AF596536

The Standard Model

This innovative textbook has been designed with approachability and engagement at its forefront, using language reminiscent of a live lecture and interspersing the main text with useful advice and expansions. Striking a balance between theoretical- and experimental-led approaches, this book immediately immerses the reader in charge and neutral currents, which are at the core of the Standard Model, before presenting the gauge fields, allowing the introduction of Feynman diagram calculations at an early stage. This novel and effective approach gives readers a head start in understanding the Model's predictions, stoking interest early on.

With in-chapter problem sessions which help readers to build their mastery of the subject, clarifying notes on equations, end-of-chapter exercises to consolidate learning, and marginal comments to guide readers through the complexities of the Standard Model, this is the ideal book for graduate students studying high-energy physics.

Marco Fabbrichesi is a researcher in theoretical high-energy physics at the Istituto Nazionale di Fisica Nucleare, Trieste, Italy. His work spans areas including string theory, supersymmetry and flavor physics. With a focus on making physics accessible, he has taught a wide variety of classes and has written two books for non-specialists. He studied at the University of Virginia and has been a fellow at the JINR, Dubna, the Neils Bohr Institute, Copenhagen, and the Theory Division at Cern, Geneva.

"Fabbrichesi has written an accessible, readable and up-to-date introduction to the Standard Model which guides the reader in understanding the theory, develops their skills and insights and introduces modern methods, tools and experiments."

Professor Alan Barr, University of Oxford

"This textbook brings the Standard Model to modern life. Professor Fabbrichesi's friendly, first-person narrative feels like an in-person lecture, guiding you from core concepts to lively problems. The step-by-step computations and blackboard-style explanations make complex topics accessible, providing an indispensable and engaging guide. A delightful and refreshing resource for particle physics learners and researchers."

Dr. Zhen Liu, University of Minnesota

"This book offers an exquisitely insightful treatment of the physics of the Standard Model to readers already familiar with the basic principles of relativistic quantum field theory and quantum electrodynamics.

Crammed full of 'problem sessions,' pedagogic tours through calculations of quantities of experimental interest, this unconventional text helps uncover the 'whys' as well as the 'whats' of Standard Model calculations."

Professor David Craig, Oregon State University

The Standard Model

A Practical Step-by-Step Guide

Marco Fabbrichesi
INFN – National Institute of Nuclear Physics, Italy

Shaftesbury Road, Cambridge CB2 8EA, United Kingdom

One Liberty Plaza, 20th Floor, New York, NY 10006, USA

477 Williamstown Road, Port Melbourne, VIC 3207, Australia

314–321, 3rd Floor, Plot 3, Splendor Forum, Jasola District Centre, New Delhi - 110025, India

103 Penang Road, #05–06/07, Visioncrest Commercial, Singapore 238467

Cambridge University Press is part of Cambridge University Press & Assessment, a department of the University of Cambridge.

We share the University's mission to contribute to society through the pursuit of education, learning and research at the highest international levels of excellence.

www.cambridge.org
Information on this title: www.cambridge.org/highereducation/isbn/9781009313872
DOI: 10.1017/9781009313865

When citing this work, please include a reference to the DOI 10.1017/9781009313865

First published 2026

A catalogue record for this publication is available from the British Library

Library of Congress Cataloging-in-Publication Data
Names: Fabbrichesi, Marco author
Title: The standard model : a practical step-by-step guide / Marco Fabbrichesi, Italian National Institute of Nuclear Physics.
Description: Cambridge, United Kingdom ; New York, NY : Cambridge University Press, 2026. | Includes bibliographical references and index.
Identifiers: LCCN 2025007667 (print) | LCCN 2025007668 (ebook) | ISBN 9781009313872 hardback | ISBN 9781009313865 epub
Subjects: LCSH: Standard model (Nuclear physics)–Textbooks
Classification: LCC QC794.6.S75 F33 2026 (print) | LCC QC794.6.S75 (ebook)
LC record available at https://lccn.loc.gov/2025007667
LC ebook record available at https://lccn.loc.gov/2025007668

ISBN 978-1-009-31387-2 Hardback

Additional resources for this publication at www.cambridge.org/Fabbrichesi

Contents

Preface

Origin

As with many a textbook before, this one too grew out of a set of lecture notes written for students, and it is based on my class on the "Standard Model of Fundamental Interactions." This course, comprising both experimental and theoretical curricula, was designed for first-year students enrolled in the Master's physics program. Most of these students had little previous exposure to high-energy physics and seemed to benefit from the classroom-and-blackboard kind of approach I utilize throughout the book.

Although the original course was only a semester long, as I was writing the book I found myself exploring additional topics and as a result expanding certain sections in the book, most of which are indicated by an asterisk after the heading. These extra sections give the possibility of selecting topics for a more advanced study. Instructors can assign or omit these topics depending on their needs.

Approach

Physics is mostly taught by example. We first follow our mentor around just watching, then we are given some simple tasks to do, and these are followed by more challenging ones. As they say in medical school: "watch one, do one, teach one." Accordingly, many concepts in the book are developed as worked examples in problem sessions, ensuring the reader is presented with the many technical manipulations that are at the heart of doing physics from an early stage. In my experience, students appreciate this style, of switching from more formal lectures to practical problem sessions and then back again.

I decided to keep the format of the book as close as possible to the actual lectures. For this reason, I chose to write in the first person rather than the more usual third person, and drafted equations with arrows and other comments as if they were written on the blackboard. Similarly, the many numbered notes are intended to mimic the flow of a live lecture, with its side comments, answers to questions, parallel lines of reasoning and assonances. I have addressed many of the questions we ask ourselves as we learn, providing the answers to these as well as related ideas that cross over with my work in particle physics. I wanted to share with the readers – as they follow the discussion about equations and experiments – my understanding of these concepts as a practicing researcher. Idiosyncratic though this approach might be, it strives to convey the meaning of the Standard Model – why this rather than that – and to make the physics behind its formulation clear, or at least clearer than in a plainer presentation.

Throughout the book I take a step-by-step approach, and I have included several important computations in detail in the main text and the problem sessions. After all, "the devil is in the detail", as the proverb goes. Our physical intuition comes from the memory of all the problems we have previously solved and the minute details these solutions have taught us. Faced by a new question, we fall back on these examples; without them, we do not know where to start. Familiarity with these solutions builds up our competence – the first step toward mastery.

Coverage

This book tackles an area somewhat neglected by current textbooks, which are either at a lower level (mostly keyed for experimentalists) or at a higher level (developing the full quantum field theory technology). An intermediate coverage assumes that most of the basis of quantum field theory has been already learned (or at least glanced at) and moves on to discuss, in terms of Lagrangians, the computation of the many observables (cross sections and decay widths) that provide our current understanding of high-energy physics. With this method, the reader is dropped right into the action from the very beginning and hopefully is not dragged down by too many preliminary mathematical questions nor left wondering because of being deprived of a more theoretical framework.

The first chapter is dedicated to recalling and reviewing all the prerequisite knowledge needed to understand the rest of the book. I take the opportunity in this chapter to explain and answer more than one question I had as a student which was not addressed by the textbook I used. The rest of the book is then organized into two parts, the first of which introduces the model and all its different components (first the electroweak and then the strong interactions). In the second part, the model is put to work in more challenging settings. I start out by presenting neutral and charged currents rather than the gauge fields, because it is the currents that you see in experiments and the fields only come as a construction to help us to organize the way in which these currents arise and are connected to each other. Whenever possible I present some of the essential steps in an historical context to remind the reader how that particular result was originally derived and that nothing of what we learn today was obvious when it was first discovered.

Difficulty

Learning about the Standard Model takes time. Though it requires, as any subject in physics, an effort, every step makes us feel more alert – energized by the pleasure of a heightened awareness. It is hard but rewarding work that causes a change in the state of our mind. That change may seem small but it is real. It is what reading fiction does. What art does. Anything requiring our undivided attention brings that change of mind – whereas what comes easily does not.

1 What is This Book About?

The theory called the **Standard Model** represents the state of the art of our understanding of the physics of elementary particles and their interactions. It is, as explained in the following chapters, a **gauge** theory, that is, a theory in which the interactions among the matter constituents are determined by gauge invariance and carried by gauge bosons. Everything else in physics (except perhaps gravitational phenomena at very high energies), as well as everything in the Universe, from chemistry to biology, is but an application (admittedly, very convoluted in most cases) of this model.

I see a raised hand: "Do you really believe in deducing the behavior of the brain from Dirac's equation?" To be sure, like many a sweeping statement, the one above is also open to criticism. Many practitioners think that such a complete reduction to fundamental physics might not be the best way to look at the world. Be that as it may, the reductionist point of view has proved to be very successful in the past and should not be abandoned too quickly. I suggest we try to bear it in mind, at least in the context of this book, because it underlines the importance of the Standard Model not only in high-energy physics but in the whole field of physics and in all the branches of science that make use of it.

The careful reader may have already at this point noticed the contradictory usage of the terms "theory" and "model" in the first line of this page.[1.1] In physics, the term "theory" indicates an established set of equations – think of Maxwell's electromagnetism – while the term "models" a rougher set of assumptions, whose working is limited and open to future improvements. For this reason it has been proposed that the Standard Model be called the "Standard Theory" – which already tells you the recognition and prestige enjoyed by it – even though the terminology is by now established and would be hard to change.

1.1 The popular usage of "in theory" for something that may actually not be true adds a further layer of confusion.

The Standard Model is a theory that has been verified in, literally, thousands of experiments. It is a theory of something. I intend this as an endorsement. A good physical theory is always a theory of (a

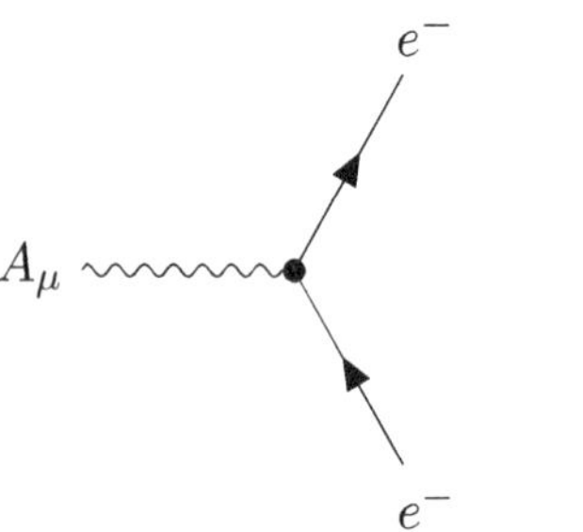

Figure 1.1 Feynman diagram for the interaction vertex between a fermion (here, an electron) and a photon.

Figure 1.2 The analogy of the basketball players: Bouncing the ball back and forth keeps the players near each other as does the exchanging of a photon between electron and positron [Aleksandrar Georgiev/Getty Images].

well-defined and specific) something. Contemplating this something – the many experimental facts and their interconnections – is a bit like looking at some powerful engine, a shining marvel in which every part sets another part in motion. There is no way to stop it. Being in its presence – humbled by its complexity, thrilled by its power – makes us feel part of it. Makes us feel somehow better.

For reasons I review in the next chapter, a theory of the fundamental interactions is necessarily a theory of elementary particles. We do not choose to study them because they are pretty and interesting (they are not, they are kind of boring compared to things like plants and animals); they choose us, in the sense that we are forced to use elementary particles if we want to have a relativistically invariant quantum theory.

And yet the Standard Model is not primarily about these elementary particles. It is a theory of the fundamental constituents of matter but it is first and foremost about the way these elementary particles interact. Quarks and leptons are used because they, by being the simplest system we can think of, can better reveal the fundamental interactions in Nature. Simple systems, and only simple systems, teach us about the fundamental laws of the world. Elementary particles play in modern physics the role played by atomic spectra in the physics of the first half of the last century and by the motion of the planets in the night sky in the history of physics from Ptolemy to Newton. One could say that the behavior of elementary particles led to gauge theories in the same way as the retrograde motion of the planets eventually led to the elliptic orbits dictated by Newton's law of gravity.

The history of the Standard Model begins where most of modern physics begins: from electromagnetism. It is through Maxwell's equations that we become familiar with seeing the interaction of charged particles as mediated by a field. At each point x of space-time there is a field quantity, the four-vector $A_\mu(x)$, and charged particles such as the electron interact with this field vector rather than directly among themselves. This process is depicted by a **Feynman diagram**, in which an electron is represented by a straight line and a photon by a wiggly one, as in Figure 1.1.

In quantum mechanics, the field described by the four-vector is quantized and the electromagnetic force itself is understood as a particle (the photon) that two charged particles exchange, like the basketball between the two players in Figure 1.2.

The above apparently simple statement is the final result of a long-standing debate about how two "bodies" interact. In Newtonian physics the gravitational interaction takes place "at a distance" and instantaneously.[1.2] The origin and actual working of this interaction is not really explained (Newton's "*Hypotheses non fingo*", "I feign no hypotheses"). In Maxwell's electromagnetism the interaction is "local" because the force is mediated

[1.2] What would a theory of gravity based on a mechanical model look like? An example is the model proposed by Le Sage in which bodies are constantly hit by a cloud of tiny particles coming from all directions, and gravitational attraction is due to the screening effect on these particles due to the presence of a nearby planet. This screening, interestingly enough, gives rise to a force – bodies tend to move toward regions where there are fewer particles hitting them – decreasing as the inverse of the square of the relative distance, as in Newton's law. Le Sage's theory provides a completely mechanical explanation of gravity for which "Hypotheses fingo". Unfortunately it does not really work – or does it?

Figure 1.3 Whereas a hand by itself remains unaffected by a rotation of 360°, when it is attached to an arm, as in the figures above, the two come back to the initial positions only after a rotation of 720° about a vertical axis. The author's forearm is the "internal" structure revealed by this simple experiment.

by the field. This description admits a more mechanical visualization with respect to action at a distance: the charged particle feels the electromagnetic field here and now and the equations of motion are driven by the value of the field, a quantity that is locally determined. The debate about how to properly describe a force has gone back and forth but it is now firmly on the side of local interactions.

The actual story is more complicated (and also more interesting) because the interacting "bodies" are made of fermions, and fermions are not as point-like as Newton might have liked them to be. Instead, a fermion has some "internal" structure, its spin, as indicated by the spinor index attached to its wave function.

To convince yourself of the existence of such a structure just think of rotating a fermion, described by its spinor wave function Ψ_α, by 360 degrees. This can be achieved by taking a fermion sealed in a box in an external magnetic field and rotating the box.[1] The mathematics of spinors tells us that instead of going back into itself, as one might expect, the spinor Ψ_α comes back (surprisingly) with a change of sign:

$$\Psi_\alpha \xrightarrow{\text{rotation of } 360^\circ} -\Psi_\alpha \, .$$

If you have two of these fermions in two boxes and rotate one of them, leaving the other alone, when they are brought back into contact and the boxes opened, you will see an interference caused by the change of sign in the wave function of the fermion that has gone around.

Fermions are not the only objects to do this trick. Also, the palm of your hand does not go back to its initial state after a rotation of 360 degrees. Try it! As shown in Figure 1.3, you need to go around by 720 degrees to come back to where you started.

---

[1] Y. Aharonov and L. Susskind, *Physical Review* **158** (1968) 1237.

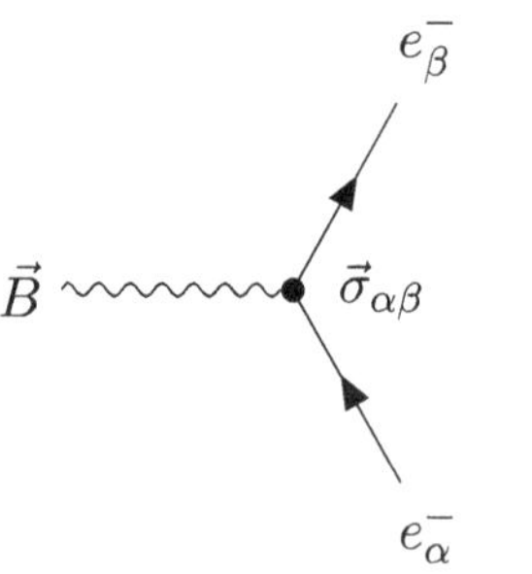

Figure 1.4 Feynman diagram for the interaction vertex between the spin of a fermion and the magnetic field of the photon.

The spin of a fermion, tracked by the spinor index α or β of its wave function, can change because it is linked to the particle's magnetic moment – proportional to the Pauli's matrices $\vec{\sigma}_{\alpha\beta}$ – which is rotated by the presence of an external magnetic field $\vec{B}$, as shown in Figure 1.4.

The process can be seen as the motion of the particle in its **internal** space (of the spin states, that is, of the indices α and β) driven by the electromagnetic interaction.[1.3] In the Standard Model this point of view is generalized by including other charges beside the electric charge and also more complicated "charges," which, like the spin, have more than one possible value. These charges add new degrees of freedom to the elementary particles, which come to live simultaneously in various internal spaces defined at each and every point in space-time. These degrees of freedom are encoded in indices that the fermion fields carry – which, mathematically speaking, track the representations of the group symmetries describing the internal spaces. The Standard Model is the theory of these degrees of freedom.

[1.3] No reason to be alarmed if something here (or later) is not immediately clear. The book is about explaining these concepts and they should become more understandable as I proceed. Pushing on – not being daunted even when not every step in some reasoning is completely clear – is the hallmark of real learning.

There are spaces as simple as the phase of complex number (for the electric charge) that do not even need an index, and there are spaces that are more complicated (for isospin or color) or even more complicated, where the index stands for four-vectors and spinors in space-time (taken to be Minkowski). Altogether, the notation looks like this (EW stands for electroweak),

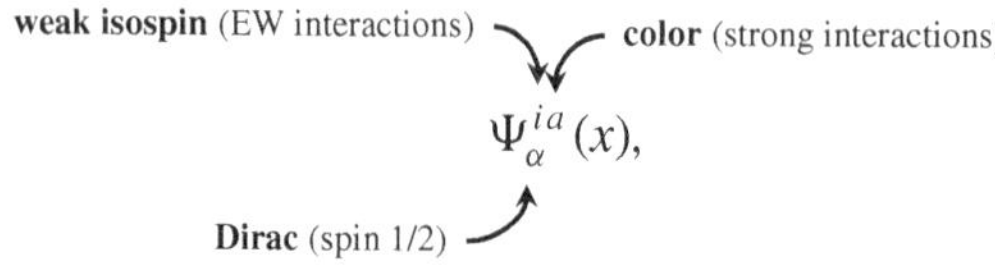

for a fermion (of spin 1/2), with "charge" due to the electroweak and strong interactions, and like this,

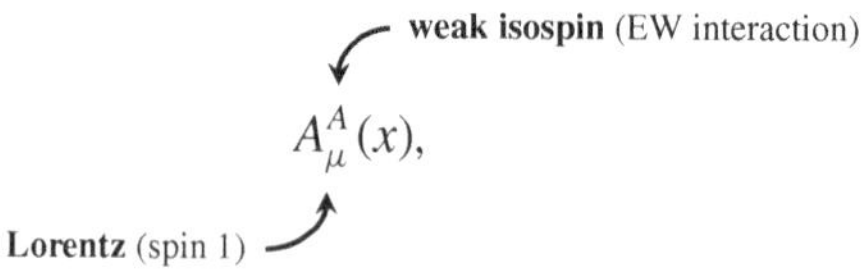

for a boson (of spin 1) charged under the electroweak interaction (but it could also have been the strong interaction).

These internal spaces, labeled by their indices, are not just a bookkeeping device. Like the spin of an electron in a magnetic field, the degrees of freedom in these spaces are endowed with their own dynamics by the gauge interactions, as they come to describe all (well – almost all, as we shall see) the interactions of the Standard Model.

The origin of the idea of describing all the fundamental interactions as a gauge theory, and therefore the origin of what came to be known as the Standard Model, can be traced to the collaboration of two physicists, Chen

Ning Yang and Robert Laurence Mills, in the spring of 1954. It is actually possible to point to the publication of a single paper where the idea is spelled out for the first time. The article is surprisingly readable and the basic concept – of making the symmetry a local one – is expressed in the following paragraph:

> The conservation of isotopic spin is identical with the requirement of invariance of all interactions under isotopic spin rotation. This means that when electromagnetic interactions can be neglected, as we shall hereafter assume to be the case, the orientation of the isotopic spin is of no physical significance. The differentiation between a neutron and a proton is then a purely arbitrary process. As usually conceived, however, this arbitrariness is subject to the following limitation: once one chooses what to call a proton, what a neutron, at one space-time point, one is then not free to make any choices at other space-time points. It seems that this is not consistent with the localized field concept that underlies the usual physical theories. In the present paper we wish to explore the possibility of requiring all interactions to be invariant under independent rotations of the isotopic spin at all space-time points, so that the relative orientation of the isotopic spin at two space-time points becomes a physically meaningless quantity (the electromagnetic field being neglected).[2]

The quoted excerpt contains the main idea behind the Standard Model: interactions affect the motion of particles in their internal spaces (the isospin in the Yang–Mills paper); these spaces are defined at each point of space-time and, because of this locality, the motion in the internal space becomes motion in space-time, that is, what in mechanics is produced by a force. In the following years, more internal spaces were added (so as to include the strong, weak and electromagnetic interactions) and new features about how the symmetries were to be realized were better understood – a process giving rise eventually to the Standard Model as we know it today.

---

[2] C.N. Yang and R.L. Mills, *Physical Review* **96** (1954) 191.

2 Review of What the Reader Already Knows

Contents

The discussion of any new topic necessarily makes use of knowledge that is to some extent assumed to have been already acquired. One cannot start from the very beginning and teach the whole of physics every time something new is to be introduced – even though the Landau and Lifshitz series of books comes close to pulling it off. More pragmatically, I would like to make sure that we are all on the proverbial same page with some of the basic notions. And where, you may ask, are these basic notions learned? I have in mind what can be called the **canon** of physics, that is, the books where we, as students, first studied the basic concepts and equations, the books that everyone has read to study, say, classical mechanics, electromagnetism, quantum mechanics, thermodynamics and statistical mechanics – as well as the mathematical tools necessary to comprehend the equations and the statistics to make sense of the data analysis. I have them (most of them, at least) in my office, and so do most of the physicists I know. The covers of some of them are shown in Figure 2.1.

Such a collection of books will vary from place to place but probably some of those I include in Figure 2.1 are known to many physics students. There is a recognition that the subject covered by each of these books has been presented in a definitive manner. I still leaf through these books, from time to time, for the simple pleasure of re-reading some of their chapters, following through their explanations and revisiting the places where I first learned about the concepts explained therein.

Whereas books such as those in Figure 2.1 provide a shared common ground and indispensable tools for thinking about physics, there are more specific preliminaries that have to do with concepts that are proper to the Standard Model. These topics, where I assume some previous knowledge, go a little beyond the canon and the consensus about where they should be studied is perhaps less unanimous. Be that as it may, there are just three such topics; the covers of the corresponding books are shown in Figure 2.2:

Figure 2.1 A selection of textbooks in the canon for the practitioner of physics.

Figure 2.2 Covers of three books mentioned in the main text.

- Some notions of **group theory**, Lie groups in particular, and their representations. I do sketch the main concepts in this chapter but under the implicit assumption that they have been seen before in, for instance, Zee's book.[1]
- A first course in **quantum field theory** (covering quantum electrodynamics as in the book of Landau and Lifshitz[2]) at the level of knowing that fermions are described by the Dirac equation and that particles are understood in terms of creation and annihilation operators in Fock space; some familiarity with covariant perturbation theory (that is, Feynman's diagrams) is also taken for granted.
- Some knowledge about **elementary particle physics** at the level of, say, Perkins' book:[3] the catalog of particles discovered in the 1960s and 1970s, their charges and organization into isospin (or larger group) multiplets in the quark model and the experimental setup (to accelerate and detect particles) utilized in high-energy physics.

This background material is usually provided in the first few years of a curriculum in physics at most universities. It is not really possible to proceed and discuss the Standard Model at the level of this book without

[1] A. Zee, *Group Theory in a Nutshell for Physicists* (Princeton University Press, 2016).
[2] L.D. Landau and E.M. Lifshitz, *Quantum Electrodynamics* (Pergamon Press, 1971).
[3] D. Perkins, *Introduction to High Energy Physics* (Cambridge University Press, 2000).

having first acquired these notions. Take it as a kind of disclaimer: if you find it difficult to understand what I am talking about, the reason may be that you have not acquired sufficient background from taking those courses or reading those books.

Having said that, I can move on and go through the actual preliminaries alluded to in the chapter's title.

2.1 Why Study Elementary Particles?

Physics students, at a certain point in their studies, learn that they live in a **Hilbert space**. A physical state is not a point in phase space but a vector in Hilbert space. Classical mechanics is done with. Goodbye to coordinates and momenta, and welcome to the quirky world of quantum-mechanics operators!

What exactly a Hilbert space is – let alone living in it – takes a little explaining. Luckily, I do not have to do this explaining because it has already been done in your quantum-mechanics class. I can just recall the final results.

A Hilbert space is a complex linear space made of vectors (well, they are "rays", but never mind), the physical observables being operators on these vectors. These vectors contain all the information on the physical system in which we are interested, and the rules of quantum mechanics tell us how to extract this information.

It is best not to think of the vectors (or the wave functions) as physical quantities (I mean, real objects out there with well-defined values before being measured) because this line of reasoning leads to never-ending problems and inconsistencies with, and in the implementation of, the rules in actual experiments. People sometime wonder about this question – most famously Bohr and Einstein in the early days of quantum mechanics – but in this book I just take the rules of quantum mechanics at face value. Whatever your interpretation of quantum mechanics is, you will get the same results as I do – at least for all the computations in this book.

Quantum mechanics can be summarized in five rules.[2.1] They fit into a short table, as in Table 2.1.

[2.1] Classical mechanics is even simpler. The observables are the position and momentum of the particles, and the time evolution is given by Newton's law $\vec{F} = m\vec{a}$. Measurement of the observables just gives their value, and there is no Born's rule nor probabilistic outcomes.

That is all there is! Admittedly, these rules by themselves are not enough. Only the many exercises and assigned problems we have to solve in our quantum-mechanics classes give life to the rules and provide our actual understanding of quantum mechanics and, through it, of the world around us.

By duly applying the rules, I can study any physical system – elementary particles included.

The Hilbert space is complex. It would be simpler to work in a space made of real numbers as we do in classical mechanics. Can we do quantum

Table 2.1 The five rules of quantum mechanics.

	Rule
1. Physical states	Physical states are vectors $\|\psi_n\rangle$ in a Hilbert space.
2. Observables	Observables are Hermitian operators $\hat{A}$, acting as follows: $$\hat{A}\|\psi_n\rangle = \alpha_n\|\psi_n\rangle,$$ with $\hat{A}^\dagger = \hat{A}$. The measurement of an observable gives one of its eigenvalues α_n.
3. Probability (Born's rule)	A generic state $\|\psi\rangle$ has a probability $\|\langle\psi\|\psi_a\rangle\|^2$ of being found in a given state $\|\psi_a\rangle$.
4. Time evolution	The evolution in time of an observable $\hat{A}$ is given by the Schrödinger equation $$i\hbar\frac{\partial\hat{A}}{\partial t} = [\hat{A},\hat{H}]\,.$$
5. Symmetry	The symmetry operators $\hat{U}$ are linear and either unitary ($\hat{U}^\dagger = \hat{U}^{-1}$) or anti-unitary.

mechanics in real space? It seems that we cannot. Here are three basic reasons:

— A commutator like that between position and momentum, $[\hat{x},\hat{p}] = \hat{x}\hat{p} - \hat{p}\hat{x}$, is an anti-Hermitian operator and therefore it must be equal to i times the reduced Planck's constant $\hbar(= h/2\pi)$ in order to change sign under conjugation. The same applies to the commutation relation between angular momentum components.

— The norm of a state $|\psi\rangle$ in the time evolution

$$i\hbar\frac{\mathrm{d}}{\mathrm{d}t}|\psi\rangle = \hat{H}|\psi\rangle$$

is conserved only if the state $|\psi\rangle$ is a complex quantity.

— The wave function of a free particle must be complex in order to carry only positive frequencies and to represent the time evolution for that particle, whereas in classical mechanics a wave is made up of both positive and negative frequencies (think of an electromagnetic wave) and is a real function of space-time. Since the wave function is thus given by only positive frequencies, the probability of finding the particle at a given point in space is a constant and the particle in a given eigenvector of the momentum is delocalized in space. Had the wave function been real, the particle would be localized in space.

Necessary though they are, complex numbers appear only in the wave function, and the eigenvalues of the observables are – as they must be— real numbers (rule number **2**).

Given that we live in a Hilbert space, the first ingredient of any theory of fundamental interactions is **quantum mechanics**: the theory must be a quantum theory.

The other requirement of a theory of fundamental interactions is that it must satisfy **relativistic invariance** – which we know to be the other essential property of the world in which we live. The latter requirement means that the physical states $|\psi\rangle$ in rule **1** must be representations of the Poincaré group and transform as

$$|\psi\rangle \rightarrow \hat{U}(\Lambda, a)|\psi\rangle, \tag{2.1}$$

where the tensor Λ and quadrivector a are defined by their effect on space-time coordinates:

$$x^\mu \rightarrow \underbrace{\Lambda^\mu_\nu}_{\text{Lorentz}} x^\nu + \overbrace{a^\mu}^{\text{translation}}. \tag{2.2}$$

The matrix Λ^μ_ν implements a (homogeneous) Lorentz transformation and a^μ is a translation in space-time.

As I am going to discuss, these irreducible representations of the Poincaré group are one-particle states. The Hilbert space is built out of these states and is called **Fock space**. The basic observables of a theory marrying quantum mechanics and relativity are obtained from these states. Such one-particle states are what we call **elementary particles** and this is the reason why we study them.[4]

Fock space contains an infinite collection of harmonic oscillators. Elementary particles are the excitations of these oscillators as expressed in terms of creation and annihilation operators. Seeing the world as a collection of harmonic oscillators is a firmly established way of doing physics, by no means peculiar to particle physics.[2.2] As long as I study small perturbations away from the equilibrium position, everything looks like (or should I say is?) a harmonic oscillator. This is a great feature because the harmonic oscillator is one of that handful of problems of which we know an exact solution – and having an exact solution is very valuable because it brings many benefits to our understanding and more power to model building.

2.2 "The career of a young theoretical physicist consists of treating the harmonic oscillator in ever-increasing levels of abstraction." S. Coleman

[4] This point has been made most authoritatively by S. Weinberg in his textbook *Quantum Field Theory* (Cambridge University Press, 1995).

Because of the Fock space construction in terms of creation and annihilation operators, experiments separated in space (and time) do not influence each other – what is called **cluster decomposition** – and we can discuss the physics of an experiment here and now, say, on Earth without having to worry about what is going to happen tomorrow on the Moon.

Elaborating the implications of requiring our theory to be a quantum theory that is relativistically invariant provides a gentle way of reviewing some of the tools necessary to deal with the Standard Model – which is what I am going to do for the remainder of this chapter.

2.2 A Little Bit of Group Theory

A **symmetry** is an expression of the simplicity of an object. Take an orange like that in Figure 2.3: because of its rotational symmetry, the entire fruit is defined once one of its slices is given.

Figure 2.3 A nicely symmetric slice of an orange [Xinzheng/Getty Imges].

Symmetries are studied by means of that branch of mathematics that goes under the name of **group theory**. A group – let me just recall the definition – is a set of elements with an internal composition law (think of the multiplication of two numbers) among its elements which is associative and for which there is an identity and inverse. The idea of symmetry seems to permeate the world described by physics and it is because symmetry arguments can be framed in the language of group theory that they are so important in physics.

Though probably few self-respecting physicists would admit to not knowing group theory, I will go briefly through some definitions – mainly to stress the use of group representations, because of the crucial role they play in the Standard Model and in preparation for their use in the study of the Poincaré group.

A finite group has a finite number of elements.[2.3] I can collect the effect of the composition rule in a table like that in Table 2.2.

The concept of a group is fine in the abstract but of little practical help. The way a group is actually used in physics is by means of **representations**, that is, mappings of the group into a vector space: the homomorphisms from the group into the group of automorphisms of the

[2.3] For group elements a, b and c, I have that the composition law satisfies

$$(a * b) * c = a * (b * c)$$
$$\mathbb{1} * a = a * \mathbb{1} = a$$
$$a^{-1} * a = a * a^{-1} = \mathbb{1}.$$

If, in addition,

$$a * b = b * a,$$

the group is called **Abelian**.

[2.4] The **order** of a group is its number of elements.

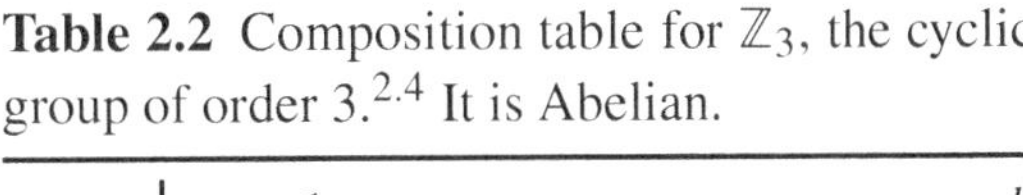

Table 2.2 Composition table for $\mathbb{Z}_3$, the cyclic group of order 3.[2.4] It is Abelian.

	$\mathbb{1}$	a	b
$\mathbb{1}$	$\mathbb{1}$	a	b
a	a	b	$\mathbb{1}$
b	b	$\mathbb{1}$	a

vector space. In plain words: a matrix (or just a number) is associated to each operation of the group.

A one-dimensional representation of the group $\mathbb{Z}_3$ is given by

$$D(\mathbb{1}) = 1, \quad D(a) = e^{2\pi i/3}, \quad D(b) = e^{4\pi i/3}. \tag{2.3}$$

Another representation, one with the dimension equal to the order of the group, is

$$D(\mathbb{1}) = \begin{pmatrix} 1 & 0 & 0 \\ 0 & 1 & 0 \\ 0 & 0 & 1 \end{pmatrix}, \quad D(a) = \begin{pmatrix} 0 & 0 & 1 \\ 1 & 0 & 0 \\ 0 & 1 & 0 \end{pmatrix}, \quad D(b) = \begin{pmatrix} 0 & 1 & 0 \\ 0 & 0 & 1 \\ 1 & 0 & 0 \end{pmatrix}. \tag{2.4}$$

Rotations in two dimensions provide another simple example of a symmetry and its related group. The elements of this group are the rotations $R(\theta)$ on a plane, which are characterized by a single parameter: the angle of rotation θ. This is perhaps the simplest example of a group characterized by a continuous parameter, known as a **Lie group**. The groups are Lie groups for most applications within the Standard Model.

In physics we write all equations in terms of vectors (or scalar) quantities: force and the velocity are vectors while mass is a scalar. These are objects with well-defined transformation laws under rotations. In this way, the invariance of the equations under rotations is guaranteed. If you were to insert into an equation a quantity which is neither a vector or a scalar, the invariance of the equation would have to be checked.

The representations of the group of rotations on the plane are given by the 2×2 orthogonal matrices $M(\theta)$ with unit determinant. They can be written as

$$M(\theta) = \begin{pmatrix} \cos\theta & -\sin\theta \\ \sin\theta & \cos\theta \end{pmatrix}. \tag{2.5}$$

This group is indicated as $SO(2)$.[2.5]

The composition of two rotations, as for instance one of 15° and another of 30°, written as

$$R(15^\circ)R(30^\circ) = R(45^\circ), \tag{2.6}$$

becomes, in terms of the representation matrices[2.6]

$$\begin{aligned} M(15^\circ)M(30^\circ) &= \begin{pmatrix} \cos 15^\circ & -\sin 15^\circ \\ \sin 15 & \cos 15 \end{pmatrix} \begin{pmatrix} \cos 30^\circ & -\sin 30^\circ \\ \sin 30^\circ & \cos 30^\circ \end{pmatrix} \\ &= \begin{pmatrix} \cos 45^\circ & -\sin 45^\circ \\ \sin 45^\circ & \cos 45^\circ \end{pmatrix} = M(45^\circ). \end{aligned} \tag{2.7}$$

The representation mimics all the group operations by embodying them in the corresponding matrices.

2.5 The 2×2 matrices providing the representation are orthogonal matrices, as indicated by the "O"; the "S" indicates that they have determinants equal to 1. An orthogonal matrix is a square matrix whose columns and rows are orthogonal unit vectors.

2.6 Mind your trigonometry: $\cos 15^\circ \cos 30^\circ - \sin 15^\circ \sin 45^\circ = \cos(15^\circ + 45^\circ)$ etc.

If I consider the matrix $M(\theta)$ for very small values of the parameter θ, I can Taylor-expand it as

$$M(\theta) = \mathbb{1} + \underbrace{\theta M'(0)}_{i\theta X} + \frac{\theta^2}{2} M''(0) + \cdots , \tag{2.8}$$

where the primes stand for derivatives and

$$X = -iM'(0) = -i\frac{\mathrm{d}}{\mathrm{d}\theta}\begin{pmatrix} \cos\theta & -\sin\theta \\ \sin\theta & \cos\theta \end{pmatrix}\Bigg|_{\theta=0} = \begin{pmatrix} 0 & i \\ -i & 0 \end{pmatrix}; \tag{2.9}$$

X is called the **generator** of the group transformation.[2.7]

2.7 The **rank** of a group is the number of its generators that can be simultaneously diagonalized.

The group $SO(2)$ being very simple has only one generator. A more interesting example is provided by rotations in three dimensions and the group $SO(3)$. It has three generators:

$$X_1 = \begin{pmatrix} 0 & 0 & 0 \\ 0 & 0 & -i \\ 0 & i & 0 \end{pmatrix} \quad X_2 = \begin{pmatrix} 0 & 0 & i \\ 0 & 0 & 0 \\ -i & 0 & 0 \end{pmatrix} \quad X_3 = \begin{pmatrix} 0 & -i & 0 \\ i & 0 & 0 \\ 0 & 0 & 0 \end{pmatrix}, \tag{2.10}$$

which satisfy the relations

$$[X_i, X_j] = i\epsilon_{ijk} X_k , \tag{2.11}$$

where the Levi–Civita symbol ϵ_{ijk} provides the **structure constants**. The relationship in Eq. (2.11) is called the **Lie algebra** of the Lie group.

The same group can have different representations; $SO(3)$ can also be represented by the Pauli matrices (divided by 2)

$$\sigma_1 = \frac{1}{2}\begin{pmatrix} 0 & 1 \\ -1 & 0 \end{pmatrix} \quad \sigma_2 = \frac{1}{2}\begin{pmatrix} 0 & -i \\ i & 0 \end{pmatrix} \quad \sigma_3 = \frac{1}{2}\begin{pmatrix} 1 & 0 \\ 0 & -1 \end{pmatrix}, \tag{2.12}$$

which satisfy the same algebra,

$$[\sigma_i, \sigma_j] = i\epsilon_{ij}^{k}\sigma_k . \tag{2.13}$$

These matrices also provide a representation of the group $SU(2)$.[2.8] The group $SU(2)$ is isomorphic to $SO(3)$ and its representations provide a double covering of $SO(3)$ (which has to do with the fact that spinorial representations go back to the identity only after a rotation of 720 degrees, as mentioned in Chapter 1).

2.8 The "U" in the notation of these groups indicates that they are complex and unitary in addition to having unit determinants, which is, as before, encoded in the "S".

The generators and their Lie algebra tell me all I need to know about a group. The Standard Model is defined by the Lie groups according to which we classify the elementary particles taking part in the interactions. They are the unitary groups $U(1)$, $SU(2)$ and $SU(3)$.

2.3 *Problem Session*: Wigner *D*-Matrices

Equation (2.10) gives the generators of rotations in space. These rotations can be represented by matrices that are different in different contexts. The differences arise because of the choices of angles with which the rotations are to be described. A common choice consists in taking three angles α, β and γ, the Euler angles usually first introduced in classical mechanics to describe the orientation of a rigid body. The motion of the body can then be described as changing of the Euler angles about the three principal axes as the body rolls, pitches and yaws.

By reversing Eq. (2.8), we see that a rotation about the y-axis is given by[2.9]

[2.9] The identity follows from the Taylor expansion and that $X_3^3 = X_3, X_3^4 = X_3^2$ and so on.

$$e^{-i\beta X_2} = \mathbb{1} + \sin\beta X_2 + (\cos\beta - 1)X_3^2\,, \tag{2.14}$$

which is a 3×3 matrix that is called after the physicist E. Wigner: $M^{(1)}_{m,m'}(\beta)$. More generally, an arbitrary element of the group $SO(3)$ is represented by a $(2j+1) \times (2j+1)$ matrix[2.10]

[2.10] *D* stands for *Darstellung*, which means representation in German.

$$D^{(j)}(\alpha,\beta,\gamma) = e^{-im'\alpha}\, e^{-im\gamma}\, M^{(j)}(\beta)_{m,m'}\,, \tag{2.15}$$

in which j tracks the representation and m and m' the positions of the matrix elements. The first two Wigner matrices are

$$M^{(1/2)}(\beta) = \begin{pmatrix} \cos\beta/2 & -\sin\beta/2 \\ \sin\beta/2 & \cos\beta/2 \end{pmatrix} \tag{2.16}$$

for the spin $j = 1/2$ representation, and

$$M^{(1)}(\beta) = \begin{pmatrix} \frac{1}{2}(1+\cos\beta) & -\frac{1}{\sqrt{2}}\sin\beta & \frac{1}{2}(1-\cos\beta) \\ \frac{1}{\sqrt{2}}\sin\beta & \cos\beta & -\frac{1}{\sqrt{2}}\sin\beta \\ \frac{1}{2}(1-\cos\beta) & \frac{1}{\sqrt{2}}\sin\beta & \frac{1}{2}(1+\cos\beta) \end{pmatrix} \tag{2.17}$$

for the spin $j = 1$ representation.

The Wigner D-matrices are the equivalent in space of the matrices in Eq. (2.5) on the plane. They are particularly useful because their absolute square $|D^{(j)}(\alpha,\beta,\gamma)|^2$ gives the probability for a system with spin j prepared in a state with spin projection m along some direction to be measured as having spin projection m' along a second direction. In particle physics they give the relations between amplitudes with helicity defined along different directions.

2.4 *Problem Session*: The Hydrogen Atom

I first want to dispel a common misconception about the classical model of the atom: the hydrogenic atom as a miniature solar system, with electrons orbiting around the nucleus, failed because the electrons would have spiraled down by emitting radiation; enter Bohr's model.

The problem of orbiting electrons radiating away their energy exists but only if there is a single electron, and the classical model of the hydrogenic atom did not have just a single electron. If you have more than one electron orbiting around the nucleus, they can be arranged so as to have the light irradiated by them interfere destructively and no residual radiation is emitted. The atom is stable.

This point is made clear in one of the exercises of Chapter 14 of Jackson's book *Classical Electrodynamics* – of physics canon fame. The power emitted by N electrons moving on a ring is proportional to

$$\left| \sum_{n=1}^{N} e^{-in\theta_i} \right|^2 , \tag{2.18}$$

where θ_i are the angular positions of the electrons on the ring. The power decreases exponentially for an increasing number N of electrons. The effect is only possible if the positions remain fixed (as they are when constrained in the metal of a conducting loop). If the electrons are randomly distributed, the destructive interference disappears, as is the case, for example (and unfortunately) for a circular collider. This feature explains why the electrons in the wiring of my house appliances do not radiate: a steady electric current in a circuit shaped as a small loop does not radiate even though the electrons circulating in the loop are accelerated as they go round in a circle.

Figure 2.4 The yellow light from the D-lines emission of sodium [mjrodafotografia/Getty images].

Most historical examples utilized in physics textbooks should always be taken with a grain of salt (those in this book no exception!). The motivation for a quantum theory of the atom is not the instability of the classical model of the hydrogen atom. The motivation comes from that very peculiar feature of atomic spectra of being monochromatic rather than having a broad distribution: if you heat up a sample of sodium in a dark room you will see the glimmer of two well-defined emission lines in the yellow part of the visible light spectrum, the famous D-lines of the transition from the $3p$ to the $2s$ level (see Figure 2.4). What is mysterious about atoms, and cannot be explained by a classical model, is that they emit light at discrete and fixed frequencies.

Think now of the beginning of quantum mechanics. Think of the days before Schrödinger's equation, when there was only Heisenberg's matrix mechanics. How did Pauli compute the energy levels of the hydrogen atom? Did he really manipulate the infinite-dimensional matrices that are the position and momentum in the Heisenberg formalism of quantum mechanics? He did not. He used group theory by performing a computation based only on the symmetries of the problem and not on dynamical equations.[2.11]

[2.11] Using just the symmetry of the problem is actually a common procedure when there are no known equations of motion and the dynamics is unknown – as was the case at the beginning of particle physics in the middle of last century, before the rise of the Standard Model, in the dark ages of "current algebra."

But before revisiting Pauli's computation, a little anticipation to whet your appetite.

The energy levels of a hydrogen-like atom are given by

$$E_n = -\frac{m\, e^4 Z^2}{8\, h^2 \epsilon_0^2\, n^2} \quad \text{for} \quad n = 1, 2, 3, \dots, \tag{2.19}$$

where h is Planck's constant, ϵ_0 is the vacuum permittivity, Z is the atomic number and e and m are the electron's charge and mass. While each energy level is characterized by three quantum numbers (the principal quantum number n, the angular momentum quantum number ℓ and its projection m), the energy depends only on n. Given the angular momentum ℓ of the orbiting electron, this remarkable result implies a degeneracy of each energy level equal to

$$\sum_{\ell=0}^{n-1} (2\ell + 1) = n^2, \tag{2.20}$$

with each level composed of $2\ell + 1$ different projections of the angular momentum.

Degeneracy means symmetry. The $(2\ell+1)$-fold degeneracy of each state with angular momentum ℓ follows from the invariance under rotations of the system. What is the symmetry behind the full n^2-fold degeneracy? This question remains unanswered in the usual treatment, in terms of Schrödinger's equation.

The computation *à la* Pauli by means of conserved quantities is a beautiful application of symmetry arguments in solving a problem. The power of group theory is in full display. In addition it makes clear the origin of the n^2-fold degeneracy.

Here goes.[5]

The Kepler problem has three (yes, three not two) conserved quantities, they are: the energy

$$E = \frac{\vec{p}^{\,2}}{2m} - \frac{k}{r}, \tag{2.21}$$

where the coefficient $k = Ze^2/4\pi\varepsilon_0$, the angular momentum

$$\vec{L} = \vec{r} \times \vec{p}, \tag{2.22}$$

and the Runge–Lenz vector[2.12]

2.12 The vector product between $\vec{L}$ and $\vec{p}$ is symmetrized in preparation for using the quantum mechanics operators for the angular momentum and momentum.

$$\vec{N} = \frac{1}{2}\Big[\vec{p} \times \vec{L} - \vec{L} \times \vec{p}\Big] - m k \frac{\vec{r}}{r}. \tag{2.23}$$

The Runge–Lenz vector at each point in the orbit is the same as the vector pointing from the focus to the periapsis of the ellipse.

[5] In this problem session I mostly follow Chapter 9 of the beautiful book *Shattered Symmetry* by P. Thyssen and A. Ceulemans (Oxford University Press, 2017).

Whereas energy and angular momentum are conserved for all forms of central potentials, the Runge–Lenz vector is conserved only for potentials with a $1/r$ dependence on the distance r from the center. If the potential is not strictly $1/r$, as it is for the Earth and all planets in the solar system, the Runge–Lenz vector will no longer be fixed, and the orbit will show a precession.

Since the energy is conserved and E is just a constant, I can redefine the Runge–Lenz vector as

$$\vec{M} = \frac{\vec{N}}{\sqrt{-2mE}} \tag{2.24}$$

for each solution with a given energy.

In quantum mechanics the vector in Eq. (2.24) and the angular momentum $\vec{L}$ are operators and satisfy commutation relations, which can be written by components as

$$[L_i, L_j] = i\hbar\epsilon_{ijk}L_k\,, \quad [M_i, L_j] = i\hbar\epsilon_{ijk}M_k \quad \text{and} \quad [M_i, M_j] = i\hbar\epsilon_{ijk}M_k\,. \tag{2.25}$$

The first commutator on the left is the Lie algebra of the group $SO(3)$; by adding the operator $\vec{M}$ to the algebra in Eq. (2.25), the Lie algebra of the group $SO(4)$ is obtained.

This group can be factorized by introducing the new operators

$$\vec{J}_1 = \frac{1}{2}\left(\vec{L} + \vec{M}\right) \quad \text{and} \quad \vec{J}_2 = \frac{1}{2}\left(\vec{L} - \vec{M}\right) \tag{2.26}$$

with commutation relations

$$[J_{1i}, J_{1j}] = i\hbar\epsilon_{ijk}J_{1k}\,, \quad [J_{2i}, J_{2j}] = i\hbar\epsilon_{ijk}J_{2k} \quad \text{and} \quad [J_{1i}, J_{2j}] = 0, \tag{2.27}$$

which show that the group $SO(4)$ is isomorphic to the product of two $SU(2)$ groups.

After introducing these definitions, I have two copies of the usual system of eigenvalues and eigenvectors of the angular momentum. As such, I can simply work out the solution by what I remember from the quantum mechanics of atomic systems. The two operator vectors $\vec{J}_1$ and $\vec{J}_2$, when acting on the eigenvectors $|j_1\, j_2\,, m_{j1},\, m_{j2}\rangle$ – defined as the direct product of the two angular momentum j_1 and j_2 eigenstates and their respective projections m_{j1} and m_{j2} – have eigenvalues $\hbar m_{j1}$ and $\hbar m_{j2}$, that is,

$$\begin{aligned} \vec{J}_1|j_1, j_2\,, m_{j1},\, m_{j2}\rangle &= \hbar m_{j1}|j_1, j_2\,, m_{j1},\, m_{j2}\rangle \\ \vec{J}_2|j_1, j_2\,, m_{j1},\, m_{j2}\rangle &= \hbar m_{j2}|j_1, j_2\,, m_{j1},\, m_{j2}\rangle\,. \end{aligned} \tag{2.28}$$

Two operators $\vec{J}_1^2$ and $\vec{J}_2^2$ can be defined with eigenvalues on the same states:

$$\begin{aligned} \vec{J}_1^2|j_1, j_2\,, m_{j1},\, m_{j2}\rangle &= \hbar^2 j_1(j_1+1)|j_1, j_2\,, m_{j1},\, m_{j2}\rangle \\ \vec{J}_2^2|j_1, j_2\,, m_{j1},\, m_{j2}\rangle &= \hbar^2 j_2(j_2+1)|j_1, j_2\,, m_{j1},\, m_{j2}\rangle\,. \end{aligned} \tag{2.29}$$

These operators commute with all the other operators in the problem; they are **Casimir operators** for the group $SO(4)$ and can be used to solve the problem of finding the energy levels.[2.13]

Now a bit of algebraical manipulation.

In the hydrogen-like atom the angular momentum and the Runge–Lenz vector are always orthogonal, that is,

$$\vec{N} \cdot \vec{L} = \vec{L} \cdot \vec{N} = 0\,, \tag{2.30}$$

and therefore the dot products of $\vec{L}$ and $\vec{M}$ vanish and the two Casimir operators in Eq. (2.29) are equal: $\vec{J}_1^{\,2} = \vec{J}_2^{\,2} = \vec{J}^{\,2}$.

The orthogonality of $\vec{N}$ and $\vec{L}$ can be used to write the square of the Runge–Lenz vector as

$$\vec{N}^2 = 2mE(\vec{L}^2 + \hbar^2) + m^2k^2\,, \tag{2.31}$$

which in turn can be rewritten in terms of the Casimir operator $\vec{J}^{\,2}$ as

$$4\vec{J}^{\,2} = -\hbar^2 - \frac{mk^2}{2E}\,. \tag{2.32}$$

Equation (2.32) contains the energy, the eigenvalues of which I would like to find. The Casimir operator can be replaced by its eigenvalues on the eigenstates to give

$$4j(j+1)\hbar^2 = -\frac{mk^2}{2E} - \hbar^2\,, \tag{2.33}$$

which is an algebraic equation that can be readily solved to provide the energy levels

$$E_j = -\frac{mk^2}{2\hbar^2(2j+1)^2} \quad \text{for } j = 0,\ \frac{1}{2},\ 1,\ \frac{3}{2}, \ldots \tag{2.34}$$

in which only the combination $n = 2j + 1$ appears. Replacing the coefficient k with its value[2.14] gives back the energy levels as written at the beginning in Eq. (2.19).

A backward glance shows how the energy levels are obtained by a purely algebraical argument without any recourse to the dynamical equations. Compare this, please, with the way we learn in basic quantum-mechanics courses about how to find the energy levels of the hydrogen atom by solving the Schrödinger equation – and consider how much simpler is the approach described above. Pauli's is a powerful, virtuosic computation, one that fills us lesser physicists (this one, anyway) with envy, even though we might take some consolation from the thought that it was done with group theory to which we too have access.

This algebraical computation brings an additional gift. I now have the key to understanding the n^2 degeneracy of the energy levels.

[2.13] This problem session completes my short review of group theory by introducing the celebrated **Schur's lemma**: if I have some n-dimensional irreducible representation, the matrices of which commute with the Hamiltonian, then the Hamiltonian is equal to the $n \times n$ identity matrix (times some constant). In the language of physics, the lemma implies the degeneracy of the energy of all states connected by a symmetry transformation. In the hydrogen-like atom, the symmetry is, as just discussed, $SO(4)$ and the degeneracy is that of the $n = 2j + 1$ levels connected by the generator $\vec{J}_1$ times that of the $n = 2j + 1$ levels connected by the generator $\vec{J}_2$.

[2.14] Recall: $k = Ze^2/4\pi\epsilon_0$.

The definition of the Lenz–Runge vector in Eq. (2.23) can be written as

$$\vec{p} \times \vec{L} - \vec{N} = m\,k\,\frac{\vec{r}}{r}\,. \tag{2.35}$$

Take both sides and dot-multiply them by themselves to obtain

$$m^2k^2 = (\vec{p} \times \vec{L}) \cdot (\vec{p} \times \vec{L}) - 2\vec{L} \cdot (\vec{N} \times \vec{p}) + \vec{N}^2\,. \tag{2.36}$$

The motion takes place in a plane (angular momentum is conserved) and I can write the vector components of the angular momentum and the momentum can be written as

$$\vec{L} = (0,0,L) \quad \text{and} \quad \vec{p} = (\pi_x, \pi_y, 0)\,. \tag{2.37}$$

In these components, Eq. (2.36) is written as

$$\left(\frac{mk}{L}\right)^2 = \pi_x^2 + \left(\pi_y - \frac{N}{L}\right)^2\,. \tag{2.38}$$

Equation (2.38) is the equation of a circle.[2.15] The momentum vector $\vec{p}$, for a given value of N and E, traces a circle in momentum space, what Hamilton called a **hodograph**, centered at the point $(0, N/L)$ on the (π_x, π_y) plane and with radius mk/L (see Figure 2.5).

2.15 Equation 2.38 is a nice result. It recovers, albeit in momentum rather than ordinary space, the circular motion so much loved by Ptolemy and that was lost with Kepler's ellipses. The ancients were right, after all: planets do move in circles. They only got wrong the space in which these circles are!

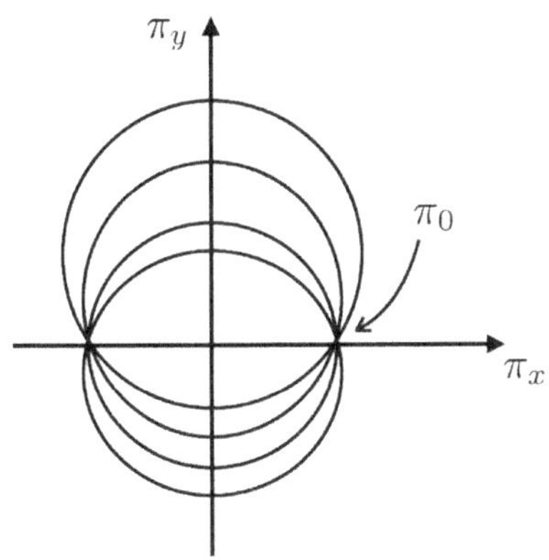

Figure 2.5 Hodograph orbits in momentum space.

All orbits in momentum space cross the π_x-axis at two points $\pi_0 = \pm\sqrt{-2mE}$ that are symmetrical about the π_y-axis. All hodographs with different values of L and M but which cross at the same points have the same energy. They are a degenerate set.

The mapping of one hodograph into another in the degenerate set is best seen by introducing a fourth component π_t for the momentum and going to four dimensions. The four-dimensional momenta describe a hypersphere

$$\pi_x^2 + \pi_y^2 + \pi_z^2 + \pi_t^2 = \pi_0^2 \tag{2.39}$$

with radius $\pi_0 = \sqrt{-2mE}$. Just as a sphere makes the $SO(3)$ symmetry manifest, the symmetry $SO(4)$ is made manifest by the hypersphere.

The degenerate set of hodographs comprises great circles on the hypersphere. The stereographic projection[2.16] of the hodographs on a hyperplane, which maps circles into circles, reproduces the collection of degenerate hodographs. By acting with the generators of the $SO(4)$ group, I can rotate one of these great circles into another. These rotations span the full n^2 degeneracy of each energy level.

2.16 An example of stereographic projection is the projection of a three-dimensional sphere onto the two-dimensional plane as is done when drawing maps of the whole Earth.

The same stereographic projection makes it possible to produce a picture showing how orbitals with different angular momenta can be rotated into each other, in the same way as an ordinary rotation in three dimensions sends orbitals with different components of the angular momentum, like the orbitals p_0, p_1 and p_{-1}, into one another.

Figure 2.6 shows how a spherically symmetric orbital s can be rotated into a rotationally asymmetric orbital like p_0 in a four-dimensional

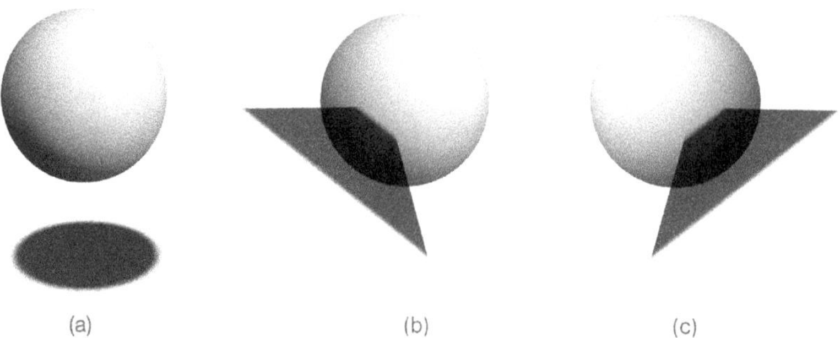

Figure 2.6 Wave functions with different j as well as m quantum numbers are related in the four-dimensional space of the momenta in Eq. (2.39). Analogy in the stereographic projection in three dimensions: an s-orbital (the sphere (a) on the left, the orbital shape shown by the projected shadow) is rotated into a p-orbital (the other spheres (b) and (c), the shadow depicting the asymmetric orbital) by changing the angle of the projection in the extra dimension.

construction. The picture is as suggestive as that provided by the rotation of a three-dimensional sphere to connect orbitals with different quantum numbers p. As states with different components of the angular momentum are related by rotations in three-dimensional space – which explains their energy degeneracy – rotations of the hypersphere in four dimensions relate states with different angular momentum – which explains the puzzling n^2-fold degeneracy of the hydrogen-like atom.

2.5 Magical Units: $c = \hbar = 1$

Now, before proceeding any further, is the time to adopt a sensible system of units of measure. Let us agree that we take the speed of light c and the reduced Planck constant $\hbar$ as both equal to 1. We can always do this by simply deciding to measure any speed in units of the light speed and any action in units of the reduced Planck constant. In the usual SI units, these two quantities have the values[2.17]

$$c = 299\,792\,458 \text{ m s}^{-1}$$
$$\hbar = 1.054\,571\,81\ldots \times 10^{-27} \text{ kg m}^2\text{s}^{-1}. \tag{2.40}$$

After this choice of natural units, there is only one dimensionful quantity: the energy, which I can measure in electron volts:[2.18]

$$1 \text{ eV} = 1.602\,176\,634 \times 10^{-19} \text{ kg m}^2\text{s}^{-2}. \tag{2.41}$$

All dimensionful quantities can be expressed in terms of eV.[2.19] I have

2.17 The values of c and $\hbar$ are exact numbers because of the way they are defined in the updated (2018) version of the SI. The dots in $\hbar$ come from the number of digits of π you want to keep.

2.18 Again, it is an exact number because so is the elementary electric charge.

2.19 The rest mass of the proton is about 1 GeV $= 10^9$ eV $\simeq 1.8 \times 10^{-27}$ kg.

$$[\text{mass}] \simeq [\text{energy}] \simeq [\text{momentum}] \simeq \text{eV}$$
$$[\text{time}] \simeq [\text{length}] \simeq \frac{1}{\text{eV}}. \tag{2.42}$$

Also, the space-time infinitesimal volume

$$\left[\mathrm{d}^4 x\right] \simeq \text{eV}^{-4} \tag{2.43}$$

and therefore the density of the Lagrangian will have to scale as eV^4 in order to make the action a pure number, which it has to be since it is proportional to $\hbar$, which is now equal to the number 1. Fermion (ψ) and boson fields (spin-1 A_μ and spin-0 ϕ) scale as follows:

$$[\psi] \simeq \text{eV}^{3/2} \quad \text{and} \quad \left[A_\mu\right],\ [\phi] \simeq \text{eV}. \tag{2.44}$$

The unit of charge is defined by taking the permittivity in vacuum $\varepsilon_0 = 1$ (and accordingly $\mu_0 = 1$), so that the charge of the electron is dimensionless (and the fine structure constant $\alpha = e^2/4\pi$).

It is useful to remember that

$$1 = \hbar c \simeq 197 \text{ MeV fm}, \tag{2.45}$$

where 1 fm = 10^{-13} cm $\simeq 5 \text{ GeV}^{-1}$. Equation (2.45) helps in going back to the usual units the few times when you want, or need, to do that.

This choice of units is not only a helpful trick for reducing the number of mistakes in writing down formulas.[2.20] Having only one dimensionful unit makes it possible to divide the world into small and large parts in a simple manner. Just set a scale and the characteristic energy of every process will fall either below or above it. As long as you remain above that energy scale, you need to worry only about the physics involving energies above that scale.

This approach of having telescoping theories separated by energy scales of increasing size brings back the ultimate reductionist dream of organizing the whole world in a single physical theory. Elementary-particle physics is the physics at energies above, say, 1 GeV, nuclear physics takes place between 1 GeV and 1 MeV, atomic physics deals with processes around 10 keV, molecular chemistry with those just below that and most of biology occurs at energies below 1 eV.[2.21]

2.20 It would be nice to be also able to set $2 = \pi = 1$!

2.21 From here on:

$c - 1$ and $\hbar = 1$.

2.6 Representations of the Poincaré Group

In the same way as we make our equations covariant under rotation by writing physical quantities as vectors (in three-dimensional Euclidean space), we make them relativistically covariant by writing them in terms of four-vectors in four-dimensional **Minkowski space**. This is a great advantage. I do not have to check for every equation what happens under a Lorentz transformation.

The scalar product of two four-vectors x and y in Minkowski space-time is defined by

$$x \cdot y = g_{\mu\nu} x^{\mu} y^{\nu} , \tag{2.46}$$

where the metric tensor $g_{\mu\nu}$ is given by the matrix[2.22]

2.22 The world is divided into people who use the same metric as I am using and those who use the one with the $(-1, 1, 1, 1)$ signature. Before starting a discussion with your colleague, to be safe, check whether you are both using the same metric!

$$\begin{pmatrix} 1 & 0 & 0 & 0 \\ 0 & -1 & 0 & 0 \\ 0 & 0 & -1 & 0 \\ 0 & 0 & 0 & -1 \end{pmatrix} \tag{2.47}$$

in the convention I follow.

Lorentz transformations are going to be, in this notation, 4×4 matrices Λ that leave the scalar product in Eq. (2.46) invariant, namely

$$(\Lambda x) \cdot (\Lambda y) = x \cdot y , \tag{2.48}$$

which in components gives

$$g_{\alpha\beta} = g_{\mu\nu} \, \Lambda^{\mu}_{\alpha} \Lambda^{\nu}_{\beta} . \tag{2.49}$$

Lorentz transformations are controlled by the parameters given by the elements of the matrix Λ, which is a 4×4 matrix and has 16 components – minus the components of the metric tensor (a symmetric 4×4 matrix with therefore 10 independent components) which has to be preserved. I thus count

$$4 \times 4 - 10 = 6 \tag{2.50}$$

independent parameters.

The group leaving the scalar product invariant is indicated by

$$\underset{\text{space} \nearrow}{O(3, \overset{\swarrow \text{time}}{1})} , \tag{2.51}$$

and is the group of orthogonal rotations in $3 + 1$ space-time dimensions. Time is unified to space in Minkowski space but still retains its own specificity, as discernible by the opposite sign in the metric component.

There is an additional condition originating in the fact that $\Lambda^{\mathrm{T}} g \Lambda = g$ implies that

$$(\det \Lambda)^2 = 1 \quad \Longrightarrow \quad \det \Lambda = \pm 1 . \tag{2.52}$$

The proper, orthochronous Lorentz group $SO(3, 1)$ is defined as having $\det \Lambda = 1$ and $\Lambda^0_0 \geq 1$. It preserves the time-like (or space-like) nature of four-vectors.

In practice, the Lorentz transformations in $SO(3, 1)$ are organized into **boosts** and **rotations** as follows:

$$\Lambda = L(\vec{\beta})\, R(\vec{\alpha}) = \begin{pmatrix} \gamma & \gamma\beta\vec{n}^{\mathrm{T}} \\ \gamma\beta\vec{n} & \mathbb{1} + (\gamma - 1)\vec{n}\cdot\vec{n}^{\mathrm{T}} \end{pmatrix} \begin{pmatrix} 1 & \vec{0}^{\mathrm{T}} \\ \vec{0} & R(\vec{\alpha}) \end{pmatrix}, \tag{2.53}$$

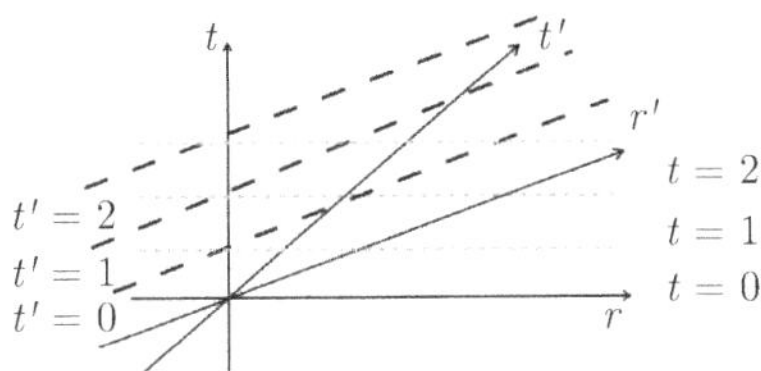

Figure 2.7 A boost in Minkowski space corresponds to tilting the axes of the space-time diagram. Many a seemingly paradoxical result in special relativity can be understood by using such diagrams.

where β is the speed of the boost, and $\gamma = (1 - \beta^2)^{-1/2}$. The boosts $L(\vec{\beta})$ are Lorentz transformations moving the frame of reference to another frame in motion with speed β in the direction $\vec{n}$ (three parameters) with respect to the original frame. They are represented by a tilting of the axes in Minkowski space, as in Figure 2.7. The rotations $R(\vec{\alpha})$ are, well, just rotations in three-dimensional space by an angle $\vec{\alpha}$ (three more parameters). You boost a rocket to its escape velocity; you rotate your head to look at it rising in the sky.

The components of a vector in different frames of reference are easily computed. Take a vector $\vec{a}$ in space at rest; its four-vector is given by

$$a = (0, \vec{a})\,. \tag{2.54}$$

A boost to a system moving with velocity $\vec{\beta}$ (for simplicity, the direction $\vec{n}$ is included in the speed $\vec{\beta}$ now) will give you the four-vector

$$\left(\gamma(\vec{\beta}\cdot\vec{a}),\, \vec{a} + \frac{\gamma^2}{1+\gamma}(\vec{\beta}\cdot\vec{a})\,\vec{\beta}\right). \tag{2.55}$$

The full invariance in Minkowski space is actually larger than that described by the Lorentz group because the scalar product in Eq. (2.46) is also invariant under the inhomogeneous transformation

$$x^{\mu\prime} = \Lambda^{\mu}_{\ \nu}\, x^{\nu} + a^{\mu}\,, \tag{2.56}$$

where a^μ are translations in space-time. The group has 10 parameters; it is the Lie group $ISO(3, 1)$, which is called the **Poincaré group**.

A representation $U(\Lambda, a)$ of the Poincaré group in some linear space can be written as[2.23]

$$U(\Lambda, a) = e^{-\frac{i}{2}\omega_{\mu\nu}M^{\mu\nu}}\, e^{-ia_\mu P^\mu}\,, \tag{2.57}$$

with $M^{\mu\nu}$ and P^μ the generators (the explicit form of which depends on the representation). The Lie algebra of the generators is

$$\begin{aligned} [M^{\mu\nu}, M^{\rho\sigma}] &= i\left(g^{\mu\sigma}M^{\nu\rho} + g^{\nu\rho}M^{\mu\sigma} - g^{\mu\rho}M^{\nu\sigma} - g^{\nu\sigma}M^{\mu\rho}\right) \\ [P^\mu, M^{\rho\sigma}] &= i\left(g^{\mu\rho}P^\sigma - g^{\mu\sigma}P^\rho\right) \\ [P^\mu, P^\nu] &= 0\,. \end{aligned} \tag{2.58}$$

These generators can be rewritten in terms of the generators J^i of the rotations and K^i of the boosts, which are defined as

$$J^i = \frac{1}{2}\epsilon^{ijk}M^{jk} \quad \text{and} \quad K^i = M^{0i} \tag{2.59}$$

2.23 The transformations with infinitesimally small a^μ and with

$$\Lambda^{\mu}_{\ \nu} = \delta^{\mu}_{\nu} + \omega^{\mu}_{\ \nu} \quad (\omega_{\mu\nu} = -\omega_{\nu\mu})$$

give rise to

$$U(\Lambda, a) \simeq -\frac{i}{2}\omega_{\mu\nu}M^{\mu\nu} - ia_\rho P^\rho + \cdots\,.$$

The factor one-half in front of the Lorentz part compensates for summing over all pairs of indices instead of only pairs with $\mu < \nu$.

for the Lorentz subgroup, and the Hamiltonian $P^0 = H$ and the momentum P_i for the space-time translations, yielding the following algebra[2.24]

2.24 The boosts do not form a subgroup!

$$\begin{aligned}
[J_i, J_j] &= i\epsilon_{ijk} J_k \\
[J_i, K_j] &= i\epsilon_{ijk} K_k \\
[K_i, K_j] &= -i\epsilon_{ijk} J_k \\
[J_i, P_j] &= i\epsilon_{ijk} P_k \\
[K_i, P_j] &= iH\delta_{ij} \\
[K_i, H] &= iP_i \\
[P_i, P_j] &= [J_i, H] = [H, H] = [P_i, H] = 0\,.
\end{aligned} \tag{2.60}$$

A representation U of the Poincaré group is given in terms of these generators as[2.25]

2.25 Since $\frac{1}{2}\omega_{\mu\nu}M^{\mu\nu} = \omega_{12}M^{12} + \omega_{13}M^{13} + \sum_{i=1}^{3}\omega_{0i}M^{i0} = \vec{\theta}\cdot\vec{J} - \vec{\omega}\cdot\vec{K}$.

$$U(\vec{\theta}, \vec{\omega}, \vec{a})^{\mu}_{\nu} = \left[e^{-i\vec{\theta}\cdot\vec{J} + i\vec{\omega}\cdot\vec{K} + i\vec{a}\cdot\vec{P}}\right]^{\mu}_{\nu}\,. \tag{2.61}$$

The specific representation is defined by the values for the three parameters $\vec{\theta}$, $\vec{\omega}$ and $\vec{a}$.

The representations in Eq. (2.61) act on the basis of the fields. For the Lorentz subgroup and for a scalar field ϕ,

$$U(\vec{\theta}, \vec{\omega}, \vec{0}) = \mathbb{1}_{1\times 1} \quad \text{and} \quad \phi \xrightarrow{\Lambda} \phi \tag{2.62}$$

while, for a vector field V^{μ},

$$U(\vec{\theta}, \vec{\omega}, \vec{0})^{\mu}_{\nu} = \Lambda^{\mu}_{\nu} \quad \text{and} \quad V^{\mu} \xrightarrow{\Lambda} \Lambda^{\mu}_{\nu} V^{\nu}\,, \tag{2.63}$$

in which $\Lambda = \exp(-i\omega_{\mu\nu}M^{\mu\nu}/2)$ with $(M^{\mu\nu})^{\alpha}_{\beta} = i(g^{\mu\alpha}\delta^{\nu}_{\beta} - g^{\nu\alpha}\delta^{\mu}_{\beta})$. For a spinor (Dirac) field ψ_a,

$$U(\vec{\theta}, \vec{\omega}, \vec{0})_{ab} = S_{ab} \quad \text{and} \quad \psi_a \xrightarrow{\Lambda} S_{ab}\psi_b\,, \tag{2.64}$$

in which $S = \exp(-i\omega_{\mu\nu}\sigma^{\mu\nu}/4)$ with $\sigma_{\mu\nu} = i[\gamma_\mu, \gamma_\nu]/2$. The matrices $\sigma_{\mu\nu}$ satisfy the Lie algebra in Eq. (2.58).[2.26]

2.26 The fields are here taken as being independent of the space-time coordinates. If the Lorentz transformation is applied also to the coordinates I have, for a generic field $\phi_a(x)$,

$$\begin{aligned}
\phi_a(x) &\xrightarrow{\Lambda} U_{ab}(\Lambda)\phi_b(\Lambda^{-1}x) \\
&= \left(e^{-i\omega_{\mu\nu}J^{\mu\nu}}\right)_{ab}\phi_b\,,
\end{aligned} \tag{2.65}$$

where $J^{\mu\nu} = M^{\mu\nu} + L^{\mu\nu}$ with $L^{\mu\nu} = i(x^{\mu}\partial^{\nu} - x^{\nu}\partial^{\mu})$. These representations are infinite dimensional.

Before the mathematical notation takes over and becomes jargon, it is useful to look at a specific example and actually compute the matrices of the representation. Take the case of the vector representation, for which the generators are

$$(M^{\mu\nu})^{\alpha}_{\beta} = i(g^{\mu\alpha}\delta^{\nu}_{\beta} - g^{\nu\alpha}\delta^{\mu}_{\beta})\,. \tag{2.66}$$

Explicit expressions for the boosts and the rotations can be computed by plugging in the 0s and 1s in the metric tensor according to the definition. The generators are then

$$K_1 = -i\begin{pmatrix}0&1&0&0\\1&0&0&0\\0&0&0&0\\0&0&0&0\end{pmatrix} \qquad J_1 = -i\begin{pmatrix}0&0&0&0\\0&0&0&0\\0&0&0&1\\0&0&-1&0\end{pmatrix}$$

$$K_2 = -i\begin{pmatrix}0&0&1&0\\0&0&0&0\\1&0&0&0\\0&0&0&0\end{pmatrix} \qquad J_2 = -i\begin{pmatrix}0&0&0&0\\0&0&0&-1\\0&0&0&0\\0&1&0&0\end{pmatrix} \tag{2.67}$$

$$K_3 = -i\begin{pmatrix}0&0&0&1\\0&0&0&0\\0&0&0&0\\1&0&0&0\end{pmatrix} \qquad J_3 = -i\begin{pmatrix}0&0&0&0\\0&0&-1&0\\0&-1&0&0\\0&0&0&0\end{pmatrix}.$$

The representation itself is obtained from these generators by using Eq. (2.61). For instance, a rotation about the z-axis is generated by the matrix[2.27]

2.27 Expand the exponential and use a Taylor series for the trigonometric functions. Notice that $(iJ_3)^3 = -iJ_3$ and $(iJ_3)^4 = iJ_3^2$ so that the even and odd powers in the series are multiplied by $(iJ_3)^2$ and iJ_3 respectively (the same happens for K_3, except that there is no minus sign).

$$e^{i\theta J_3} = \sum_{n=0}^{\infty} \frac{1}{n!}(iJ_3)^n\theta^n = \begin{pmatrix}1&0&0&0\\0&\cos\theta&-\sin\theta&0\\0&\sin\theta&\cos\theta&0\\0&0&0&1\end{pmatrix}, \tag{2.68}$$

and similarly a boost in the z-direction is generated by the matrix

$$e^{i\omega K_3} = \sum_{n=0}^{\infty} \frac{1}{n!}(iK_3)^n\omega^n = \begin{pmatrix}\cosh\omega&0&0&\sinh\omega\\0&1&0&0\\0&0&1&0\\\sinh\omega&0&0&\cosh\omega\end{pmatrix}. \tag{2.69}$$

Back to the general formalism. I can write the Casimir operators of the Lorentz group in terms of the generators in Eq. (2.58) as[2.28]

2.28 Notice the similarity to the invariants built out of the electric $\vec{E}$ and magnetic $\vec{B}$ fields: $\vec{E}\cdot\vec{B}$ and $\vec{E}^2 - \vec{B}^2$.

$$\frac{1}{2}M^{\mu\nu}M_{\mu\nu} = \vec{J}^2 - \vec{K}^2 \tag{2.70}$$

$$\frac{1}{2}\epsilon_{\mu\nu\alpha\beta}M^{\alpha\beta}M_{\mu\nu} = 2\vec{J}\cdot\vec{K}\,. \tag{2.71}$$

Classification of the various representations of the Lorentz group is easier using the two vectors

$$\vec{A} = \frac{1}{2}(\vec{J} - i\vec{K}) \quad \text{and} \quad \vec{B} = \frac{1}{2}(\vec{J} + i\vec{K}), \tag{2.72}$$

the algebra of which is

$$\begin{aligned}[][A_i, A_j] &= i\epsilon_{ijk}A_k\\ [B_i, B_j] &= i\epsilon_{ijk}B_k\\ [A_i, B_j] &= 0\,,\end{aligned} \tag{2.73}$$

and is therefore that of two (commuting) copies of $SU(2)$.[2.29]

2.29 The group $SO(3,1)$ is locally isomorphic to $SU(2) \oplus SU(2)$.

$(A,B)=(0,0)$ **Trivial rep. (scalars)** $U(\Lambda,0)^\alpha_\beta=\delta^\alpha_\beta$	$M^\mu_\nu=0$
$(1/2,1/2)$ **Fundamental rep. (vectors)** $U(\Lambda,0)^\alpha_\beta=\Lambda^\alpha_\beta$	$(M^{\mu\nu})^\alpha_\beta=-i(g^{\mu\alpha}\delta^\nu_\beta-g^{\nu\alpha}\delta^\mu_\beta)/2$
$(0,1/2)$ **Spinor rep. (right-handed Weyl)**	$\vec{J}=\frac{1}{2}\vec{\sigma},\quad \vec{K}=-i\frac{1}{2}\vec{\sigma}$
$(1/2,0)$ **Spinor rep. (left-handed Weyl)**	$\vec{J}=\frac{1}{2}\vec{\sigma},\quad \vec{K}=+i\frac{1}{2}\vec{\sigma}$
$(1/2,0)\oplus(0,1/2)$ **Spinor rep. (Dirac)**	$M^{\mu\nu}=\sigma^{\mu\nu}=\frac{i}{2}[\gamma^\mu\,\gamma^\nu]$

Figure 2.8 The irreducible representations of the Lorentz group in the field basis.

To sort out the irreducible representations I need the Casimir operators. The eigenvalues of the two Casimir operators $\vec{A}^2$ and $\vec{B}^2$ classify the irreducible representations $D^{(A,B)}$ of the Lorentz group by their eigenvalues A and B, as in Figure 2.8.

Figure 2.8 shows some of the possible bases and representations in terms of fields. We can have boson particles with spin equal to zero (or one). We can have fermion particles with spin one half – which come in two varieties as Weyl and Dirac spinors. Weyl spinors describe massless fermions. These are the fundamental fields of the Standard Model.

Moving to the full Poincaré group, the Casimir operators are

$$\begin{aligned} P^\mu P_\mu &= P^2 \\ W^\mu W_\mu &= W^2\,, \end{aligned} \tag{2.74}$$

where $W_\mu=-\epsilon_{\mu\rho\sigma\lambda}M^{\rho\sigma}P^\lambda/2$ is the Pauli–Lubanski vector,[2.30] which is related to rotations and boosts by the identities

2.30 The Pauli–Lubanski vector corresponds in four dimensions to what I called the Lenz–Runge vector in the problem session on the hydrogen atom.

$$W^0=\vec{P}\cdot\vec{J}\quad\text{and}\quad\vec{W}=P^0\vec{J}+\vec{P}\times\vec{K}\,. \tag{2.75}$$

The eigenvalues of the Casimir operators in Eq. (2.74) can be used to classify the irreducible representations of the Poincaré group in term of mass and spin (of the one-particle states) as

$$\begin{aligned} P^\mu|\vec{p},\sigma\rangle &= p^\mu|\vec{p},\sigma\rangle \\ \vec{J}^{\,2}|\vec{p},\sigma\rangle &= \sigma(\sigma+1)|\vec{p},\sigma\rangle \\ J_3|\vec{p},\sigma\rangle &= \sigma|\vec{p},\sigma\rangle\,, \end{aligned} \tag{2.76}$$

with $P^2=m^2$ and $W^2=-m^2\vec{J}^{\,2}=-m^2\sigma(\sigma+1)$.

The representations on the basis formed by one-particle states are organized as follows:

- by having P^μ diagonal, and
- by means of the **little group**.

Translations are therefore represented as

$$U(\mathbb{1},\vec{a})\,|\vec{p},\,\sigma\rangle = e^{-i\vec{p}\cdot\vec{a}}\,|\vec{p},\,\sigma\rangle\,. \tag{2.77}$$

The little group is defined by those Lorentz transformations λ that leave the momentum four-vector p^μ of the particle invariant:[2.31]

2.31 In the rest frame of the particle,

$$P^\mu = (m,\vec{0}) \text{ and } W^\mu = (0, m\vec{J})$$

and the little group gives the transformations of the spin of the particle under the Lorentz group – from which I can extract the representation to which the particle belongs.

$$\lambda^\mu_{\ \nu} p^\nu = p^\mu\,. \tag{2.78}$$

Under the little group, a one-particle state transforms as

$$U(\lambda,\vec{0})\,|\vec{p},\,\sigma\rangle = \sum_{\sigma',\,\sigma} D_{\sigma',\,\sigma}(\lambda)\,|\vec{p},\,\sigma\rangle\,, \tag{2.79}$$

which provides the remaining part of the representation. These representations are what we call **elementary particles**.[2.32] Elementary particles are their mass and spin. They embody the representations of the Poincaré group and our way of looking at the world through quantum mechanics and relativity.

2.32 This point of view is a subtle conceptual shift from the more usual one in which elementary particles are just discovered by their behavior and are classified by their mass and spin afterwards.

We are already using the mass and the spin of the elementary particles to classify them. Now it is clear from where these quantum numbers come: they are the eigenvalues of the Casimir operators of the Poincaré group, as shown in Table 2.3.

In the case of massless particles, the little group reduces to the two-dimensional space $E(2)$ of the helicity. The **helicity** is the projection (either left- or right-handed) of the spin of that particle along the direction of motion (given by its momentum),[2.33] as shown in Figure 2.9.

2.33 The helicity is left unchanged by Lorentz boosts.

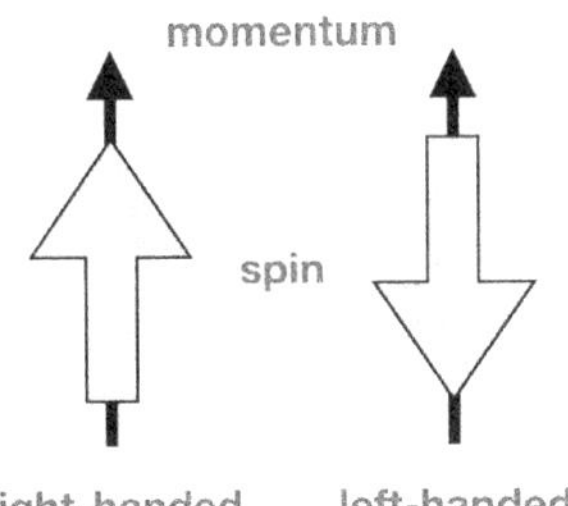

Figure 2.9 Helicity and handedness.

In this language, elementary-particle states are classified by two quantities, spin and momentum:

$$|\vec{p},\sigma\rangle\,. \tag{2.80}$$

(momentum → $\vec{p}$; spin → σ)

It is the space built out of these states that is called Fock space. It starts out empty: the **vacuum**. This space is filled with the population

Table 2.3 Possible eigenvalues of the Casimir operators of the Poincaré group and the corresponding one-particle state basis.

vacuum ($P^\mu = 0$)	$p^\mu = (0,0,0,0)$	little group: $SO(3,1)$
massive particle ($P^2 = m^2$)	$p^\mu = (m,0,0,0)$	little group: $SO(3)$, $J_i = 0, 1/2, 1$
massless particle ($P^2 = 0$)	$p^\mu = (\omega,0,0,\omega)$	little group: $ISO(2) \simeq E(2)$, J_3 (helicity)

of particles that we want include in our model by repeated action of the creation operator on the vacuum state, as in

$$|\vec{p},\sigma\rangle = \hat{a}^{\dagger}(\vec{p},\sigma)|0\rangle\,, \tag{2.81}$$

for a single particle. In a more general manner, the Fock state is defined as

$$|\Phi\rangle = \prod_{i=1}^{N} \hat{a}^{\dagger}(\vec{p}_i,\sigma_i)|0\rangle\,, \tag{2.82}$$

for N particles. I am going to use the Fock state when we come to the S-matrix and Feynman diagrams.

2.7 Particles and Fields*

This problem session is a bit on the formal side of things. The purpose is to point the reader toward the origin of the equations we use in quantum field theory and to why we use fields when the physics is actually given by the particles.

The representations $D^{(A,B)}$, introduced in the previous section, provide the transformations of the particle creation and annihilator operators under the Lorentz group. For instance, for the annihilator operator $\hat{a}(\vec{p},\sigma)$,

$$U[\Lambda,a]\,\hat{a}(\vec{p},\sigma)\,U^{-1}[\Lambda,a] = e^{(\Lambda p)\cdot a}\sqrt{\frac{(\Lambda p)^0}{p^0}}D^{(A,B)}[\lambda^{-1}(\Lambda,\vec{p})]^{\sigma}_{\rho}\,\hat{a}(\Lambda\vec{p},\rho)\,, \tag{2.83}$$

where the factor in the square root comes from the covariant normalization and $\lambda(\Lambda,\vec{p})$ represents a Lorentz transformation belonging to the little group.[6]

The representations $D^{(A,B)}$ are **unitary** but **infinite** dimensional. The impossibility of having finite unitary representations comes from the non-compactness of the group.[2.34] The non-covariant transformation of the creation and annihilator operators (due to the exponential factor coming from translations) makes writing invariant operators directly in terms of them difficult.

2.34 A group is **compact** if the corresponding manifold is topologically compact. The group $U(1)$ is compact because its manifold is the circle S. The manifold of the Lorentz boosts in a given direction is $\mathbb{R}$, which is not compact. Same for the translations in the Poincaré group.

The quantum operator representing the field

$$\psi^{\alpha}(x) = \sum_{\sigma}\int \mathrm{d}p\,\big[\underbrace{\hat{a}(\vec{p},\sigma)}_{\text{particle}}u^{\alpha}(\vec{p},\sigma)\,e^{-ip\cdot x} + \underbrace{\hat{b}^{\dagger}(\vec{p},\sigma)}_{\text{antiparticle}}v^{\alpha}(\vec{p},\sigma)\,e^{+ip\cdot x}\big] \tag{2.84}$$

transforms, under the same Poincaré group, as

$$U[\Lambda,a]\psi^{\alpha}(x)U^{-1}[\Lambda,a] = D[\Lambda^{-1}]^{\alpha}_{\beta}\,\psi^{\beta}(\Lambda x + a)\,, \tag{2.85}$$

[6] The reference for this problem session is, again, S. Weinberg, *Quantum Field Theory* (Cambridge University Press, 1995).

where the representations $D[\Lambda^{-1}]^{\alpha}_{\beta}$ are **finite** dimensional but **non-unitary**. The finiteness of these representations together with the covariant transformation property is the reason why fields are introduced in the first place. They make it possible to write the Lagrangians we use in defining the Standard Model.

The connection between the unitary (infinite) and the non-unitary (finite) representations,

$$D[\Lambda^{-1}]^{\alpha}_{\beta}\, u^{\beta}(\vec{p},\sigma)\, e^{ip\cdot(\Lambda x+b)} = \sqrt{\frac{(\Lambda p)^0}{p^0}} e^{(\Lambda p)\cdot a}\, D^{(A,B)}[\lambda^{-1}(\Lambda,\vec{p})]^{\sigma}_{\rho} u^{\alpha}(\Lambda\vec{p},\rho) e^{ip\cdot x}, \quad (2.86)$$

with a similar relation for the wave function $v^{\beta}(\vec{p},\sigma)$ of the antiparticle, provides a set of constraints on the wave functions u and v of the field operator.

Equation (2.86) with its many indices is quite a handful, because it is very general. Let us take it now for the case of a boost $L(q)$ from the rest frame (of a particle of mass m) to a momentum $\vec{q}$ for which $\lambda^{-1}(\Lambda,\vec{p}\,) = 1$, so that Eq. (2.86) simplifies to

$$u^{\beta}(\vec{q},\sigma) = \sqrt{\frac{m}{q^0}} D[L(q)]^{\beta}_{\alpha} u^{\alpha}(0,\sigma)\,. \quad (2.87)$$

Similarly, for a rotation R, $\lambda^{-1}(\Lambda,\vec{p}\,) = R$ and Eq. (2.86) is just

$$u^{\beta}(0,\sigma)\vec{J}_{\sigma,\rho} = D[R]^{\beta}_{\alpha} u^{\alpha}(0,\rho)\,, \quad (2.88)$$

where the $\vec{J}_{\sigma,\rho}$ form the angular-momentum (spin-dependent) representation of the rotation.

The constraints in Eqs. (2.87) and (2.88) can be solved to obtain the equations of motion of the particles, namely, depending on the spin, the Klein–Gordon, Dirac and Maxwell equations.

For example, in the simplest case, namely that of a scalar particle, for which $D(\Lambda) = 1$ and $J_{\sigma,\rho} = 0$, Eq. (2.88) has only a constant solution for the wave function:

$$u(\vec{p}) = \frac{1}{\sqrt{2p^0}}\,, \quad (2.89)$$

with a similar one for the antiparticle wave function $v(\vec{p}\,)$. The wave function has no index and satisfies the Klein–Gordon equation

$$(\Box^2 + m^2)\,\frac{e^{ip\cdot x}}{\sqrt{2p^0}} = 0\,. \quad (2.90)$$

The field is given by

$$\psi(x) = \frac{1}{\sqrt{2p^0}} \int \mathrm{d}p \left[\hat{a}(\vec{p}\,)\,e^{-ip\cdot x} + \hat{b}^\dagger(\vec{p}\,)\,e^{+ip\cdot x}\right], \tag{2.91}$$

which is the quantum operator for a complex scalar (spin-0) field.[2.35]

2.35 The same steps can be followed in the case of the spin-1 and Dirac fermions – only with more complicated expressions and more work – which the dedicated reader is encouraged to explore. A derivation of the Dirac equation from the spinor representation is presented in Appendix D.

Historically it was the other way around, with first the discovery of the equations of motion of the various particles and then the study of the elementary-particle representations. Nevertheless, the most logical path starts with the particle representations; the equations satisfied by the wave function in the field operators then follow as constraints that these representations must satisfy.

2.8 Discrete Symmetries

There are three **discrete** transformations that are of interest in particle physics. Two act on space-time coordinates. They are the **parity** transformation $\hat{P}$, which changes the sign of the space coordinates:

$$\hat{P} \,:\, (\vec{x}, t) \to (-\vec{x}, t)\,, \tag{2.92}$$

and the **time reversal** transformation $\hat{T}$, which changes the sign of the time coordinate:

$$\hat{T} \,:\, (\vec{x}, t) \to (\vec{x}, -t). \tag{2.93}$$

Because of the negative sign of their corresponding determinants, $\hat{P}$ and $\hat{T}$ belong to the group $O(3,1)$ rather than $SO(3,1)$.

A third discrete transformation, the **charge conjugation** transformation $\hat{C}$, exchanges particles with antiparticles, thus reversing the charge (let me denote it by e):[2.36]

2.36 Symmetry under charge conjugation does not imply charge conservation (which comes from invariance under a **continuous** symmetry) and *vice versa* a violation of charge conjugation does not imply that the charge is not conserved.

$$\hat{C} \,:\, e \to -e\,. \tag{2.94}$$

These discrete symmetries were found to be important in the old days of particle physics and they are still useful in characterizing interactions.

To discuss the transformation properties of a specific process it is essential to keep track of the helicity of the particles. A parity transformation changes the direction of the momentum, leaving the spin unchanged. The helicity is therefore switched. A charge conjugation leaves the momentum alone but switches the spin and therefore the helicity.

How do these discrete transformations act on matter, that is, on the spinor fields describing elementary particles? To get there I need to start from the electromagnetic fields.

The transformation properties of the electromagnetic fields $\vec{E}$ and $\vec{B}$ can be deduced from the form of the Lorentz force,

$$\frac{\mathrm{d}\vec{p}}{\mathrm{d}t} = e\left(\vec{E} + \vec{v} \times \vec{B}\right), \tag{2.95}$$

to which it is easy to apply the discrete symmetry transformations.

Beginning with parity, it transforms the coordinates $\vec{x}$ into $-\vec{x}$ and therefore changes the sign of the momentum $\vec{p}$. For Eq. (2.95) to remain valid $\vec{E}$ must flip sign while $\vec{B}$ remains unchanged since it is $\vec{v}$ that changes sign. A charge conjugation flips the sign of e but leaves the left-hand side of Eq. (2.95) unchanged. Accordingly, both $\vec{E}$ and $\vec{B}$ must flip sign. Finally, under a time reversal, both the momentum $\vec{p}$ and the coordinate t change sign so that the force is left unchanged and therefore $\vec{E}$ must remain the same while $\vec{B}$, whose term contains the velocity $\vec{v}$, which changes sign, must flip sign. Thus the actions of the discrete transformations are as follows:

$$\vec{E}(\vec{x},t) \quad \Longrightarrow \quad \begin{cases} \stackrel{\hat{P}}{\rightarrow} & -\vec{E}(-\vec{x},t) \\ \stackrel{\hat{T}}{\rightarrow} & \vec{E}(\vec{x},\,-t) \\ \stackrel{\hat{C}}{\rightarrow} & -\vec{E}(\vec{x},t) \end{cases} \tag{2.96}$$

and

$$\vec{B}(\vec{x},t) \quad \Longrightarrow \quad \begin{cases} \stackrel{\hat{P}}{\rightarrow} & \vec{B}(-\vec{x},t) \\ \stackrel{\hat{T}}{\rightarrow} & -\vec{B}(\vec{x},\,-t) \\ \stackrel{\hat{C}}{\rightarrow} & -\vec{B}(\vec{x},t)\,. \end{cases} \tag{2.97}$$

To study the transformation properties of spinors I actually need those of the electromagnetic potentials φ and $\vec{A}$.[2.37] Their properties follow from their definition in terms of the electromagnetic fields and are as follows:

2.37 $\vec{E} = -\vec{\nabla}\varphi - \dfrac{\partial \vec{A}}{\partial t}$
$\vec{B} = \vec{\nabla} \times \vec{A}\,.$

$$\vec{A}(\vec{x},t) \quad \Longrightarrow \quad \begin{cases} \stackrel{\hat{P}}{\rightarrow} & -\vec{A}(-\vec{x},t) \\ \stackrel{\hat{T}}{\rightarrow} & -\vec{A}(\vec{x},\,-t) \\ \stackrel{\hat{C}}{\rightarrow} & -\vec{A}(\vec{x},t) \end{cases} \tag{2.98}$$

and

$$\phi(\vec{x},t) \quad \Longrightarrow \quad \begin{cases} \stackrel{\hat{P}}{\rightarrow} & \phi(-\vec{x},t) \\ \stackrel{\hat{T}}{\rightarrow} & \phi(\vec{x},\,-t) \\ \stackrel{\hat{C}}{\rightarrow} & -\phi(\vec{x},t)\,. \end{cases} \tag{2.99}$$

These transformation rules can be used in the Dirac equation (see Appendix D for a derivation) to determine how the Dirac spinor Ψ transforms under the same operations. The explicit form of this equation depends on the representation of the Dirac matrices that is chosen; here I use

$$\gamma^0 = \begin{pmatrix} \mathbb{0} & \mathbb{1} \\ -\mathbb{1} & \mathbb{0} \end{pmatrix} \quad \text{and} \quad \gamma^i = \begin{pmatrix} \mathbb{0} & -\sigma^i \\ \sigma^i & \mathbb{0} \end{pmatrix}. \tag{2.100}$$

Since the discrete transformations of spinors are non-trivial, unlike those for scalar and vector quantities, I will go through the derivation in detail.

Consider the Dirac equation in an external electromagnetic field and separate the time and space components:

$$\left\{\gamma^0\left[i\frac{\partial}{\partial t}-e\phi(\vec{x},t)\right]-\gamma^i\left[i\frac{\partial}{\partial x^i}-eA_i(\vec{x},t)\right]-m\right\}\Psi(\vec{x},t)=0\,. \tag{2.101}$$

Let us begin with the parity transformation. Under a parity transformation $\hat{P}$, Eq. (2.101) becomes

$$\left\{\gamma^0\left[i\frac{\partial}{\partial t}-e\phi(-\vec{x},t)\right]-\gamma^i\left[i\frac{\partial}{\partial(-x^i)}+eA_i(-\vec{x},t)\right]-m\right\}\Psi(-\vec{x},t)=0\,, \tag{2.102}$$

which can be rewritten as (multiply by γ^0 on the left, and then move the γ^0 across[2.38])

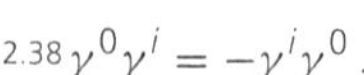
2.38 $\gamma^0\gamma^i=-\gamma^i\gamma^0$.

$$\left\{\gamma^0\left[i\frac{\partial}{\partial t}-e\phi(-\vec{x},t)\right]-\gamma^i\left[i\frac{\partial}{\partial x^i}-eA_i(-\vec{x},t)\right]-m\right\}\underbrace{\gamma^0\Psi(-\vec{x},t)}_{\text{solution}}=0\,. \tag{2.103}$$

The quantity outside the braces in Eq. (2.103) is the solution of the parity-transformed equation.

The parity transformation of $\Psi(\vec{x},t)$ is not just a change in the argument, as in $\Psi(-\vec{x},t)$, because the spinor is also multiplied by γ^0. For the Dirac equation to be invariant under parity (Figure 2.10), the spinor must transform as follows:

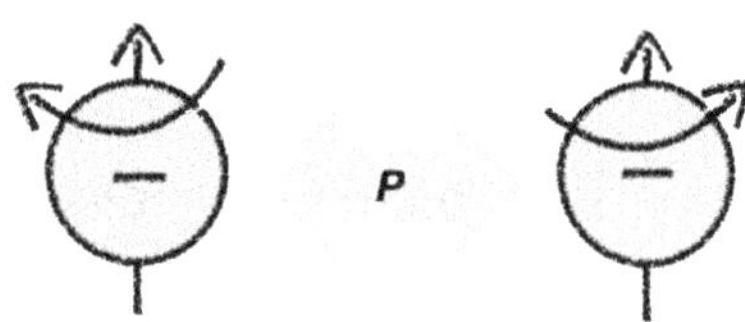

Figure 2.10 A parity transformation.

$$\boxed{\hat{P}:\ \Psi(\vec{x},t)\xrightarrow{\hat{P}}\eta_P\boldsymbol{\gamma}^{\mathbf{0}}\Psi(-\vec{x},t)\,,} \tag{2.104}$$

with an arbitrary phase factor η_P of unit magnitude. This phase plays no physical role.

Because $\{\gamma^0,\gamma^5\}=0$, the chiral spinors

$$\psi_L=\frac{1}{2}(1-\gamma^5)\psi\quad\text{and}\quad\psi_R=\frac{1}{2}(1+\gamma^5)\psi \tag{2.105}$$

are exchanged under the parity transformation:

$$\hat{P}:\ \psi_L\to\psi_R\quad\text{and}\quad\hat{P}:\ \psi_R\to\psi_L\,. \tag{2.106}$$

Moving on to charge transformation: it sends particles into antiparticles, as in Figure 2.11; I must conjugate the equation, change $e\to -e$ and transform the potentials according to Eq. (2.98) and Eq. (2.99):

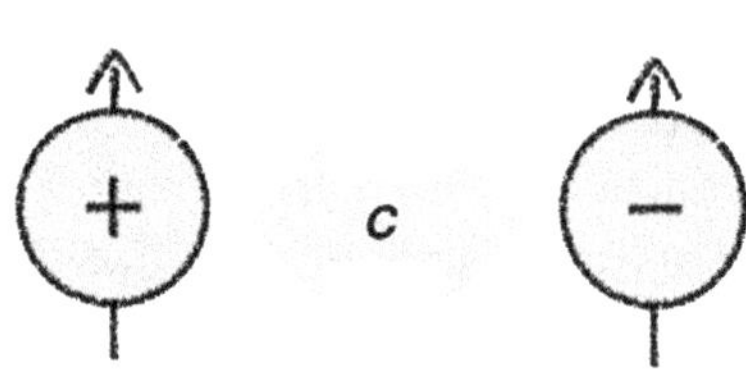

Figure 2.11 Charge conjugation.

$$\begin{aligned}&\left\{-\gamma^0\left[i\frac{\partial}{\partial t}+e\phi(\vec{x},t)\right]+\gamma^1\left[i\frac{\partial}{\partial x^1}+eA_1(\vec{x},t)\right]\right.\\&\left.-\gamma^2\left[i\frac{\partial}{\partial x^2}+eA_2(\vec{x},t)\right]+\gamma^3\left[i\frac{\partial}{\partial x^3}+eA_3(\vec{x},t)\right]-m\right\}\Psi^*(\vec{x},t)=0\,.\end{aligned} \tag{2.107}$$

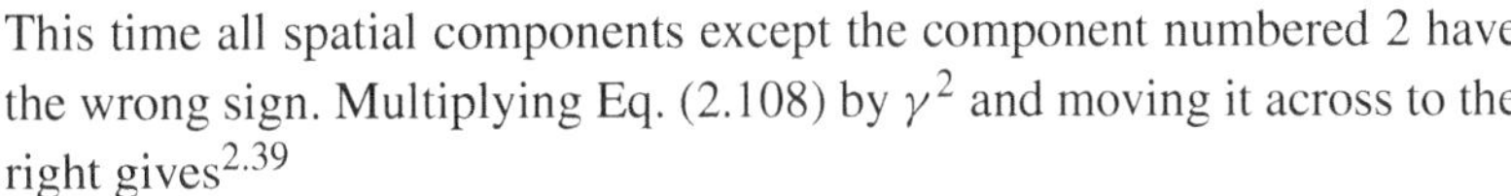

This time all spatial components except the component numbered 2 have the wrong sign. Multiplying Eq. (2.108) by γ^2 and moving it across to the right gives[2.39]

$$\left\{\gamma^0\left[i\frac{\partial}{\partial t}+e\phi(\vec{x},t)\right]-\gamma^i\left[i\frac{\partial}{\partial x^i}+eA_i(\vec{x},t)\right]-m\right\}\underbrace{\gamma^2\Psi^*(\vec{x},t)}_{\text{solution}}=0\,. \tag{2.108}$$

2.39 $\gamma^{0*}=\gamma^0$, $\gamma^{1*}=\gamma^1$, $\gamma^{2*}=-\gamma^2$, $\gamma^{3*}=\gamma^3$.

In other words, for the Dirac equation to be invariant under charge conjugation, the Dirac spinor must transform as[2.40]

2.40 Again, η_C is a phase factor.

$$\boxed{\hat{C}:\ \Psi(\vec{x},t)\xrightarrow{\hat{C}}\eta_C\boldsymbol{\gamma^2}\Psi^*(\vec{x},t)\,,} \tag{2.109}$$

which can also be written using $\bar{\Psi}^T=\gamma^0\Psi^*$ as

$$\eta_C\gamma^0\gamma^2\bar{\Psi}^T(\vec{x},t)\,. \tag{2.110}$$

I can define spinors that are self-conjugating under $\hat{C}$ and for which

$$\eta_C\gamma^0\gamma^2\bar{\Psi}(\vec{x},t)^T=\Psi(\vec{x},t)\,. \tag{2.111}$$

They are called **Majorana spinors** and describe truly neutral particles. They will come in handy when I discuss neutrinos in the last chapter.

Finally, the time-reversal transformation, Figure 2.12, consists in conjugation, $t\to -t$ exchange and transformation of the potentials according to Eq. (2.98) and Eq. (2.99). It changes the Dirac equation to[2.41]

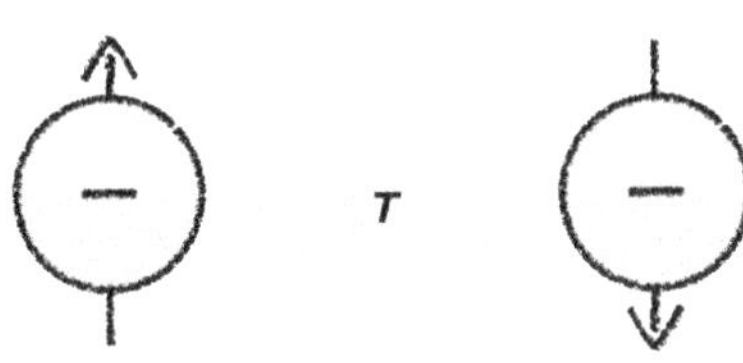

Figure 2.12 Time reversal.

2.41 Time reversal exchanges the final and initial states and conjugates the fields.

$$\begin{aligned}\Big\{-\gamma^0\left[i\frac{\partial}{\partial(-t)}+e\phi(\vec{x},-t)\right]+\gamma^1\left[i\frac{\partial}{\partial x^1}-eA_1(\vec{x},-t)\right]\\ -\gamma^2\left[i\frac{\partial}{\partial x^2}-eA_2(\vec{x},-t)\right]+\gamma^3\left[i\frac{\partial}{\partial x^3}-eA_3(\vec{x},-t)\right]-m\Big\}\ \Psi^*(\vec{x},-t)=0\,.\end{aligned} \tag{2.112}$$

The spatial components numbered 1 and 2 have the wrong sign this time. Multiplying on the left by $\gamma^1\gamma^3$ gives

$$\left\{\gamma^0\left[i\frac{\partial}{\partial t}-e\phi(\vec{x},-t)\right]-\gamma^i\left[i\frac{\partial}{\partial x^i}-eA_i(\vec{x},-t)\right]-m\right\}\underbrace{\gamma^1\gamma^3\Psi^*(\vec{x},-t)}_{\text{solution}}=0, \tag{2.113}$$

and therefore the Dirac spinor must transform as

$$\boxed{\hat{T}:\ \Psi(\vec{x},t)\xrightarrow{\hat{T}}\eta_T\boldsymbol{\gamma^1\gamma^3}\Psi^*(\vec{x},-t)\,,} \tag{2.114}$$

with η_T again a phase factor.

An electron with an electric dipole moment violates time-reversal symmetry. Both the electric dipole moment and the magnetic moment of the electron lie along the same axis as the electron spin. The operation of time reversal reverses the magnetic moment but does not affect the electric dipole moment; therefore an electron with a non-zero electric dipole moment violates time-reversal symmetry.

The time transformation is anti-unitary. Acting on the commutator $[\hat{x}, \hat{p}] = i\hbar$ results in $\hat{T}\hat{x}\hat{T}^{-1} = \hat{x}$, while $\hat{T}\hat{p}\hat{T}^{-1} = -\hat{p}$, and therefore $\hat{T}i\hat{T}^{-1} = -i$ must also hold.

It is worth dwelling for a moment on these particulars because they illuminate the relationship between quantum mechanics and time reversal. To time-reverse the evolution of a system we must have

$$\hat{T}^{-1}e^{-iHt}\hat{T} = e^{iHt}. \tag{2.115}$$

Equation (2.115) implies that (take the time derivative at $t = 0$)

$$\hat{T}^{-1}(-iHt)\hat{T} = iHt, \tag{2.116}$$

which for a unitary operator would imply $\hat{T}^{-1}(-H)\hat{T} = H$. This is wrong – the operator H cannot be unitarily related to $-H$ because that would imply (since the spectra of unitarily related operators are the same) that the spectrum of the Hamiltonian H is not bounded from below (given that its spectrum is not bounded from above). The operator $\hat{T}$ must then be anti-unitary so that $\hat{T}^{-1}(H)\hat{T} = H$. Unitary and anti-unitary operators are the only operators preserving probabilities.

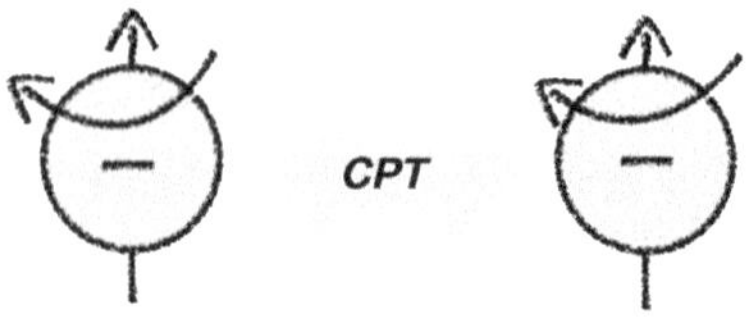

The combined operation $\hat{C}\hat{P}\hat{T}$, as in Figure 2.13, is always a symmetry in quantum field theory. This can be understood by considering the combined action of $\hat{P}\hat{T}$ – which is the transformation in the Lorentz group given by a boost with rapidity $i\pi$. $\hat{P}\hat{T}$ is equivalent to $\hat{C}$ and therefore $\hat{C}\hat{P}\hat{T} = \hat{C}^2 = \mathbb{1}$ and $\hat{C}\hat{P}\hat{T}$ symmetry follows in quantum field theory from Lorentz invariance.

CPT

Figure 2.13 *CPT* transformation.

The validity of this symmetry is currently experimentally confirmed by comparing the mass and lifetime of a particle with that of its antiparticle. For instance, for neutral kaons,[7]

$$\left|\frac{m_{K^0} - m_{\bar{K}^0}}{m_{K^0} + m_{\bar{K}^0}}\right| < 3 \times 10^{-19}, \tag{2.117}$$

and for their widths,

$$\left|\frac{\Gamma_{K^0} - \Gamma_{\bar{K}^0}}{\Gamma_{K^0} + \Gamma_{\bar{K}^0}}\right| < (4 \pm 4) \times 10^{-18}. \tag{2.118}$$

Also, the charges of particle and antiparticle are expected to be of opposite sign but the same size. For example, for the electron and positron,

$$q_{e^+} + q_{e^-} < 4 \times 10^{-8}\, e. \tag{2.119}$$

Can $\hat{C}\hat{P}\hat{T}$ symmetry be broken? Though it is easier to not tread along that path but to always assume that $\hat{C}\hat{P}\hat{T}$ is a good symmetry (because of

[7] Data from R.L. Workman *et al.* (Particle Data Group), *Prog. Theor. Exp. Phys.* **2022** (2022) 083C01 http://pdg.lbl.gov.

its intimate relation to Lorentz invariance and locality), in fact it can be broken without necessarily breaking Lorentz invariance. For instance,[2.42] if you introduce a $\hat{C}\hat{P}\hat{T}$-non-invariant interaction, the widths of particles and antiparticles will no longer be equal whereas their masses will remain the same. *Vice versa*, a symmetry-breaking mass term (as could arise for a bound state) will lead to different masses for particles and antiparticles while leaving their widths the same.

2.42 For the brave of heart.

2.9 *Problem Session*: The Hydrogen Atom, Again

I want to go back to the hydrogen atom and discuss the effect of having a relativistically invariant equation for the electron. I could just solve the Dirac equation in the central potential of the nucleus but it is more instructive to see how the components of the Dirac spinor separate to give rise to the relativistic corrections.

This discussion should give a more tangible feeling to the Dirac equation, a feeling that will be helpful when I move on to the computation of Standard Model amplitudes.

First, take the Dirac equation for a free electron. The Hamiltonian is given by

$$H = \vec{\alpha} \cdot \vec{p} + \beta m\,, \tag{2.120}$$

with

$$\beta = \begin{pmatrix} \mathbb{1} & \mathbb{0} \\ \mathbb{0} & -\mathbb{1} \end{pmatrix} \quad \text{and} \quad \vec{\alpha} = \begin{pmatrix} \mathbb{0} & \vec{\sigma} \\ \vec{\sigma} & \mathbb{0} \end{pmatrix} \tag{2.121}$$

in the Dirac representation. The Dirac wave function

$$\psi = \begin{pmatrix} \varphi \\ \chi \end{pmatrix} \tag{2.122}$$

has two elements: φ which has two "large" components, and χ which has two "small" components. The small components are small because the free Dirac equation (with the speed of light c reintroduced explicitly for the next two equations) couples φ and χ:

$$\begin{aligned} E\varphi &= c\,\vec{\sigma} \cdot \vec{p}\,\chi + mc^2\varphi \\ E\chi &= c\,\vec{\sigma} \cdot \vec{p}\,\varphi - mc^2\chi\,. \end{aligned} \tag{2.123}$$

Equation (2.123) can be solved approximately (using $E \simeq mc^2$) to give

$$\chi \simeq \frac{\vec{\sigma} \cdot \vec{p}}{2mc}\varphi\,, \tag{2.124}$$

which shows that the components of χ are c times smaller than those in φ.

2.43 The Foldy–Wouthyusen transformation.

I want to find a transformation of the wave function[2.43]

$$\psi' = e^{iS}\psi \tag{2.125}$$

such that the new Hamiltonian H' contains only **even** operators, that is, operators that act only between the large (or between the small) components. Operators mixing the large and small components are called **odd**. The Dirac matrix β is even while the matrices $\vec{\alpha}$ are odd. The desired transformation can be achieved by taking

$$S = \beta\frac{\vec{\alpha}\cdot\vec{p}}{|\vec{p}\,|}\theta \tag{2.126}$$

and writing the transformed Hamiltonian

$$\begin{aligned} H' = e^{iS}He^{-iS} &= \left(\cos\theta + \beta\frac{\vec{\alpha}\cdot\vec{p}}{|\vec{p}\,|}\sin\theta\right) \\ &\quad \times(\vec{\alpha}\cdot\vec{p}+\beta m)\left(\cos\theta - \beta\frac{\vec{\alpha}\cdot\vec{p}}{|\vec{p}\,|}\sin\theta\right) \\ &= (\vec{\alpha}\cdot\vec{p}+\beta m)\left(\cos\theta - \beta\frac{\vec{\alpha}\cdot\vec{p}}{|\vec{p}\,|}\sin\theta\right)^2 \\ &= \vec{\alpha}\cdot\vec{p}\left(\cos 2\theta - \frac{m}{|\vec{p}\,|}\sin 2\theta\right) + \beta\left(m\cos 2\theta - |\vec{p}\,|\sin 2\theta\right). \end{aligned} \tag{2.127}$$

Choosing $\tan\theta = |\vec{p}\,|/m$ gives

$$H' = \beta\sqrt{m^2 + |\vec{p}\,|^2}, \tag{2.128}$$

which acts separately on the small and large components and has the usual form of the relativistic energy (for a free electron).

For the hydrogen atom, the electron is no longer free and the Hamiltonian is given by

$$H = c\vec{\alpha}\cdot\left(\vec{p} - e\vec{A}\right) + \beta m + e\,\Phi, \tag{2.129}$$

where $\vec{A}$ is the vector potential and Φ the scalar potential of the field of the nucleus.

The "odd" operator is

$$\hat{O} = \vec{\alpha}\cdot\left(\vec{p} - e\vec{A}\right) \tag{2.130}$$

while

$$\hat{E} = \beta m + e\,\Phi \tag{2.131}$$

is even.

To eliminate the odd operator, the same kind of transformation can be tried as in the free case by taking

$$S = -i\beta\frac{\hat{O}}{2m}; \tag{2.132}$$

acting on the Hamiltonian H gives[2.44]

[2.44] Use

$$e^A B e^A = \sum_{n=0}^{\infty} \frac{1}{n!}[A,[A,[\ldots[A,B]]]\ldots].$$

$$\begin{aligned} H' = e^{iS} H e^{-iS} &= H + i[S,H] - \frac{1}{2}[S,[S,H]] - \frac{i}{6}[S,[S,[S,H]]] \\ &= +\frac{1}{24}[S,[S,[S,\beta m]]] - \dot{S} - \frac{i}{2}[S,\dot{S}] + \frac{1}{6}[S,[S,\dot{S}]] + O\left(\frac{1}{m^4}\right), \end{aligned} \tag{2.133}$$

where the time derivatives (indicated by overdots) come from the Schrödinger equation. There is quite a handful of terms in Eq. (2.133):

$$\begin{aligned} i[S,H] &= -\hat{O} + \frac{\beta}{2m}[\hat{O},\hat{E}] = \frac{1}{m}\beta\hat{O}^2 \\ \frac{i^2}{2}[S,[S,H]] &= \frac{\beta\hat{O}^2}{2m} - \frac{1}{8m^2}[\hat{O},[\hat{O},\hat{E}]] - \frac{\hat{O}^2}{2m^2} \\ \frac{i^3}{3!}[S,[S,[S,H]]] &= \frac{\hat{O}^3}{6m^2} - \frac{1}{6m^2}\beta\hat{O}^4 \\ -\dot{S} &= \frac{i}{2m}\beta\hat{\dot{O}}^2 \\ \frac{i^4}{4!}[S,[S,[S,[S,H]]]] &= \frac{i}{24m^3}\beta\hat{O}^4 \\ -\frac{i}{2}[S,\dot{S}] &= -\frac{i}{2m^2}[\hat{O},\dot{\hat{O}}], \end{aligned} \tag{2.134}$$

which, when gathered up, gives

$$\begin{aligned} H' &= \beta\left(m + \frac{\hat{O}^2}{2m} - \frac{\hat{O}^4}{8m^3}\right) + \hat{E} \\ &\quad - \frac{1}{8m^2}[\hat{O},[\hat{O},\hat{E}]] - \frac{i}{8m^2}[\hat{O},\dot{\hat{O}}] \\ &\quad + \frac{\beta}{2m}[\hat{O},\hat{E}] - \frac{\hat{O}^3}{3m^2} + i\frac{\beta\dot{\hat{O}}}{2m} \\ &= \beta m + \hat{E}' + \hat{O}'. \end{aligned} \tag{2.135}$$

The first two lines of Eq. (2.135) define the even operator $\hat{E}'$ and the last one the odd operator $\hat{O}'$. Since there is still an odd operator in the Hamiltonian I must persist and eliminate $\hat{O}'$ with another transformation,

$$H'' = e^{iS'} H e^{-iS'}, \tag{2.136}$$

where $S' = -i\beta\hat{O}'/2m$.[2.45] This, as it turns out, is still not sufficient and the Hamiltonian

[2.45] With O' given by the last line in Eq. (2.135).

$$H'' = \beta m + \hat{E}' + \hat{O}'' \tag{2.137}$$

still has an odd part, which is $O(1/m^2)$. Performing another (and last) transformation by means of $S'' = -i\beta\hat{O}''/2m$ finally brings the Hamiltonian to a fully even form:[2.46]

[2.46] With E' given by the two first lines in Eq. (2.135).

$$
\begin{aligned}
H''' &= e^{iS''} H e^{-iS''} = \beta m + \hat{E}' \\
&= \beta\left(m + \frac{\hat{O}^2}{2m} - \frac{\hat{O}^4}{8m^3}\right) + \hat{E} - \frac{1}{8m^2}[\hat{O},[\hat{O},\hat{E}]] - \frac{i}{8m^2}[\hat{O},\dot{\hat{O}}] .
\end{aligned}
\tag{2.138}
$$

The terms in Eq. (2.139) can be written out to give

$$
\begin{aligned}
\frac{\hat{O}^2}{2m} &= \frac{\left[\vec{\alpha}\cdot\left(\vec{p} - e\vec{A}\right)\right]^2}{2m} \\
\frac{1}{8m^2}\left([\hat{O},\hat{E}] + i\dot{\hat{O}}\right) &= \frac{e}{8m^2}\left(-i\vec{\alpha}\cdot\vec{\nabla}\Phi - i\vec{\alpha}\cdot\dot{\vec{A}}\right) = \frac{ie}{8m^2}\vec{\alpha}\cdot\vec{E} \\
\left[\hat{O}, \frac{ie}{6m^3}\vec{\alpha}\cdot\vec{E}\right] &= \frac{ie}{8m^2}[\vec{\alpha}\cdot\vec{p},\vec{\alpha}\cdot\vec{E}] .
\end{aligned}
\tag{2.139}
$$

The last line in Eq. (2.138) can be written as

$$
\frac{e}{8m^2}(\vec{\nabla}\cdot\vec{E}) + \frac{ie}{8m^2}\vec{\Sigma}\cdot(\vec{\nabla}\times\vec{E}) + \frac{e}{4m^2}\vec{\Sigma}\cdot(\vec{E}\times\vec{p}) ,
\tag{2.140}
$$

where $\Sigma = \begin{pmatrix} \vec{\sigma} & 0 \\ 0 & \vec{\sigma} \end{pmatrix}$ is the spin matrix.

The Hamiltonian is now

$$
\begin{aligned}
H''' = \beta &\left[m + \frac{\left(\vec{p} - e\vec{A}\right)^2}{2m} - \underbrace{\frac{\vec{p}^{\,4}}{8m^3}}_{\text{KE correction}} \right] \\
&+ \frac{e}{c}\Phi - \underbrace{\frac{e\hbar}{2m}\beta\vec{\sigma}\cdot\vec{B}}_{\text{Pauli term}} - \underbrace{\frac{ie}{8m^2}\vec{\sigma}\cdot\left(\nabla\times\vec{E}\right)}_{\text{spin–orbit, I}} \\
&- \underbrace{\frac{e}{4m^2}\vec{\sigma}\cdot\left(\vec{E}\times\vec{p}\right)}_{\text{spin–orbit, II}} - \frac{e}{8m^2}\underbrace{\left(\nabla\cdot\vec{E}\right)}_{\text{Darwin term}} ,
\end{aligned}
\tag{2.141}
$$

which contains five relativistic correction terms.

For the particular case of the hydrogen-like atom, for which

$$
e\Phi = -\frac{Ze^2}{4\pi r} \quad \text{and} \quad \vec{A} = 0 ,
\tag{2.142}
$$

The second spin–orbit term[2.47] is given by

$$
\vec{\sigma}\cdot\left(\nabla\times\vec{E}\right) = -\frac{1}{r}\frac{\partial\Phi}{\partial r}\vec{\sigma}\cdot(\vec{r}\times\vec{p}) = -\frac{1}{r}\frac{\partial\Phi}{\partial r}\vec{\sigma}\cdot\vec{L} ,
\tag{2.143}
$$

[2.47] The first spin–orbit term is zero for a spherically symmetric potential, for which $\nabla\times\vec{E} = 0$. We also have that $\vec{E} = -\nabla\Phi$.

and, for the Darwin term,

$$
\frac{e}{8m^2}\left(\nabla\cdot\vec{E}\right) = \frac{Ze^2}{8m^2}\delta^3(\vec{r}) .
\tag{2.144}
$$

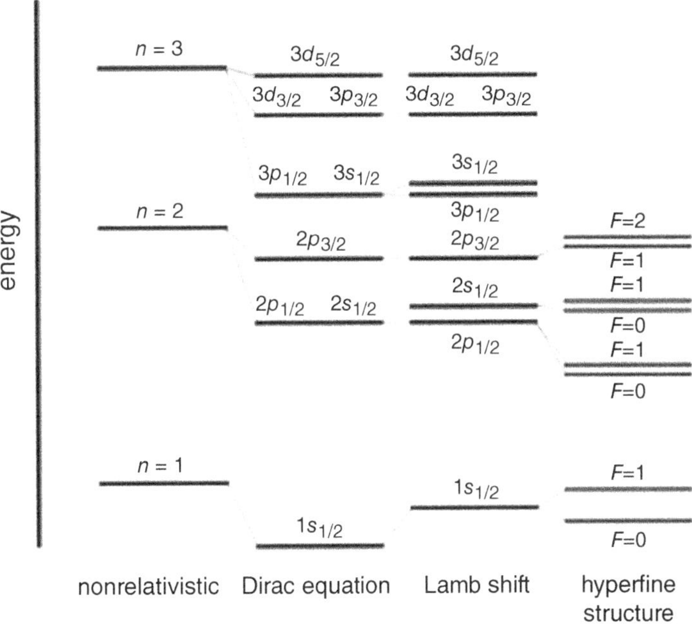

Figure 2.14 Energy levels of the hydrogen-like atom.

That was long but interesting, I hope. I am now ready to discuss the relativistic corrections to the energy levels of the hydrogen-like atom. See Figure 2.14.

The expectation values of the terms of the Hamiltonian in Eq. (2.141) are to be evaluated on the wave functions $|\psi_{n\ell s}\rangle$ of the nonrelativistic hydrogen atom as perturbations. This gives three corrections:[2.48]

1. Darwin: $\langle\psi_{n\ell s}|\dfrac{Ze^2}{8m^2}\delta^3(\vec{r})|\psi_{n\ell s}\rangle = \dfrac{m}{2}\dfrac{(Z\alpha)^4}{n^3}$;
2. spin–orbit : $\langle\psi_{n\ell s}|\dfrac{Z\alpha}{2m^2}\dfrac{1}{r^3}\vec{L}\cdot\vec{\sigma})|\psi_{n\ell s}\rangle = \dfrac{m}{4}\dfrac{(Z\alpha)^4}{n^3}\dfrac{j(j+1)-\ell(\ell+1)-3/4}{\ell(\ell+1/2)(\ell+1)}$;
3. KE : $\langle\psi_{n\ell s}| -\dfrac{\vec{p}^{\,4}}{8m^3}|\psi_{n\ell s}\rangle = \dfrac{m}{2}\dfrac{(Z\alpha)^4}{n^4}\left(\dfrac{3}{4}-\dfrac{n}{\ell+1/2}\right)$.

2.48 Recall that:

$$\langle\psi_{n\ell s}|1/r|\psi_{n\ell s}\rangle = Z\alpha m/n^2$$

$$\langle\psi_{n\ell s}|1/r^2|\psi_{n\ell s}\rangle = Z\alpha m^2/(\ell+1/2)n^2$$

$$\langle\psi_{n\ell s}|1/r^3|\psi_{n\ell s}\rangle = Z\alpha m^3/(\ell+1/2)n^3.$$

The $\vec{p}^{\,4}$ correction is computed using $\vec{p}^{\,4} = 4m^2(E-V(r))^2$. The expectation value of $\vec{L}\cdot\vec{\sigma}$ is equal to half that of $\vec{J}^2-\vec{L}^2-\vec{S}^2$.

By collecting all terms, and using $j=\ell\pm 1/2$, the relativistic correction to the energy levels is obtained:

$$\Delta E_{nj} = \frac{m}{2}\frac{(Z\alpha)^4}{n^4}\left(\frac{3}{4}-\frac{n}{j+1/2}\right). \tag{2.145}$$

The result in Eq. (2.145) can be compared with the exact solution of the Dirac equation for the hydrogen-like atom, which is given by[2.49]

2.49 What happens when $Z\alpha > j+1/2$? The energy turns into a complex number, signalling that the atom has become unstable.

$$E_{nj} = m\left[1 + \left(\frac{Z\alpha}{n - j - \frac{1}{2} + \sqrt{(j+\frac{1}{2})^2 - (Z\alpha)^2}}\right)^2\right]^{-1/2}$$

$$\simeq m\left[1 - \underbrace{\frac{(Z\alpha)^2}{2n^2}}_{\text{Bohr}} + \underbrace{\frac{(Z\alpha)^4(6j+3-8n)}{8(2j+1)n^4}}_{\Delta E_{nj}\ \text{in Eq. (2.145)}} + \cdots\right]. \tag{2.146}$$

The energy levels in Eq. (2.145) depend only on j! This means that two levels with the same j will remain degenerate even after the relativistic corrections. However, levels with different values of j, for instance, $j = 3/2$ and $j = 1/2$, are separated by

$$\Delta E(l = 1, j = 3/2) - \Delta E(l = 1, j = 1/2) = \frac{Z^4\alpha^4 mc^2}{4n^3}, \tag{2.147}$$

the degeneracy being lifted by the spin–orbit interaction. The energy levels thus obtained are the **fine structure** of the atom. The remaining degeneracy is eventually lifted by the Lamb shift. The origin of this correction is the combined effect of the self-energy (numerically the largest), the vacuum polarization and the vertex corrections in quantum electrodynamics, which together produce the energy shift

$$\Delta E_{\text{LS}} = \underset{\text{number to be computed}}{\#}\left(\frac{\alpha}{\pi}\right)\frac{\alpha^2 Z^4}{n^3}. \tag{2.148}$$

One should add to the correction in Eq. (2.148) the comparable corrections from the finite size and recoil of the nucleus to obtain the **hyperfine structure** of the energy levels of the hydrogen-like atom in Figure 2.14.

2.10 Why Use Quantum Field Theory?

Usually, one starts from a diagram – like the one in Figure 2.15 – in which the regions of validity of classical, relativistic and quantum mechanics are plotted versus the inverse of the speed of light c and Planck's constant $\hbar$ (both momentarily brought back as different from 1).

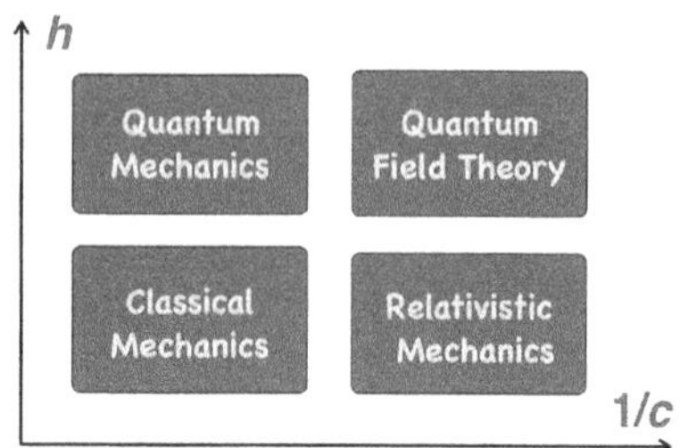

Figure 2.15 Theory space.

When neither $\hbar$ nor $1/c$ can be neglected, we have quantum field theory, which is quantum mechanics made relativistically invariant. It is a theory with a variable number of particles. No theory in which the number of particles is fixed – like the Dirac equation – is consistent with relativity. Efforts to "make do" using a single-particle theory always lead to trouble and awkward concepts like Dirac's "sea" of electrons.

This point can be made clear – and in the process we learn a few interesting things – by considering the propagation of a particle.

Consider first quantum mechanics. The state $|\vec{x}_0\rangle$ describing a particle at point $\vec{x}_0$ evolves to the state $|\vec{x}\rangle$, where the same particle is now at point $\vec{x}$, through Schrödinger's equation (rule number **4** of quantum mechanics)

$$i\frac{\partial}{\partial t}|\vec{x}_0\rangle = \hat{H}|\vec{x}_0\rangle \tag{2.149}$$

whose solution is

$$e^{-iHt}|\vec{x}_0\rangle \,. \tag{2.150}$$

The amplitude for this process is given (rule number **1**) by[2.50]

$$\langle\vec{x}|e^{-iHt}|\vec{x}_0\rangle = \int \frac{\mathrm{d}^3\vec{p}}{(2\pi)^3} e^{-i\vec{p}\cdot(\vec{x}-\vec{x}_0)} e^{-it\vec{p}^{\,2}/2m} \tag{2.151}$$

for $H = \vec{p}^{\,2}/2m$.

The integral in Eq. (2.151) can be computed to give

$$\left(\frac{m}{2\pi i t}\right)^{3/2} e^{i(m/2t)(\vec{x}-\vec{x}_0)^2} \,. \tag{2.152}$$

If a space-like interval, with $|\vec{x}-\vec{x}_0| > t$, is considered then the amplitude does not vanish as it should in relativity. The particle has a finite probability of going from $\vec{x}_0$ to $\vec{x}$ even when the two points are separated by a space-like interval. To do so, the particle must move faster than light, which it cannot.

2.50 Use the normalization

$$\langle\vec{p}|\vec{x}\rangle = \frac{1}{(2\pi)^3} e^{-i\vec{p}\cdot\vec{x}}$$

and the completeness relation

$$|\vec{x}\rangle = \int \mathrm{d}^2\vec{p}\,|\vec{p}\,\rangle\langle\vec{p}\,|\vec{x}\rangle$$

of the states. The integral can be brought into the standard form

$$I = \int_{-\infty}^{+\infty} e^{-ax^2}\,\mathrm{d}x,$$

which is integrated by writing its square as a two-dimensional integral and going into polar coordinates, to yield $I = \sqrt{\pi}$. A neat trick!

Maybe the problem is the nonrelativistic Hamiltonian I have used. It can be replaced with $H = \sqrt{\vec{p}^{\,2} + m^2}$ and the computation of the amplitude redone. The amplitude is then

$$\langle\vec{x}|e^{-iHt}|\vec{x}_0\rangle = \int \frac{\mathrm{d}^3\vec{p}}{(2\pi)^3} e^{-i\vec{p}\cdot(\vec{x}-\vec{x}_0)} e^{-it\sqrt{\vec{p}^{\,2}+m^2}} \tag{2.153}$$

and the integral is a bit more elaborate. First I integrate the angular variables in

$$\langle\vec{x}|e^{-iHt}|\vec{x}_0\rangle = \frac{1}{(2\pi)^3}\int \mathrm{d}\phi \int \mathrm{d}\cos\theta \int p^2 \mathrm{d}p\, e^{-ipr\cos\theta} \,, \tag{2.154}$$

setting $r = |\vec{x}-\vec{x}_0|$ and $p = |\vec{p}\,|$, and obtain

$$\frac{i}{4\pi^2 r}\int_{-\infty}^{+\infty} \mathrm{d}p\, p\, e^{-ipr-it\sqrt{p^2+m^2}} \,, \tag{2.155}$$

which can be written as

$$-\frac{1}{4\pi^2 r}\frac{\partial}{\partial r}\left[\int_{-\infty}^{+\infty} \mathrm{d}p^2 e^{-ipr-it\sqrt{p^2+m^2}}\right] \tag{2.156}$$

by getting rid of the p in the numerator (at the price of having a derivative). The integral in Eq. (2.156) is complicated. It can be approximated by the **saddle point** technique to give

$$g(r,t)e^{-m\sqrt{r^2+m^2}} \,, \tag{2.157}$$

where

$$g(r,t) = \sqrt{\frac{mt^2}{(r^2-t^2)^{3/2}} \frac{2mr^2(r^2-t^2)+\sqrt{r^2-t^2}(5r^2-2t^2)}{4\sqrt{2}\pi^{3/2}r^2(r^2-t^2)\sqrt{r^2-t^2}}}. \tag{2.158}$$

If you do not know about the saddle point method, do not worry. The result is that this amplitude is still non-zero for $r > t$ even though there is some improvement and it is only exponentially small outside the future space-time cone, modulated by an exponential factor e^{-mr} which suppresses propagation beyond a Compton wavelength ($\lambda_C = 1/m$).

Mathematically speaking, this result is due to the fact that, in a Fourier transformation such as

$$f(t) = \int_{-\infty}^{+\infty} \mathrm{d}t\, e^{-i\omega t} F(\omega), \tag{2.159}$$

the function $F(\omega)$ is different from zero in any interval of t unless it is identically zero everywhere. To have $F = 0$ on some finite interval it is necessary to have both positive **and** negative frequencies. I will come back to this point in a moment.

Conclusion: it appears impossible to write a relativistic quantum theory – at least by just writing a relativistic equation of motion.

As shown by the computation above, the worst possible case occurs for a massless particle, for which the entire space-like region is accessible. On the bright side, in this case I can do the integration exactly because the Hamiltonian simplifies to $H = p$. Then

$$\langle \vec{x}|e^{-iHt}|\vec{x}_0\rangle = -\frac{1}{4\pi^2 r}\frac{\partial}{\partial t}\int_{-\infty}^{+\infty} \mathrm{d}p \left[e^{-ip(t+r)} - e^{-ip(t-r)}\right], \tag{2.160}$$

which is divergent. To regularize it, use the principal-value prescription, in which, for an arbitrary function $f(x)$,[2.51]

2.51 Chaucy's principal value is usually studied as some arcane bit of mathematics but here it takes a life of its own.

$$\text{p.v.}\int_a^b \mathrm{d}x \frac{f(x)}{x} = \frac{1}{2}\left[\lim_{\epsilon\to 0}\int_{a-i\epsilon}^{b-i\epsilon} \mathrm{d}x \frac{f(x)}{x} + \lim_{\epsilon\to 0}\int_{a+i\epsilon}^{b+i\epsilon} \mathrm{d}x \frac{f(x)}{x}\right]; \tag{2.161}$$

thus taking the average of the paths in the complex plane, one above and one below the singularity. The two singularities in x cancel out, leaving half the residue at $x = 0$, that is,

$$\int_{-\epsilon}^{+\epsilon} \delta(x) f(x) \mathrm{d}x = f(0), \tag{2.162}$$

which gives, in terms of distributions (that is, things that have a meaning only after integration)[2.52]

2.52 $\delta(t-r) - \delta(t+r) = 2r\delta(t^2 - r^2)$

$$\text{p.v.}\frac{1}{t-r} - \text{p.v.}\frac{1}{t+r} = \frac{2r}{t^2-r^2}.$$

$$\frac{1}{x - i\epsilon} = \text{p.v.}\frac{1}{x} + i\pi\delta(x), \tag{2.163}$$

and the integral in Eq. (2.165) can be computed to be

$$\int_0^\infty \mathrm{d}p \left[e^{-ip(t+r)} - e^{-ip(t-r)}\right] = i\left(\frac{-1}{t+r-i\epsilon} + \frac{1}{t-r-i\epsilon}\right). \quad (2.164)$$

The propagator is given by

$$\langle \vec{x}|e^{-iHt}|\vec{x}_0\rangle = \frac{\partial}{\partial t}\frac{1}{2\pi^2}\left[-i\pi\,\delta(t^2-r^2) + \text{p.v.}\frac{1}{t^2-r^2}\right]. \quad (2.165)$$

Equation (2.165) says that the massless particle evolves in space-time in two simultaneous manners:

— at the speed of light on its light-cone where $r = t$;
— as a principal value that is different from zero everywhere and for which the particle propagates faster than light (and violates causality).[2.53]

2.53 If you were wondering, causality in quantum field theory is always preserved by having the commutator (or anticommutator) vanish for fields taken at two points separated by a space-like interval.

This is the same problematic result as for the massive case – only much worse, with the principal value different from zero everywhere. Yet amidst all this inconsistency the solution arises, because for a space-like interval there is no way of knowing whether the position $\vec{x}$ comes after $\vec{x}_0$. A moving observer may see the order reversed. Since there is no way of having a definite order of the events, I must simultaneously include the propagation from point $\vec{x}$ to point $\vec{x}_0$:

$$\begin{aligned}\langle \vec{x}_0|e^{iHt}|\vec{x}\,\rangle &= \int \frac{\mathrm{d}^3\vec{p}}{(2\pi)^3} e^{-i\vec{p}\cdot(\vec{x}_0-\vec{x})}e^{+itp}\\ &= \frac{\partial}{\partial t}\frac{1}{2\pi^2}\left[-i\pi\,\delta(t^2-r^2) - \text{p.v.}\frac{1}{t^2-r^2}\right]. \end{aligned} \quad (2.166)$$

A backward glance shows that the principal value part now has the opposite sign with respect to Eq. (2.165), and when I sum the amplitude for the two possible time orderings, I obtain

$$\langle \vec{x}_0|e^{iHt}|\vec{x}\rangle + \langle \vec{x}|e^{-iHt}|\vec{x}_0\rangle = \frac{\partial}{\partial t}\frac{1}{\pi^2}\left[-i\pi\,\delta(t^2-r^2)\right], \quad (2.167)$$

where the principal value part has gone, together with the propagation outside the cone.

The backward propagation of a particle from $\vec{x}$ to point $\vec{x}_0$ is interpreted in quantum field theory as the propagation of an antiparticle, that is, a particle with the same mass but opposite charge. The separation into particles and antiparticles is not a well-defined concept in relativistic dynamics.[2.54] Whether there is a particle or an antiparticle at a given

2.54 Pair creation within the Compton length suggests that the position eigenfunctions are not a good basis in the relativistic theory. For this reason, in quantum field theory we go to a momentum basis.

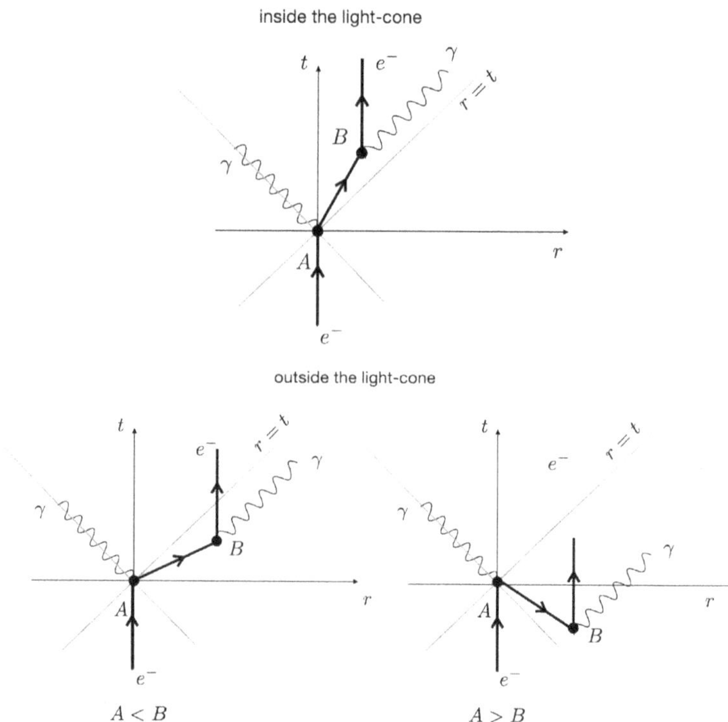

Figure 2.16 Propagation of an electron as seen by two moving observers. The electron emits a photon at *A* and *B*. If the propagation is outside the light-cone, the order of events *A* and *B* can be reversed for a moving observer. What looks like just one electron to one observer may look like one electron plus an electron–positron pair to the other observer.

moment of time depends on the motion of the observer, in the same way as in relativity the simultaneity of two events is ill defined for space-like intervals.[2.55]

[2.55] A pole vaulter, carrying a 4-meter long pole, runs across a barn that is 3.5-meter long from door to door. If she runs fast enough (in relativistic terms!) the Lorentz contraction of the pole will allow the two doors of the barn to close simultaneously as she runs across and then open again to let her through. This is what happens in the frame of reference of the barn. What happens in the frame of reference of the running pole vaulter? In that frame it is the barn that shrinks. How can the doors close simultaneously now that the barn is even shorter than before? Can they?

What occurs for the propagation of a single particle is true in general. The same process, taking place outside the light-cone, is shown in Figure 2.16 as seen by two observers moving with respect to each other. For the first observer event A takes place before event B; the opposite is true for the second observer, for whom it is event B that takes place before event A.

The second observer ($B < A$) sees a particle, an electron, which begins to move backward in time before moving forward again. If I take a slice in time, after event B, the initial electron appears together with an electron–positron pair created by the photon in B. The electron and the positron are annihilated in A to create the other photon.

There is no process that can be consistently described by a fixed number of particles. We always need a variable number of particles to be present. We need quantum field theory where the field operator of, for instance, a scalar field is given by

$$\hat{\varphi}(x) = \int \frac{\mathrm{d}^4\vec{p}}{\sqrt{2E_{\vec{p}}}} \left(\hat{a}_{\vec{p}}\, \varphi(\vec{p}\,)\, e^{-ip\cdot x} + \hat{b}^{\dagger}_{\vec{p}}\, \varphi^{\dagger}(\vec{p}\,)\, e^{+ip\cdot x} \right) \tag{2.168}$$

and contains simultaneously the operator for the particle, $\hat{a}$, and for the antiparticle, $\hat{b}^{\dagger}$. Repeated insertions of the field operator generate multiple particles and antiparticles.

2.11 The *S*-Matrix Toolbox: Decays and Cross Sections

In a typical scattering experiment, Figure 2.17, the particles are taken to be free – sufficiently far from each other to not interact – at time $t \to -\infty$, and they end up again free at time $t \to +\infty$. As they come close to each other they interact.[2.56]

2.56 Long-distance interactions must die off sufficiently fast to preserve the (almost) free state of the incoming and outgoing particles. Strictly speaking, the S-matrix is well defined only if all particles partaking in the scattering process have non-vanishing mass.

An operator $\hat{S}$, the **scattering matrix**, or S-matrix for short, is defined such that

$$\underbrace{\langle f|}_{\text{OUT states}}\hat{S}\underbrace{|i\rangle}_{\text{IN states}}. \tag{2.169}$$

It is unitary ($\hat{S}^{\dagger}\hat{S} = \hat{S}\hat{S}^{\dagger} = \mathbb{1}$) and thus preserves orthonormal states. The S-matrix connects the incoming states with the outgoing states. This is all I need and all that can be observed. The interaction itself is a black box in which, for the moment, I am not interested and do not specify. The entire process is written (as in the Hebrew or Arabic languages) from right to left because of the ordering in quantum mechanics.

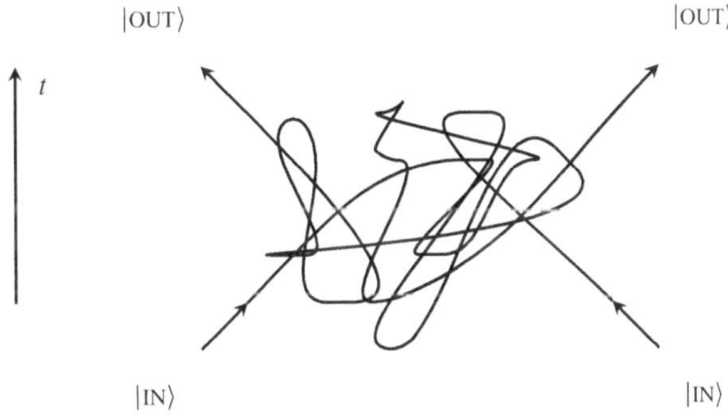

Figure 2.17 Particle scattering in S-matrix language. Particles come in from far away in the far past (the IN states) and fly off again so that they are far from one another in the far future (the OUT states). In between, they interact in a manner that can be very complicated.

The element S_{fi} of the S-matrix is the amplitude for the initial state to go into the final state. This part is computed in perturbation theory by means of quantum field theory.

To compute decays and cross sections I need the (properly normalized) differential probability

$$\mathrm{d}P = \frac{|\langle f|\hat{S}|i\rangle|^2}{\langle f|f\rangle\langle i|i\rangle}\mathrm{d}\Pi \tag{2.170}$$

for the process to take place. The differential factor $\mathrm{d}\Pi$ is the phase space element, which is given by

$$\mathrm{d}\Pi = \prod_f \frac{V}{(2\pi)^3}\mathrm{d}^3\vec{p}_f\,. \tag{2.171}$$

The integral over phase space in Eq. (2.171) is over the final states of the scattering process, which are labeled by their momenta $\vec{p}_f$.

I need to say a few words about the normalization in Eq. (2.170). All states are labeled by their momenta $\vec{p}$ and spin σ. The initial and final states are normalized by the condition

$$\langle\vec{p},\sigma|\vec{p}\,',\sigma'\rangle = (2\pi)^3 2E_{\vec{p}}\,\delta^{(3)}(\vec{p}-\vec{p}\,')\delta_{\sigma,\sigma'}\,, \tag{2.172}$$

which is convenient because it also works for massless particles. The condition in Eq. (2.172) is different from the usual one in non-relativistic quantum mechanics because of the energy factor. This factor is there to make the normalization transform as a density, the time component of a four-vector. Following the normalization in Eq. (2.172), I have

$$\langle f|f\rangle = \prod_f(2E_f V) \quad \text{and} \quad \langle i|i\rangle = \prod_f(2E_i V)\,, \tag{2.173}$$

because

$$\delta^{(3)}(0) = \frac{1}{(2\pi)^3}\int \mathrm{d}^3\vec{x} = \frac{V}{(2\pi)^3}\,, \tag{2.174}$$

given the definition of δ-function

$$\delta^{(3)}(\vec{p}) = \frac{1}{(2\pi)^3}\int \mathrm{d}^3\vec{x}\; e^{i\vec{p}\cdot\vec{x}}\,. \tag{2.175}$$

The phase-space integral is normalized with a density of one particle with momentum $\vec{p}_f$ for each unit of volume V of space.

I like to think of the scattering process of particles as a car crash, Figure 2.18. I stand on the side of the road (as detector) and as the crash takes place I look at the fragments of the cars coming my way. The cars could come from opposite directions and hit each other head on. In this case the collision takes place in the center of mass (CM, for short) frame. Instead, a single car could hit a wall. In this case the collision takes place in the laboratory frame. In both cases, I must reconstruct from the flying debris the make, model and year of the cars. It does not seem simple and it is not.

Figure 2.18 Scattering of cars (dummies onboard) [Toru Yamanaka/Getty Images].

Yet this is what is done for all collisions in particle accelerators where, instead of cars, we tear apart particles to look at their insides.

A little bit of refinement. I am not interested in the collision amplitude when nothing happens. It can be subtracted from the S-matrix, writing

$$\langle f|\hat{S} - \mathbb{1}|i\rangle = i(2\pi)^4\delta^{(4)}\left(\sum_i p_i^\mu - \sum_f p_f^\mu\right)\langle f|\hat{\mathcal{M}}|i\rangle\,, \tag{2.176}$$

where $\langle f|\hat{\mathcal{M}}|i\rangle$ are the matrix elements for something to happen.

I need to take the complex conjugate of the quantity in Eq. (2.176) and multiply it by the original expression to obtain the probability:

$$\begin{aligned}|\langle f|\hat{S} - \mathbb{1}|i\rangle|^2 &= i(2\pi)^8\delta^{(4)}(0)\delta^{(4)}\left(\sum_i p_i^\mu - \sum_f p_f^\mu\right)|\langle f|\hat{\mathcal{M}}|i\rangle|^2\\ &= TV(2\pi)^4\delta^{(4)}\left(\sum_i p_i^\mu - \sum_f p_f^\mu\right)|\langle f|\hat{\mathcal{M}}|i\rangle|^2\,,\end{aligned} \tag{2.177}$$

where I have used one of the two δ-functions in the product to set to zero the argument of the other and then taken its value over the entire spacetime to be TV. The transition probability is then

$$\begin{aligned}\mathrm{d}P &= \frac{TV(2\pi)^4\delta^{(4)}\left(\sum_i p_i^\mu - \sum_f p_f^\mu\right)}{\prod_i(2E_iV)}\frac{|\langle f|\hat{\mathcal{M}}|i\rangle|^2}{\prod_f(2E_fV)}\prod_f\frac{V}{(2\pi)^3}\mathrm{d}^3\vec{p}_f\\ &= \frac{TV}{\prod_i(2E_iV)}|\langle f|\hat{\mathcal{M}}|i\rangle|^2\mathrm{d}\Phi_{\mathrm{LIPS}}\,,\end{aligned} \tag{2.178}$$

where the **Lorentz-invariant phase space** (LIPS) element is defined by

$$\boxed{\mathrm{d}\Phi_{\mathrm{LIPS}} = \prod_f\frac{\mathrm{d}^3\vec{p}_f}{(2\pi)^3}\frac{1}{2E_f}(2\pi)^4\delta^{(4)}\left(\sum_i p_i^\mu - \sum_f p_f^\mu\right).} \tag{2.179}$$

The transition probability in Eq. (2.178) seems ill defined because of the infinite factors V and T; these factors will drop out for the transitions in which we are interested: decays and cross sections.

For a decay in which the initial state consists of a single particle and the final states are the decay products, the probability is

$$dP = \frac{T}{2E_1} |\langle f|\hat{\mathcal{M}}|i\rangle|^2 d\Phi_{\text{LIPS}} , \tag{2.180}$$

which divided by T gives the differential width,

$$\boxed{d\Gamma = \frac{dP}{T} = \frac{1}{2E_1} |\langle f|\hat{\mathcal{M}}|i\rangle|^2 d\Phi_{\text{LIPS}} .} \tag{2.181}$$

Equation (2.181) gives the probability of decay per unit time.

Given an initial number $N(0)$ of unstable particles, their population is depleted by a factor $e^{-\Gamma t}$ as more and more of the particles decay:

$$N(t) = N(0)\, e^{-\Gamma t} . \tag{2.182}$$

The lifetime τ of the decaying particles is defined by

$$\tau = \frac{\displaystyle\int_0^\infty t\, N(t) dt}{\displaystyle\int_0^\infty N(t) dt} = \frac{\displaystyle\int_0^\infty t e^{-\Gamma t} dt}{\displaystyle\int_0^\infty e^{-\Gamma t} dt} = \frac{1}{\Gamma} . \tag{2.183}$$

The general formula for the decay width in Eq. (2.181) is usually applied to the cases of two and three final particles. For two final particles, now labeled as 1 and 2, I have, in the rest frame of the decayed particle with, say, mass M,

$$d\Gamma = \frac{(2\pi)^4}{2M} |\langle f|\hat{\mathcal{M}}|i\rangle|^2 \frac{d^3\vec{p}_1}{(2\pi)^3} \frac{1}{2E_1} \frac{d^3\vec{p}_2}{(2\pi)^3} \frac{1}{2E_2} \delta^{(3)}(\vec{p}_1 + \vec{p}_2)\delta(M - E_1 - E_2) . \tag{2.184}$$

The three-dimensional δ-function kills one of the momentum integrations, so that[2.57]

2.57 I have used the property of the δ-function

$$\delta[f(p_1)] dp_1 = \left|\frac{df}{dp_1}\right|^{-1}_{p^*} \delta(p_1 - p^*)$$

with

$$f(p_1) = M - \sqrt{|\vec{p}_1|^2 + m_1^2} - \sqrt{|\vec{p}_2|^2 + m_2^2} .$$

$$d\Gamma = \frac{1}{32\pi^2 M} |\langle f|\hat{\mathcal{M}}|i\rangle|^2 \frac{|\vec{p}_1|^2 d|\vec{p}_1| d\Omega}{E_1 E_2} \delta(M - E_1 - E_2) , \tag{2.185}$$

which can be further simplified by means of the surviving δ-function; this imposes that the variable $|\vec{p}_1|$ is equal to p^*, where

$$p^* = \frac{1}{2M} \sqrt{[M^2 - (m_1 + m_2)^2][M^2 - (m_1 - m_2)^2]} ; \tag{2.186}$$

thus by the vanishing of the argument of the δ-function in Eq. (2.185) it follows that

$$M - \sqrt{|\vec{p}_1|^2 + m_1^2} - \sqrt{|\vec{p}_2|^2 + m_2^2} = 0 . \tag{2.187}$$

In this way I find

$$d\Gamma = \frac{1}{32\pi^2 M^2} p^* |\langle f|\hat{\mathcal{M}}|i\rangle|^2 d\Omega\,. \tag{2.188}$$

Things are more complicated when there are three particles in the final state, Figure 2.19, because the δ-function cannot be used to kill all the integrations but one. The phase space is effectively two dimensional, and whereas in the two-body decay the kinematics is fully described by a single variable – be it the scattering angle θ or the Mandelstam variable t – here two variables are needed. These are usually taken to be the Mandelstam-like variables

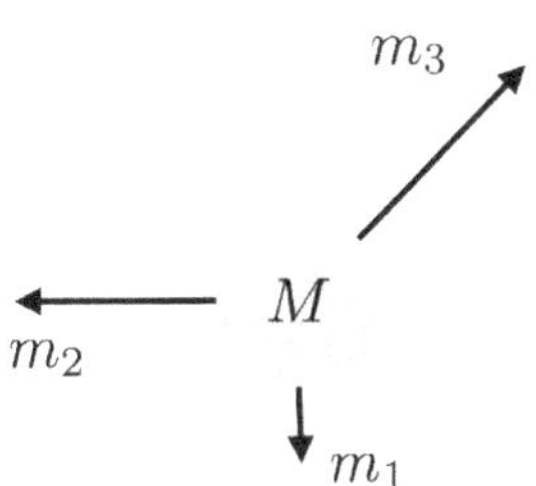

Figure 2.19 Three-body configuration.

$$\begin{aligned} m_{12}^2 &= (p_1 - p_2)^2 = (E_1 + E_2)^2 - (\vec{p}_1 - \vec{p}_2)^2 \\ m_{23}^2 &= (p_2 - p_3)^2 = (E_2 + E_3)^2 - (\vec{p}_2 - \vec{p}_3)^2\,. \end{aligned} \tag{2.189}$$

The two variables can be plotted in what is called a Dalitz plot, as in Figure 2.20.

The differential decay width is now given by

$$d\Gamma(m_{12}^2, m_{23}^2) = \frac{1}{(2\pi)^3 32 M^2} |\langle f|\hat{\mathcal{M}}|i\rangle|^2 dm_{12}^2 dm_{23}^2\,. \tag{2.190}$$

Even though all computations on the phase space become more complicated, there is a beauty to it because by a careful study of the Dalitz plot one can infer some properties of the decaying particles such as, for instance, the presence of resonances.

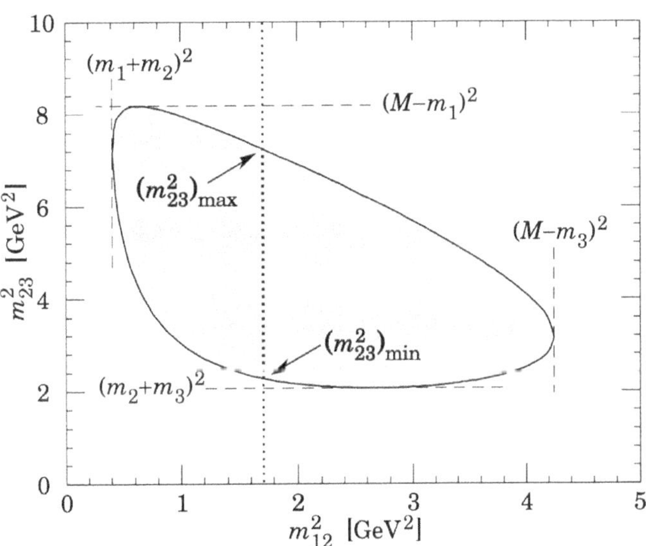

Figure 2.20 Dalitz plot of the two variables m_{12}^2 and m_{23}^2 [R.L. Workman *et al.* (Particle Data Group), *Prog. Theor. Exp. Phys.* **2022** (2022) 083C01 http://pdg.lbl.gov].

For a differential cross section in which two particles are in the initial state, the probability of their scattering, for unit flux of the incoming particles, is given by

$$\boxed{\mathrm{d}\sigma = \frac{\mathrm{d}P}{T \times \text{flux}} = \frac{1}{(2E_1)(2E_2)} \frac{1}{|\vec{v}_1 - \vec{v}_2|} |\langle f|\hat{\mathcal{M}}|i\rangle|^2 \mathrm{d}\Phi_{\text{LIPS}}\,,} \tag{2.191}$$

where in momentum space the flux is given by $|\vec{v}_1 - \vec{v}_2|/V$. The denominator in Eq. (2.191) can be rewritten in a form that is relativistically invariant, since

$$(2E_1)(2E_2)|\vec{v}_1 - \vec{v}_2| = 4\sqrt{(p_1 \cdot p_2)^2 - m_1^2 m_2^2}\,. \tag{2.192}$$

I leave it as an exercise for you to convince yourself that the above identity is indeed correct.

The general formula in Eq. (2.191) for the cross section is used in most cases where there are just two particles, labeled 3 and 4, in the final state. In this case Eq. (2.191) simplifies:

$$\mathrm{d}\Phi_{\text{LIPS}} = (2\pi)^4 \delta^{(4)}\,(p_1 + p_2 - p_3 - p_4)\,\frac{\mathrm{d}^3\vec{p}_3}{(2\pi)^3}\frac{1}{2E_3}\frac{\mathrm{d}^3\vec{p}_4}{(2\pi)^3}\frac{1}{2E_4}\,, \tag{2.193}$$

which can be further simplified in the center of mass frame, where $\vec{p}_1 + \vec{p}_2 = \vec{p}_3 + \vec{p}_4$, to

$$\begin{aligned} p_i = |\vec{p}_1| = |\vec{p}_2| \quad &\text{and} \quad p_f = |\vec{p}_3| = |\vec{p}_4| \\ E_1 + E_2 = E_3 + E_4 &= E_{\text{CM}}\,, \end{aligned} \tag{2.194}$$

so that the integration over $\vec{p}_4$ can be carried out using a δ-function to obtain

$$\frac{1}{16\pi^2}\mathrm{d}\Omega \int \mathrm{d}p_f \frac{p_f^2}{E_3 E_4}\delta(E_3 + E_4 - E_{\text{CM}})\,, \tag{2.195}$$

where $E_3 = \sqrt{|\vec{p}_3|^2 + m_3^2}$ and $E_4 = \sqrt{|\vec{p}_4|^2 + m_4^2}$.

The remaining integral is dispatched by the change of variable $p_f \to x - E_3 - E_4 - E_{\text{CM}}$ with Jacobian $\mathrm{d}x/\mathrm{d}p_f = (E_3 + E_4)p_f/E_3E_4$, so that (see Figure 2.21)

$$\frac{1}{16\pi^2}\mathrm{d}\Omega \int_{m_3+m_4-E_{\text{CM}}}^{\infty} \mathrm{d}x \frac{p_f}{E_{\text{CM}}}\delta(x) = \frac{1}{16\pi^2}\mathrm{d}\Omega \frac{p_f}{E_{\text{CM}}}\theta(E_{\text{CM}} - m_3 - m_4)\,. \tag{2.196}$$

After these manipulations, the two-body differential cross section is given by[2.58]

2.58 Heaviside function:

$$\theta(x) = \begin{cases} 1 & x > 0 \\ 0 & x \leq 0\,. \end{cases}$$

$$\left(\frac{\mathrm{d}\sigma}{\mathrm{d}\Omega}\right)_{\text{CM}} = \frac{1}{64\pi^2 E_{\text{CM}}^2}\frac{p_f}{p_i}|\langle f|\hat{\mathcal{M}}|i\rangle|^2 \theta(E_{\text{CM}} - m_3 - m_4)\,, \tag{2.197}$$

since

$$|\vec{v}_1 - \vec{v}_2| = \left|\frac{|\vec{p}_1|}{E_1} + \frac{|\vec{p}_2|}{E_2}\right| = p_i \frac{E_{\text{CM}}}{E_1 E_2}\,. \tag{2.198}$$

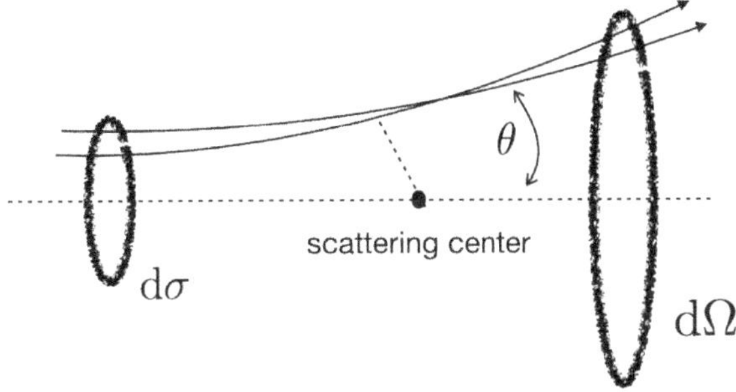

Figure 2.21 Geometry and variables in the definition of the cross section.

The expression in Eq. (2.197) gives the threshold $E_{\rm CM} > m_3 + m_4$ necessary for the process to take place. You cannot create two particles if you do not have at least as much energy as the sum of their masses.

In the elastic case[2.59] for which $m_1 = m_3$ and $m_2 = m_4$ it follows that $p_f = p_i$, and Eq. (2.197) simplifies further to

$$\left(\frac{\mathrm{d}\sigma}{\mathrm{d}\Omega}\right)_{\rm CM} = \frac{1}{64\pi^2 E_{\rm CM}^2} |\langle f|\hat{\mathcal{M}}|i\rangle|^2 \theta(E_{\rm CM} - m_3 - m_4)\,. \tag{2.199}$$

2.59 For non-elastic scattering for which there are different masses in the initial ($m_1 = m_2 = m_i$) and final states ($m_3 = m_4 = m_f$), we must keep the factor

$$\frac{p_f}{p_i} = \frac{\sqrt{1 - 4m_f^2/E_{\rm CM}^2}}{\sqrt{1 - 4m_i^2/E_{\rm CM}^2}}$$

in Eq. (2.197).

I may as well hit you with another change of variables and introduce those of Mandelstam. These variables consist of relativistically invariant quantities. For the two-body scattering process they are given by

$$\begin{aligned} s &= (p_1 + p_2)^2 \\ t &= (p_1 - p_3)^2 \\ u &= (p_1 - p_4)^2 \end{aligned} \tag{2.200}$$

with $s + t + u = m_1^2 + m_2^2 + m_3^2 + m_4^2$ and $E_{\rm CM} = \sqrt{s}$.

I can exchange the integration over the angular variables in Eq. (2.197) for one over the Mandelstam variable t by using the relation

$$t = (p_1 - p_3)^2 = m_1^2 + m_3^2 - 2E_1E_3 + 2|\vec{p}_1||\vec{p}_3|\cos\theta\,, \tag{2.201}$$

because

$$\mathrm{d}t = 2|\vec{p}_1||\vec{p}_3|\mathrm{d}\cos\theta = p_i p_f \frac{\mathrm{d}\Omega}{\pi} \tag{2.202}$$

and therefore

$$\left(\frac{\mathrm{d}\sigma}{\mathrm{d}\Omega}\right)_{\rm CM} = \frac{\mathrm{d}\sigma}{\mathrm{d}t} = \frac{1}{64\pi s}\frac{1}{p_i^2} |\langle f|\hat{\mathcal{M}}|i\rangle|^2\,. \tag{2.203}$$

2.12 Relativistically Invariant Perturbation Theory

Before moving on to actual computations it is useful to put some perspective on what I am doing when computing the matrix elements $\langle f|\hat{M}|i\rangle$ by means of Feynman diagrams.

What is happening is that, first, the free fields are reduced to (infinite) collections of harmonic oscillators. The oscillators are then (weakly) coupled by the interactions. And these interactions are treated by means of the relativistic generalization of time-dependent perturbation theory (Fermi's "golden rule") to obtain widths and cross sections.

Feynman diagrams are an effective book-keeping device to write down these various contributions in the perturbative expansion. As temping as it might be, the idea that Feynman diagrams correspond to actual physical processes is **wrong**.[2.60] There are no electrons coming in and deciding to go around a little bit before emitting a single photon or whatever. The diagrams are only a shorthand for terms in the perturbative expansion of an exponential (containing the Lagrangian density and its creation and annihilation operators) and its expectation value on the external (free) one-particle states. The diagrams stand for the various terms of this expansion and they help us to write them down without having to go back each time to that expansion and the contractions among the numerous creation and annihilation operators. It is useful to repeat this to ourselves from time to time because the notation and the shorthand are so compelling that we may forget this and end up believing that elementary particles in Nature behave as in the Feynman diagrams.

[2.60] Kind of obvious but nevertheless important word of caution!

To better understand this point and see how the Feynman rules and diagrams emerge from perturbation theory, I need to go into some of the details.

The evolution in time of the Fock state, defined in Eq. (2.82), is given by quantum mechanics as

$$i\frac{\partial}{\partial t}|\Phi\rangle = (H_0 + V(t))|\Phi\rangle\,, \tag{2.204}$$

where H_0 represents the unperturbed Hamiltonian and $V(t)$ the time-dependent perturbation. In the interaction picture,

$$V(t) = e^{iH_0 t} V(t) e^{-iH_0 t} \tag{2.205}$$

and therefore Eq. (2.204) becomes

$$i\frac{\partial}{\partial t}|\Phi\rangle = V(t)|\Phi\rangle\,, \tag{2.206}$$

which can be solved to give

$$|\Phi(t_f)\rangle = T\exp\left\{-i\int_{t_i}^{t_f} V(t)\,\mathrm{d}t\right\}|\Phi(t_i)\rangle\,. \tag{2.207}$$

In Eq. (2.207), T stands for the ordering in time of the perturbations $V(t)$ entering the integral; it is necessary because perturbations at different times do not necessarily commute.

If I take the initial time t_i and final time t_f to be minus and plus infinity, respectively, I can write the S-matrix connecting the incoming to the outgoing state,

$$|\Phi(+\infty)\rangle = \hat{S}\,|\Phi(-\infty)\rangle\,, \tag{2.208}$$

in terms of the expression in Eq. (2.207) as[2.61]

2.61 Here I am glossing over several steps that are discussed in QFT textbooks under the heading of "LSZ reduction formulas".

$$\begin{aligned}\hat{S} &= T\exp\left\{-i\int_{-\infty}^{\infty} V(t)\mathrm{d}t\right\} \\ &= \sum_{k=0}^{\infty}\frac{(-i)^k}{k!}\int_{-\infty}^{\infty}\mathrm{d}t_1\int_{-\infty}^{\infty}\mathrm{d}t_2\cdots\int_{-\infty}^{\infty}\mathrm{d}t_k\; T\left\{V(t_1)V(t_2)\cdots V(t_k)\right\}\,,\end{aligned} \tag{2.209}$$

in which the chronological ordering is from right to left. The expectation value of the S-matrix in Eq. (2.209) between the initial and final Fock states in which we are interested corresponds to the relativistically invariant generalization of Fermi's "golden rule" of time-dependent perturbation theory.

To put some flesh on Eq. (2.209), let me use a definite expression for the interaction:

$$V(t) = e\int \mathrm{d}^3x\, j_\mu(x)A^\mu(x), \tag{2.210}$$

which is the interaction for quantum electrodynamics (see the next section). Inserting this into Eq. (2.209), the first non-vanishing term is

$$\begin{aligned}\hat{S}^{(2)} &= \frac{e^2}{2!}\int\!\!\int \mathrm{d}^4x_1\mathrm{d}^4x_2\; T\left\{j^\mu(x_1)A_\mu(x_1)j^\nu(x_2)A_\nu(x_2)\right\} \\ &= \frac{e^2}{2!}\int\!\!\int \mathrm{d}^4x_1\mathrm{d}^4x_2\; T\left\{j^\mu(x_1)j^\nu(x_2)\right\}\; T\left\{A_\mu(x_1)A_\nu(x_2)\right\}\,,\end{aligned} \tag{2.211}$$

where I can write the second line of Eq. (2.211) because the currents (made out of electron operators) commute with the potentials (consisting of photon operators).

In preparation for following problem sessions, let us take, as an example, the expectation value of the S-matrix in Eq. (2.211) between the Fock states for two initial electrons and zero photons and two final electrons and, again, zero photons, that is,

$$|\Phi^{(i)}\rangle = a^\dagger_{\vec{p}_1,\sigma_1}\,a^\dagger_{\vec{p}_2,\sigma_2}|0\rangle \quad\text{and}\quad \langle 0|a_{\vec{p}_3,\sigma_3}\,a_{\vec{p}_4,\sigma_4} = \langle\Phi^{(f)}|\,. \tag{2.212}$$

Then

$$\langle\Phi^{(f)}|\hat{S}^{(2)}|\Phi^{(i)}\rangle = \frac{e^2}{2!}\int\int \mathrm{d}^4x_1\mathrm{d}^4x_2 \times \langle 0|T\left\{A_\mu(x_1)A_\nu(x_2)\right\}|0\rangle \times \langle 0|a_{\vec{p}_3,\sigma_3}\,a_{\vec{p}_4,\sigma_4}T\left\{j^\mu(x_1)j^\nu(x_2)\right\}a^\dagger_{\vec{p}_1,\sigma_1}\,a^\dagger_{\vec{p}_2,\sigma_2}|0\rangle \tag{2.213}$$

in which the photon operator is given by

$$A_\mu(x) = \frac{1}{(2\pi)^{3/2}}\int\frac{\mathrm{d}^3\vec{p}}{\sqrt{2p^0}}\sum_\sigma\left[a_{\vec{p},\sigma}\varepsilon_\mu(\vec{p},\sigma)\,e^{-ip\cdot x} + a^\dagger_{\vec{p},\sigma}\varepsilon^*_\mu(\vec{p},\sigma)\,e^{+ip\cdot x}\right], \tag{2.214}$$

with $\varepsilon_\mu(\vec{p},\sigma)$ the polarization wave function and σ running over the photon polarizations. The currents

$$j^\mu(x) = \bar{\psi}(x)\gamma^\mu\psi(x) \tag{2.215}$$

contain the electron operators, each given by

$$\psi(x) = \frac{1}{(2\pi)^{3/2}}\int\frac{\mathrm{d}^3\vec{p}}{\sqrt{2p^0}}\sum_\sigma\left[a_{\vec{p},\sigma}u(\vec{p},\sigma)\,e^{-ip\cdot x} + b^\dagger_{\vec{p},\sigma}v(\vec{p},\sigma)\,e^{+ip\cdot x}\right], \tag{2.216}$$

with $u(\vec{p},\sigma)$ and $v(\vec{p},\sigma)$, respectively, the electron and positron wave functions, $a_{\vec{p},\sigma}$ the annihilation operator for an electron and $b^\dagger_{\vec{p},\sigma}$ the creation operator for a positron. The parameter σ runs over the two polarizations of the fermions.

Taking the first expectation value in Eq. (2.213), $\langle 0|T\left\{A_\mu(x_1)A_\nu(x_2)\right\}|0\rangle$, it can contain only terms proportional to

$$\langle 0|a_{\vec{p},\sigma}a^\dagger_{\vec{p},\sigma}|0\rangle = 1\,, \tag{2.217}$$

in which a photon is created at x_1 and destroyed at x_2, the time x_2^0 being after x_1^0, as dictated by the chronological ordering, or *vice versa*. The two chronologically ordered terms give the **propagator** of the photon:

$$-iD_{\mu\nu}(x_2-x_1) = \frac{1}{(2\pi)^3}\int\frac{\mathrm{d}^3\vec{p}}{2p^0}\sum_\sigma\varepsilon_\mu(\vec{p},\sigma)\varepsilon^*_\nu(\vec{p},\sigma)\left[e^{ip\cdot(x_1-x_2)}\theta(x_1^0-x_2^0)+e^{ip\cdot(x_2-x_1)}\theta(x_2^0-x_1^0)\right], \tag{2.218}$$

in which the θ-functions take care of the time ordering of the events. All other terms vanish because of the effect of the creation and annihilation operators acting on the vacuum.

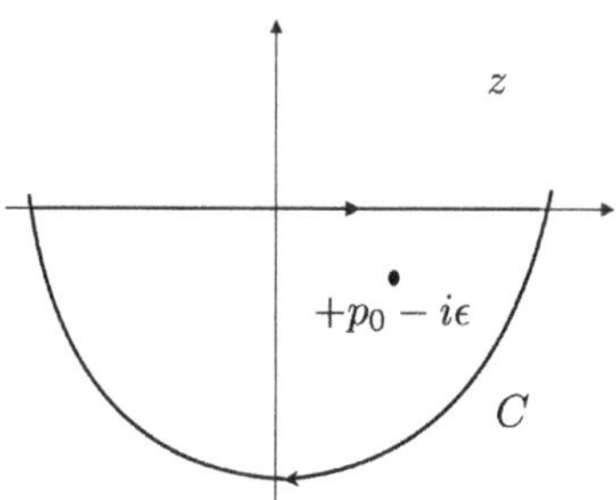

Figure 2.22 The path C in the complex plane.

The photon propagator can be put into the form that is actually used in the Feynman rules by noticing that the first term in Eq. (2.218) can be rewritten as an integral over the complex path C of Figure 2.22 (running from left to right along the real axis of z and closing in the lower plane, where the imaginary part of z is negative, as indicated by the $+i\epsilon$ factor in Eq. (2.219)):[2.62]

[2.62] To go from Eq. (2.218) to Eq. (2.219), use the residue theorem and

$$z^2 - \vec{p}^{\,2} + i\epsilon = (z - |\vec{p}|_i\epsilon')(z + |\vec{p}|_i\epsilon')$$

with $2i\epsilon'|\vec{p}\,|$ identified with ϵ.

$$\theta(x_1^0 - x_2^0) \int \frac{\mathrm{d}^3\vec{p}}{(2\pi)^3} \Big[\sum_\sigma \varepsilon_\mu(\vec{p},\sigma)\varepsilon_\nu^*(\vec{p},\sigma) \Big] e^{i\vec{p}\cdot(\vec{x}_1-\vec{x}_2)} \frac{1}{2\pi} \int_C \mathrm{d}z \frac{e^{-iz(x_1^0 - x_2^0)}}{z^2 - \vec{p}^{\,2} + i\epsilon}, \tag{2.219}$$

which becomes an integral over d^4p, identifying $z = p_0$ on the real axis. The same happens for the second term in Eq. (2.218), only now the path C is from right to left and closes on the upper plane. Together they therefore give

$$-iD_{\mu\nu}(x_2 - x_1) = \frac{1}{(2\pi)^4} \int \mathrm{d}^4p \, \frac{\sum_\sigma \varepsilon_\mu(\vec{p},\sigma)\varepsilon_\nu^*(\vec{p},\sigma)}{p^2 + i\epsilon} \, e^{ip\cdot(x_2 - x_1)} . \tag{2.220}$$

The Fourier transform of Eq. (2.220) gives the propagator of the photon in momentum space. The sum over the polarizations $\sum_\sigma \varepsilon_\mu(\vec{p},\sigma)\varepsilon_\nu^*(\vec{p},\sigma)$ depends on the choice of gauge. More on this when we come to discuss gauge fields in Chapter 4.

The second expectation value in Eq. (2.213),

$$\langle 0 | a_{\vec{p}_3,\sigma_3}\, a_{\vec{p}_4,\sigma_4} T\left\{ j^\mu(x_1) j^\nu(x_2) \right\} a^\dagger_{\vec{p}_1,\sigma_1}\, a^\dagger_{\vec{p}_2,\sigma_2} | 0 \rangle , \tag{2.221}$$

is more complicated because the creation and annihilation operators in the fermion fields Eq. (2.216) can meet those operators acting on the vacuum in more than one way. There are four possible contractions (as they are called) in which the creation and annihilation operators walk through to find their opposites and give 1, as in Eq. (2.217), namely

$$\begin{aligned}
&\langle 0 | a_{\vec{p}_3,\sigma_3}\, a_{\vec{p}_4,\sigma_4} (\bar{\psi}(x_1)\, \gamma^\mu \psi(x_1)) (\bar{\psi}(x_2) \gamma^\nu\, \psi(x_2)) a^\dagger_{\vec{p}_1,\sigma_1}\, a^\dagger_{\vec{p}_2,\sigma_2} | 0 \rangle \\
&+ \langle 0 |\, a_{\vec{p}_3,\sigma_3}\, a_{\vec{p}_4,\sigma_4} (\bar{\psi}(x_1)\, \gamma^\mu \psi(x_1)) (\bar{\psi}(x_2) \gamma^\nu\, \psi(x_2)) a^\dagger_{\vec{p}_1,\sigma_1}\, a^\dagger_{\vec{p}_2,\sigma_2}\, | 0 \rangle \\
&+ \langle 0 | a_{\vec{p}_3,\sigma_3}\, a_{\vec{p}_4,\sigma_4} (\bar{\psi}(x_1)\, \gamma^\mu \psi(x_1)) (\bar{\psi}(x_2) \gamma^\nu\, \psi(x_2)) a^\dagger_{\vec{p}_1,\sigma_1}\, a^\dagger_{\vec{p}_2,\sigma_2} | 0 \rangle \\
&+ \langle 0 | a_{\vec{p}_3,\sigma_3}\, a_{\vec{p}_4,\sigma_4} (\bar{\psi}(x_1)\, \gamma^\mu \psi(x_1)) (\bar{\psi}(x_2) \gamma^\nu\, \psi(x_2)) a^\dagger_{\vec{p}_1,\sigma_1}\, a^\dagger_{\vec{p}_2,\sigma_2}\, | 0 \rangle ,
\end{aligned} \tag{2.222}$$

in which the braces over and under the expectation values indicate the contractions yielding the non-vanishing products of the annihilation and creation operators.

After taking expectation values, we have[2.63]

$$\begin{aligned}&\langle 0|a_{\vec{p}_3,\sigma_3}\, a_{\vec{p}_4,\sigma_4} T\left\{j^\mu(x_1) j^\nu(x_2)\right\} a^\dagger_{\vec{p}_1,\sigma_1}\, a^\dagger_{\vec{p}_2,\sigma_2}|0\rangle\\ &\quad= (\bar{\psi}_4\gamma^\mu\psi_2)(\bar{\psi}_3\gamma^\nu\psi_1) + (\bar{\psi}_3\gamma^\mu\psi_1)(\bar{\psi}_4\gamma^\nu\psi_2)\\ &\qquad - (\bar{\psi}_3\gamma^\mu\psi_2)(\bar{\psi}_4\gamma^\nu\psi_1) - (\bar{\psi}_4\gamma^\mu\psi_1)(\bar{\psi}_3\gamma^\nu\psi_2)\end{aligned} \tag{2.223}$$

2.63 The minus signs come from the anticommuting of the creation and annihilation operators as they walk through.

in which the labels on the fermion operators stand for the incoming and outgoing electrons. Inserting this result into Eq. (2.213) gives[2.64]

$$\begin{aligned}\langle\Phi^{(f)}|S^{(2)}|\Phi^{(i)}\rangle &= i\,e^2\int\!\!\int \mathrm{d}^4x_1\mathrm{d}^4x_2\, D_{\mu\nu}(x_1-x_2)\\ &\quad\times\Big[(\bar{\psi}_4\gamma^\mu\psi_2)(\bar{\psi}_3\gamma^\nu\psi_1) - (\bar{\psi}_4\gamma^\mu\psi_1)(\bar{\psi}_3\gamma^\nu\psi_2)\Big],\end{aligned} \tag{2.224}$$

2.64 Remember that the currents $\bar{\psi}_i\gamma^\mu\psi_j$ contains the exponential factors $e^{-i(p_j-p_i)\cdot x}$ (see Eq. (2.216)), which, after the integration, give the conservation of momentum in the factor $\delta^4(p_3+p_4-p_1-p_2)$ that is factorized out in defining the amplitude M_{fi}. Two of the four terms in Eq. (2.223) give, after integration, results equal to the other two.

which, after performing the space-time integration, results in the amplitude

$$\begin{aligned}\langle f|\hat{S}^{(2)}-1|\rangle \equiv \mathcal{M}^{(2)}_{fi} &= e^2\Big[(\bar{u}_4\gamma^\mu u_2)D_{\mu\nu}(p_2-p_4)(\bar{u}_3\gamma^\nu u_1)\\ &\quad - (\bar{u}_4\gamma^\mu u_1)D_{\mu\nu}(p_2-p_3)(\bar{u}_3\gamma^\nu u_2)\Big].\end{aligned} \tag{2.225}$$

Equation (2.225) represents the amplitude for the interaction of two electrons at first order in the perturbative expansion. Each of the two terms in Eq. (2.225) can be represented by a diagram. The first one is shown in Figure 2.23 together with the mathematical expression it represents: the vertices are associated to the currents and the internal line to the photon propagator.

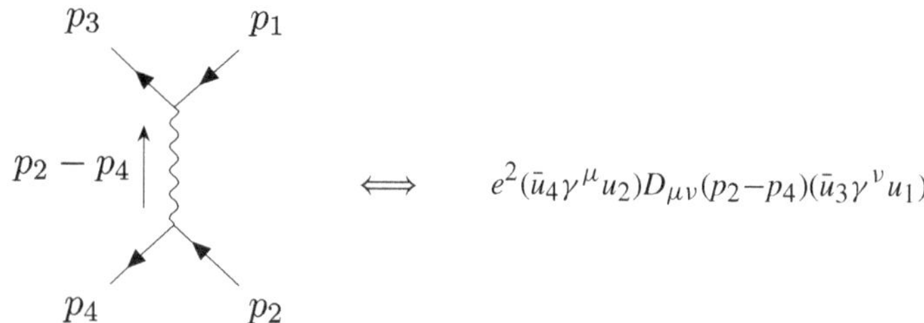

Figure 2.23 First term in Eq. (2.225).

The second term is shown in Figure 2.24.

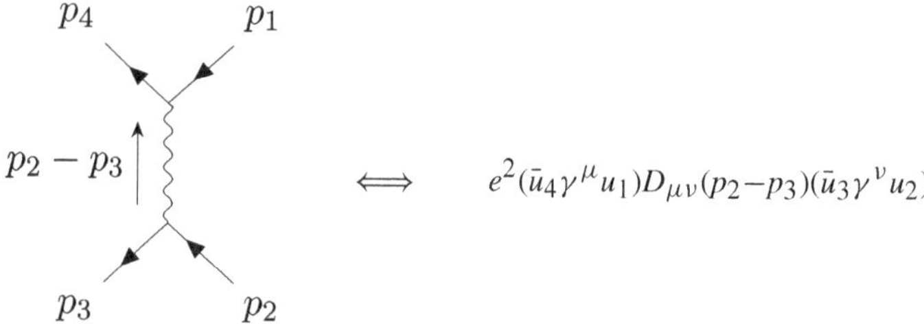

Figure 2.24 Second term in Eq. (2.225).

The terms in the perturbative expansion can easily be reconstructed by drawing the possible diagrams and writing the corresponding expressions using the **Feynman rules** – as I am about to do in the next problem session.

2.13 *Problem Session*: Pair Annihilation

This problem session is the first of three dedicated to computing cross sections in quantum electrodynamics, QED for short. They serve the purpose of reviewing the tools necessary to derive the cross section from the Feynman diagrams.

To compute the amplitude we need the Feynman rules. Some of these are collected in appendix section A.6 for the reader's convenience.

The Feynman rules can be extracted from the Lagrangian density by considering the action

$$S = i \int \overbrace{L}^{L_{cl}\,(+L_{g.f.} + L_{gh})} \, \mathrm{d}^4 x \tag{2.226}$$

(where I have put between parentheses terms that are not essential in QED – I shall come back to these terms in Chapter 3) and[2.65]

2.65 Notation: $\not{X} \equiv \gamma^\mu X_\mu$.

$$L_{cl} = -\frac{1}{4} F^{\mu\nu} F_{\mu\nu} + \bar{\psi}_e (i\not{\partial} - e\not{A} - m)\psi_e \,. \tag{2.227}$$

The Lagrangian density (let me call it just the Lagrangian from now since it is understood that we are always dealing with fields) must be organized in terms bilinear in the fields (the free part) and all the rest (the interactions):

$$L = \underbrace{L_0}_{\bar{\psi}_e(i\gamma^\mu \partial_\mu - m)\psi_e - \frac{1}{4}F^{\mu\nu}F_{\mu\nu}} + \underbrace{L_I}_{-e\bar{\psi}_e \gamma^\mu A_\mu \psi_e} \,. \tag{2.228}$$

All terms must be written in momentum space. The bilinear terms give me the propagators, the other terms the interactions.

The rule is to take the terms in the Lagrangian, after removing the fields, and multiplying by i (because of the way the Lagrangian enters the action).

For the internal lines and vertices the propagators and interaction vertices are given in Figure 2.25. The photon propagator depends on the gauge parameter ξ, as I will discuss in Chapter 3. The previous section explains how these rules embody the relativistically invariant perturbative expansion.

Table 2.4 Feynman rules for the wave functions in QED.

	IN	OUT
fermion	$u^s(p)$	$\bar{u}^s(p)$
antifermion	$\bar{v}^s(p)$	$v^s(p)$
photon	$\varepsilon_\lambda^\mu(k)$	$\varepsilon_\lambda^{\mu *}(k)$

vertex

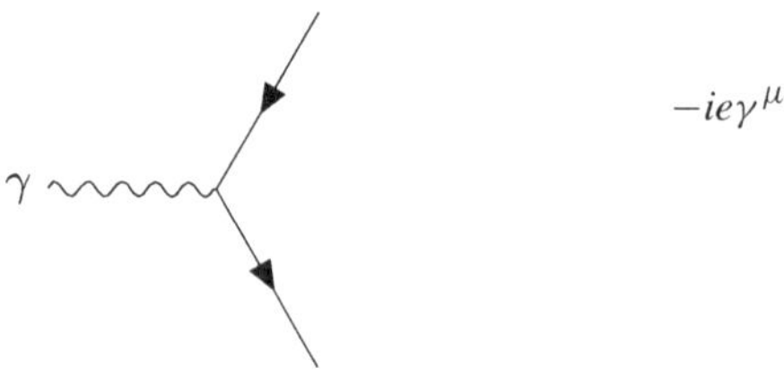

$-ie\gamma^\mu$

fermion propagator

$$\left(\frac{i}{\not{p} - m - i\epsilon}\right)_{ab}$$

photon propagator

$$\frac{ig_{\mu\nu} - i(1-\xi)\frac{q_\mu q_\nu}{q^2}}{q^2 + i\epsilon}$$

Figure 2.25 Feynman rules for QED.[2.66]

2.66 "Like the silicon chip of more recent years, the Feynman diagram was bringing computation to the masses." J. Schwinger

For the external incoming and outgoing states, the wave functions as described in Table 2.4 must be inserted, with attention being paid to what is going in and what is coming out.

The Dirac matrices satisfy anticommutation relations:

$$\{\gamma^\mu, \gamma^\nu\} = 2g^{\mu\nu} . \tag{2.229}$$

These matrices can be written in various representations. I will be using three of them. Beside the standard (Dirac) representation, there is the Weyl representation in which the Dirac spinor is split into left- and right-handed components:

$$\psi_L = \frac{1}{2}(1-\gamma^5)\psi \quad \text{and} \quad \psi_R = \frac{1}{2}(1+\gamma^5)\psi , \tag{2.230}$$

where

$$\frac{1+\gamma^5}{2} = \begin{pmatrix} \mathbb{0} & \mathbb{0} \\ \mathbb{0} & \mathbb{1} \end{pmatrix} \quad \text{and} \quad \frac{1-\gamma^5}{2} = \begin{pmatrix} \mathbb{1} & \mathbb{0} \\ \mathbb{0} & \mathbb{0} \end{pmatrix} \tag{2.231}$$

are the chirality projectors ($\mathbb{1}$ and $\mathbb{0}$ are 2×2 matrices).

I can go from one representation to the other by the transformations

$$\gamma_\mu^{\text{Weyl}} = S\,\gamma_\mu^{\text{Dirac}}\,S^{-1}$$
$$\psi_{\text{Weyl}} = S\,\psi_{\text{Dirac}}\,, \tag{2.232}$$

where

$$S = \frac{1}{\sqrt{2}}(1+\gamma^5\gamma^0) = \frac{1}{\sqrt{2}}\begin{pmatrix} \mathbb{1} & -\mathbb{1} \\ \mathbb{1} & \mathbb{1} \end{pmatrix}. \tag{2.233}$$

In the Weyl basis, the Dirac equation becomes[2.67]

2.67 The Dirac equation is introduced in Appendix D.

$$(i\gamma^\mu\partial_\mu - m)\psi = \begin{pmatrix} -m & i\sigma^\mu\partial_\mu \\ i\bar{\sigma}^\mu\partial_\mu & -m \end{pmatrix}\begin{pmatrix} \psi_L \\ \psi_R \end{pmatrix} = 0\,, \tag{2.234}$$

defining $\sigma^\mu = (1, \vec{\sigma})$ and $\bar{\sigma}^\mu = (1, -\vec{\sigma})$; σ^i are the Pauli matrices I have already introduced.

This basis is particular useful in the massless case $m = 0$ because of the factorization of the Dirac equation, which becomes

$$\underbrace{i\bar{\sigma}^\mu\partial_\mu\psi_R}_{(E+\vec{\sigma}\cdot\vec{p})\psi_R} = \underbrace{i\sigma^\mu\partial_\mu\psi_L}_{(E-\vec{\sigma}\cdot\vec{p})\psi_L} = 0\,; \tag{2.235}$$

this shows how the chiral spinors $\psi_{L,R}$ are eigenstates of the projection of the spin on the momentum, that is, the helicity:

$$\frac{\vec{s}\cdot\vec{p}}{|\vec{p}|} = \pm 1. \tag{2.236}$$

($+$: *R*-handed; $-$: *L*-handed)

In addition there is a representation in which the spinor is real (and represents a neutral particle such as the neutrino); it is called the Majorana representation. I can go from from the standard representation to the Majorana representation by the transformations

$$\gamma_\mu^{\text{Majorana}} = S\,\gamma_\mu^{\text{Dirac}}\,S^{-1}$$
$$\psi_{\text{Majorana}} = S\,\psi_{\text{Dirac}}\,, \tag{2.237}$$

where

$$S = \frac{1}{\sqrt{2}}\begin{pmatrix} \mathbb{1} & \sigma^2 \\ \sigma^2 & -\mathbb{1} \end{pmatrix}. \tag{2.238}$$

In the Majorana representation, the two components of the spinor

$$\psi = \begin{pmatrix} \varphi \\ \chi \end{pmatrix} \tag{2.239}$$

are not independent. Given φ, we know χ, which is

$$\chi = -i\sigma_2\varphi^*\,. \tag{2.240}$$

Standard (Dirac) rep.

$$\gamma^0 = \begin{pmatrix} \mathbb{1} & \mathbb{0} \\ \mathbb{0} & -\mathbb{1} \end{pmatrix} \quad \gamma^i = \begin{pmatrix} \mathbb{0} & \sigma^i \\ -\sigma^i & \mathbb{0} \end{pmatrix} \quad \gamma^5 = \begin{pmatrix} \mathbb{0} & \mathbb{1} \\ \mathbb{1} & \mathbb{0} \end{pmatrix}$$

Weyl rep.

$$\gamma^0 = \begin{pmatrix} \mathbb{0} & \mathbb{1} \\ \mathbb{1} & \mathbb{0} \end{pmatrix} \quad \gamma^i = \begin{pmatrix} \mathbb{0} & \sigma^i \\ -\sigma^i & \mathbb{0} \end{pmatrix} \quad \gamma^5 = \begin{pmatrix} -\mathbb{1} & \mathbb{0} \\ \mathbb{0} & \mathbb{1} \end{pmatrix}$$

Majorana rep.

$$\gamma^0 = \begin{pmatrix} \mathbb{0} & \sigma^2 \\ \sigma^2 & \mathbb{0} \end{pmatrix} \quad \gamma^1 = i\begin{pmatrix} \sigma^3 & \mathbb{0} \\ \mathbb{0} & \sigma^3 \end{pmatrix} \quad \gamma^2 = \begin{pmatrix} \mathbb{0} & -\sigma^2 \\ \sigma^2 & \mathbb{0} \end{pmatrix} \quad \gamma^3 = i\begin{pmatrix} -\sigma^1 & \mathbb{0} \\ \mathbb{0} & -\sigma^1 \end{pmatrix}$$

Figure 2.26 Representations of the Dirac matrices.

The Dirac equation is in this case

$$\sigma^\mu \partial_\mu \varphi + m\sigma_2 \varphi^* = 0\,, \tag{2.241}$$

and, under charge conjugation,

$$\psi^c = i\gamma^2 \psi^* = \psi\,. \tag{2.242}$$

For this reason, the Majorana fermions are truly neutral particles.

It is important to know about these representations of the Dirac matrices (see Figure 2.26) even though they rarely appear in the actual computation of cross sections – in which the Dirac matrices are usually traced over and yield only products of momenta.

Because the application of the Feynman rules produces strings of Dirac matrices, the identities in Figure 2.27 for these matrices are going to be needed over and over again.

Now consider the annihilation of an electron and positron pair to give, say, a muon and antimuon pair. In more than one way, this is the simplest process in QED.

By reading the diagram in Figure 2.28 from right to left I can substitute for the external lines, vertices and internal lines their expressions according the the rules of QED I have just summarized. The amplitude is obtained as

$$\langle f|\hat{\mathcal{M}}|i\rangle \equiv i\hat{\mathcal{M}}_{fi}$$

$$= \left[\underbrace{\bar{u}(p_3)}_{\mu^-\ \text{OUT}}\ \overbrace{(-ie\gamma^\nu)}^{\text{vertex}}\ \underbrace{v(p_4)}_{\mu^+\ \text{OUT}}\right] \overbrace{\frac{-ig_{\mu\nu} + i(1-\xi)\frac{q_\mu q_\nu}{q^2}}{q^2}}^{\text{photon propagator}} \left[\underbrace{\bar{v}(p_2)}_{e^+\ \text{IN}}\overbrace{(-ie\gamma^\mu)}^{\text{vertex}}\underbrace{u(p_1)}_{e^-\ \text{IN}}\right].$$

Before plunging in, we need to check that this is indeed the only diagram that needs to be considered. The other diagram would be in the t-channel

Diracology

$$\mathrm{Tr}\,\mathbb{1} = 4$$

$$\mathrm{Tr}\ (\text{odd number of } \gamma) = 0$$

$$\mathrm{Tr}\,\gamma^\mu\gamma^\nu = 4g^{\mu\nu}$$

$$\mathrm{Tr}\,\gamma^\mu\gamma^\nu\gamma^\rho\gamma^\sigma = 4(g^{\mu\nu}g^{\rho\sigma} - g^{\mu\rho}g^{\mu\sigma} + g^{\mu\sigma}g^{\mu\rho})$$

$$\gamma^\mu\gamma_\mu = 4$$

$$\gamma^\mu\gamma^\nu\gamma_\mu = -2\gamma^\nu$$

$$\gamma^\mu\gamma^\nu\gamma^\rho\gamma_\mu = 4g^{\nu\rho}$$

$$\mathrm{Tr}\,\gamma^\mu\gamma^\nu\gamma^\rho\gamma^\sigma\gamma_5 = -4i\epsilon^{\mu\nu\rho\sigma}$$

$$\mathrm{Tr}\,\gamma_5 = \mathrm{Tr}\,\gamma_5\gamma^\mu\gamma^\nu = 0$$

Figure 2.27 Useful formulas and identities of Dirac matrices.

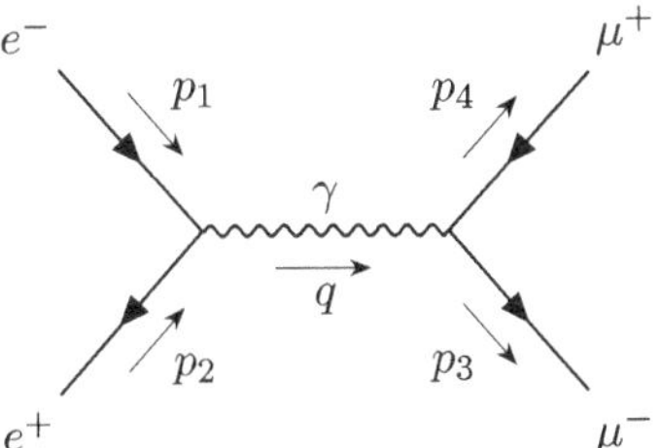

Figure 2.28 Feynman diagram for pair annihilation.

but that is simply not there because an electron cannot go into a muon by emitting a photon. I am also happy to notice that the gauge-dependent part of the photon propagator can be dropped because, for $q = p' - p$,

$$q_\mu \bar{u}(p')\gamma^\mu u(p) = u(p')\not{p}' u(p) - \bar{u}(p')\not{p} u(p) = -m\,\bar{u}u + m\,\bar{u}u = 0\,, \quad (2.243)$$

which is nothing but the conservation of the electromagnetic currents.

The amplitude M_{fi} can be written as a current–current interaction:

$$-\frac{e^2}{s}\bar{v}(p_2)\gamma^\mu u(p_1)\bar{u}(p_3)\gamma_\mu v(p_4) = -\frac{e^2}{s} j^e \cdot j^\mu\,, \quad (2.244)$$

where $s = -q^2$. The total cross section can be obtained from this relation.

To make it more interesting, I want to compute separately the contributions of electrons of different chirality to see how the partial cross sections depend on the angular variable and how they combine to give the total result.

To select the electron chirality, the electron wave function simply needs to be projected as

$$u_R(p) = \frac{1}{2}(1+\gamma^5)u(p) \quad \text{and} \quad u_L(p) = \frac{1}{2}(1-\gamma^5)u(p)\,. \tag{2.245}$$

The two possibilities for the initial states are $e^-_R e^+_L$ and $e^-_L e^+_R$ because, in the high-energy limit (which is equivalent to vanishing masses for both electrons and muons), the amplitude vanishes unless the spinors have the same helicity and left-handed electrons correspond to right-handed positrons and *vice versa*. For the same conservation of helicity, the electron-pair states can only end up as $\mu^-_L \mu^+_R$ and $\mu^-_R \mu^+_L$. Altogether there are four amplitudes to be computed.

For the first case $e^-_R e^+_L$ I obtain

$$j^e_\mu j^{e*}_\nu = \sum_{s,s'} \bar{v}^s(p_2)\gamma^\mu \frac{1+\gamma^5}{2} u^{s'}(p_1)\bar{u}^{s'}(p_1)\gamma^\nu \frac{1+\gamma^5}{2} v^s(p_2)\,. \tag{2.246}$$

Perhaps that left out too many steps. Here is the same computation in slow motion. First, take the Hermitian conjugate of the current

$$j^e = \bar{v}^s(p_2)\gamma^\mu \frac{1+\gamma^5}{2} u^{s'}(p_1), \tag{2.247}$$

that is,

$$j^{e*}_\nu = \left(\bar{v}^s(p_2)\gamma^\mu \frac{1+\gamma^5}{2} u^{s'}(p_1)\right)^\dagger, \tag{2.248}$$

which is, remembering that conjugation reverses the order of the operators,[2.68]

2.68 Remember that $\gamma^{5\dagger} = \gamma^5$ and $\gamma^5\gamma^\mu = -\gamma^\mu\gamma^5$.

$$u^{s'\dagger}(p_1)\frac{1+\gamma^5}{2}\gamma^{\mu\dagger}(\bar{v}^s(p_2))^\dagger\,. \tag{2.249}$$

Now insert $\gamma^0\gamma^0 = 1$ next to the chiral projector as follows:

$$u^{s'\dagger}(p_1)\gamma^0\gamma^0\frac{1+\gamma^5}{2}\gamma^{\mu\dagger}(\bar{v}^s(p_2))^\dagger, \tag{2.250}$$

to obtain (one γ^0 goes left and is combined with the spinor, the other goes right, across the chiral projector)[2.69]

2.69 Use the definition of the barred spinor: $\bar{v}^s(p_2) = v^s(p_2)^\dagger\gamma^0$ and $(\bar{v}^s(p_2))^\dagger = \gamma^0 v^s(p_2)$,

$$\bar{u}^{s'}(p_1)\frac{1-\gamma^5}{2}\gamma^0\gamma^{\mu\dagger}\gamma^0 v^s(p_2), \tag{2.251}$$

so that[2.70]

2.70 Use $\gamma^{\mu\dagger} = \gamma^0\gamma^\mu\gamma^0$.

$$\bar{u}^{s'}(p_1)\frac{1-\gamma^5}{2}\gamma^\nu v^s(p_2) = \bar{u}^{s'}(p_1)\gamma^\nu\frac{1+\gamma^5}{2} v^s(p_2)\,, \tag{2.252}$$

which is a term in Eq. (2.246). In practice, this step-by-step derivation should convince you that Hermitian conjugation takes the string of

Dirac spinors and matrices, reverses their order, turns barred spinors into unbarred one spinors, and *vice versa*, and flips the chirality of the chiral projectors.

Back to normal speed. The sum over the polarizations of the external particles is done according to the formula

$$\boxed{|\hat{\mathcal{M}}_{fi}|^2 = \underbrace{\prod_i \frac{1}{2s_i+1}}_{\substack{\text{average over}\\ \text{initial states}}} \underbrace{\sum_{\text{pol}}}_{\substack{\text{sum over}\\ \text{final states}}} |\langle f|\hat{\mathcal{M}}|i\rangle|^2 ,} \tag{2.253}$$

where all possible final states (and therefore sum over their polarizations) are considered and, not knowing the polarizations of the initial states, an average is taken.[2.71]

In practice, I first use the sum over the fermion polarizations given by

$$\boxed{\sum_{s=1,2} u^s(p)\bar{u}^s(p) = \not{p} + m} \quad \text{and} \quad \boxed{\sum_{s=1,2} v^s(p)\bar{v}^s(p) = \not{p} - m} \tag{2.254}$$

and then rewrite the sum as the trace over the Dirac structure, thanks to a certain manipulation, a trick named after the physicist H. Casimir. To make manifest why the Casimir trick works, write the Dirac indices explicitly, as in

$$\Big[v(p_2)\bar{v}(p_2)\Big]_{da}\left[\gamma^\mu \frac{1+\gamma^5}{2}\right]_{ab}\Big[u(p_1)\bar{u}(p_1)\Big]_{bc}\left[\gamma^\mu \frac{1+\gamma^5}{2}\right]_{cd}, \tag{2.255}$$

which shows that you are just multiplying matrices in the manner dictated by taking a trace,

$$\mathrm{Tr}\,(ABCD) = A_{da}B_{ab}C_{bc}D_{cd}\,. \tag{2.256}$$

Now I do the trick and obtain[2.72]

$$\begin{aligned} j^e_\mu j^{e*}_\nu &= \mathrm{Tr}\left[\not{p}_2\gamma^\mu \frac{1+\gamma^5}{2}\not{p}_1\gamma^\nu \frac{1+\gamma^5}{2}\right] = \mathrm{Tr}\left[\not{p}_2\gamma^\mu \not{p}_1\gamma^\nu \frac{1+\gamma^5}{2}\right] \\ &= 2\left(p_2^\mu p_1^\nu + p_1^\mu p_2^\nu - g^{\mu\nu} p_1\cdot p_2 - i\epsilon^{\alpha\mu\beta\nu}p_{2}^{\alpha} p_{1}^{\beta}\right). \end{aligned} \tag{2.257}$$

At the other end of the diagram, there is the meson current. The spinors must conserve the chirality and the amplitude with initial states $e^-_R e^+_L$ must have final states $\mu^-_R\mu^+_L$ and is, neglecting all the masses, the amplitude I just computed for the electrons. Accordingly, the square of the amplitude is just the product of $j^e_\mu j^{e*}_\nu$ times $j^\mu_\mu j^{\mu *}_\nu$, that is,[2.73]

$$\begin{aligned} |\overline{\mathcal{M}}_{fi}|^2 &= \frac{e^2}{s^2}\left[2(p_1\cdot p_3)(p_2\cdot p_4) + 2(p_1\cdot p_4)(p_2\cdot p_3) - \epsilon^{\alpha\mu\beta\nu}\epsilon_{\rho\mu\sigma\nu}p_2^\alpha p_1^\beta p_3^\rho p_4^\sigma\right] \\ &= \frac{4e^2}{s^2}(p_1\cdot p_4)(p_2\cdot p_3)\,. \end{aligned} \tag{2.258}$$

2.71 Ignorance of the initial polarizations is not necessarily always the case and I could, for instance, study a process in which I take an initial state with a known polarization. The rule must then accordingly change. I discuss this case in the next chapter.

2.72 The projector $(1+\gamma^5)/2$ simply goes across with the γ^5 anticommuting with any Dirac matrix it encounters, and after an even number of anticommutations it remains the same.

2.73 I have multiplied by 1/4 for the average over the initial polarizations and have used the relation $\epsilon^{\alpha\mu\beta\nu}\epsilon_{\rho\mu\sigma\nu} = -2(g^{\alpha\rho}g^{\beta\sigma} - g^{\alpha\sigma}g^{\beta\rho})$.

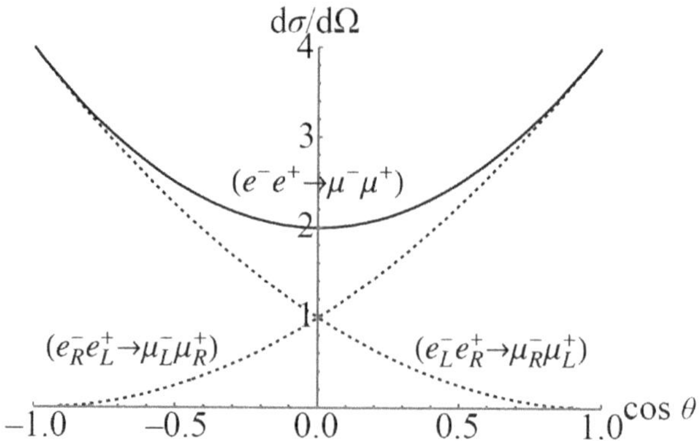

Figure 2.29 The differential cross sections for the chiral contributions in the processes $e_R^- e_L^+ \to \mu_L^- \mu_R^+$ and $e_L^- e_R^+ \to \mu_R^- \mu_L^+$ (dotted curves) and the sum of the chiral contributions (solid curve).

I obtain the differential cross section

$$\frac{d\sigma}{d\Omega} = \frac{|\overline{\mathcal{M}}_{fi}|^2}{64\pi^2 E^2} \tag{2.259}$$

by replacing the momenta with the energy and scattering angle (in the center of mass frame, where $s = 4E^2$).[2.74]

The cross section for $e_R^- e_L^+ \to \mu_R^- \mu_L^+$ is given by[2.75]

$$\frac{d\sigma}{d\Omega}(e_R^- e_L^+ \to \mu_R^- \mu_L^+) = \frac{\alpha^2}{16E^2}(1 + \cos\theta)^2 \,. \tag{2.260}$$

By parity invariance, the cross section for $e_L^- e_R^+ \to \mu_L^- \mu_R^+$ must be the same as in Eq. (2.260).

The cross section for $e_R^- e_L^+ \to \mu_L^- \mu_R^+$ can be obtained from the one just computed because it is the same apart for the switch of chirality projectors, which implies a minus sign:

$$\frac{d\sigma}{d\Omega}(e_R^- e_L^+ \to \mu_L^- \mu_R^+) = \frac{\alpha^2}{16E^2}(1 - \cos\theta)^2 \,. \tag{2.261}$$

Again, because of parity invariance, the cross section for $e_L^- e_R^+ \to \mu_R^- \mu_L^+$ is the same as in Eq. (2.261).

The sum of the four contributions is

$$\frac{d\sigma}{d\Omega} = \frac{\alpha^2}{4E^2}(1 + \cos^2\theta) \,. \tag{2.262}$$

Figure 2.29 shows the partial differential cross sections for the fixed chiralities as well as their sum, as given by Eq. (2.262).

[2.74] Components of the four-momenta in the center of mass frame:

$$\begin{aligned} p_1 &= (E, 0, 0, p) \\ p_2 &= (E, 0, 0, -p) \\ p_3 &= (E, E\sin\theta, 0, E\cos\theta) \\ p_4 &= (E, -E\sin\theta, 0, -E\cos\theta) \,. \end{aligned}$$

[2.75] Note that $(p_1 \cdot p_4)(p_2 \cdot p_3) = E^2(1 + \cos\theta)^2$.

2.14 Where is the "Quantum" in Quantum Field Theory?

In the computations in the previous problem session there seems to be little or no trace of the paraphernalia of quantum mechanics: wave function collapse, particle–wave duality and cats simultaneously dead and alive. Indeed, the S-matrix approach apparently avoids having to address these problems at all. Granted, the cross section is only providing the probability for a particle to leave a track in the detector with a given momentum. The fundamental probabilistic nature of quantum interactions is there. But otherwise the computation seems to proceed mainly along classical lines.

What is going on? Has quantum field theory removed all the quirky aspects of quantum theory (but for the probabilistic character of the interactions)?

For one thing, the S-matrix is utilized in a preferred basis of states.[2.76] Momentum (and spin) are the only bases that can be used because of the Poincaré representations. There is no possibility of changing to, say, a position-dependent set of states and comparing the same process in the two bases. The localization of the wave function collapse in space is removed from the formalism altogether.

[2.76] Pair production makes localization in space impossible below the Compton length of a particle.

In addition, all observables are chosen from among commuting operators such as the occupation number and the momentum of the one-particle states. We do not have the problems that are necessarily encountered in atomic physics with non-commuting operators.

One can say that the S-matrix setup is a clever way to avoid many (but not all) of the issues arising from quantum theory. Many a discussion on the interpretation of quantum mechanics cannot even be formulated in the language of quantum field theory.

This setup is part of a shift of emphasis consisting of moving away from quantum states and energy levels, which quantum mechanics is all about, toward the scattering process as described by the S-matrix in terms of the momenta and number of particles in the (asymptotically distant and free) initial and final states. It is a change of point of view initiated in the 1950s and has become nowadays accepted, in high-energy physics at least.[8]

Nevertheless, two very quantum aspects are preserved in quantum field theory.

First, loop diagrams (I discuss one of them in the problem session on the anomalous magnetic moment of the electron) represent the uncertainty in energy. They are there because even in a vacuum the quantum theory

[8] Read A. S. Blum, The state is not abolished, it withers away, *Stud. His. Sci.* **B60** (2017) 46 [arXiv:2011.05908], for a recent discussion.

allows (indeed, prescribes) the presence of particles. In the perturbative expansion of the Feynman diagrams, the tree diagrams represent the classical process (as long as there is no interference) while the loop diagrams give the quantum corrections. This point is made clear in Chapter 6, where the quantum corrections to the electroweak classical Lagrangian of the Standard Model are discussed.

Second, the prescription that, in the presence of different possible Feynman diagrams representing the same scattering process, we must sum their amplitudes (a case in point is Compton scattering, discussed in the next problem section), and only after that take the square of the modulus to compute the probability, leads to the presence of interference terms that make the computation of the probability of the cross section non-classical.

The most striking consequence of quantum mechanics is however more subtle. By looking carefully at the angular distributions of the final particles in a scattering process, I can find correlations which, for some of the possible processes, are maximally entangled (to be explained below) and provide a testing ground for the violation of the **Bell inequalities** – the crucial test of the non-local nature of quantum theory.

Consider the measurement of an observable (take the spin, assume it to be 1/2) for two particles originating from the decay of some state (with vanishing total angular momentum) and then freely moving to two far away positions in space. Call $\mathcal{P}(-;\ \uparrow_{\hat{n}_2})$ the probability of measuring the second particle with spin up along the direction $\hat{n}_2$ and $\mathcal{P}(\uparrow_{\hat{n}_1};-)$ the probability of measuring the first particle with spin up along the direction $\hat{n}_1$. In this notation, $\mathcal{P}(\uparrow_{\hat{n}_i};\ \uparrow_{\hat{n}_j})$ is the probability of measuring spin up along the direction $\hat{n}_i$ for the first particle and simultaneously measuring spin up along the direction $\hat{n}_j$ for the second particle.

Correlations along only one direction in a given frame of reference can only probe classical properties, as in the case of angular momentum conservation – the correlation in this case being that if, say, spin up for one particle is measured in one direction, necessarily spin down will be measured in the same direction for the other particle.[2.77] It is only by the simultaneous measurement of correlations along more axes (or different bases) that we can probe quantum correlations, in particular, by measuring non-commuting quantities for the two particles, as in the case when the spin is along the $\hat{n}_1 = \hat{z}$ direction for the first particle and along the $\hat{n}_2 = \hat{x}$ direction for the second particle.

2.77 This case is identical to that in which I have two boxes and I put a diamond in one of the two. Then I send one box to one location and the other to another (far away) location. If I open the box in one of the two locations and find it empty, I will know that the box at the other location contains the diamond. This is a case in which quantum and classical physics predict the same outcome.

The spins of the two particles are (classically) correlated in the sense that conservation of the angular momentum dictates that, after having measured the spin component in a given direction to be up for one of the two particles, it must be down (with respect to the same direction) for the other particle. If I measure along the same direction in both locations I will always find that when the spin is up in one location it will be down in the other: $\mathcal{P}(\uparrow_{\hat{n}_i}\ ;\ \downarrow_{\hat{n}_i}) = 1$. If instead I measure the spin along a direction orthogonal to the direction used in the first location, the outcome is completely undetermined and the correlation is equal to 0. This is also true in the quantum case.

With complex questions, it is helpful to start by figuring out what, exactly, we want to know. The difference between a classical and a quantum correlation is all about locality, that is, the possibility of factorizing the probabilities of the two spin measurements at the two distant locations. In the classical case I can factorize them; in the quantum case I cannot. If I consider only measurements along the same direction (or orthogonal to each other) this difference in factorizing the probabilities does not appear. I need to consider a more general situation in which the direction of the measurements can point in any direction.

Let me then consider the case of two measurements of spin along arbitrary directions at two distant locations.

The composite probability can be factorized as[2.78]

$$\mathcal{P}(\uparrow_{\hat{n}_i}\ ;\ \uparrow_{\hat{n}_j}) = \mathcal{P}(-;\ \uparrow_{\hat{n}_j})\mathcal{P}(\uparrow_{\hat{n}_i}\ ;-) \tag{2.263}$$

for two measurements taken at far distant locations, if I assume that they behave like classical probabilities.

Under that assumption,[2.79] I can use the generic property of four non-negative numbers x_1, x_2, x_3 and x_4 which are less than or equal to one (as probabilities must be) of satisfying the inequality

$$x_1x_2 - x_1x_4 + x_3x_2 + x_3x_4 \leq x_3 + x_2\,, \tag{2.264}$$

to write[2.80]

$$\mathcal{P}(\uparrow_{\hat{n}_1};\ \uparrow_{\hat{n}_2}) - \mathcal{P}(\uparrow_{\hat{n}_1};\ \uparrow_{\hat{n}_4}) + \mathcal{P}(\uparrow_{\hat{n}_3};\ \uparrow_{\hat{n}_2}) + \mathcal{P}(\uparrow_{\hat{n}_3};\ \uparrow_{\hat{n}_4}) \\ \leq \mathcal{P}(\uparrow_{\hat{n}_3};-) + \mathcal{P}(-;\ \uparrow_{\hat{n}_2})\,, \tag{2.265}$$

which is the Bell inequality in this case. What is crucial in deriving Eq. (2.265) is to be able to factorize the composite probabilities as in Eq. (2.263). This property makes the theory **local**.

The inequality in Eq. (2.265) is necessarily satisfied if I assume that the probabilities are obtained as an average over some hidden variable λ (which completely fixes the state of the particle and follows a distribution of values $\eta(\lambda)$) and are given, for instance, by

$$\mathcal{P}(\uparrow_{\hat{n}_i}\ ;\ \uparrow_{\hat{n}_j}) = \int \mathrm{d}\lambda\ \eta(\lambda)\, p_\lambda(\uparrow_{\hat{n}_i}\ ;\ \uparrow_{\hat{n}_j})\,, \tag{2.266}$$

where the probabilities $p_\lambda(\uparrow_{\hat{n}_i}\ ;\ \uparrow_{\hat{n}_j})$ are classical and factorize,

$$p_\lambda(\uparrow_{\hat{n}_i}\ ;\ \uparrow_{\hat{n}_j}) = p_\lambda(\uparrow_{\hat{n}_i}\ ;-)\, p_\lambda(-;\ \uparrow_{\hat{n}_j}), \tag{2.267}$$

2.78 Recall that I called $\mathcal{P}(\uparrow_{\hat{n}_i};\uparrow_{\hat{n}_j})$ the probability of measuring spin up along the direction $\hat{n}_i$ for the first particle and simultaneously measuring spin down along the direction $\hat{n}_j$ for the second particle.

2.79 Take $x_1 = \mathcal{P}(\uparrow_{\hat{n}_1};-)$, $x_2 = \mathcal{P}(-;\ \uparrow_{\hat{n}_2})$, $x_3 = \mathcal{P}(\uparrow_{\hat{n}_3};-)$ and $x_4 = \mathcal{P}(-;\ \uparrow_{\hat{n}_2})$, and use Eq. (2.263).

2.80 This form of the Bell inequality goes under the name of the physicists J. Clauser, M. Horne, A. Shimony and R. Holt – CHSH for short.

for measurements taken either near or far apart, it does not matter which. This is the gist of Bell's argument to characterize correlations in theories with hidden variables that are local.

If I now instead use quantum mechanics, for which the state of the two particles is the superposition[2.81]

$$|\Psi\rangle = \frac{1}{\sqrt{2}}\Big(|\uparrow\rangle|\downarrow\rangle + |\downarrow\rangle|\uparrow\rangle\Big), \tag{2.268}$$

for an arbitrary direction in space, the composite probabilities, which are not factorizable, are given by (the measurement is represented by Pauli matrices as in a Stern–Gerlach experiment)[2.82]

$$\begin{aligned} \mathcal{P}(\uparrow_{\hat{n}_i}\,;\,\uparrow_{\hat{n}_j}) &= \frac{1}{4}\langle\Psi|(\mathbb{1} + \hat{n}_i\cdot\vec{\sigma})\otimes(\mathbb{1} + \hat{n}_j\cdot\vec{\sigma})|\Psi\rangle \\ &= \frac{1}{4}\left(\hat{n}_i^x\hat{n}_j^x + \hat{n}_i^y\hat{n}_j^y - \hat{n}_i^z\hat{n}_j^z\right). \end{aligned} \tag{2.269}$$

Taking for the four directions

$$\hat{n}_1 = \hat{z}, \quad \hat{n}_2 = \frac{-1}{\sqrt{2}}(\hat{z}+\hat{x}), \quad \hat{n}_3 = -\hat{x} \quad \text{and} \quad \hat{n}_4 = \frac{1}{\sqrt{2}}(\hat{z}-\hat{x}), \tag{2.270}$$

it follows that

$$\mathcal{P}(\uparrow_{\hat{n}_1};\,\uparrow_{\hat{n}_2}) - \mathcal{P}(\uparrow_{\hat{n}_1};\,\uparrow_{\hat{n}_4}) + \mathcal{P}(\uparrow_{\hat{n}_3};\,\uparrow_{\hat{n}_2}) + \mathcal{P}(\uparrow_{\hat{n}_3};\,\uparrow_{\hat{n}_4}) = \frac{1}{2} + \frac{\sqrt{2}}{2}, \tag{2.271}$$

while $\mathcal{P}(\uparrow_{\hat{n}_i}\,;-) = \frac{1}{2}\langle\Psi|(\mathbb{1} + \hat{n}_i\cdot\vec{\sigma})\otimes\mathbb{1}|\Psi\rangle = \frac{1}{2}$ and

$$\mathcal{P}(\uparrow_{\hat{n}_3};-) + \mathcal{P}(-;\,\uparrow_{\hat{n}_2}) = 1, \tag{2.272}$$

with a manifest violation of the inequality in Eq. (2.265).

The violation of the Bell inequalities shows that quantum correlations are different from, and stronger than in, the classical case. Since the crucial assumption is the factorization in Eq. (2.263), there must be some **non-locality** built into the quantum theory.[2.83]

Figure 2.30 depicts the correlation between observations of the spin of the two particles as a function of the relative angle between the detectors used in the two distant locations. For the case when the two detectors in the two locations are measuring along the same direction (with a relative angle of 0° or 180°), the classical and quantum correlations coincide. The same equivalence between classical and quantum descriptions is true if the measurement is done with respect to two orthogonal directions (with a relative angle of 90°) except that now the result of the measurement in the second location is completely random with no correlation.

The equivalence between classical and quantum correlations is no longer true if the relative angle is anything in between these values – as in the case defined in Eq. (2.270).

2.81 The pure state $|\Psi\rangle^T = (0, 1, 1, 0)/\sqrt{2}$ in Eq. (2.268) corresponds to maximum entanglement between the spins of the two particles.

2.82 It is a double experiment with one polarimeter, the operator $\vec{\sigma}$, in the direction $\hat{n}_i$ and the other in the direction $\hat{n}_j$. Each polarimeter is represented by an operator $\frac{1}{2}(1 + \hat{n}\cdot\vec{\sigma})$.

2.83 If you are wondering: Bell inequalities have been tested in numerous experiments and found to be violated. The world is indeed quantum. However, the non-locality of the correlation does not imply a violation of relativistic invariance and there is no faster-than-light exchange of information.

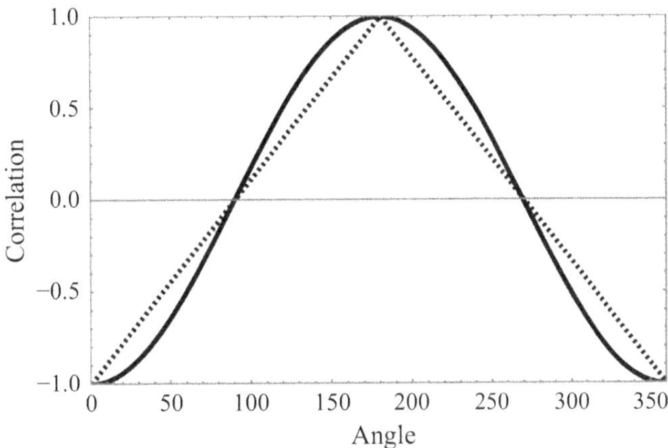

Figure 2.30 Classical (dotted line) and quantum (solid line) correlation as a function of the relative angle between the detectors.

The Bell inequalities set a limit given by the classical correlations which is violated by the quantum correlations at these intermediate values of the relative angle between the two detectors – where the quantum correlation is substantially larger because of the non-local nature of the quantum phenomena and the two observables are maximally entangled – as is the case for the angle between $\hat{n}_1$ and $\hat{n}_2$ (135°) and between $\hat{n}_1$ and $\hat{n}_4$ (45°).

Quantum field theory contains maximally entangled states whose measurement would lead to violations of Bell inequalities. This is unavoidable since the theory is the marriage of relativity and quantum mechanics. If I want to find the "quantum" in quantum field theory I must go after these correlations (in the cross sections) and the non-local effects they entail.

Taking the pair annihilation process of the previous problem session as an example, the final state produced by the incoming left-handed electrons and right-handed positrons can be written in terms of the possible chiral states of the final muons and antimuons as

$$\zeta_1|\mu_L^- \mu_L^+\rangle + \zeta_2|\mu_R^- \mu_L^+\rangle + \zeta_3|\mu_L^- \mu_R^+\rangle + \zeta_4|\mu_R^- \mu_R^+\rangle \tag{2.273}$$

with $\sum_i |\zeta_i|^2 = 1$. Among the many possible measures of the entanglement of the state, the quantity

$$2\,|\zeta_1\zeta_4 - \zeta_2\zeta_3|\,, \tag{2.274}$$

which is equal to 1 for maximal entanglement and 0 for a product state (with no entanglement) suffices for the purpose of the present discussion. It is called the **concurrence**.

Same chirality states are suppressed in the high-energy limit and vanish when neglecting the masses. Therefore only two out of the four states in Eq. (2.273) remain.

The states in Eq. (2.273) can be written for the final muon momenta along an arbitrary direction by rotating the chiral states – which are here defined along the beam direction – to that direction by means of the Wigner D-matrix for the $l = 1$ representation (because of the spin-1 photon). The element of the Wigner D-matrix to be used depends on the helicity of the state. The rotated state is[2.84]

2.84 $M^{(1)}_{1\,1}(\theta) = \frac{1}{2}(1+\cos\theta)$

$M^{(1)}_{1\,-1}(\theta) = \frac{1}{2}(1-\cos\theta).$

$$M^{(1)}_{1\,1}(\theta)\,|\mu^-_R\mu^+_L\rangle + M^{(1)}_{1\,-1}(\theta)\,|\mu^-_L\mu^+_R\rangle\,, \tag{2.275}$$

because the state $|\mu^-_R\mu^+_L\rangle$ has the two helicities pointing in the direction of the momentum (and therefore the total helicity is $+1$) and the state $|\mu^-_L\mu^+_R\rangle$ has opposite helicities (and therefore the total helicity is -1).

After normalizing the coefficients, the final state of the two muons is found to be (at very large energies)

$$\frac{1}{\sqrt{2}}\frac{1+\cos\theta}{\sqrt{1+\cos^2\theta}}\,|\mu^-_R\mu^+_L\rangle + \frac{1}{\sqrt{2}}\frac{1-\cos\theta}{\sqrt{1+\cos^2\theta}}\,|\mu^-_L\mu^+_R\rangle\,. \tag{2.276}$$

Therefore, the entanglement is given by[2.85]

2.85 $\zeta_1 = \zeta_4 = 0$

$\zeta_2 = (1+\cos\theta)/\sqrt{2\,(1+\cos^2\theta)}$

$\zeta_3 = (1-\cos\theta)/\sqrt{2\,(1+\cos^2\theta)}.$

$$2\,|-\zeta_2\zeta_3| = \frac{(1-\cos\theta)(1+\cos\theta)}{1+\cos^2\theta} = \frac{\sin^2\theta}{1+\cos^2\theta}\,. \tag{2.277}$$

The state of the final muons has maximal entanglement for a scattering angle $\theta = \pi/2$ – when the muon pair comes out perpendicular to the direction of the electron pair – and is given by

$$\frac{1}{\sqrt{2}}\Big(|\mu^-_R\mu^+_L\rangle + |\mu^-_L\mu^+_R\rangle\Big)\,, \tag{2.278}$$

which is, like the state in Eq. (2.268), maximally entangled. The reason for the entanglement is that the photon sees the two opposite chiral states of the final muons equally and cannot distinguish one from the other.

2.15 *Problem Session*: Compton Scattering

This problem session uses the same Feynman rules as those utilized in the previous session but adds two new features: first, the amplitude consists the sum of two Feynman diagrams; second, two of the external particles are photons. This provides the chance to review some additional machinery.

The Compton scattering of electrons and light is a very famous process. It is almost impossible to graduate in physics without having encountered it in one form or another.

To apply perturbation theory, for a given scattering process I have to consider the initial and final states and draw all inequivalent Feynman diagrams connecting these states. In the case of Compton scattering, there are two diagrams, those shown in Figure 2.31.

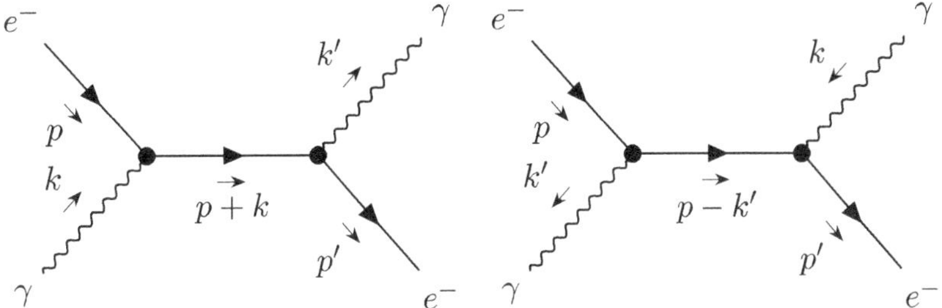

Figure 2.31 Feynman diagrams for Compton scattering. The final photons are indistinguishable and the two possible momentum and polarization identifications must be included by summing the two diagrams.

The amplitude for the scattering process is the sum of these two diagrams:

$$\mathcal{M}_{fi} = \mathcal{M}^a + \mathcal{M}^b , \tag{2.279}$$

where

$$i\mathcal{M}^a = \Bigg[\underbrace{\bar{u}(p')}_{e^-\ \text{OUT}}\ \overbrace{(-ie\gamma^\mu)}^{\text{vertex}}\ \underbrace{\varepsilon_\mu^{*\lambda'}(k')}_{\gamma\ \text{OUT}}\Bigg]\ \overbrace{\frac{i(\not{p}+\not{k}+m)}{(p+k)^2-m^2}}^{\text{electron propagator}}\ \Bigg[\underbrace{u(p)}_{e^-\ \text{IN}}\ \overbrace{(-ie\gamma^\nu)}^{\text{vertex}}\ \underbrace{\varepsilon_\mu^{\lambda}(k)}_{\gamma\ \text{IN}}\Bigg]$$

and

$$i\mathcal{M}^b = \Bigg[\underbrace{\bar{u}(p')}_{e^-\ \text{OUT}}\ \overbrace{(-ie\gamma^\mu)}^{\text{vertex}}\ \underbrace{\varepsilon_\mu^{\lambda}(k)}_{\gamma\ \text{OUT}}\Bigg]\ \overbrace{\frac{i(\not{p}-\not{k}'+m)}{(p-k')^2-m^2}}^{\text{electron propagator}}\ \Bigg[\underbrace{u(p)}_{e^-\ \text{IN}}\ \overbrace{(-ie\gamma^\nu)}^{\text{vertex}}\ \underbrace{\varepsilon_\mu^{*\lambda'}(k')}_{\gamma\ \text{IN}}\Bigg] .$$

The photons enter with their polarization four-vectors $\varepsilon_\mu^\lambda(k)$. There are only two polarizations for a massless spin-1 particle. For a photon moving in the z-direction, they are given by

$$\varepsilon_\mu^{\lambda=1}(\hat{z}) = \frac{-1}{\sqrt{2}}\begin{pmatrix}0\\1\\i\\0\end{pmatrix} \quad \text{and} \quad \varepsilon_\mu^{\lambda=2}(\hat{z}) = \frac{1}{\sqrt{2}}\begin{pmatrix}0\\+1\\-i\\0\end{pmatrix} , \tag{2.280}$$

in a circular polarization basis. These explicit expressions are not needed in this computation but it is useful to bear them in mind as I write the various equations.

The presence of two diagrams gives rise to a specifically quantum mechanical interference term and there are three terms to compute, namely, we obtain

$$|\mathcal{M}_{fi}|^2 = |\mathcal{M}_a|^2 + |\mathcal{M}_b|^2 + 2\,\mathrm{Re}\,\mathcal{M}_a\mathcal{M}_b^* . \tag{2.281}$$

Beginning with M_a, write

$$i\mathcal{M}_a = -ie^2\varepsilon_\mu'^{*}\varepsilon_\nu\,\bar{u}(p')\frac{\gamma^\mu(\not{p}+\not{k}'+m)\gamma^\nu}{(p+k)^2-m^2}u(p) , \tag{2.282}$$

where

$$(p+k)^2 - m^2 = p^2 + 2p\cdot k + k^2 - m^2 = 2p\cdot k\,, \tag{2.283}$$

because of the mass-shell conditions for the electron ($p^2 = m^2$) and the massless photon ($k^2 = 0$). Now simplify the numerator by means of the Dirac equation and write[2.86]

2.86 Use the anticommutation of the Dirac matrices $\gamma^\alpha\gamma^\nu = 2g^{\alpha\nu} + \gamma^\nu\gamma^\alpha$ and the Dirac equation $(\not{p} - m)u = 0$.

$$\begin{aligned}(\not{p}+m)\gamma^\nu u(p) &= (\gamma^\alpha\gamma^\nu p_\alpha + \gamma^\nu m)u(p)\\ &= (2p^\nu - \gamma^\nu\not{p} + \gamma^\nu m)u(p) = 2p^\nu u(p)\end{aligned} \tag{2.284}$$

so that

$$|\mathcal{M}_a|^2 = -ie^2\varepsilon'^*_\mu\varepsilon'_\rho\varepsilon^*_\mu\varepsilon_\rho\,\bar{u}(p')\frac{2\gamma^\mu p^\nu + \gamma^\mu\not{k}\gamma^\nu}{2p\cdot k}u(p)\,\bar{u}(p')\frac{2\gamma^\sigma p^\rho + \gamma^\sigma\not{k}\gamma^\rho}{2p\cdot k}u(p)\,. \tag{2.285}$$

What I need to compute is

$$|\overline{\mathcal{M}}_a|^2 = \frac{1}{4}\sum_{s,s',\lambda,\lambda'}|\mathcal{M}_a|^2\,, \tag{2.286}$$

where the sum over the polarizations of the product of the wave functions is given by Eq. (2.254) of the previous problem section. The factor 1/4 comes from averaging over the polarizations of the initial states.

The sum over the photon polarizations is given by

$$\boxed{\sum_{\lambda=1,2,3,4}\varepsilon^{\lambda*}_\mu(k)\varepsilon^\lambda_\nu(k) = -g_{\mu\nu}\,.} \tag{2.287}$$

The reason why I can sum over all polarizations and yet still only the two physical ones contribute is explained in Chapter 3.

I can use the Casimir trick to sum over the polarizations. I have that

$$|\overline{\mathcal{M}}_a|^2 = \frac{e^4}{4}\frac{1}{(2p\cdot k)^2}\mathrm{Tr}\left[(\not{p}'+m)(2\gamma^\mu p^\nu + \gamma^\mu\not{k}\gamma^\nu)(\not{p}+m)(2\gamma_\mu p_\nu + \gamma_\nu\not{k}\gamma_\mu)\right]. \tag{2.288}$$

The expansion of the trace in Eq. (2.288) unfortunately contains 16 terms! There is nothing I can do but keep going and doggedly compute them.[2.87]

2.87 Only seven traces out of the 16 turn out to be non-vanishing:

$$\begin{aligned}\mathrm{Tr}\,[\not{p}'\gamma^\mu\not{k}'\gamma^\nu\not{p}\gamma_\nu\not{k}\gamma_\mu] &= 32\,(p\cdot k)(p'\cdot k)\,,\\ \mathrm{Tr}\,[\not{p}'\gamma^\mu\not{k}'\gamma^\nu\not{p}2\gamma_\nu p_\mu] &= -16\,m^2(p'\cdot k)\,,\\ \mathrm{Tr}\,[\not{p}'2\gamma^\mu p^\nu\not{p}\gamma_\nu\not{k}\gamma_\mu] &= -16\,m^2(p'\cdot k)\,,\\ \mathrm{Tr}\,[\not{p}'2\gamma^\mu p^\nu\not{p}2\gamma_\mu p_\nu] &= -32\,m^2(p'\cdot p)\,,\\ \mathrm{Tr}\,[\gamma^\mu p^\nu\not{k}\gamma_\nu\gamma_\mu p_\nu] &= 32\,m(k\cdot p)\,,\\ \mathrm{Tr}\,[\gamma^\mu p^\nu\gamma_\nu\not{k}\gamma_\mu] &= 16\,(p\cdot k)\,,\\ \mathrm{Tr}\,[\gamma^\mu p^\nu\gamma_\mu p_\nu] &= 16\,m^2\,.\end{aligned}$$

Trace-taking and Diracology gives

$$|\overline{\mathcal{M}}_a|^2 = \frac{32\,e^4}{4\,(2p\cdot k)^2}\left[2\,m^4 + 2m^2(p\cdot k) - m^2(p\cdot p') - m^2(p'\cdot k) + (p\cdot k)(p'\cdot k)\right], \tag{2.289}$$

which can be rewritten in terms of the Mandelstam variables, given in this case by

$$\begin{aligned}s &= (p+k)^2 = 2p\cdot k + m^2\\ t &= (p'-p)^2 = -2p\cdot p' + 2m^2\\ u &= (p'-k)^2 = -2p'\cdot k + m^2\,,\end{aligned} \tag{2.290}$$

as

$$|\overline{\mathcal{M}_a}|^2 = 2\,e^4 \left[\frac{4m^4}{(s-m^2)^2} + \frac{2m^2}{(s-m^2)} + \frac{u-m^2}{(s-m^2)} \right] . \tag{2.291}$$

Before starting work on the next amplitude, let's go back and notice that $\mathcal{M}_b$ is just like $\mathcal{M}_a$ except that the variables s and u are interchanged. This is called crossing symmetry. I do not have to redo the computation for $\mathcal{M}_b$, which is just given by

$$|\overline{\mathcal{M}_a}|^2 = 2\,e^4 \left[\frac{4m^4}{(u-m^2)^2} + \frac{2m^2}{(u-m^2)} + \frac{s-m^2}{(u-m^2)} \right] . \tag{2.292}$$

Let us push on – not to be daunted. The remaining interference term takes another bit of manipulation, which in the end yields

$$2\overline{\mathcal{M}_a \mathcal{M}_b^*} = 2\,e^4 \left[\frac{8m^4}{(s-m^2)(u-m^2)} + \frac{2m^2}{(u-m^2)} - \frac{2m^2}{(s-m^2)} \right] . \tag{2.293}$$

Finally, combining the three terms,

$$|\overline{\mathcal{M}}|^2 = 2\,e^4 \left[4m^4 \left(\frac{1}{s-m^2} + \frac{1}{u-m^2} \right)^2 + 4m^2 \left(\frac{1}{s-m^2} + \frac{1}{u-m^2} \right) \right. \\ \left. - \frac{u-m^2}{s-m^2} - \frac{s-m^2}{u-m^2} \right] . \tag{2.294}$$

In the center of mass frame, see Figure 2.32, the momenta of the electrons and photons are given by

$$\begin{aligned} k &= (\omega,\, 0,\, 0,\, \omega) \\ k' &= (\omega,\, \omega \sin\theta,\, 0,\, \omega\cos\theta) \\ p &= (E,\, 0,\, 0,\, -\omega) \\ p' &= (E,\, -\omega\sin\theta,\, 0,\, -\omega\cos\theta)\,, \end{aligned} \tag{2.295}$$

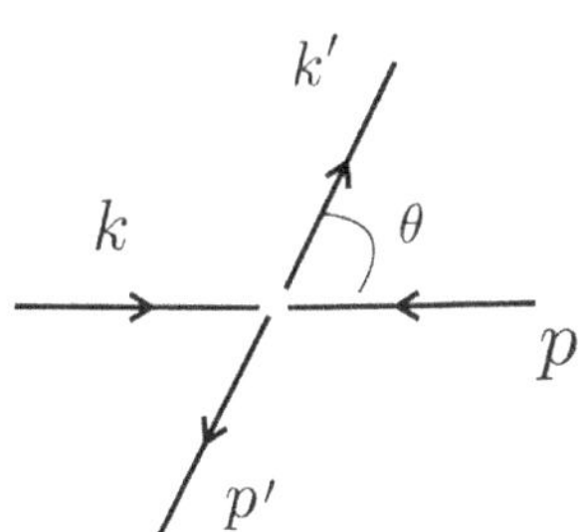

Figure 2.32 Momenta in the center of mass frame.

where ω is the photon energy, E that of the electron and θ the scattering angle. The Mandelstam variables in this frame of reference are given by

$$\begin{aligned} s &= m^2 + 2\,\omega\,(E+\omega) \\ t &= 2\,\omega^2(\cos\theta - 1) \\ u &= -2\,\omega\,(E + \omega\cos\theta) + m^2\,. \end{aligned} \tag{2.296}$$

The differential cross section is obtained by inserting the momenta in Eq. (2.295) and Eq. (2.296) into the general formula to obtain

$$\mathrm{d}\sigma = \frac{1}{(2E)(2\omega)\,|1+\omega/E|} |\overline{\mathcal{M}}|^2 \mathrm{d}\Pi_2 = \frac{1}{2\,(s-m^2)} |\overline{\mathcal{M}}|^2 \mathrm{d}\Pi_2\,. \tag{2.297}$$

The integrated phase space is given by

$$\int \mathrm{d}\Pi_2 = \int \frac{\mathrm{d}^3\vec{p}\,'}{(2\pi)^3}\frac{1}{2E_{\vec{p}\,'}}\int \frac{\mathrm{d}^3\vec{k}'}{(2\pi)^3}\frac{1}{2E_{\vec{k}'}}(2\pi)^4\delta^{(4)}(p+k-p'-k')$$
$$= \frac{1}{(2\pi)^2}\int \frac{(k')^2\mathrm{d}k'\mathrm{d}\cos\theta\,\mathrm{d}\phi}{4E_{\vec{p}'}E_{\vec{k}'}}\delta(E+\omega-E_{\vec{p}'}-E_{\vec{k}'}), \quad (2.298)$$

where the second line in Eq. (2.298) is obtained after the space δ-function $\delta^{(3)}(\vec{p}+\vec{k}-\vec{p}\,'-\vec{k}')$ in the phase-space integral removes the integration over $\vec{p}\,'$ by enforcing the conservation of momentum:

$$\vec{p}+\vec{k}=\vec{p}\,'+\vec{k}'\,. \quad (2.299)$$

This relation makes it possible to write

$$\underbrace{(p')^2}_{=m^2} = (p+k-k')^2 = m^2+2p\cdot(k-k')-2k\cdot k'+\underbrace{k^2+(k')^2}_{=0} \quad (2.300)$$

and, in the laboratory frame,

$$m^2 = m^2+2m(\omega-\omega')+2\omega\omega'(\cos\theta-1), \quad (2.301)$$

that is,

$$\frac{1}{\omega'}-\frac{1}{\omega}=\frac{1}{m}(1-\cos\theta), \quad (2.302)$$

which gives the relationship between the energies of the outgoing and incoming photons and the scattering angle θ.

The remaining δ-function in Eq. (2.298) can be written, using the property of this function, as

$$\delta(E+\omega-E_{\vec{p}'}-E_{\vec{k}'}) = \delta(k-k_0)\left|\frac{\partial(E+\omega-E_{\vec{p}'}-E_{\vec{k}'})}{\partial k'}\right|^{-1}, \quad (2.303)$$

where k_0 is the root of the equation

$$E+\omega-E_{\vec{p}'}-E_{\vec{k}'}=0 \quad (2.304)$$

and

$$\frac{\partial(E+\omega-E_{\vec{p}'}-E_{\vec{k}'})}{\partial k'} = \frac{-2k'}{2E_{\vec{p}'}}-1=-\left(\frac{k'}{E_{\vec{p}'}}+\frac{k'}{E_{\vec{k}'}}\right). \quad (2.305)$$

Therefore the phase-space factor in Eq. (2.298) can be written as a differential of the scattering angle:

$$\mathrm{d}\Pi_2 = \mathrm{d}\cos\theta\left(\frac{k'}{E_{\vec{p}'}+E_{\vec{k}'}}\right)$$
$$= \frac{\mathrm{d}\cos\theta}{8\pi}\frac{(\omega')^2}{m\,\omega} = \frac{\mathrm{d}\cos\theta}{8\pi}\frac{\omega}{\underbrace{E+\omega}_{=\frac{s-m^2}{2\omega}}}\,. \quad (2.306)$$

At this point, I may as well exchange the scattering angle for the Mandelstam variable t, the differential of which is $\mathrm{d}t = 2\omega^2 \mathrm{d}\cos\theta$, and finally write the cross section as

$$\frac{\mathrm{d}\sigma}{\mathrm{d}t} = \frac{1}{16\pi(s - m^2)} |\overline{\mathcal{M}}|^2 , \tag{2.307}$$

where $|\overline{\mathcal{M}}|^2$ is given by Eq. (2.294).

The Compton cross section in Eq. (2.294) can also be computed in the laboratory frame of reference, Figure 2.33, where

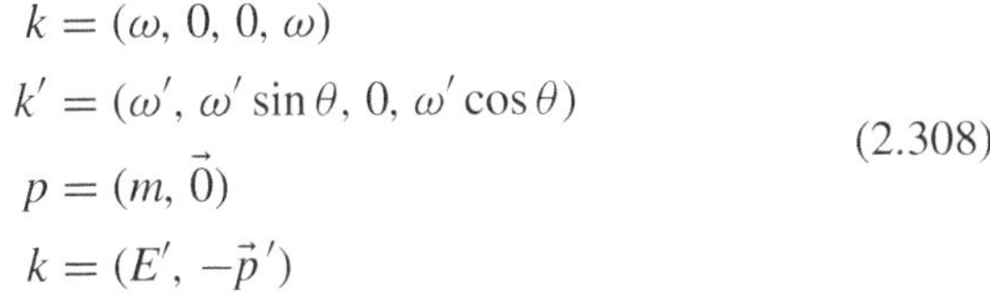

$$\begin{aligned} k &= (\omega, 0, 0, \omega) \\ k' &= (\omega', \omega' \sin\theta, 0, \omega' \cos\theta) \\ p &= (m, \vec{0}) \\ k &= (E', -\vec{p}\,') \end{aligned} \tag{2.308}$$

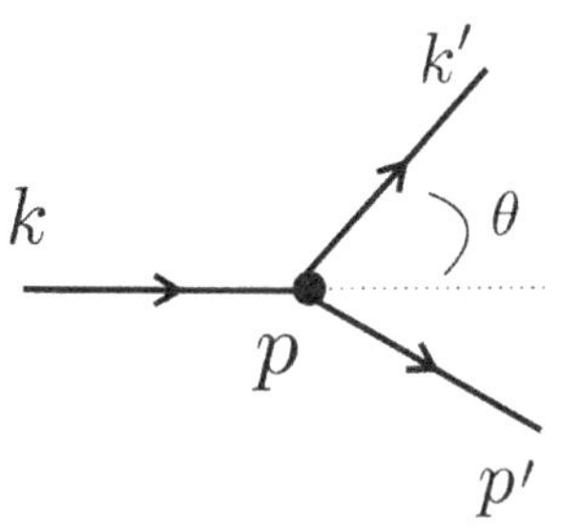

Figure 2.33 Momenta in the laboratory frame.

and

$$\begin{aligned} s &= 2\, m\, \omega + m^2 \\ t &= 2\, \omega\, \omega' (\cos\theta - 1) \\ u &= -2\, m\, \omega' + m^2 . \end{aligned} \tag{2.309}$$

Taking the Mandelstam variables in Eq. (2.309) and inserting them in Eq. (2.294) gives

$$|\overline{\mathcal{M}}|^2 = 2\, e^4 \left(\frac{\omega'}{\omega} + \frac{\omega}{\omega'} - \sin^2\theta \right) . \tag{2.310}$$

The phase-space factor in this frame of reference is

$$\begin{aligned} \mathrm{d}\Pi_2 &= \frac{\mathrm{d}\cos\theta\, \omega' \mathrm{d}\omega'}{8\pi E'} \left(1 + \frac{\omega' - \omega\cos\theta}{E'} \right)^{-1} \delta(\omega' - \omega_0') \\ &= \frac{\mathrm{d}\cos\theta}{8\pi} \frac{\omega'}{\underbrace{E'}_{E' = m + \omega - \omega'} + \omega' - \omega\cos\theta} = \frac{\mathrm{d}\cos\theta}{8\pi} \frac{(\omega')^2}{m\omega} , \end{aligned} \tag{2.311}$$

and therefore

$$\frac{\mathrm{d}\sigma}{\mathrm{d}\cos\theta} = \frac{1}{8\pi} \frac{(\omega')^2}{m\omega} \frac{1}{2\underbrace{(2m\omega)}_{= s - m^2}} |\overline{\mathcal{M}}|^2 , \tag{2.312}$$

which gives, after replacing the matrix element with its value in the laboratory frame, as in Eq. (2.310),

$$\frac{\mathrm{d}\sigma}{\mathrm{d}\cos\theta} = \frac{e^4}{16\pi m^2} \left(\frac{\omega'}{\omega} \right) \left(\frac{\omega'}{\omega} + \frac{\omega}{\omega'} - \sin^2\theta \right) . \tag{2.313}$$

Equation (2.313) is the celebrated **Klein–Nishima formula** for Compton scattering.

As a useful check, going to low energies, where the initial and final energies of the electron are about the same ($\omega'/\omega \to 1$), I correctly obtain the classical Thomson cross section for the scattering of light from electrons:

$$\frac{\mathrm{d}\sigma}{\mathrm{d}\cos\theta} = \frac{\pi\alpha^2}{m^2}\left(1 + \cos^2\theta\right). \tag{2.314}$$

2.16 Unitarity and the S-Matrix

The S-matrix gives the probability for a given scattering (or decay) to take place. From the properties of the probability of an event we can learn something about the S-matrix in general without having to assume anything about the scattering process involved.[2.88]

An important relation is obtained by enforcing the fact that the probabilities for any scattering process must sum to one. I cannot create more than I have put in, that is, the S-matrix is unitary and

2.88 This idea of finding the dynamics by only enforcing properties like unitarity, analyticity and crossing on the S-matrix – often dubbed the **bootstrap**, a nod to Baron Munchausen – is part of a research program that periodically makes its way to the forefront of particle physics.

$$\hat{S}\hat{S}^\dagger = 1. \tag{2.315}$$

By writing the S-matrix in terms of the matrix $\hat{M}$ (after subtracting scattering in which nothing happens) I have that

$$\hat{S}\hat{S}^\dagger = (1 + i\hat{\mathcal{M}})(1 - i\hat{\mathcal{M}}^\dagger) = 1 + i(\hat{\mathcal{M}} - \hat{\mathcal{M}}^\dagger) + \hat{\mathcal{M}}\hat{\mathcal{M}}^\dagger = 1, \tag{2.316}$$

from which I obtain that

$$-i(\hat{\mathcal{M}} - \hat{\mathcal{M}}^\dagger) = 2\,\mathrm{Im}\,\hat{\mathcal{M}} = \hat{\mathcal{M}}\hat{\mathcal{M}}^\dagger. \tag{2.317}$$

This relation can be written in the case of elastic forward scattering[2.89] as (now going to the components and using the completeness relation)

2.89 Remember that elastic scattering means that the particles in the initial and final states are the same.

$$2\,\mathrm{Im}\,\mathcal{M}_{ii} = (\mathcal{M}\mathcal{M}^\dagger)_{ii} = \sum_k \mathcal{M}_{ik}\mathcal{M}^*_{ki}. \tag{2.318}$$

The total cross section (see for example Figure 2.34 below) is given by

$$\sigma_T = \frac{4\pi}{s}\sum_k |\mathcal{M}_{ik}|^2 \tag{2.319}$$

and therefore[2.90]

2.90 This identity tells me that loops (see the next section) are necessary because the imaginary part of the amplitude is to be found – except for a delta function – in the loops.

$$\boxed{\sigma_T = \frac{8\pi}{s}\mathrm{Im}\,\mathcal{M}_{ii},} \tag{2.320}$$

which is the **optical theorem**; it relates the imaginary part of the elastic amplitude in the forward direction to the total cross section.

The matrix elements M_{ii} can be written in terms of partial waves in the angular momentum J.[2.91] Then M_{ii} is a finite complex function that can be written as[2.92]

$$\mathcal{M}_{ii}(s,\theta) = \sum_J (2J+1)\, a_J(s) P_J(\cos\theta)\,. \tag{2.321}$$

In Eq. (2.321) there is a complex function $a_J(s)$ for which it is true that

$$2\,\mathrm{Im}\, a_J(s) \geq [\mathrm{Im}\, a_J(s)]^2 + [\mathrm{Re}\, a_J(s)]^2\,, \tag{2.322}$$

which implies

$$0 \leq \mathrm{Im}\, a_J(s) \leq 2 \tag{2.323}$$

and therefore the bound

$$\mathrm{Im}\,\mathcal{M}_{ii}(s,\theta) = \sum_J (2J+1)\,\mathrm{Im}\, a_J(s) P_J(\cos\theta) \leq 2\sum_J (2J+1)\,, \tag{2.324}$$

since $|P_J(\cos\theta)| \leq 1$.

The optical theorem in Eq. (2.320) together with the bound in Eq. (2.324) on the imaginary part of the forward amplitude gives an upper bound on the size of the total cross section:

$$\sigma_T \leq \frac{16\pi}{s} \sum_J (2J+1)\,. \tag{2.325}$$

The limit in Eq. (2.325) is particularly simple when the energy is sufficiently small that waves with $J \geq 1$ can be neglected to obtain

$$\sigma_T \leq \frac{16\pi}{s}\,. \tag{2.326}$$

The sum over all angular momenta J goes up to the largest value $J^{\max} = \mathrm{const.} \times \sqrt{s}\ln s$.[2.93]

Now summing over all the partial waves,

$$\sum_0^{J^{\max}} (2J+1) = \mathrm{const.} \times (1 + \sqrt{s}\ln s)^2 \simeq \mathrm{const.} \times s\,(\log s)^2\,, \tag{2.327}$$

which, inserted into Eq. (2.325), gives the **Froissart bound** on the total cross section,

$$\sigma_T \leq \mathrm{const.} \times (\log s)^2\,; \tag{2.328}$$

this allows for a slow (logarithmic) increase of the cross section with the center of mass energy. Cross sections growing faster, for example rising linearly with s, are not allowed. If a model produces such a cross section it will violate unitarity and I will have to look for a better model.

2.91 The matrix element $\mathcal{M}_{ii}$ in terms of the (incoming and outgoing) momenta requires an infinite normalization and cannot be used. Had it been possible, the unitary bound would have been simply Eq. (2.326) below.

2.92 Remember that the Legendre polynomials $P_J(\cos\theta)$ provide a orthonormal basis. They are real functions. The inequality (2.322) requires $\mathrm{Im}\, a_J$ to be within the circle of radius 1 going through the origin of coordinates and having equation $(\mathrm{Im}\, a_J)^2 + (\mathrm{Re}\, a_J)^2 - 2\,\mathrm{Im}\, a_J = 0$.

2.93 A detailed analysis is required to argue that $J^{\max}$ is indeed the largest angular momentum contributing to the scattering. Let me take the result from the literature.

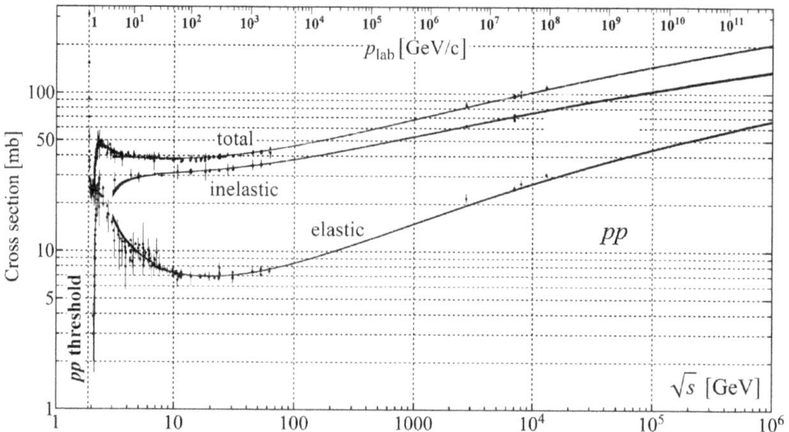

Figure 2.34 Cross section for *pp* scattering [R.L. Workman *et al.* (Particle Data Group), *Prog. Theor. Exp. Phys.* **2022** (2022) 083C01 http://pdg.lbl.gov].

2.17 *Problem Session*: Anomalous Magnetic Moment*

In this problem session I want to examine the case for which the Feynman diagram consists of lines closed in a loop. This kind of process adds the final bit of machinery, the loop integral, to the toolbox of perturbative quantum field theory.

The example of the anomalous magnetic moment is simple because there are no divergent integrations. It also played a tremendous historical role in convincing the community of the power of QED. The value of the magnetic moment of the electron thus computed agrees within 10 significant digits with its experimental value. You are not offered every day the possibility of computing something with this kind of precision!

To set the stage, recall that the magnetic moment of an electron is defined as

spin

$$\vec{\mu} = g \, \frac{e}{2m} \, \vec{s} \, . \tag{2.329}$$

gyromagnetic ratio

From the problem session on the hydrogen atom I know that the spin is $\vec{s} = \frac{1}{2}\vec{\sigma}$ and therefore the gyromagnetic ratio of the electron is $g = 2$.

Lorentz covariance can be used to write a generic vertex between the electron and the photon as

$$\Gamma^\mu = \Gamma_1(q^2)\gamma^\mu + \Gamma_2(q^2)(p+p')^\mu + \Gamma_3(q^2)(p-p')^\mu \, . \tag{2.330}$$

2.94 $(p-p')^\mu \Gamma_\mu = 0$.

Because of the conservation of the electromagnetic current,[2.94] the function $\Gamma_3(q^2)$ must vanish.

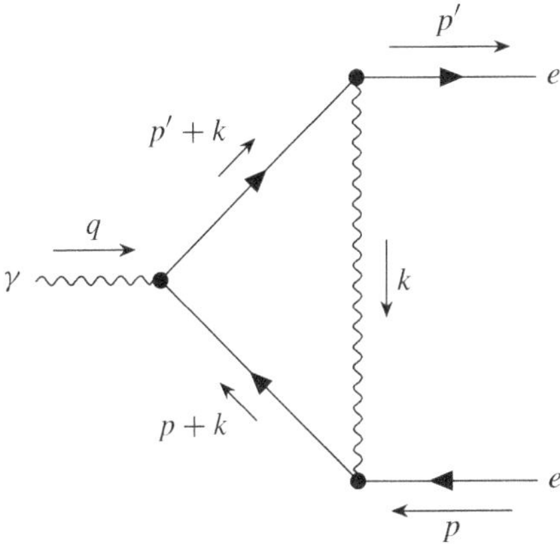

Figure 2.35 The Feynman diagram for the anomalous magnetic moment of the electron.

The **Gordon identity**

$$\bar{u}(p')\gamma^{\mu}u(p) = \frac{1}{2m}\bar{u}(p')(p + p')^{\mu}u(p) + \frac{i}{2m}\bar{u}(p')\sigma^{\mu\nu}(p - p')_{\nu}u(p)$$

makes it possible to recast the functions $\Gamma_1(q^2)$ and $\Gamma_2(q^2)$ in terms of two form factors as

$$\Gamma^{\mu} = F_1(q^2)\gamma^{\mu} + \frac{i}{2m}F_2(q^2)\sigma^{\mu\nu}q_{\nu}\,, \tag{2.331}$$

where F_1 is called the electric part and F_2 the magnetic part. At the tree level – with the coupling just given by the Dirac matrix – the form factors are given by $F_1(0) = 1$ and $F_2(0) = 0$. At the one-loop level again $F_1(0) = 1$, because the electric charge does not renormalize, and

$$F_2(0) = (g - 2)/2\,, \tag{2.332}$$

which is the required anomalous magnetic moment of the electron.

I can apply the Feynman rules of QED for the **loop diagram** in Figure 2.35 by going around the loop, obtaining the amplitude

$$\bar{u}(p')i\Gamma^{\mu}u(p) = e^3 \underbrace{\int \frac{\mathrm{d}^4k}{(2\pi)^4}}_{\substack{\text{integration}\\ \text{over}\\ \text{internal}\\ \text{momentum!}}} \frac{\bar{u}(p')\left[\gamma^{\rho}(\not{p}' + \not{k} + m)\gamma^{\mu}(\not{p} + \not{k} + m)\gamma_{\rho}\right]u(p)}{[(p' + k)^2 - m^2]k^2[(p + k)^2 - m^2]}\,. \tag{2.333}$$

Let us first do the usual Diracology and then come back to the integral. Take the numerator in Eq. (2.333):

$$\bar{u}(p')\Big[\gamma^{\rho}(\not{p}' + \not{k} + m)\gamma^{\mu}(\not{p} + \not{k} + m)\gamma_{\rho}\Big]u(p)\,. \tag{2.334}$$

I can rewrite it by means of the equations of motion and by letting some of the Dirac matrices go from right to left. To be specific, the term in Eq. (2.334) with only Dirac matrices gives[2.95]

$$\gamma^\rho(\not{p}' + \not{k})\gamma^\mu(\not{p} + \not{k})\gamma_\rho = -2(\not{p} + \not{k})\gamma^\mu(\not{p}' + \not{k}), \tag{2.335}$$

while the term with mass can be rewritten as

$$\bar{u}(p')\Big[m\gamma^\rho(\not{p}' + \not{k})\gamma^\mu\gamma_\rho + m\gamma^\rho\gamma^\mu(\not{p} + \not{k})\gamma_\rho\Big]u(p) = 4m(p' + p + 2k)^\mu \tag{2.336}$$

and the term with m^2 as

$$m^2\gamma^\rho\gamma^\mu\gamma_\rho = -2m^2\gamma^\mu . \tag{2.337}$$

Further simplifications can be implemented by using the equations of motion to tackle the individual parts of Eq. (2.334). At the end I obtain for the numerator in Eq. (2.333) the expression[2.96]

$$\bar{u}(p')\Big[2\gamma^\mu k^2 - 4k^\mu\not{k} + 4(p'\cdot p)\gamma^\mu + 4(p+p')\cdot k\gamma^\mu + 4mk^\mu - 4(p+p')^\mu\not{k}\Big]u(p), \tag{2.338}$$

which is the most convenient form for integration in the variable k.

The denominator of Eq. (2.333) is unyielding to integration and is best transformed by using the identity[2.97]

$$\boxed{\frac{1}{ABC} = 2\int_0^1 \mathrm{d}x \int_0^1 \mathrm{d}y \int_0^1 \mathrm{d}z \frac{\delta(1-x-y-z)}{[xA + yB + zC]^3},} \tag{2.339}$$

which can be easily verified for any quantities A, B and C.

I therefore rewrite the denominator in Eq. (2.333) as follows:

$$\begin{aligned}&\frac{1}{[(p'+k)^2 - m^2]k^2[(p+k)^2 - m^2]}\\ &\quad = 2\int_0^1 \mathrm{d}x \int_0^1 \mathrm{d}y \int_0^1 \mathrm{d}z \frac{\delta(1-x-y-z)}{[(k^2 + 2k\cdot p'x + 2k\cdot pz]^3}.\end{aligned} \tag{2.340}$$

Using a change of variables, Eq. (2.340) can be brought into a simpler form, by introducing a new momentum

$$\tilde{k} = k + p'x + pz . \tag{2.341}$$

This change of integration variable eliminates all the terms linear in the momentum k in the denominator on the right-hand side of Eq. (2.340), which becomes

$$\left[\tilde{k}^2 + q^2xz - m^2(x+z)^2\right]^3 . \tag{2.342}$$

2.95 $$\gamma^\rho\gamma^\mu\gamma_\rho = -2\gamma^\mu$$ $$\gamma^\rho\gamma^\nu\gamma^\mu\gamma_\rho = 4g^{\mu\nu}$$ $$\gamma^\rho\gamma^\nu\gamma^\mu\gamma^\sigma\gamma_\rho = -2\gamma^\sigma\gamma^\mu\gamma^\nu .$$

2.96 $$\not{p}u(p) = mu(p)$$ $$\bar{u}(p')\not{p}' = m\bar{u}(p') .$$

2.97 This trick is so neat that it has a name: Feynman's parametrization. There are other parametrizations for these integrals but this one is sufficient here.

This is fine but now I have to shift to the variable $\tilde{k}$ in the numerator and that is going to be painful. The reason for doing this shift is that now all the integrals over $\tilde{k}$ can be performed using the formula[2.98]

$$\int \frac{d^4\tilde{k}}{(2\pi)^4} \frac{(\tilde{k}^2)^r}{[\tilde{k}^2 - R^2]^3} = i(R^2)^{r-1} \frac{(-1)^{r-3}}{(4\pi)^2} \frac{\Gamma(r+2)}{\Gamma(2)} \frac{\Gamma(1-r)}{\Gamma(3)} . \quad (2.343)$$

2.98 This formula is derived by doing the integral in complex space after a Wick rotation of the time component. I am well aware that the integral is divergent for $r > 2$ because of the poles of the Γ-function. Fortunately, only $r = 0$ is needed and the integral is then finite. In general, these divergent integrals are a problem that leads to having to renormalize the theory, as discussed at length in any quantum field theory class.

So let me do the shift, and obtain that the numerator in Eq. (2.333) is now

$$\begin{aligned} \bar{u}(p')\Big[& 2\gamma^\mu \tilde{k}^2 + 2\gamma^\mu (p'x + px)^2 - 4(\tilde{k} - p'x - pz)^\mu (\tilde{\not{k}} - \not{p}'x - \not{p}z) \\ & + 4(p' \cdot p)\gamma^\mu + 4(p' + p) \cdot (\tilde{k} - p'x - pz)\gamma^\mu \\ & + 4m\tilde{k}^\mu - 4m(p' + p)^\mu - 4(p' + p)^\mu (\tilde{\not{k}} - \not{p}'x - \not{p}z)\Big] u(p) . \end{aligned} \quad (2.344)$$

It is getting complicated, but physics, for better or for worse, is made by those who never give up.

Terms linear in $\tilde{k}$ give zero because of the symmetric integration. Anyway, of all the terms in Eq. (2.344) I want only those proportional to p'^μ and p^μ because they give rise to the magnetic moment part through the Gordon decomposition.

There are only three of these terms, namely

$$-4m(x+z)(xp'^\mu + zp^\mu) - 4m(xp' + xp)^\mu - 4m(p' + p)^\mu)(x + z) . \quad (2.345)$$

Together they give[2.99]

$$2m\Big[(x + z) - (x + z)^2\Big](p' + p)^\mu . \quad (2.346)$$

2.99 I collect the terms and use the symmetry for the reciprocal exchange of x with z in the denominator to write both $x^2 + xz + z$ and $z^2 + xz - x$ as $\frac{1}{2}[(x + z)^2 - (x + z)]$.

Therefore,

$$\bar{u}(p')i\Gamma^\mu u(p) = 4me^2 \int_0^1 \mathrm{d}x\mathrm{d}z \int_0^{1-x} \frac{\mathrm{d}\tilde{k}}{(2\pi)^4} \frac{(x + z) - (x + z)^2}{[\tilde{k}^2 - R^2]^3} (p' + p)^\mu , \quad (2.347)$$

with $R^2 = m^2(x + z)^2 - q^2 xz$. I can use the formula in Eq. (2.343) to do the integration over $\tilde{k}$, which gives

$$\frac{(-1)}{(4\pi)^2} \frac{i}{2R^2} , \quad (2.348)$$

and, using the Gordon decomposition, I find that

$$F_2(q^2) = \left(\frac{\alpha}{4\pi}\right) \int_0^1 \mathrm{d}x \int_0^{1-x} \mathrm{d}z \, \frac{(x + z)^2 - (x + z)}{m^2(x + z)^2 - q^2 xz} (-4m^2) , \quad (2.349)$$

where $\alpha = e^2/4\pi$. Equation (2.349) at $q^2 = 0$ gives (after relabeling with $y = x + z$)

$$F_2(0) = -4\left(\frac{\alpha}{4\pi}\right) \int_0^1 \mathrm{d}x \int_x^1 \mathrm{d}y \, \frac{y^2 - y}{y^2} = \frac{\alpha}{2\pi} . \quad (2.350)$$

Going back to the definition of $F_2(0)$, the anomalous part of the magnetic moment of the electron is

$$a = \frac{g-2}{2} = F_2(0) = \frac{\alpha}{2\pi} \tag{2.351}$$

at one-loop order. This number is so important that it has been computed to $O(\alpha^5)$ to give

$$a_{th} = 0.001\,159\,652\,181\,643(764)\,, \tag{2.352}$$

which can be compared with the experimental number

$$a_{exp} = 0.001\,159\,652\,180\,73(28) \tag{2.353}$$

with an uncertainty of only 1 part per billion!

I am always amazed by how long computations – from the amplitude, through the Diracology of all those funny looking 4×4 matrices to the integration of the phase space to the final cross section – leads to a number that is actually what is found in the experiment. There is a bit of magic here in the way in which our mental constructs, quantum field theory and the Standard Model are really in the world out there.

2.18 Resonances

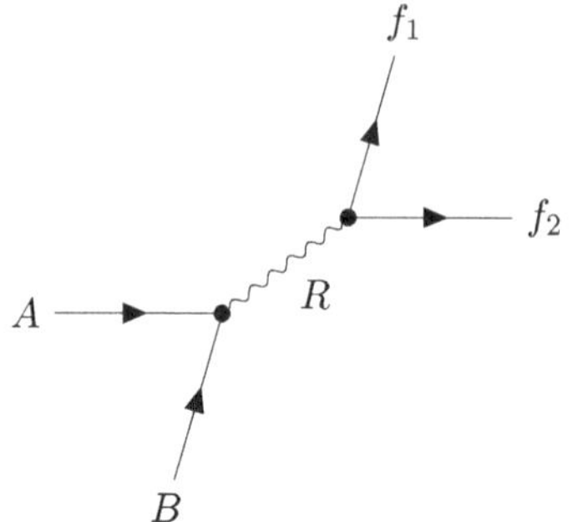

Figure 2.36 Diagram for the process in which the initial particles A and B go into the final particles f_1 and f_2 through the formation of a resonance R with finite width.

Looking at all the possible processes to be described in the S-matrix language, there remains one case not discussed yet. How do we write the amplitude for a process in which the initial states (say, A and B) go into an intermediate state (R) which then decays into some final states f? See Figure 2.36. This kind of processes take place when **resonances** are produced in strong interactions or in the production and decay of gauge bosons and other weakly interacting heavy states.

The difference here from the processes discussed up to this point is that a propagator of intermediate resonance cannot be inserted because the particle is on its mass shell. If such a resonance is sufficiently long lived, we could put the momentum in the propagator equal to the mass of the resonance. The cross section is enhanced – this is the resonant behavior giving its name to the phenomenon – but too much so since then an ill-defined (divergent) quantity is obtained.

The solution comes from considering the next orders in the perturbative expansion of quantum field theory. I must introduce radiative corrections (collectively indicated by a "blob" in the Feynman diagram) to the (scalar, for simplicity) propagator and write

$$\frac{i}{p^2 - m^2} + \frac{i}{p^2 - m^2}\, i\,\Pi(p^2) \frac{i}{p^2 - m^2} + \cdots\,, \tag{2.354}$$

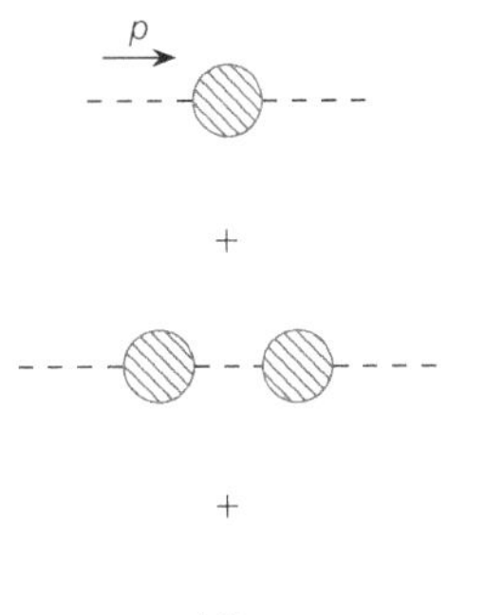

Figure 2.37 A geometric series of diagrams.

which can be re-summed as a geometric series, whose first two terms are depicted in Figure 2.37, to give

$$\frac{i}{p^2 - m^2 + \Pi(p^2)} \,. \tag{2.355}$$

The pole of this propagator gives the mass of the resonance m_R as the solution of the equation

$$m_R^2 - m^2 + \Pi(m_R) = 0 \,. \tag{2.356}$$

The function $\Pi(p^2)$ can be computed order by order in perturbation theory and contains an imaginary part (because of the optical theorem in Eq. (2.320))

$$\text{Im}\ \Pi(p^2) = \frac{1}{2} \sum_f \int \mathrm{d}\Phi_f |M(R \to f)|^2 = \sqrt{p^2}\, \Gamma_R \,, \tag{2.357}$$

where Γ_R is the width for the resonance to decay to the final states f.

For energies $s - p^2$ close to the resonance mass, the propagator can be written (neglecting quartic terms) as

$$\frac{i}{p^2 - m_R^2 + i p^2\, \Gamma_R / m_R} \,, \tag{2.358}$$

for a scalar resonance – and similar expressions enter the denominators of the propagators of spin-1/2 and spin-1 resonances.

The effect of introducing this modified propagator for the decaying resonance is that the cross section will show a characteristic shape as a function of s: a peak centered around the resonance mass m_R and width proportional to Γ_R:

$$\sigma(s) \simeq \frac{g_A g_B}{(s - m_R^2)^2 + m_R^2\, \Gamma_R^2} \,, \tag{2.359}$$

where I have introduced the factors g_X to account for the degeneration of the initial states (from spin or other degrees of freedom). Equation (2.359) is named after the physicists G. Breit and E. Wigner.

In the case of a very narrow width $\Gamma \ll m_R$, the resonance can only be produced on-shell and Eq. (2.359) can be approximated by

$$g_A g_B \frac{\pi}{m_R \Gamma_R} \delta(p^2 - m_R^2) \,. \tag{2.360}$$

In this narrow-width approximation, the cross section for the whole process becomes the product of the cross section for producing the resonance times the branching ratio[2.100] for the decay of the resonance to the final states in which we are interested:

2.100 The **branching ratio** for a given decay channel is defined as the width of that channel over the total decay width.

$$\sigma(A+B\to R\to f) = \sigma(A+B\to R)\times\frac{\Gamma_R(R\to f)}{\Gamma_{\text{tot}}}$$
$$= \sigma(A+B\to R)\times \text{BR}\,(R\to f)\,, \qquad (2.361)$$

which is the template for the formula that is most often used.

2.19 External (Classical) Fields and the *S*-Matrix*

In many an application, we want to study the interaction of quantum particles with external fields that are classical. Notable examples are electrons going through matter (and experiencing the Coulomb fields of the nuclei) and Compton scattering in the presence of photons from a laser. How do these classical fields enter in the S-matrix formalism?

I could try to imagine the classical field as being just a state in the Fock space with a very large occupation number n of photons. But this cannot be correct because the expectation value of any field $A_\mu(x)$ between states with a certain number n of photons is given by

$$\langle n|A_\mu(x)|n\rangle$$
$$= \sum_\lambda \int \mathrm{d}k\left[\langle n|\hat{a}(\vec{k},\lambda)|n\rangle\,\varepsilon^\lambda_\mu(\vec{k})\,e^{ik\cdot x} + \langle n|\hat{a}^\dagger(\vec{k},\lambda)|n\rangle\,\varepsilon^{\lambda *}_\mu(\vec{k})\,e^{-ik\cdot x}\right],$$

which vanishes because both $\langle n|\hat{a}^\dagger(\vec{k},\lambda)|n\rangle$ and $\langle n|\hat{a}(\vec{k},\lambda)|n\rangle$ vanish.

The classical field is not a state with a fixed, albeit large, number of photons. States with fixed occupation number do not have a classical limit.

The classical electromagnetic field is a state with a variable number of photons. This can be understood by looking at just one of the infinite number of harmonic oscillators making up the photon field. This state (characterized by the momentum k and polarization λ of the photon) is a superposition of states with different numbers of photons:

$$|\alpha^\lambda_k\rangle = \sum_n c_n|n\rangle\,. \qquad (2.362)$$

The coefficient $c_n = \langle n|\alpha^\lambda_k\rangle$ can be written as

$$c_n = \frac{(\alpha^\lambda_k)^n}{\sqrt{n!}}\langle 0|\alpha\rangle \qquad (2.363)$$

by repeatedly acting on the vacuum: $|n\rangle = (a^\dagger)^n|0\rangle/\sqrt{n!}$.

The value of $\langle n|\alpha^\lambda_k\rangle$ can be determined by the overall normalization, that is,

$$1 = \sum_n\langle\alpha^\lambda_k|n\rangle\langle n|\alpha^\lambda_k\rangle = |\langle\alpha^\lambda_k|0\rangle|^2\sum_n\frac{|\alpha^\lambda_k|^2}{n!} = |\langle\alpha^\lambda_k|0\rangle|^2 e^{|\alpha^\lambda_k|^2} \qquad (2.364)$$

for each of the modes identified by λ and k. The state under discussion is therefore given by

$$|\alpha_k^\lambda\rangle = e^{-\frac{1}{2}|\alpha_k^\lambda|^2} \sum_n \frac{\alpha_k^\lambda}{\sqrt{n!}} = e^{-\frac{1}{2}|\alpha_k^\lambda|^2 + \alpha_k^\lambda a^\dagger(\vec{k},\lambda)}|0\rangle, \quad (2.365)$$

which I can also write as[2.101]

2.101 Use Hausdorff's formula for two non-commuting operators $\hat{A}$ and $\hat{B}$:

$$e^{\hat{A}+\hat{B}} = e^{\hat{A}} e^{\hat{B}} e^{-\frac{1}{2}[\hat{A},\hat{B}]} .$$

$$|\alpha_k^\lambda\rangle = e^{\alpha_k^\lambda a^\dagger(\vec{k},\lambda) + \alpha_k^{\lambda *} a(\vec{k},\lambda)}|0\rangle . \quad (2.366)$$

How many photons are there in the state $|\alpha_k^\lambda\rangle$? The probability of having n photons (with polarization λ and momentum k) is

$$|\langle \alpha_k^\lambda | n\rangle|^2 = e^{-|\alpha_k^\lambda|^2} \frac{|\alpha_k^\lambda|^2}{n!}, \quad (2.367)$$

which tells me that the photons follow a Poisson distribution. Their number is undefined, the phases are fixed: they are **coherent states** of the photon field.[2.102] These states are as classical as they can be in the sense that the product of the uncertainties in the values of the field and its momentum is equal to $\hbar/2$, the smallest possible value.

2.102 Coherent states are defined by their being eigenstates of the annihilation operator, that is,

$$a^\dagger(\vec{k},\lambda)|\alpha_k^\lambda\rangle = \alpha_k^\lambda|\alpha_k^\lambda\rangle .$$

I can now go back and compute the expectation value of the photon field $A_\mu(x)$ on these coherent states and find

$$\begin{aligned}
&\langle \alpha | A_\mu(x) | \alpha\rangle \\
&= \sum_\lambda \int \mathrm{d}k \left[\langle \alpha_k^\lambda | \hat{a}(\vec{k},\lambda) | \alpha_k^\lambda\rangle\, \varepsilon_\mu^\lambda(\vec{k})\, e^{ik\cdot x} + \langle \alpha_k^\lambda | \hat{a}^\dagger(\vec{k},\lambda)|\alpha_k^\lambda\rangle\, \varepsilon_\mu^{\lambda *}(\vec{k})\, e^{-ik\cdot x} \right] \\
&= \sum_\lambda \int \mathrm{d}k \left[\alpha_k^\lambda\, \varepsilon_\mu^\lambda(\vec{k})\, e^{ik\cdot x} + \alpha_k^{\lambda *}\, \varepsilon_\mu^{\lambda *}(\vec{k})\, e^{-ik\cdot x} \right],
\end{aligned}$$

which is indeed the value of the classical field.

To insert such a classical field as an external source into the S-matrix, I use the coherent state and write, for a generic process in which an initial electron with momentum p_i goes into a final electron with momentum p_f plus a photon with momentum k,

$$\langle p_f, k, \alpha_q^\lambda | \hat{S} | p_i, \alpha_q^\lambda\rangle = \langle p_f, k | \hat{S}(A_\mu) | p_i\rangle . \quad (2.368)$$

In Eq. (2.368) the coherent state finds the annihilation operator and produces a S-matrix $\hat{S}(A_\mu)$ in which I can use the value of the classical field A_μ and its coupling to the quantum particle as a vertex.

For instance, for an electron traveling through matter,[2.103] the external (classical) field of the nuclei is just the Fourier transform of the Coulomb potential $A_0 = -Ze/|\vec{q}|^2$ (with $\vec{A} = 0$). The polarization $\varepsilon_\mu^\lambda(\vec{k})$ selects the time-like component, whose size is given by $\alpha_q^\lambda = A_0$.

2.103 Without the external potential the electron cannot emit a photon without violating the conservation of energy.

The amplitude for the electron to change momentum from p_i to p_f while emitting a photon with momentum k is given by[2.104]

2.104 Because of the external Coulomb field, only the energy is conserved in Eq. (2.369).

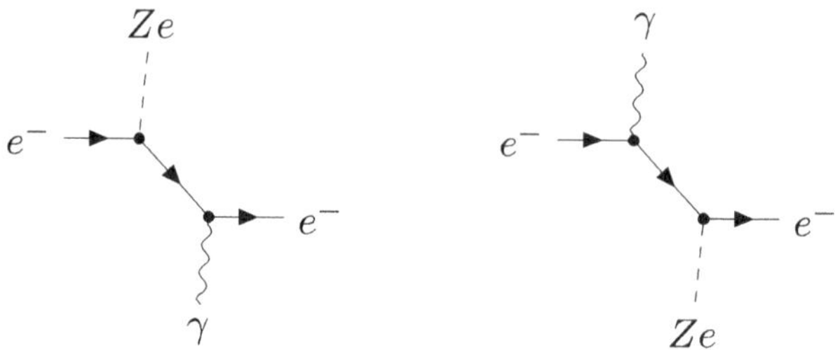

Figure 2.38 Diagrams for electron Bremsstrahlung. The dashed line represents the external Coulomb field.

$$\mathcal{M}_{fi} = \frac{ie^2}{|\vec{p}_f - \vec{p}_i + \vec{k}|^2} 2\pi\, \delta(p_i^0 - p_f^0 - k^0)$$

Coulomb field insertion

$$\times\, \bar{u}(p_f) \left[\not{\epsilon} \frac{Ze}{\not{p}_f + \not{k} - m} \gamma^0 + \gamma^0 \frac{Ze}{\not{p}_i - \not{k} - m} \not{\epsilon} \right] u(p_i) \,. \qquad (2.369)$$

The first denominator in Eq. (2.369) comes from the Fourier transform of the Coulomb potential with $\vec{q} = \vec{p}_f - \vec{p}_i + \vec{k}$; the factor γ^0 in the amplitude represents the coupling of the electron to the time component of the external potential. The second term in the square brackets comes from the inverse ordering of the interactions.

The amplitude in Eq. (2.369) represents the *Bremsstrahlung*, that is, the emission of a photon by an electron moving through matter. (See Figure 2.38). There is just one emitted photon but those making up the Coulomb field are indefinite in number and are organized in a coherent state – which is effectively described in Eq. (2.369) by the single insertion of the classical Coulomb field.

Part I

Where the Model is Explained

3 *In Medias Res*: Charged and Neutral Currents

Contents

Figure 3.1 The Lagrangian of the Standard Model on a mug.

A remarkable booklet published by the *Particle Data Group* (PDG for short) contains everything I need to know about the Standard Model – plus a lot more, of which I only need to remember that this is the place where it can be found. Inside its pages I find the electroweak Lagrangian:

$$\mathcal{L}_\psi = \sum_i \bar{\psi}_i \left(i\partial\!\!\!/ - m_i\right) \psi_i \tag{3.1}$$

$$- \sum_i Q^i \bar{\psi}_i \gamma^\mu \psi_i A_\mu \tag{3.2}$$

$$- \frac{g}{2\sqrt{2}} \sum_i \bar{\psi}_i^{\mathrm{D}} \gamma^\mu \left(1 - \gamma_5\right) \left(T^+ W_\mu^+ + T^- W_\mu^-\right) \psi_i^{\mathrm{D}} \tag{3.3}$$

$$- \frac{g}{2\cos\theta_W} \sum_i \bar{\psi}_i \gamma^\mu \left(g_V^i - g_A^i \gamma_5\right) \psi_i Z_\mu \,. \tag{3.4}$$

The first line, Eq. (3.1), is the Dirac Lagrangian for free fermions. The second line, Eq. (3.2), is the electromagnetic current. I will take it that these two terms are already known (I discussed them briefly in Chapter 1).

The last two lines provide new information: the charged and neutral currents of the Standard Model. They contain terms with fermions ψ_i, gauge bosons $W_\mu^\pm$ and Z_μ.

The electroweak Lagrangian is not the whole Standard Model – there are additional terms for the gauge and Higgs bosons and the strong interactions, depicted in Figure 3.1, which I discuss in later chapters. Equations (3.3) and (3.4) are nevertheless a good starting point because they show the currents – how matter interacts – and these currents are what is actually measured in most experiments.

The first thing I need to know is the indices of the matter fields, that is, of the Dirac fermions. There can be eight kinds of fermions for each

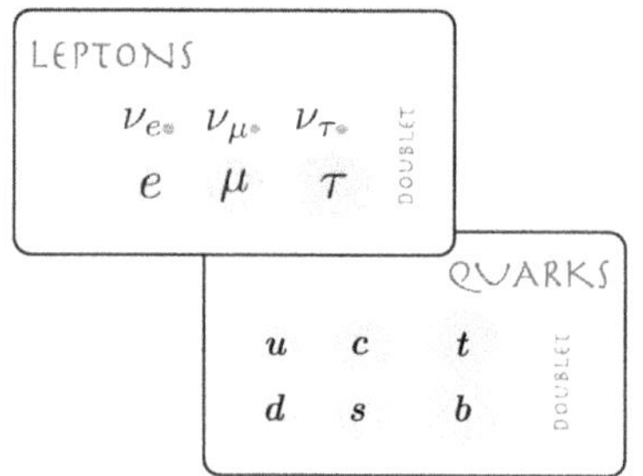

Figure 3.2 Often, a lot is made of these charge assignments in terms of quarks and leptons. True, they are the building blocks of all other particles but they are not by themselves the Standard Model – they are only the starting point of the Standard Model.

generation (and there are three generations); one keeps track of them by their quantum numbers.

For the first generation there are four Dirac fermions, two quarks $i = u, d$, the electron $i = e$ and its neutrino $i = \nu_e$, which means that there are twice as many chiral fermions (one left-handed and one right-handed) – except for the neutrino, which is going to be only left-handed (ν_R is not part of the Standard Model, at least to begin with). It is best to keep these chiral components separated because they interact differently.

There are two more generations but they are exact copies of this first one, as in the cartoon in Figure 3.2.

Lest it be thought that I am forgetful, I do not mention the masses of these fermions because they do not enter explicitly at the level of the currents. Masses are the origin of the mixing of the quark flavors in the charged currents (more about this in the next section); they are crucial in flavor physics but for the moment they are not needed.

Altogether the fermions come in a variety of **flavors**. The quantum numbers of the first generation, as far as the electroweak interactions are concerned, are given by the **weak isospin** (the third component T^3 of the $SU(2)$ group generators) and the **hypercharge** (Y).[3.1] I summarize them in Table 3.1.

3.1 Weak isospin and hypercharge are related to the electric charge:

$$Q = T^3 + \frac{Y}{2}.$$

The taxonomy is at first bewildering. There is no way around it: one has to memorize these charges. They provide the basis of the rich phenomenology of the Standard Model.

The crucial difference between Eq. (3.4) and Eq. (3.3) is that the fermions in the charged currents appear as weak **isospin doublets**,

$$\psi_i^{\mathrm{D}} = \begin{pmatrix} u_i \\ d_i' \end{pmatrix} \quad \begin{matrix} \leftarrow T^3 = 1/2 \\ \leftarrow T^3 = -1/2 \end{matrix} \tag{3.5}$$

where $d_i' = V_{ij}^{\mathrm{CKM}} d_j$ (more about this rotation shortly), whereas the fermions enter as an isospin singlet ψ_i in the neutral currents. The charged current changes the charges of the fermions taking part in the interaction,

Table 3.1 The charge assignments for the first generation of quarks and leptons.

	u_L	d_L	u_R	d_R	ν_L	e_L	ν_R	e_R
T^3	1/2	−1/2	0	0	1/2	−1/2	0	0
Y	1/3	1/3	4/3	−2/3	−1	−1	0	−2
Q	2/3	−1/3	2/3	−1/3	0	−1	0	−1

which is the reason why it is called charged. The neutral current leaves the charge unchanged, and that why it is called neutral.

How this Lagrangian was conceived is a long story. It was not the work of a single physicist. Many experimental clues needed to be patched together before the final formulation as we know it today was reached. It is an interesting example of how science is actually done. In stark contrast with, for instance, general relativity (which was the work of a single physicist), the Standard Model is an example of the most common kind of scientific theory, those developed by many different persons working independently, or in collaboration, and over the course of many years. It is hard, if not impossible, to single out an "aha!" moment, a change of paradigm.[3.2] It has always seemed to me the triumph – to use T. Kuhn's terminology – of normal science over the idealized model of science development going from revolution to revolution so popular among science writers and so scarce in the actual history of science.

[3.2] There is no single crucial insight, no Newton's apple or Kepler's ellipses for the Standard Model.

3.1 Charged Currents

The Lagrangian[3.3] in Eq. (3.3) describes physics in three spaces simultaneously. First, the index μ lives in space-time. It tells me that the interaction is between a current and a vector field. Second, the spinor index keeps track of the spin space of the fermion fields. Third, the matrices $T^{\pm}$ and the doublet components of the fermions are defined in weak isospin space.

[3.3] Why do we always write Lagrangians in particle physics? I think that there are two reasons. First, we are used to it. While studying classical mechanics we begin to frame all problems not in terms of their equations of motion but as Lagrangians (or Hamiltonians). And once we are used to doing something it is difficult to change – especially if it works well. Second, it is a compact way of encoding the degrees of freedom, which helps in connecting the symmetries of the problem with the observables. The invariance of the Lagrangian implies that of the equations of motion and the conserved quantities are easily identified.

The interaction has to do with the change in the charge of the particles because it acts on isospin doublets by flipping the isospin T^3 component, with a net change of one unit of isospin (carried away by the gauge boson $W_\mu^{\mp}$). The change in isospin implies a change in electric charge:

$$\Delta Q = \Delta T^3 \,, \tag{3.6}$$

assuming a constant hypercharge Y.

To better understand Eq. (3.3), it is useful to deconstruct its terms in the three spaces in which they are defined (Lorentz, weak isospin and spin):

$$\mathcal{L}^{cc} = -\underbrace{\frac{g}{2\sqrt{2}}}_{\text{coupling}} \sum_i \underset{\substack{\text{isospin}\\\text{doublet}}}{\bar{\psi}_i^{D}} \overbrace{\gamma^\mu (1-\gamma_5)}^{\text{chiral structure}} \underbrace{\left(T^+ W_\mu^+ + T^- W_\mu^-\right)}_{\text{isospin space}} \underset{\substack{\text{isospin}\\\text{doublet}}}{\psi_i^{D}} \,. \tag{3.7}$$

This Lagrangian was first brought forth as Fermi's theory of β-decay. Fermi knew that in the β-decay of the neutron n, the final states were a proton p, an electron e and its antineutrino $\bar{\nu}_e$. He therefore knew that the interaction vertex had to contain these four fermion fields and, by following the electrodynamics, he combined them as a product of two currents.[3.4] The corresponding Feynman diagram is displayed in Figure 3.3. Its mathematical expression is

[3.4] Notice here the revolutionary idea that the electron (and the neutrino) are not part of what we call a nucleon. They are created in the process of the nucleon's decay; an idea only possible by fully embracing quantum field theory.

$$\frac{G_F}{\sqrt{2}}\left[\bar{\psi}_p\gamma^\mu\psi_n\right]\left[\bar{\psi}_e\gamma_\mu\psi_{\bar{\nu}_e}\right]. \tag{3.8}$$

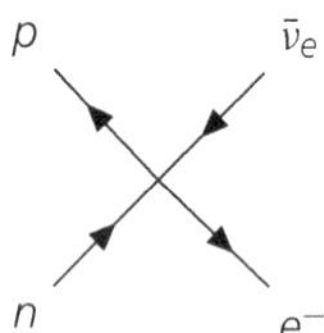

Figure 3.3 Fermi interaction for β-decay.

[3.5] Each fermion field has dimension $[E]^{3/2}$. You do the math.

The dimension of this product of two currents,[3.5] which is $[E]^6$, fixes the dimensions of the constant G_F in front to be $[E]^{-2}$, because the interaction as a whole must have dimension $[E]^4$. Why $[E]^4$? Because the action is dimensionless (in natural units) and in this case it is going to be the Fermi interaction, which is a Lagrangian density times the space-time integration, which has dimension $[E]^{-4}$.

Before proceeding, we can pause and admire the elegance of Fermi's theory. A neutron goes into a proton in the interaction of Eq. (3.8). They are states of the same multiplet of isospin – as long as we forget about the electric charge (and their tiny mass difference). The neutron is the component pointing up, the proton that pointing down. The Fermi interaction rotates one component into the other.

The structure in isospin space is controlled by the $SU(2)$ matrices $T^\pm$, which are projectors given by

$$T^+ = \begin{pmatrix} 0 & 0 \\ 1 & 0 \end{pmatrix} \quad \text{and} \quad T^- = \begin{pmatrix} 0 & 1 \\ 0 & 0 \end{pmatrix}. \tag{3.9}$$

These matrices are built from the Pauli matrices, which are traditionally renamed as T^1, T^2 and T^3, as

$$T^\pm = \frac{1}{2}(T^1 \mp iT^2). \tag{3.10}$$

The effects of the matrices $T^\pm$ (a change of 1 in the isospin) on the isospin doublet ψ^{D} of the first generation of quark fields are

$$\left(\bar{u}, \bar{d}\right) T^+ \begin{pmatrix} u \\ d \end{pmatrix} = \bar{d}u \quad \text{and} \quad \left(\bar{u}, \bar{d}\right) T^- \begin{pmatrix} u \\ d \end{pmatrix} = \bar{u}d, \tag{3.11}$$

respectively. This projection gives the desired transition (in terms of quark fields) of a neutron going into a proton. An analogous transition can be written for the lepton fields.

Equation (3.3) seems right in isospin space. Yet its chiral structure cannot be quite correct. The vector vertex γ^μ conserves parity – which is not conserved by the weak interactions. This fundamental notion was first shown by the study of the weak decay of cobalt into nickel:

$$^{60}_{27}\mathrm{Co} \rightarrow ^{60}_{28}\mathrm{Ni} + e^- + \bar{\nu}_e + 2\gamma\,. \tag{3.12}$$

One neutron in the atom of cobalt decays into a proton, emitting an electron and its antineutrino, and the resulting atom of nickel is an excited state, which decays into the ground state by emitting two photons.

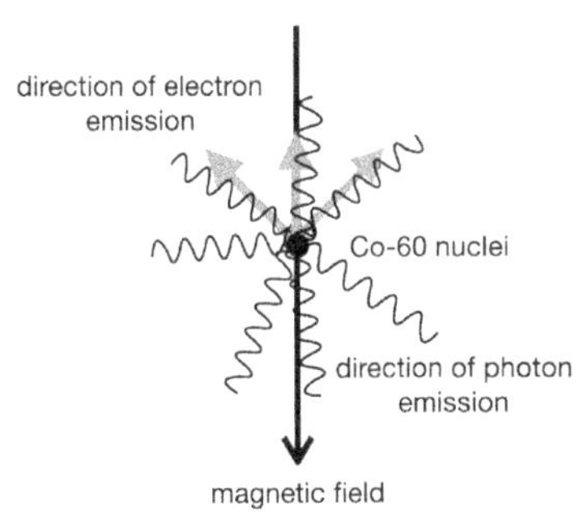

Figure 3.4 Schematics of the experiment with cobalt nuclei.

By putting the atoms of cobalt into a magnetic field $\vec{B}$, as in Figure 3.4, one can tell the momentum of the electrons that come out. While the emitted photons are isotropically distributed, more electrons are seen with momentum in the opposite direction with respect to the magnetic field than in the same direction. Parity is not conserved. That was a shocker at the time; it is now a fundamental property embedded in the Standard Model.

A parity transformation (as discussed in Chapter 1) transforms a space point $\vec{x}$ into $-\vec{x}$. A spinor transforms under a parity transformation as

$$\psi \xrightarrow{\hat{P}} \gamma^0\psi \quad \text{and} \quad \bar{\psi} \xrightarrow{\hat{P}} \bar{\psi}\gamma^0\,, \tag{3.13}$$

and a vector current transforms as

$$J^\mu = \bar{\psi}\gamma^\mu\psi \xrightarrow{\hat{P}} \bar{\psi}\gamma^0\gamma^\mu\gamma^0\psi\,, \tag{3.14}$$

which depends on which component is considered:[3.6]

3.6 $\gamma^0\gamma^0\gamma^0 = \gamma^0$
$\gamma^0\gamma^i\gamma^0 = -\gamma^i\,.$

$$\begin{aligned} J^0 &\xrightarrow{\hat{P}} J^0 \\ J^i &\xrightarrow{\hat{P}} -J^i\,. \end{aligned} \tag{3.15}$$

Under a parity transformation, the vector current behaves as a polar vector.

However, an axial current like $J^\mu_A = \bar{\psi}\gamma^\mu\gamma_5\psi$ transforms as

$$J^\mu_A = \bar{\psi}\gamma^\mu\gamma_5\psi \xrightarrow{\hat{P}} -\bar{\psi}\gamma^0\gamma^\mu\gamma_5\gamma^0\psi \tag{3.16}$$

and therefore[3.7]

3.7 $\gamma^0\gamma^0\gamma^0\gamma_5 = \gamma^0\gamma_5$
$\gamma^0\gamma^i\gamma^0\gamma_5 = -\gamma^i\gamma_5$
and $\{\gamma^0, \gamma_5\} = 0\,.$

$$\begin{aligned} J^0_A &\xrightarrow{\hat{P}} -J^0_A \\ J^i_A &\xrightarrow{\hat{P}} J^i_A\,. \end{aligned} \tag{3.17}$$

Under a parity transformation, the axial current behaves as an axial vector.

If I take a product of two vector or polar currents, the result will turn out to be parity invariant, whereas an arbitrary combination of the two currents will violate parity. If I take such a mixed combination of currents,

$$J^\mu = \bar{\psi}(g_V\gamma^\mu + g_A\gamma^\mu\gamma_5)\psi = g_V J^\mu_V + g_A J^\mu_A\,, \tag{3.18}$$

the product of the currents is then

$$J_\mu J^\mu = g_V^2 J_\mu^V J_V^\mu + g_A^2 J_\mu^A J_A^\mu + g_V g_A (J_\mu^V J_A^\mu + J_\mu^A J_V^\mu), \tag{3.19}$$

where the first two terms conserve parity and the last one does not. The relative weight of the parity-violating term with respect to the parity-conserving terms is given by

$$\frac{g_V g_A}{g_V^2 + g_A^2} \tag{3.20}$$

and is maximal for $|g_V| = |g_A|$.[3.8]

Neutrinos have nearly zero mass (I come to their mass in the last chapter) and have a simple helicity structure: left-handed neutrinos have helicity -1, which means that their momentum and spin are always antiparallel. Right-handed antineutrinos have the opposite helicity. Since in β-decay the electron antineutrino is always right-handed, I must take the charged current to maximally violate parity:

[3.8] The Wu experiment tells me that charged currents violate parity. Do they do so maximally? The Goldhaber experiment (1957), on the capture of electrons by the nuclei of europium and the subsequent decay into samarium plus a neutrino and a photon, found that the photon always has helicity -1 and therefore the neutrino – which is produced back-to-back with the photon – must be left-handed and so the charged currents violate parity maximally.

$$J^\mu = \bar{\psi}\gamma^\mu(1-\gamma_5)\psi\,. \tag{3.21}$$

Fermi's interaction vertex becomes

$$\frac{G_F}{\sqrt{2}}\left[\bar{\psi}_n\gamma^\mu(1-\gamma_5)\psi_p\right]\left[\bar{\psi}_e\gamma_\mu(1-\gamma_5)\psi_{\bar{\nu}_e}\right], \tag{3.22}$$

which encodes in its chiral structure the parity violation of weak interactions and the result of Wu's experiment.

Yet the Lagrangian in Eq. (3.22) cannot be the correct one.

If I take the high-energy limit of the amplitude obtained by using Fermi's theory in the scattering of an electron and a proton – rather than in β decay – the result is that the amplitude grows as s, the square of the center of mass energy:[3.9]

[3.9] Use $\bar{u}(p_2)u(p_1) \propto \frac{\vec{p}_1\cdot\vec{p}_2}{E+m} \propto \sqrt{s}$ as long as the momenta are large.

$$\mathcal{M}(e^- p \to \nu_e n) \simeq \frac{G_F}{\sqrt{2}}s\,, \tag{3.23}$$

which for s sufficiently large ($s > 8\sqrt{2}\pi/G_F$) gives rise to a cross section violating the perturbative s-wave unitary bound in Eq. (2.325) in the previous chapter. This may not be such a big problem as long as we remain at energies very much below $1/\sqrt{G_F}$, where Fermi's theory applies, but it is a troublesome feature that is going to haunt us if we want the model to be well defined no matter what the energy of the process under consideration is.

To remedy this serious defect the four-fermion interaction is opened up and the four-fermion operator is made into two currents that interact through the exchange of an intermediate particle, the gauge boson W. The Fermi interaction becomes[3.10]

$$\underbrace{\left[\frac{g}{2\sqrt{2}}\bar{\psi}_n\gamma_\mu(1-\gamma_5)\psi_p\right]}_{\text{charged current}} \times \underbrace{\frac{-i}{p^2-m_W^2}\left[-g^{\mu\nu}-\frac{p^\mu p^\nu}{m_W^2}\right]}_{W\text{ propagator}}$$
$$\times \underbrace{\left[\frac{g}{2\sqrt{2}}\bar{\psi}_e\gamma_\nu(1-\gamma_5)\psi_{\bar{\nu}_e}\right]}_{\text{charged current}}. \tag{3.24}$$

3.10 As I discuss in the next chapter, the propagator of a gauge boson depends on the choice of gauge. The expression in Eq. (3.24) is in what is called the unitary gauge.

which goes back to the four-fermion interaction in the limit of small energies $s \ll m_W^2$, as follows:[3.11]

3.11 The term $p^\mu p^\nu/m_W^2$ cancels because of the vanishing mass of the neutrino.

$$\text{Eq. (3.24)} \overset{s\ll m_W^2}{\Longrightarrow} -\underbrace{\frac{g^2}{8m_W^2}}_{\equiv G_F/\sqrt{2}}\left[\bar{\psi}_n\gamma^\mu(1-\gamma_5)\psi_p\right]\left[\bar{\psi}_e\gamma_\mu(1-\gamma_5)\psi_{\bar{\nu}_e}\right]. \tag{3.25}$$

The limit in Eq. (3.25) provides a connection between the coupling g of the vertex between the fermions and the W boson and the Fermi constant G_F, namely

$$\frac{g^2}{8m_W^2} = \frac{G_F}{\sqrt{2}}. \tag{3.26}$$

Because of the presence of the intermediate gauge boson, the amplitude in Eq. (3.23) now behaves as

$$\mathcal{M}(e^-p\to\nu_e n) \simeq \frac{G_F}{2\sqrt{2}\pi}\frac{s}{s-m_W^2}, \tag{3.27}$$

which satisfies the unitary bound at all energies.

After this (final) tuning of Fermi's interaction, I can write the vertex for the interaction between the new bosons $W^\pm$ and the charged current, as given in Eq. (3.7), as in Figure 3.5.[3.12]

3.12 The same rule in the leptonic sector with ν_{ℓ_i}, ℓ_i as for the external fermions.

The presence of the charged combinations $W^\pm$ already tells me that the interaction is not charge-conjugation invariant. The gauge boson fields transform as

$$\hat{C}W^\pm\hat{C}^{-1} = \mp W^\pm \tag{3.28}$$

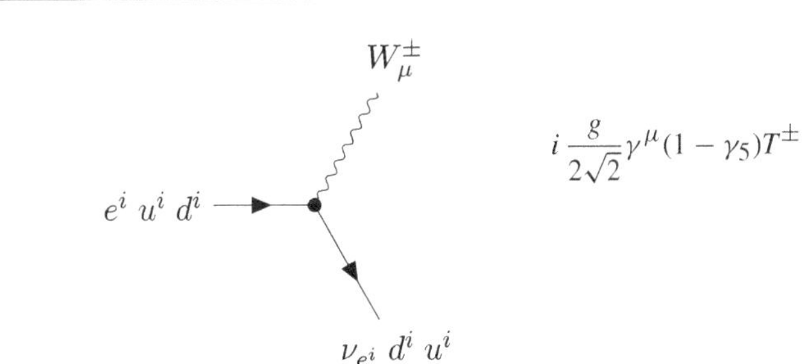

Figure 3.5 The vertex between fermions and the $W^\pm$ gauge bosons.

and the presence of both a vector and a pseudovector current makes the Lagrangian violate the symmetry. Only if charge and parity transformations are combined into CP does the Lagrangian in Eq. (3.3) become invariant.[3.13]

3.13 Only true if the Lagrangian is Hermitian. There is the possibility of violating even CP if a phase is introduced.

To sum it up: charged currents change the charges of the particles in the interaction by flipping their isospins. They are chirally fixed: the gauge bosons always couple to left-handed currents. To indicate this feature the group is denoted $SU(2)_L$.

The story is actually a bit more complicated because experimentally the β-decay of an s quark in the second generation is not equal to that, for instance, of a d quark, its cousin from the first generation. For instance, the kaon and pion weak decays have very different decay widths and it is found that

$$\frac{\Gamma(K^- \to \mu^- \bar{\nu}_\mu)}{\Gamma(\pi^- \to \mu^- \bar{\nu}_\mu)} \simeq \frac{1}{20} . \tag{3.29}$$

This indicates that charged currents have different strengths for the different generations of quarks. To account for this difference, one must introduce, as hinted in the primed notation of the down-quark fields in Eq. (3.5), a rotation in the definition of the isospin doublets:

$$\underbrace{\begin{pmatrix} d' \\ s' \\ b' \end{pmatrix}}_{\text{weak states}} = \underbrace{\begin{pmatrix} V_{ud} & V_{us} & V_{ub} \\ V_{cd} & V_{cs} & V_{cb} \\ V_{td} & V_{ts} & V_{tb} \end{pmatrix}}_{\text{CKM}} \cdot \underbrace{\begin{pmatrix} d \\ s \\ b \end{pmatrix}}_{\text{mass states}} . \tag{3.30}$$

The rotation matrix V_{CKM} (CKM comes from the names of the physicists N. Cabibbo, T. Kobayashi and M. Maskawa) is unitary and rotates the mass eigenstates into the interaction (weak) states. It is quasi-diagonal, meaning that there is a hierarchy in the size of the matrix elements, and transitions from one generation to the others are suppressed. The matrix elements V_{ud} and V_{us} modulate the strengths of the β-decays in Eq. (3.29), thus explaining their difference.

I return to the V_{CKM} matrix in Chapter 4 where I discuss flavor physics. The Feynman rule is modulated in the quark sector by the V_{ij} matrix elements and therefore given as in Figure 3.6.

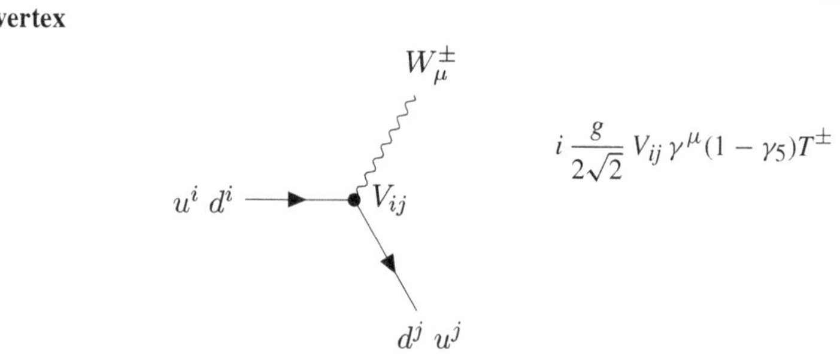

Figure 3.6 Again, the vertex between fermions and the $W^\pm$ gauge bosons, but this time with the CKM mixing matrix elements.

3.2 *Problem Session*: The β-Decay of the Muon

Consider the β-decay of the muon μ. It is conceptually the same as the β-decay of the neutron but we don't have to worry about the complications that nuclear structure brings into a problem.

The process is a three-body decay:

$$\mu^-(p) \to e^-(k) + \bar{\nu}_e(q_e) + \nu_\mu(q_\mu), \tag{3.31}$$

where both the electron and the muon neutrinos are in the final state because of the conservation of lepton number, which is true independently for each kind of lepton. I want to use Fermi's theory, where the interaction takes place at one point because the energies are much smaller than the mass of the gauge boson W. This approach is based on an effective Hamiltonian in which the gauge boson propagator (in Eq. (3.24)) is expanded as follows:

$$\begin{aligned} &\frac{-i}{p^2 - m_W^2}\left[-g^{\mu\nu} - \frac{p^\mu p^\nu}{m_W^2}\right] \\ &\leadsto \frac{i}{m_W^2}\left[g^{\mu\nu} - \frac{1}{m_W^2}(p^\mu p^\nu - g^{\mu\nu}p^2) + O\left(\frac{1}{m_W^4}\right)\right], \end{aligned} \tag{3.32}$$

and only the leading term is retained.[3.14]

3.14 The Standard Model process displayed in Figure 3.7 is represented in the effective theory as in Figure 3.8, in which the gauge boson has been "integrated out."

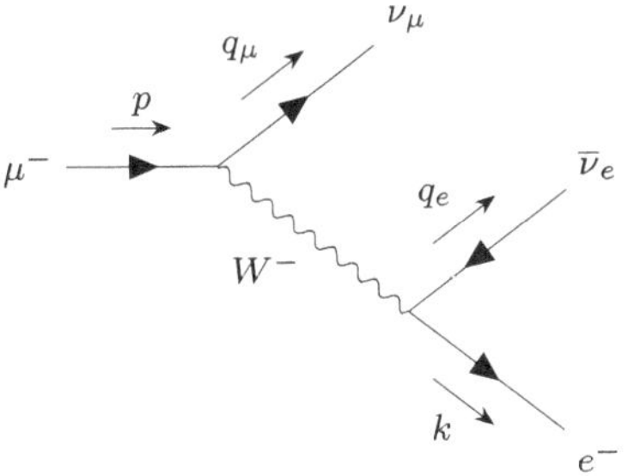

Figure 3.7 Original Feynman diagram.

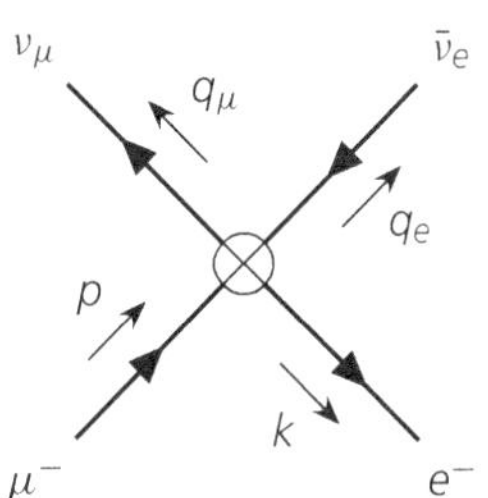

Figure 3.8 Effective-theory Feynman diagram.

I am going to use the notation in which the spinors are represented by the particles' names – it is simpler but I have to remember which spinor is a particle with a u-type wave function and which is an antiparticle with a v-type wave function. Thus

$$\mathcal{L}_{\text{eff}} = -\frac{G_F}{\sqrt{2}}\left[\bar{u}_{\nu_\mu}(q_\mu)\gamma^\alpha(1-\gamma_5)u_\mu(p)\right]\left[\bar{u}_e(k)\gamma_\alpha(1-\gamma_5)v_{\nu_e}(q_e)\right], \tag{3.33}$$

which gives directly the matrix element

$$\mathcal{M}_{fi} = i\frac{G_F}{\sqrt{2}}\left[\bar{u}_{\nu_\mu}(q_\mu)\gamma^\alpha(1-\gamma_5)u_\mu(p)\right]\left[\bar{u}_e(k)\gamma_\alpha(1-\gamma_5)v_{\nu_e}(q_e)\right], \tag{3.34}$$

the conjugate of which is[3.15]

$$\mathcal{M}_{fi}^\dagger = -i\frac{G_F}{\sqrt{2}}\left[\bar{u}_\mu(p)\gamma^\beta(1-\gamma_5)u_{\nu_\mu}(q_\mu)\right]\left[v_{\nu_e}(q_e)\gamma_\beta(1-\gamma_5)u_e(k)\right]. \tag{3.35}$$

3.15 For two fermions, say ψ_1 and ψ_2,

$$\begin{aligned} &|\bar{\psi}_1\gamma^\mu(1-\gamma_5)\psi_2|^\dagger \\ &= (\psi_1^\dagger\gamma^0\gamma^\mu(1-\gamma_5)\psi_2)^\dagger \\ &= \psi_2^\dagger(1-\gamma_5)(\gamma^0\gamma^\mu\gamma^0)\gamma^0\psi_1 \\ &= \bar{\psi}_2\gamma^\mu(1-\gamma_5)\psi_1, \end{aligned}$$

where I have used the fact that $\gamma^{\mu\dagger} = \gamma^0\gamma^\mu\gamma^0$.

Taking their product gives

$$\begin{aligned} |\mathcal{M}_{fi}|^2 = \frac{G_F^2}{2}&\left[\bar{v}_{\nu_\mu}(q_\mu)\gamma^\alpha(1-\gamma_5)u_\mu(p)\right]\left[\bar{u}_e(k)\gamma^\beta(1-\gamma_5)v_{\nu_e}(q_e)\right] \\ \times &\left[\bar{u}_\mu(p)\gamma_\alpha(1-\gamma_5)u_{\nu_\mu}(q_\mu)\right]\left[\bar{v}_{\nu_e}(q_e)\gamma_\beta(1-\gamma_5)u_e(k)\right], \end{aligned} \tag{3.36}$$

where the spinors of the muon and electron, and their neutrinos, commute with each other and so can go freely across.[3.16]

3.16 The notation is a bit unfortunate since q_μ is the momentum of the neutrino ν_μ and not a four-vector with index μ!

Now take the average over the initial polarizations of the muon and sum over those of the final electron and neutrinos.

For the muon part,[3.17]

3.17 The sum over the polarizations gives

$$\sum u_e(k)\bar{u}_e(k) \to (\not{k} + m_e),$$

$$\sum u_{\nu_i}(q_i)\bar{u}_{\nu_i}(q_i) \to \not{q}_i$$

and

$$\sum u_\mu(p)\bar{u}_\mu(p) \to (\not{p} + m_\mu)/2$$

with a factor 1/2 because it is an average.

$$\begin{aligned}&\left[\bar{u}_{\nu_\mu}(q_\mu)\gamma^\alpha(1-\gamma_5)\,u_\mu(p)\right]\left[\bar{u}_\mu(p)\,\gamma^\beta(1-\gamma_5)u_{\nu_\mu}(q_\mu)\right]\\ &\quad\to\quad \frac{1}{2}\mathrm{Tr}\left[\not{q}_\mu\gamma^\alpha(1-\gamma_5)(\not{p}+m_\mu)\gamma^\beta(1-\gamma_5)\right]\end{aligned} \tag{3.37}$$

on replacing the bracketed wave functions summed over the polarizations by the expressions in Eq. (2.254). The result can be simplified by moving the projector $(1-\gamma_5)$ across the term $(\not{p}+m_\mu)$ to obtain[3.18]

3.18 $(1-\gamma_5)(\not{p}+m_\mu)\gamma^\beta(1-\gamma_5) = \not{p}(1+\gamma_5)\gamma^\beta(1-\gamma_5) + m_\mu(1-\gamma_5)\gamma^\beta(1-\gamma_5) = 2\not{p}\gamma^\beta(1-\gamma_5)$, with the term proportional to m_μ vanishing because of the chiral projectors.

$$\mathrm{Tr}\left[\not{q}_\mu\gamma^\alpha\not{p}\gamma^\beta(1-\gamma_5)\right]. \tag{3.38}$$

A similar result is obtained for the electron part. Multiplying the two together gives

$$\sum_{\mathrm{pol}}|\mathcal{M}_{fi}|^2 = G_F^2\,\mathrm{Tr}\left[\not{q}_\mu\gamma^\alpha\not{p}\gamma^\beta(1-\gamma_5)\right]\times\mathrm{Tr}\left[\not{q}_e\gamma_\alpha\not{k}\gamma_\beta(1-\gamma_5)\right], \tag{3.39}$$

summing over the polarizations, which, after the usual Diracology,[3.19] yields

3.19 Use $\mathrm{Tr}\,(\gamma^\mu\gamma^\nu\gamma^\rho\gamma^\sigma) = 4(g^{\rho\mu}g^{\sigma\nu} - g^{\rho\sigma}g^{\mu\nu} + g^{\rho\nu}g^{\mu\sigma})$, $\mathrm{Tr}\,(\gamma^\mu\gamma^\nu\gamma^\rho\gamma^\sigma\gamma_5) = -4i\varepsilon^{\mu\nu\rho\sigma}$ and $\varepsilon^{\alpha\mu\beta\nu}\varepsilon_{\rho\mu\sigma\nu} = -2(\delta^\alpha_\rho\delta^\beta_\sigma - \delta^\alpha_\sigma\delta^\beta_\rho)$.

$$\begin{aligned}\sum\nolimits_{\mathrm{pol}}|\mathcal{M}_{fi}|^2 &= 16\,G_F^2\,p^\lambda q_\mu^\tau k^\rho q_e^\sigma\left(g_{\tau\alpha}g_{\lambda\beta} - g_{\tau\lambda}g_{\alpha\beta} + g_{\tau\beta}g_{\lambda\alpha} - i\varepsilon^{\tau\alpha\lambda\beta}\right)\\ &\quad\times\left(g_{\rho\alpha}g_{\sigma\beta} - g_{\rho\sigma}g_{\alpha\beta} + g_{\rho\beta}g_{\alpha\sigma} - i\varepsilon_{\rho\alpha\sigma\beta}\right)\\ &= 64\,G_F^2\,(p\cdot q_e)(k\cdot q_\mu).\end{aligned} \tag{3.40}$$

The differential decay width is given by

$$d\Gamma = \frac{1}{2m_\mu}\sum_{\mathrm{pol}}|M_{fi}|^2 d\Phi^{(3)}_{\mathrm{LIPS}} \tag{3.41}$$

in the rest frame of the decaying muon. Here the subscript LIPS stands for Lorentz-invariant phase space, as in Chapter 1.

I have to deal with the three-body phase space

$$d\Phi^{(3)}_{\mathrm{LIPS}} = \frac{d^3\vec{k}}{(2\pi)^3 2E_k}\frac{d^3\vec{q}_e}{(2\pi)^3 2E_e}\frac{d^3\vec{q}_\mu}{(2\pi)^3 2E_\mu}(2\pi)^4\delta^{(4)}(p-k-q_e-q_\mu), \tag{3.42}$$

which is always a bit of a challenge.

To integrate over the neutrino momenta $q_e^\alpha q_\beta^\nu$ in Eq. (3.40) I first write the relevant integral:

$$I^{\alpha\beta}(q) \equiv \int\frac{d^3\vec{q}_e}{E_e}\frac{d^3\vec{q}_\mu}{E_\nu}\,\delta^{(4)}(q-q_e-q_\mu)\,q_e^\alpha q_\mu^\beta = \mathbb{I}_1(q^2)g^{\alpha\beta} + \mathbb{I}_2(q^2)\frac{q^\alpha q^\beta}{q^2}, \tag{3.43}$$

where $q = p-k$. The last identity in the equation above is given by Lorentz invariance: a tensor with two indices, α and β, and a function of only one four-vector q must be proportional to the metric times a scalar function of q^2, which I call $\mathbb{I}_1(q^2)$, and the product of $q^\alpha q^\beta$ times another scalar function, which I call $\mathbb{I}_2(q^2)$. The required integral is obtained once these two functions have been determined.

Now computing the trace $I^\alpha_\alpha = 4\mathbb{I}_1 + \mathbb{I}_2$ and the projection $q^\alpha q^\beta I^{\alpha\beta} = q^2(\mathbb{I}_1 + \mathbb{I}_2)$, it follows that[3.20]

[3.20] Since $q_e \cdot q_\mu = q^2/2$ because of the vanishing mass of the neutrinos.

$$I^\alpha_\alpha = \int \frac{\mathrm{d}^3\vec{q}_e}{E_e}\frac{\mathrm{d}^3\vec{q}_\mu}{E_\nu}\,\delta^{(4)}(q - q_e - q_\mu)\,q_e \cdot q_\mu = \frac{1}{2}q^2 \int \frac{\mathrm{d}^3\vec{q}_e}{E_e E_\mu}\,\delta(q^0 - E_e - E_\mu), \tag{3.44}$$

where the spatial part of the δ-function has killed the integral over $\vec{q}_\mu$. The last remaining integration can be done in the rest frame of the two-neutrino system, where $E_\mu = E_e$ and $q = (q^0, \vec{0})$, to obtain[3.21]

[3.21] The angular integration is just $\int \mathrm{d}\Omega = 4\pi$. The powers of E_e simplify and I can use the property that $\delta(f(x)) = |\mathrm{d}f(x)/\mathrm{d}x|^{-1}_{x=x_0}\delta(x - x_0)$ for x_0 a zero of the function $f(x)$.

$$I^\alpha_\alpha = \frac{q^2}{2}\int \frac{E_e^2 \mathrm{d}E_e \mathrm{d}\Omega}{E_e^2}\,\delta(q^0 - 2E_e) = 2\pi q^2 \int \mathrm{d}E_e\,\frac{1}{2}\,\delta\left(E_e - \frac{q^0}{2}\right) = \pi q^2.$$

The projection

$$q_\alpha q_\beta I^{\alpha\beta} = \int \frac{\mathrm{d}^3\vec{q}_e}{E_e}\frac{\mathrm{d}^3\vec{q}_\mu}{E_\nu}\,\delta^{(4)}(q - q_e - q_\mu)\,(q \cdot q_e)(q \cdot q_\mu) \tag{3.45}$$

can be treated in a similar manner:[3.22]

[3.22] In the neutrino rest frame $q_e \cdot q = E_e q^0$ and $q_0^2 = q^2$.

$$\begin{aligned} q_\alpha q_\beta I^{\alpha\beta} &= q_0^2 \int \mathrm{d}^3\vec{q}_e \mathrm{d}^3\vec{q}_\mu\,\delta^{(4)}(q - q_e - q_\mu) \\ &= q^2 \int \mathrm{d}^3\vec{q}_e\,\delta(q^0 - 2E_e) = \frac{\pi}{2}q^4. \end{aligned} \tag{3.46}$$

The functions $\mathbb{I}_1$ and $\mathbb{I}_2$ can be found by solving the system

$$4\mathbb{I}_1 + \mathbb{I}_2 = \pi q^2 \quad \text{and} \quad q^2(\mathbb{I}_1 + \mathbb{I}_2) = \frac{\pi}{2}q^4, \tag{3.47}$$

to obtain (finally!)

$$I^{\alpha\beta}(q) = \frac{\pi}{6}(q^2\eta^{\alpha\beta} + 2q^\alpha q^\beta), \tag{3.48}$$

which can be used to compute

$$I^{\alpha\beta}p_\alpha k_\beta = \frac{\pi}{6}\Big[q^2(p \cdot k) + 2(q \cdot p)(q \cdot k)\Big]. \tag{3.49}$$

Inserting the result into Eq. (3.41), I obtain that the differential width is given by

$$\mathrm{d}\Gamma = \frac{G_F^2}{48\pi^4 m_\mu}\frac{E_k \mathrm{d}E_k \mathrm{d}\Omega}{E_k^2}\Big[q^2(p \cdot k) + 2(q \cdot p)(q \cdot k)\Big]. \tag{3.50}$$

In the rest frame of the muon, $p = (m_\mu, \vec{0})$ and so[3.23]

[3.23] For $q \cdot p = m_\mu^2 - m_\mu E_k$, $q^2 = m_\mu^2 - 2m_\mu E_k$ and $q \cdot k = m_\mu E_k$.

$$\frac{\mathrm{d}\Gamma}{\mathrm{d}E_k} = \frac{G_F^2}{12\pi^3}E_k^2\,(3m_\mu^2 - 4m_\mu E_k), \tag{3.51}$$

and, on integrating,[3.24]

[3.24] The spectrum reaches its maximum value at $E_k - m_\mu/2$.

$$\Gamma_{\text{tot}} = \int_{E_{\text{MIN}}=m_e\simeq 0}^{E_{\text{MAX}}\simeq m_\mu/2} \frac{d\Gamma}{dE_k} dE_k \simeq \frac{G_F^2 m_\mu^5}{192\pi^3}, \tag{3.52}$$

where the approximation consists of neglecting the mass of the electron.[3.25]

[3.25] The result in Eq. (3.52) can be obtained just by dimensional analysis: the width has the dimension of a mass; G_F is $[M]^{-2}$ and we have G_F^2 in the cross section. The only dimensionful parameter is m_μ, which must then come as m_μ^5. The result of the actual computation is just the number $192\pi^3$ in Eq. (3.52).

I am now in the position of being able to compute G_F once I know the mass m_μ and lifetime τ_μ of the muon. These are given by

$$\begin{aligned} m_\mu &= 105.6583715 \pm 0.0000024 \text{ MeV} \\ \tau_\mu &= (2.197019 \pm 0.000021) \times 10^{-6} \text{ s}, \end{aligned} \tag{3.53}$$

which leads to

$$G_F = 1.1663787(6) \times 10^{-5} \text{ GeV}^{-2}; \tag{3.54}$$

this is the number you will find in the PDG booklet if you look it up. Now you know where it comes from. Please just remember that it is about 10^{-5} GeV^{-2}. This is small compared with the mass of a proton ($m_p \ll 1/\sqrt{G_F} \simeq 10^{2.5}$ GeV).

Another point.

What if the initial muon is polarized? How do I proceed if I do not want to average over the muon polarizations? This is done by means of the polarization projector:[3.26]

[3.26] A useful formula to add to the toolbox.

$$\boxed{u^\sigma(p) \otimes \bar{u}^\sigma(p) = \frac{1}{2} \underbrace{(\not{p} + m)}_{\text{spin sum}} \underbrace{(\mathbb{1} - \gamma_5 \not{\sigma})}_{\text{pol. proj.}},} \tag{3.55}$$

which replaces Eq. (2.254) when the polarization is measured (with a similar expression for the antiparticles). It is another tool to be stored safely for later use.

The projection operator in Eq. (3.55) goes into the more familiar chiral projectors $(\mathbb{1} \pm \gamma_5)/2$ when the spin is entirely along the particle momentum.

The operator in Eq. (3.55) must be inserted at the appropriate position in the trace over the muon terms in Eq. (3.37):

$$\frac{1}{2}\text{Tr}\left[\not{q}\gamma^\alpha(1-\gamma_5)(\not{p} + m_\mu)\underbrace{(1-\gamma_5\not{\sigma})}_{\text{pol. proj.}}\gamma^\beta(1-\gamma_5)\right]. \tag{3.56}$$

After that, I proceed as before through the Diracology song and dance and obtain the differential decay width, which now depends on the muon's polarization four-vector:[3.27]

[3.27] $\gamma = (1-\beta^2)^{1/2}$ where $\beta = \vec{p}/|\vec{p}|$. Here σ represents just the four-vector obtained by boosting the spin vector $\vec{s}$ (as was done in Chapter 1 for an arbitrary space vector in the rest frame).

$$\sigma = \left(\gamma(\vec{\beta}\cdot\vec{s}), \vec{s} + \frac{\gamma^2}{1+\gamma}(\vec{\beta}\cdot\vec{s})\,\vec{\beta}\right), \tag{3.57}$$

3.28 $\mathrm{d}\Omega = \sin\theta\, \mathrm{d}\theta\, \mathrm{d}\phi$.

where $\vec{s}$ is the polarization in the rest frame of the muon. The decay width is now given by[3.28]

$$\frac{\mathrm{d}^2\Gamma(\mu \to e\nu_e\bar{\nu}_\mu)}{\mathrm{d}E_k\mathrm{d}\Omega} = \frac{G_F^2}{12\pi^3}\frac{1}{E_k}\Big[q^2(p\cdot k) + 2(q\cdot p)(q\cdot k)\underbrace{-m_\mu q^2(k\cdot\sigma) - 2m_\mu(k\cdot p)(q\cdot\sigma)}_{\text{new terms!}}\Big] \tag{3.58}$$

as long as the electron mass is neglected, as before.

Equation (3.58) can be written in the rest frame of the muon, where $\sigma = (0, \vec{s}\,)$, as

$$\frac{\mathrm{d}^2\Gamma(\mu \to e\nu_e\bar{\nu}_\mu)}{\mathrm{d}x_e\mathrm{d}\cos\theta} = \Gamma_{\text{tot}}\left[(3-2x_e) - P_\mu(2x_e-1)\cos\theta\right]x_e^2, \tag{3.59}$$

where $x_e \simeq 2E_k/m_\mu$, $P_\mu = |\vec{s}\,|$ is the average polarization of the decaying muons and

$$k\cdot\sigma = -q\cdot\sigma = P_\mu E_k\cos\theta \quad \text{and} \quad p\cdot\sigma = 0\,. \tag{3.60}$$

It is because I am keeping track of the polarization that the width now depends on the angle θ between the momenta of the electrons and the direction of the muon polarization. (See Figure 3.9.) This dependence can be exploited to measure the magnetic moment of the muon by having the decay take place in the presence of an external magnetic field and measuring the momenta of the final electrons, which vary following the frequency of precession of the muon magnetic moment in the magnetic field.

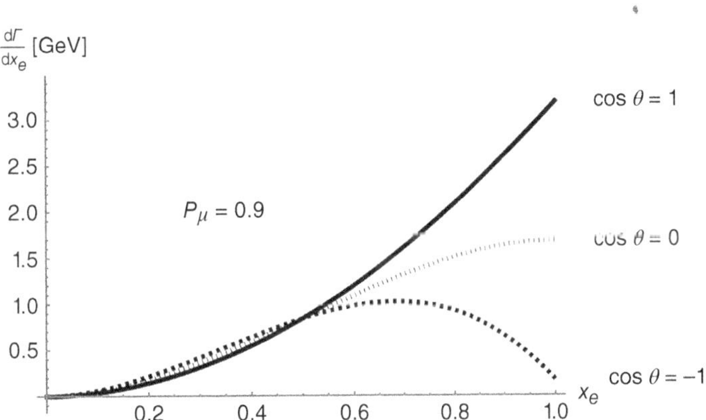

Figure 3.9 Differential cross section for the decay of a polarized muon. The mean polarization is taken as equal to 0.9 and the dependence on the electron energy fraction x_e is shown for three possible values of $\cos\theta$.

3.3 *Problem Session*: Two More Ways of Doing the Three-Body Phase-Space Integration

The integration over phase space is often the hardest part of computing a decay rate. For this reason it is useful to have different ways of doing it – the hope being that one of them is going to be well suited for the process to be computed. I introduce here two results based on the trick of rewriting the three-dimensional phase-space integration over the electron momentum as four-dimensional covariant integration. If you do not feel the urge to make this trick yours, skip this problem session. You can always come back here when the day comes that you find yourself stuck in some three-body phase-space computation.

The **first way**[1] consists of using the trick to rewrite the integral over the momentum k of the electron:

$$\begin{aligned}\int \frac{\mathrm{d}^3k}{2E_k}\,\delta^{(4)}(p-k-q_e-q_\mu) &= \int \mathrm{d}^4k\,\theta(E_k)\,\delta^{(4)}(p-k-q_e-q_\mu)\delta(k^2)\\ &= \theta(p-q_e-q_\mu)\delta\Big[(p-q_e-q_\mu)^2\Big],\end{aligned} \tag{3.61}$$

which leaves a width (in the rest frame of the muon)

$$\begin{aligned}\Gamma = {}& \frac{1}{2m_\mu(2\pi)^5}\int \frac{\mathrm{d}^3p_e}{2E_e}\frac{\mathrm{d}^3p_\mu}{2E_\mu}\\ &\times\,\theta(m_\mu-E_e-E_\mu)\delta\Big[(p-q_e-q_\mu)^2\Big]\sum_{\mathrm{pol}}|\mathcal{M}_{fi}|^2\,.\end{aligned} \tag{3.62}$$

Neglecting the neutrino masses, the remaining integrations are just over the energies and over solid angles:

$$\frac{\mathrm{d}^3p_e}{2E_e} = \frac{1}{2}E_e\mathrm{d}E_e\mathrm{d}\Omega_e \quad\text{and}\quad \frac{\mathrm{d}^3p_\mu}{2E_\mu} = \frac{1}{2}E_\mu\mathrm{d}E_\mu\mathrm{d}\Omega_\mu\,. \tag{3.63}$$

Taking as the axis the direction of p_e, the vector p_μ is defined relative to it. The angular integration over Ω_e is trivial and gives a factor 4π. I then have

$$\mathrm{d}\Omega_\mu = \overbrace{\mathrm{d}\phi_{e\mu}}^{\int \mathrm{d}\phi_{e\mu}=2\pi}\,\mathrm{d}\cos\theta_{e\mu} \tag{3.64}$$

and

$$\begin{aligned}\Gamma = {}& \frac{1}{4m_\mu(2\pi)^3}\int \mathrm{d}E_e\,E_e\int \mathrm{d}E_\mu\,E_\mu\int \mathrm{d}\cos\theta_{e\mu}\\ &\times\,\delta\Big[m_\mu^2-2m_\mu(E_e+E_\mu)+2E_eE_\mu(1-\cos\theta_{e\mu})\Big]\sum_{\mathrm{pol}}|\mathcal{M}_{fi}|^2\,.\end{aligned} \tag{3.65}$$

[1] See M.J. Savage, *Introduction to Nuclear and Particle Physics*, notes of Lecture 3, from where I learned these two ways of doing phase-space integrals.

The argument of the δ-function contains the kinematic constraints in the integration. When $\cos\theta_{e\mu} = -1$ the two neutrinos are back to back, and

$$m_\mu^2 - 2m_\mu(E_e + E_\mu) + 4E_eE_\mu = (2E_e - m_\mu)(2E_\mu - m_\mu) = 0, \quad (3.66)$$

which means that either E_e or E_μ is equal to $m_\mu/2$. This gives the maximum energy that the neutrinos can have. The other integration boundary arises when $\cos\theta_{e\mu} = 1$ and the two neutrinos are moving parallel. In this case

$$m_\mu^2 - 2m_\mu(E_e + E_\mu) = 0 \quad (3.67)$$

and therefore $E_e = E_\mu = m_\mu/2$.

Thus the width, with the integration region explicitly indicated, can be written as

$$\Gamma = \frac{1}{4m_\mu(2\pi)^3}\int_0^{m_\mu/2} \mathrm{d}E_\mu E_\mu \int_{m_\mu/2-E_\mu}^{m_\mu/2} \mathrm{d}E_e E_e \int \mathrm{d}\cos\theta_{e\mu} \quad (3.68)$$

$$\times \frac{1}{2E_eE_\mu}\,\delta\left(\cos\theta_{e\mu} - \frac{m_\mu^2 - 2m_\mu(E_e+E_\mu) + 2E_eE_\mu}{2E_eE_\mu}\right)\sum_{\text{pol}}|\mathcal{M}_{fi}|^2$$

$$= \frac{1}{8m_\mu(2\pi)^3}\int_0^{m_\mu/2} \mathrm{d}E_\mu \int_{m_\mu/2-E_\mu}^{m_\mu/2} \mathrm{d}E_e \sum_{\text{pol}}|\mathcal{M}_{fi}|^2, \quad (3.69)$$

where the squared amplitude is given by Eq. (3.40) and must be evaluated for the value of $\cos\theta_{e\mu}$ given by the δ-function in the equation above. The integrals can now be performed without further ado.

The **other way** uses the same trick of writing three-dimension integrals as four dimensional but this time for the integration over the momenta of the two neutrinos:

$$\int\frac{\mathrm{d}^3q_e}{2E_e}\int\frac{\mathrm{d}^3q_\mu}{2E_\mu} = \int\mathrm{d}^4q_e\int\mathrm{d}^4q_\mu\theta(E_e)\theta(E_\mu)\delta^{(4)}(q_e^2-m_\nu^2)\,\delta^{(4)}(q_\mu-m_\nu^2). \quad (3.70)$$

First change to new variables[3.29]

3.29 The Jacobian is unity, $p_e = (q_1+q_2)/2$ and $p_\mu = (q_1-q_2)/2$.

$$q_1 = p_e + p_\mu \quad \text{and} \quad q_2 = \frac{1}{2}(p_e - p_\mu) \quad (3.71)$$

to obtain[3.30]

3.30 Use $\delta(a+b)\delta(a-b) = \delta(2b)\delta(a-b) = \delta(b)\delta(a)/2$.

$$\int\frac{\mathrm{d}^3q_e}{2E_e}\int\frac{\mathrm{d}^3q_\mu}{2E_\mu} = \int\mathrm{d}^4q_1\int\mathrm{d}^4q_2\,\theta\left(\frac{q_1^0}{2}+q_2^0\right)\theta\left(\frac{q_1^0}{2}-q_2^0\right)$$

$$\times\,\delta\left(\frac{q_1^2}{4}+q_2^2-m_\nu^2+q_1\cdot q_2\right)\delta\left(\frac{q_1^2}{4}+q_2^2-m_\nu^2-q_1\cdot q_2\right)$$

$$= \frac{1}{2}\int\mathrm{d}^4q_1\int\mathrm{d}^4q_2\,\theta\left(\frac{q_1^0}{2}+q_2^0\right)\theta\left(\frac{q_1^0}{2}-q_2^0\right)\delta\left(\frac{q_1^2}{4}+2^2-m_\nu^2\right)$$

$$\times\,\delta(q_1\cdot q_2)\int\mathrm{d}s\,\delta(s-q_1^2), \quad (3.72)$$

where the last integral is just 1 and serves the purpose of introducing the variable $s = q_1^2$, which is the squared invariant mass of the neutrino pair. It will come in handy in a moment.

The width is at this point given by

$$\Gamma = \frac{1}{4m_\mu(2\pi)^5} \int_0^{m_\mu/2} \mathrm{d}s \int \frac{\mathrm{d}^3k}{2E_k} \int \mathrm{d}^4q_1 \int \mathrm{d}^4q_2 \sum_{\text{pol}} |\mathcal{M}_{fi}|^2$$
$$\times\ \delta(s - q_1^2)\, \theta\left(\frac{q_1^0}{2} + q_2^0\right) \theta\left(\frac{q_1^0}{2} - q_2^0\right) \delta\left(\frac{q_1^2}{4} + q_2^2 - m_\nu^2\right)$$
$$\times\ \delta(q_1 \cdot q_2)\delta^{(4)}(p - k - q_1)\,. \qquad (3.73)$$

The (Lorentz-invariant) factor

$$\int \mathrm{d}^4q_2 \sum_{\text{pol}} |\mathcal{M}_{fi}|^2\, \theta\left(\frac{q_1^0}{2} + q_2^0\right) \theta\left(\frac{q_1^0}{2} - q_2^0\right) \delta\left(\frac{q_1^2}{4} + q_2^2 - m_\nu^2\right) \delta(q_1 \cdot q_2) \qquad (3.74)$$

can be estimated in any frame, and this is simplest in the rest frame of the neutrino pair. In this frame, Eq. (3.74) is given by[3.31]

3.31 $q_1 \cdot q_2 = q_1^0\sqrt{s}$, which gives $\int \mathrm{d}q_2^0\, \delta(q_1 \cdot q_2) = 1/\sqrt{s}$ since $q_1 = (\sqrt{s}, 0, 0, 0)$.

$$\theta(q_1^0)\frac{1}{\sqrt{s}} \int \mathrm{d}^4q_2\, \delta\left(\frac{s}{4} - |\vec{q}_2|^2 - m_\nu^2\right) \sum_{\text{pol}} |\mathcal{M}_{fi}|^2$$
$$= \theta(q_1^0)\frac{1}{2\sqrt{s}} \int \mathrm{d}\Omega_{q_2}\mathrm{d}|\vec{q}_2|\, \delta\left(|\vec{q}_2| - \sqrt{\frac{s}{4} - m_\nu^2}\right) \sum_{\text{pol}} |\mathcal{M}_{fi}|^2$$
$$= \theta(q_1^0)\sqrt{\frac{s}{4} - m_\nu^2} \int \mathrm{d}\Omega_{q_2} \sum_{\text{pol}} |\mathcal{M}_{fi}|^2\,. \qquad (3.75)$$

Also, the integrations in Eq. (3.73) can be "massaged" as follows:

$$\int \mathrm{d}s \int \frac{\mathrm{d}^3k}{2E_k} \int \mathrm{d}^4q_1\, \delta(s - q_1^2)\, \delta^{(4)}(p - k - q_1)$$
$$= \int \mathrm{d}s \int \mathrm{d}^4q_1\, \delta(s - q_1^2)\, \delta\Big[(p - k)^2 - m_\nu^2\Big]$$
$$= \int \mathrm{d}s \int \frac{\mathrm{d}^3q_1}{2q_2^0}\, \delta(m_\mu^2 - m_e^2 + s - 2p \cdot q_1)\,. \qquad (3.76)$$

The integration over q_1 can be performed in the rest frame of the muon and gives

$$\int \frac{\mathrm{d}^3q_1}{2q_2^0}\, \delta(m_\mu^2 - m_e^2 + s - 2p \cdot q_1) = 2\pi \int \mathrm{d}q_1^0|\vec{q}_1|\, \delta(m_\mu^2 - m_e^2 + s - 2m_\mu q_1^0)$$
$$= \frac{\pi}{m_\mu}|\vec{q}_1|\,. \qquad (3.77)$$

From the δ-function in Eq. (3.77),

$$q_1^0 = \frac{m_\mu^2 - m_e^2 + s}{2m_\mu} \tag{3.78}$$

and therefore

$$|\vec{q}_1| = \sqrt{\left(\frac{m_\mu^2 - m_e^2 + s}{2m_\mu}\right) - s}\,. \tag{3.79}$$

The integration limits for the integration over the variable s come from the production of neutrino pairs at the threshold ($s = m_\nu^2$) or the electron at rest (in the rest frame of the muon), where $s = (m_\mu - m_2)^2$.

Pulling everything together, Eq. (3.73) can be written as

$$\Gamma = \frac{1}{2^5(2\pi)^4 m_\mu} \int_{4m_\mu^2}^{(m_\mu - m_2)^2} \mathrm{d}s \sqrt{\left(\frac{m_\mu^2 - m_e^2 + s}{2m_\mu}\right) - s}\,\sqrt{1 - \frac{4m_\mu^2}{s}} \times \sum_{\text{pol}} |\mathcal{M}_{fi}|^2 , \tag{3.80}$$

which can be integrated after inserting the expression in Eq. (3.40) for the matrix element.

3.4 Neutral Currents

Neutral currents are a bit of a surprise.

Charged currents have a structure dictated by the $SU(2)$ symmetry of the weak isospin. The first two generators of this group T^+ and T^- enter in Eq. (3.7). I can write them explicitly in terms of the time component of the respective currents and they are, in the case of the leptons,

$$T^+ = \frac{1}{2}\int \mathrm{d}^3x\, J_0^+(x) = \frac{1}{2}\int \mathrm{d}^3x\, u_{\nu_e}^\dagger (1-\gamma_5) u_e \quad \text{and} \quad T^- = (T^+)^\dagger . \tag{3.81}$$

It would seem only natural to identify the third generator T^3 with the electric charge

$$Q = \frac{1}{2}\int \mathrm{d}^3x\, u_e^\dagger u_e , \tag{3.82}$$

which would lead to a true unification of the weak and electromagnetic interactions in the group $SU(2)$.

This minimal implementation in terms of the charged and electromagnetic currents does not work, however, because the $SU(2)$ algebra

$$[T^+, T^-] = 2\,T_3 \tag{3.83}$$

gives

$$T_3 = \int \mathrm{d}^3x\, J_0^3(x) = \frac{1}{2}\int \mathrm{d}^3x \left[u_{\nu_e}^\dagger(1-\gamma_5)u_{\nu_e} - u_e^\dagger(1-\gamma_5)u_e\right] \neq Q\,. \tag{3.84}$$

For this reason the electric charge must enter as an independent group generator of an additional $U(1)$ group to be associated with the hypercharge Y, and T^3 remains as the charge of a new current, the neutral current, which even though it is as neutral as the electromagnetic current, is distinct from it. This is the origin of the odd $SU(2)_L \times U(1)_Y$ symmetry of the Standard Model and of the only partial unification between the weak and electromagnetic interactions.

That the electroweak group is $SU(2) \times U(1)$ instead of $SU(2)$ is an example of how mathematical beauty alone is not a good guiding principle in constructing physical models in high-energy physics, one is often reminded of Dirac's path to his equation, or Einstein's to general relativity, but these are exceptions. Our mind is weak and without the input from experiments often goes astray. General principles and mathematical elegance are no substitute for real data. For simplicity and the sake of unification, we would have done without the neutral currents. But they are there, and the electroweak group is the ugly product $SU(2) \times U(1)$.

The Lagrangian of the neutral currents is given by Eq. (3.4), which here I rewrite as

$$\mathcal{L}^{\text{nc}} = -\underbrace{\frac{g}{2\cos\theta_W}}_{\text{coupling}} \sum_i \underbrace{\bar{\psi}_i}_{\substack{\text{isospin}\\ \text{singlet}}} \overbrace{\gamma^\mu\left(g_V^i - g_A^i\gamma_5\right)}^{\text{chiral structure}} Z_\mu \underbrace{\psi_i}_{\substack{\text{isospin}\\ \text{singlet}}}\,. \tag{3.85}$$

The coefficients in the chiral structure in Eq. (3.85) are given in terms of the fermion quantum numbers as

$$\begin{aligned} g_V^i &= T_3^i - 2Q_i \sin^2\theta_W \\ g_A^i &= T_3^i\,. \end{aligned} \tag{3.86}$$

For instance, for the electron ($i = e$ and $T_3^e = -1/2, Q^e = -1$),

$$\begin{aligned} g_V^e &= -1/2 + 2\sin^2\theta_W \\ g_A^e &= -1/2\,. \end{aligned} \tag{3.87}$$

There are no isospin doublets here. The neutral current acts on isospin singlets.[3.32] There is no maximal parity violation either and the chirality of the interaction is modulated by the coefficients $g_{V,A}^i$. The chiral structure is made explicit by rewriting Eq. (3.85) as

[3.32] At least at tree level.

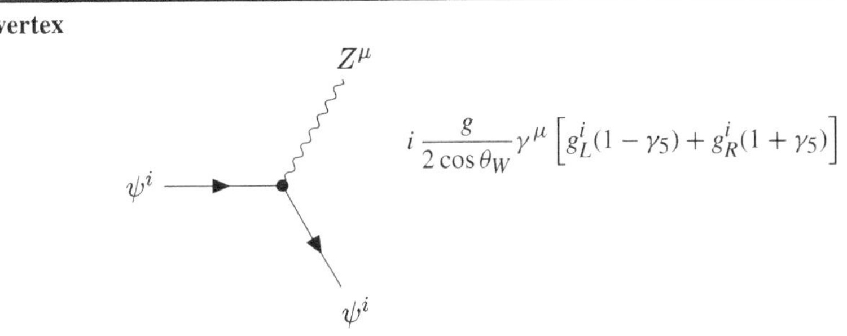

Figure 3.10 The vertex between fermions and the Z-boson.

$$\mathcal{L}^{\text{nc}} = -\frac{g}{2\cos\theta_W}\sum_i \bar{\psi}_i\gamma^\mu\left[g_L^i(1-\gamma_5)+g_R^i(1+\gamma_5)\right]\psi_i\, Z_\mu, \tag{3.88}$$

now with

$$\begin{aligned} g_L^i &= \frac{1}{2}\left(g_V^i + g_A^i\right) = T_3^i - Q^i\sin^2\theta_W \\ g_R^i &= \frac{1}{2}\left(g_V^i - g_A^i\right) = -Q^i\sin^2\theta_W\,. \end{aligned} \tag{3.89}$$

The structure of the interaction is similar to that of the charged currents, which itself is similar to the electromagnetic interaction. Yet the strengths of the coupling are different and the chiral structure is such that the interaction is not only between left-handed fields. And there is another important difference between the neutral and the charged currents: whereas the charge currents change flavor across generations, the neutral currents do not and there is no equivalent to the mixing matrix V_{CKM} either.

The Feynman rule is extracted from the vertex in Eq. (3.88) and shown in Figure 3.10.

3.5 *Problem Session*: Scattering of Neutrinos off Electrons

The scattering of muon neutrinos on electrons belonging to an atom of some material (see Figure 3.11) is described by the process

$$\nu_\mu(k) + e^-(p) \longrightarrow \nu_\mu(k') + e^-(p')\,. \tag{3.90}$$

The corresponding Feynman diagram is given in Figure 3.12. Lepton number is conserved generation by generation. Only the neutral current partakes of this interaction because only the t-channel diagram is possible.

Feynman rules for the vertices between the Z-boson and the neutrinos and the electrons are needed. It is just a matter of assigning the values of the coefficients g_L^i and g_R^i in Eq. (3.88). What I find is shown in Figure 3.13 for the neutrino, and in Figure 3.14 for the electron, for which $g_L^e + \sin^2\theta_W = 1/2$ and $g_R^e = \sin^2\theta_W$.

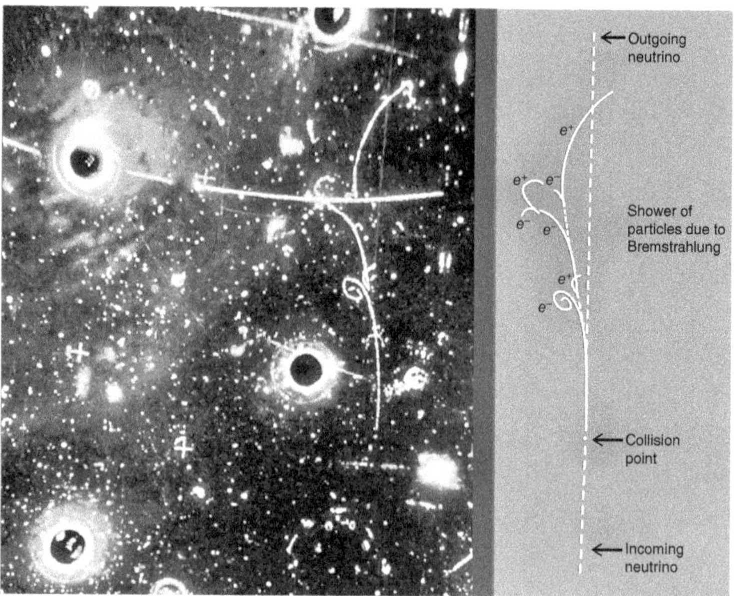

Figure 3.11 Collision of a neutrino and an electron in the *Gargamella* bubble chamber detector. This was the first evidence (July 1973) for neutral currents [© CERN].

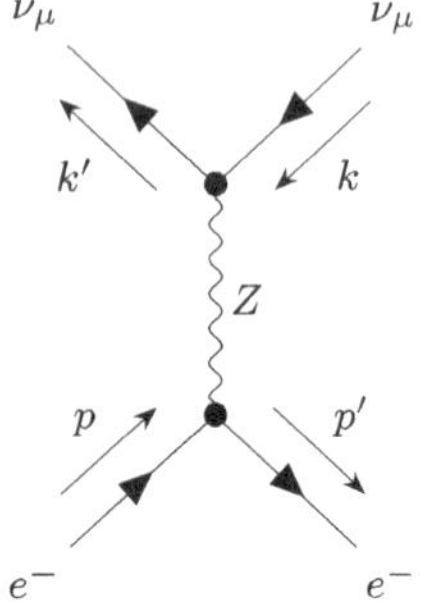

Figure 3.12 Feynman diagram for the process in Eq. (3.90).

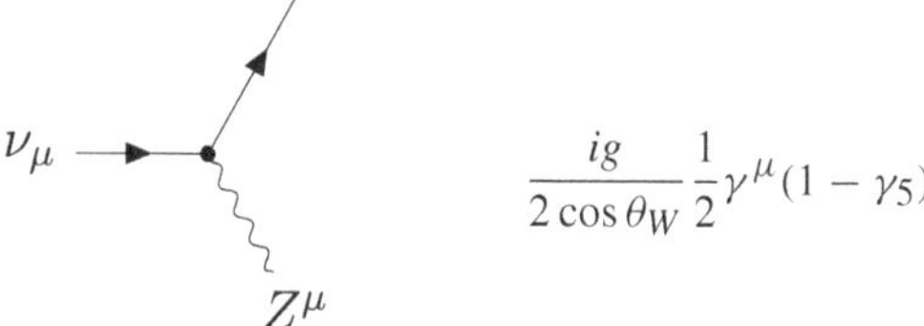

Figure 3.13 Vertex between the neutrino and the Z-boson.

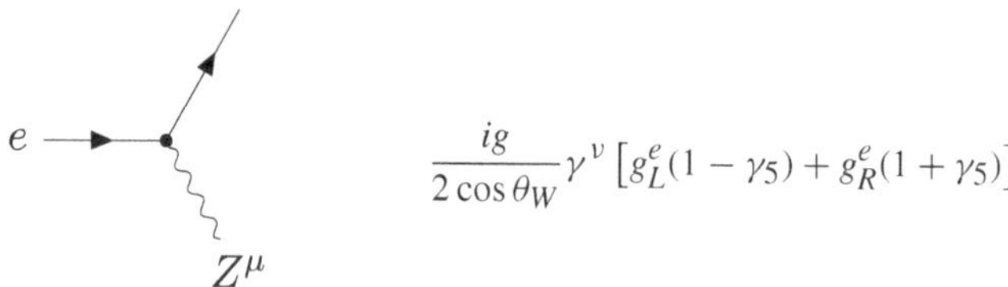

Figure 3.14 Vertex between the electron and the Z-boson.

If the energy is sufficiently smaller than the mass of the Z-boson, only the Fermi Lagrangian is needed, and the propagator of the Z-boson can be "integrated out" to give a factor

$$-\frac{1}{m_Z^2} \tag{3.91}$$

in front. The amplitude is therefore

$$\begin{aligned} \mathcal{M}_{fi} = {} & \frac{g^2}{8\cos^2\theta_W}\frac{1}{m_Z^2}\left[\bar{u}_{\nu_\mu}(k')\gamma^\alpha(1-\gamma_5)u_{\nu_\mu}(k)\right] \\ & \times\left[\bar{u}_e(p')\gamma_\alpha\Big((1-\gamma_5)g_L^e+(1+\gamma_5)g_R^e\Big)u_e(p)\right]. \end{aligned} \tag{3.92}$$

Take the conjugate of the amplitude in Eq. (3.92), multiply the two, and use the Casimir trick, in which the sum over the polarizations is turned into a trace, to obtain a long expression that is best broken down into two quantities: $\mathbb{M}_1^{\alpha\beta}$, for the neutrino part, and $\mathbb{M}_2^{\alpha\beta}$, for the electron part. The square of the amplitude is given by[3.33]

$$|\mathcal{M}_{fi}|^2 = \frac{G_F^2}{2}\mathbb{M}_{1\alpha\beta}\,\mathbb{M}_2^{\alpha\beta}\,, \tag{3.93}$$

where

$$\begin{aligned}\mathbb{M}_1^{\alpha\beta} &= 2\,\mathrm{Tr}\left\{\gamma^\alpha(1-\gamma_5)\not{k}\gamma^\beta\not{k}'\right\}\\ &= 8\left(k^\alpha k'^\beta - k\cdot k' g^{\alpha\beta} + k'^\alpha k^\beta\right) - 8i\epsilon^{\alpha\rho\beta\sigma}k_\rho k'_\sigma\,,\end{aligned} \tag{3.94}$$

and[3.34]

$$\begin{aligned}\mathbb{M}_2^{\alpha\beta} &= \frac{1}{2}\mathrm{Tr}\Big\{(\not{p}+m_e)\gamma^\alpha\left[(g_L+g_R)+(g_R-g_L)\gamma_5\right](\not{p}'+m_e)\gamma^\beta\\ &\quad\times\left[(g_L+g_R)+(g_R-g_L)\gamma_5\right]\Big\}\\ &= 4(g_L^2+g_R^2)\left[p^\alpha p'^\beta - p\cdot p' g^{\alpha\beta} + p'^\alpha p^\beta\right]\\ &\quad - 4\,i(g_R^2-g_L^2)\epsilon^{\rho\alpha\sigma\beta}p_\rho p'_\sigma + 8\,m_e^2\,g_L g_R\,g^{\alpha\beta}\,.\end{aligned} \tag{3.95}$$

Multiplying these two quantities gives a longish expression:

$$\begin{aligned}|\mathcal{M}_{fi}|^2 &= 16\,G_F^2\left\{k_\alpha k'_\beta - k\cdot k' g_{\alpha\beta} + k_\beta k'_\alpha - i\epsilon_{\alpha\rho\beta\sigma}k^\rho k'^\sigma\right\}\\ &\quad\times\Big\{(g_L^2+g_R^2)\left[p^\alpha p'^\beta - p\cdot p' g^{\alpha\beta} + p'^\alpha p^\beta\right] + 2\,m_e^2\,g_L g_R\,g^{\alpha\beta}\\ &\quad + i(g_R^2-g_L^2)\epsilon^{\rho\alpha\sigma\beta}p_\rho p'_\sigma\Big\}\,,\end{aligned} \tag{3.96}$$

which requires some massaging to be brought into a manageable form. One must be careful here because there are several terms and it is easy to miss one of them. Now multiplying the contents of the two sets of braces and simplifying the result eventually yields the compact expression

$$|\overline{\mathcal{M}}_{fi}|^2 = 64\,G_F^2\left[g_L^2(k\cdot p)(k'\cdot p') + g_R^2(k\cdot p')(k'\cdot p) - m_e^2\,g_L g_R(k\cdot k')\right]. \tag{3.97}$$

The cross section

$$\sigma = \int\frac{|\overline{\mathcal{M}}_{fi}|^2}{4\,(p\cdot k)}(2\pi)^4\delta^{(4)}(p+k-p'-k')\frac{\mathrm{d}^3\vec{p}\,'}{(2\pi)^3}\frac{1}{2p'_0}\frac{\mathrm{d}^3\vec{k}'}{(2\pi)^3}\frac{1}{2k'_0} \tag{3.98}$$

is obtained by inserting the expression for $|\overline{\mathcal{M}}_{fi}|^2$ from Eq. (3.97). After collecting the factors 2 and π, I find

$$\begin{aligned}\sigma &= \frac{G_F^2}{\pi^2}\frac{4}{(p\cdot k)}\int\left[g_L^2\,(p\cdot k)(p'\cdot k') + g_R^2\,(p'\cdot k)(p\cdot k') - m_e^2\,g_L g_R(k\cdot k')\right]\\ &\quad\times\,\delta^{(4)}(p+k-p'-k')\frac{\mathrm{d}^3\vec{p}\,'\,\mathrm{d}^3\vec{k}'}{2p'_0\;2k'_0}\,,\end{aligned} \tag{3.99}$$

where I still have to integrate over the phase space.

3.33 I have used

$$\frac{g^2}{8m_W^2} = \frac{g^2}{8m_Z^2\cos^2\theta_W} = \frac{G_F}{\sqrt{2}}.$$

3.34 The many errors I often discover afterwords, make me not too confident in what I find at the end of my computations. For this reason I like to work with others and compare results. The best collaborators are those who are both good at physics and reliable in their computations. They are like the drummer of a rock band, tapping the tempo as the other players go berserk. They are the bedrock on which physics is built – the unsung heroes, in my opinion, in an age of more extrovert personalities too often ready to get all the credit.

Looking at Eq. (3.99), the following integrals are needed:

$$I_{\alpha\beta} = \int \frac{\mathrm{d}^3\vec{p}\,'}{2p_0'}\frac{\mathrm{d}^3\vec{k}'}{2k_0'} k_\alpha' p_\beta'\, \delta^{(4)}(p+k-p'-k') \quad \text{and} \tag{3.100}$$

$$I = \int \frac{\mathrm{d}^3\vec{p}\,'}{2p_0'}\frac{\mathrm{d}^3\vec{k}'}{2k_0'} k\cdot k'\, \delta^{(4)}(p+k-p'-k')\,. \tag{3.101}$$

The good news is that the integral in Eq. (3.100) has already been computed, in the problem section on the β-decay of the muon, and I can just take the result from there:

$$I_{\alpha\beta} = \frac{\pi}{6}(q^2\eta_{\alpha\beta} + 2q_\alpha q_\beta)\,, \tag{3.102}$$

where now $q = p + k = (k_0 + p_0, \vec{0})$ in the center of mass frame.

All that is needed is to do the integral in Eq. (3.101). The space part of the δ-function can be used to kill the integration over $\vec{p}\,'$ and this gives, in the center of mass frame,[3.35]

3.35 In the center of mass frame, $\vec{p} + \vec{k} = 0$, $k_0 + p_0 = E$, $|\vec{k}'| = k'^0$, $|\vec{k}| = k^0$ and $\vec{k}\cdot\vec{k}' = k_0 k_0' \cos\theta$.

$$I = \int \frac{\mathrm{d}^3\vec{k}'}{2k_0'} k_0 k_0' (1-\cos\theta) \frac{1}{\sqrt{|\vec{k}|^2 + m_e^2}} \delta^{(3)}(\vec{p}\,' - \vec{k}')\, \delta(k_0' + p_0' - p_0 - k_0)\,, \tag{3.103}$$

which can be rewritten as[3.36]

3.36 Still in the center of mass frame, $p_0' + k_0' - k_0 - p_0 = 0$ and therefore $p_0' = k_0 + p_0 - k_0' = E - k_0'$.

$$I = k_0 \int \mathrm{d}\Omega\,(1-\cos\theta) \int_0^\infty \mathrm{d}k_0'\,(k_0')^2 \frac{\delta(\sqrt{(k_0')^2 + m_e^2} + k_0' - E)}{4\sqrt{k_0^2 + m_e^2}}\,. \tag{3.104}$$

The integral over k_0' can be done by using the remaining δ-function to give[3.37]

3.37 Remember that

$$\int f(x)\delta[g(x)] = f(x)/|g'(x)|_{g(x)=0}$$

and that the zero is in this case given by the solution of the equation $\sqrt{(k_0')^2 + m_e^2} + k_0' = E$, that is, $k_0' = (E^2 - m_e^2)/(2E)$.

$$\begin{aligned} I &= \frac{k_0}{4}\int \mathrm{d}\Omega\,(1-\cos\theta) \frac{(k_0')^2}{\sqrt{(k_0')^2 + m_e^2}} \left(\frac{\sqrt{(k_0')^2 + m_e^2}}{k_0' + \sqrt{(k_0')^2 + m_e^2}} \right) \Bigg|_{k_0' = \frac{E^2 - m_e^2}{2E}} \\ &= \frac{k_0}{4}\int \mathrm{d}\Omega\,(1-\cos\theta) \frac{(E^2 - m_e^2)^2}{4E^3}\,, \end{aligned} \tag{3.105}$$

where also performing the trivial angular integration gives

$$I = \pi \frac{k_0 E}{E^2} \frac{(E^2 - m_e^2)^2}{4E}\,. \tag{3.106}$$

I have written Eq. (3.106) in this peculiar way to make it ready to be expressed in terms of the Mandelstam variables and other invariant expressions (and ready to be written not only in the center of mass frame but in any frame of reference I may choose) as

$$I = \pi \frac{k\cdot p}{4s^2}(s - m_e^2)^2\,. \tag{3.107}$$

Having obtained the two integrals in Eq. (3.100) and Eq. (3.101), I can go back and compute the cross section, which is now given by

$$\sigma = \frac{G_F^2}{\pi^2}\frac{4}{(p\cdot k)}\Big\{g_L^2(p\cdot k)\frac{\pi}{24}(4s+2s) + g_R^2\frac{\pi}{24}\Big[s(p\cdot k) + 2(k+p)\cdot k\,(k+p)\cdot p\Big]$$
$$- m_e^2\, g_L g_R \frac{\pi}{4s^2}(p\cdot k)(s-m_e^2)^2\Big\}, \qquad (3.108)$$

which yields, after collecting the various factors,

$$\sigma = \frac{G_F^2}{\pi}s\left\{g_L^2 + \frac{1}{3}g_R^2 - g_L g_R\frac{m_e^2}{s}\right\}. \qquad (3.109)$$

And here is finally the bottom line. The total cross section in the laboratory frame is[3.38]

[3.38] $s = m_e(2E_\nu + m_e)$ in the laboratory frame. Neglect m_e with respect to E_ν within the parentheses.

$$\sigma = \frac{2G_F^2}{\pi}m_e\left[(g_L^2 + \frac{1}{3}g_R^2)E_\nu - \frac{m_e}{2}g_L g_R\right]. \qquad (3.110)$$

The linear rise of the cross section with the neutrino energy E_ν in Eq. (3.110) can be seen in Figure 3.15. It is a low-energy feature. At higher energies the exact Z-boson exchange between the vertices must be used instead of the effective interaction (as in Fermi's theory). The Z-boson propagator turns the apparent growth with the energy into a cross section decreasing with the energy in agreement with unitarity.

The two couplings g_L and g_R can be found by measuring the slope and the intercept of the cross section – after subtracting the dominant contribute of the charged currents (see Figure 3.15) – and compared with the Standard Model values.

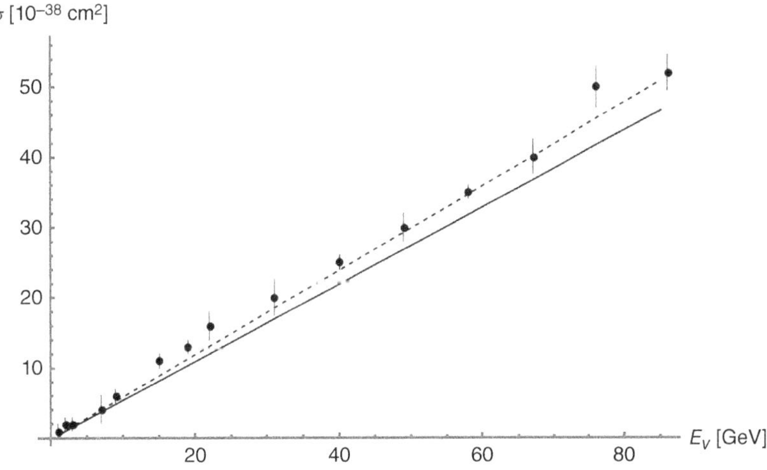

Figure 3.15 Some data for the cross section for the scattering of neutrinos off electrons. The solid line is the charged current contribution. The data show that there must be also an extra contribution, linear in the neutrino energy (the dashed line): this is the neutral current.

3.6 *Problem Session*: A Little Help (MATHEMATICA™ and FEYNCALC)

The often lengthy computations leading from the amplitude to the cross section can be broken down into a sequence of basic steps: writing the Feynman diagrams, summing over polarizations and taking traces of Dirac γ-matrices. There is no great feat of imagination here, and you only need to be careful and tenacious. It is no surprise that these steps can be automatized and that it has been done in a number of software programs. These programs give us a welcome help and spare us many a hour of computing and checking. If you do not take particular pleasure in doing the computation (many physicists actually do) these software programs are for you.

First of all a program for algebraical manipulation is needed. Over time, one program has come to dominate the market and is called MATHEMATICA™.[2] You will need it. Once you have MATHEMATICA™ on your computer you have the choice among several programs for computing cross sections. In recent years, one of them, called FEYNCALC,[3] has become perhaps the most popular. Actually, I have been mostly using a different one, which is unfortunately now becoming obsolete. Once you have put in the hours to learn how to use a specific piece of software, it is hard to change because it would require one to go again through the learning curve.

What FEYNCALC does is as follows: you write the amplitude in which you are interested by means of the Feynman rules and multiply by its complex conjugate, replace the trace by the sum over the polarizations and the program carries out the remaining computations (traces, sums and products of quadrivectors). It also does integrals over the internal momenta.

As an example, let me redo a particularly painful computation, that of the problem session on the anomalous magnetic moment of the electron in Chapter 1. It can be done in few strokes of the keyboard (and thanks to the CPU of the computer)

The traces and the multiplications of the quadrivectors, as well as all the simplifications, can be done automatically. I first initialize the package FEYNCALC within MATHEMATICA™ as shown in Figure 3.16.

The program provides the Feynman diagram to be evaluated, as in Figure 3.17.

Then the expression I wish to evaluate is inserted using the notation and grammar of the software. I also specify the on-shell conditions linking momenta and masses of the particles. The program returns what is shown in Figure 3.18.

---

[2] Website: https://www.wolfram.com/mathematica.

[3] Website: https://feyncalc.github.io/.

```
description="El -> Ga El, QED, F2(0) form factor, 1-loop";
If[ $FrontEnd === Null,
    $FeynCalcStartupMessages = False;
    Print[description];
];
If[ $Notebooks === False,
    $FeynCalcStartupMessages = False
];
$LoadAddOns={"FeynArts"};
<<FeynCalc`
$FAVerbose = 0;

FCCheckVersion[9,3,1];
```

FeynCalc 9.3.1 (stable version). For help, use the documentation center, check out the wiki or visit the forum.

To save your and our time, please check our FAQ for answers to some common FeynCalc questions.

See also the supplied examples. If you use FeynCalc in your research, please cite

- V. Shtabovenko, R. Mertig and F. Orellana, Comput.Phys.Commun. 256 (2020) 107478, arXiv:2001.04407.
- V. Shtabovenko, R. Mertig and F. Orellana, Comput.Phys.Commun. 207 (2016) 432–444, arXiv:1601.01167.
- R. Mertig, M. Böhm, and A. Denner, Comput. Phys. Commun. 64 (1991) 345–359.

FeynArts 3.11 (25 Mar 2022) patched for use with FeynCalc, for documentation see the manual or visit www.feynarts.de.

If you use FeynArts in your research, please cite

- T. Hahn, Comput. Phys. Commun., 140, 418–431, 2001, arXiv:hep-ph/0012260

Figure 3.16 Initialization of FEYNCALC.

Even the loop integral is easily performed, as shown by the output in Figure 3.19.

After adding the on-shell conditions in Figure 3.20, the final result is obtained, as shown in the screen shot in Figure 3.21.

More challenging examples can be found in the software documentation. What is important is to know that you are not alone and your computations can be checked or even performed entirely by computer thus freeing you for more important tasks.

I remember the first time I had access to one of these software programs for analytic manipulation. It was magical! Up to that time, I had used computers only for numerical computations but that day I understood that

Nicer typesetting

```
MakeBoxes[mu,TraditionalForm]:="μ";
MakeBoxes[p1,TraditionalForm]:="\!\(\*SubscriptBox[\(p\), \(1\)]\)";
MakeBoxes[p2,TraditionalForm]:="\!\(\*SubscriptBox[\(p\), \(2\)]\)";
```

```
diags = InsertFields[CreateTopologies[1, 1 → 2,
        ExcludeTopologies→{Tadpoles, WFCorrections}], {F[2,{1}]} →
        {V[1],F[2,{1}]}, InsertionLevel → {Particles},
        ExcludeParticles→{S[_],V[2|3],(S|U)[_],F[3|4],F[2,{2|3}]}];
```

```
Paint[diags, ColumnsXRows → {1, 1}, Numbering → Simple,
    SheetHeader→None,ImageSize→{256,256}];
```

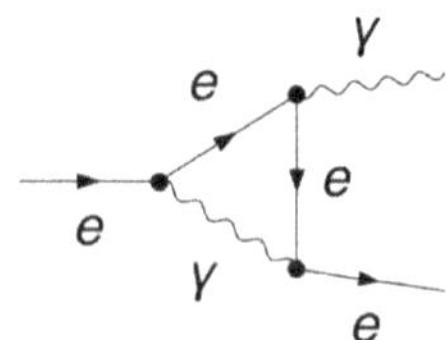

Figure 3.17 Running FeynCalc: Generation of the diagram to be evaluated.

The 1/(2Pi)^D prefactor is implicit. We need to replace e with -e to be compatible with the convention D^mu = d^mu + ie A^mu

```
amp[0] = FCFAConvert[CreateFeynAmp[diags,PreFactor→1],
    IncomingMomenta→{p1}, OutgoingMomenta→{k,p2},
    LorentzIndexNames→{mu},
    LoopMomenta→{q}, UndoChiralSplittings→True,
    ChangeDimension→D, List→False, SMP→True,
    FinalSubstitutions→ {SMP["e"]→-SMP["e"]}]/.
    k→p1-p2/.q→q+p1
```

$$\frac{i\, g^{\mathrm{Lor2}\,\mathrm{Lor3}}\, \varepsilon^{*\mu}(p_1 - p_2)\,(\varphi(p_2, m_e)).\left(-i\,\mathrm{e}\,\gamma^{\mathrm{Lor3}}\right).(m_e + \gamma\cdot(q+p_2)).(-i\,\mathrm{e}\,\gamma^{\mu}).(m_e + \gamma\cdot(q+p_1)).\left(-i\,\mathrm{e}\,\gamma^{\mathrm{Lor2}}\right).(\varphi(p_1, m_e))}{\left((p_1+q)^2 - m_e^2\right).\left((p_2+q)^2 - m_e^2\right).q^2}$$

Figure 3.18 Running FeynCalc: The beginning of the computation of the electron's anomalous magnetic moment.

```
amp[1] = (#1 /. SMP["e"]^3 → 4*Pi*SMP["e"]*SMP["alpha_fs"] & )[
    Contract[(#1 /. Pair[Momentum[Polarization[___], ___],
              ___] :→ 1 & )[amp[0]]]]
```

$$-\frac{4\,\pi\,\alpha\,\mathrm{e}\,(\varphi(p_2, m_e)).\gamma^{\mathrm{Lor3}}.(m_e+\gamma\cdot(q+p_2)).\gamma^{\mu}.(m_e+\gamma\cdot(q+p_1)).\gamma^{\mathrm{Lor3}}.(\varphi(p_1, m_e))}{((p_1+q)^2-m_e^2).((p_2+q)^2-m_e^2).q^2}$$

```
amp[2] = (Collect2[#1, Spinor] & )[DiracSimplify[
    TID[amp[1], q, ToPaVe → True]]]
```

$$8\,i\,\pi^3\,\alpha\,\mathrm{e}\,m_e\,(p_1{}^{\mu}+p_2{}^{\mu})\Big(2\,\mathrm{C}_1(m_e^2,0,m_e^2,0,m_e^2,m_e^2)+D\,\mathrm{C}_{11}(m_e^2,0,m_e^2,0,m_e^2,m_e^2)-2\,\mathrm{C}_{11}(m_e^2,0,m_e^2,0,m_e^2,m_e^2)+ D\,\mathrm{C}_{12}(m_e^2,0,m_e^2,0,m_e^2,m_e^2)-2\,\mathrm{C}_{12}(m_e^2,0,m_e^2,0,m_e^2,m_e^2)\Big)(\varphi(p_2, m_e)).(\varphi(p_1, m_e)) - 4\,i\,\pi^3\,\alpha\,\mathrm{e}\Big(D\,\mathrm{B}_0(0,m_e^2,m_e^2)-6\,\mathrm{B}_0(0,m_e^2,m_e^2)+4\,\mathrm{B}_0(m_e^2,0,m_e^2)+4\,m_e^2\,\mathrm{C}_0(0,m_e^2,m_e^2,m_e^2,m_e^2,0)- 2\,D\,\mathrm{C}_{00}(m_e^2,0,m_e^2,0,m_e^2,m_e^2)+4\,\mathrm{C}_{00}(m_e^2,0,m_e^2,0,m_e^2,m_e^2)\Big)(\varphi(p_2, m_e)).\gamma^{\mu}.(\varphi(p_1, m_e))$$

```
amp[3] = DotSimplify[(#1 /. FCI[GAD[mu]] :→ 0 & )[amp[2]]]
```

$$8\,i\,\pi^3\,\alpha\,\mathrm{e}\,m_e\,(p_1{}^{\mu}+p_2{}^{\mu})\Big(2\,\mathrm{C}_1(m_e^2,0,m_e^2,0,m_e^2,m_e^2)+D\,\mathrm{C}_{11}(m_e^2,0,m_e^2,0,m_e^2,m_e^2)-2\,\mathrm{C}_{11}(m_e^2,0,m_e^2,0,m_e^2,m_e^2)+ D\,\mathrm{C}_{12}(m_e^2,0,m_e^2,0,m_e^2,m_e^2)-2\,\mathrm{C}_{12}(m_e^2,0,m_e^2,0,m_e^2,m_e^2)\Big)(\varphi(p_2, m_e)).(\varphi(p_1, m_e))$$

```
amp[4] = amp[3] /. {PaVe[1, {ME^2, 0, ME^2}, {0, ME^2, ME^2},
        OptionsPattern[]] → 1/(32*Pi^4*ME^2),
    PaVe[1, 1, {ME^2, 0, ME^2}, {0, ME^2, ME^2},
        OptionsPattern[]] → -(1/(96*Pi^4*ME^2)),
    PaVe[1, 2, {ME^2, 0, ME^2}, {0, ME^2, ME^2},
        OptionsPattern[]] → -(1/(192*Pi^4*ME^2))}
```

$$8\,i\,\pi^3\,\alpha\,\mathrm{e}\,m_e\left(\frac{3}{32\,\pi^4\,m_e^2}-\frac{D}{64\,\pi^4\,m_e^2}\right)(p_1{}^{\mu}+p_2{}^{\mu})\,(\varphi(p_2, m_e)).(\varphi(p_1, m_e))$$

Figure 3.19 Running FeynCalc: Loop integration.

```
FCClearScalarProducts[];
ME=SMP["m_e"];
ScalarProduct[p1,p1]=ME^2;
ScalarProduct[p2,p2]=ME^2;
ScalarProduct[k,k]=0;
ScalarProduct[p1,p2]=ME^2;
```

Figure 3.20 Running FeynCalc: On-shell conditions.

```
amp[5] = (#1 /. D -> 4 & )[(ChangeDimension[#1, 4] & )[amp[4]]]
```

$$\frac{i\,\alpha\,e\,(\overline{p_1}^{\mu}+\overline{p_2}^{\mu})\,(\varphi(\overline{p_2},m_e)).(\varphi(\overline{p_1},m_e))}{4\,\pi\,m_e}$$

```
f2[0] = (#1 /. {Spinor[__] . Spinor[__] :> 1,
        FCI[FV[p1, _] + FV[p2, _]] :> 1} & )[
    amp[5]/((I*SMP["e"])/(2*ME))]
```

$$\frac{\alpha}{2\,\pi}$$

Figure 3.21 Running FeynCalc: Final result for the anomalous magnetic moment of the electron.

from then on I could use them to actually do the algebra, integrals and, well, yes, also the tedious traces of the Dirac matrices.

3.7 Neutral Currents and Pair Annihilation

Many a process that is computed in QED must be re-computed in the Standard Model because the existence of neutral currents makes possible a new diagram in which the photon is replaced by the Z-boson.

A case in point is the annihilation of an electron–positron pair into a muon–antimuon. It is easy to write the amplitude for this process as the sum of terms like (in the Feynman gauge)

$$-i\mathcal{M}_\gamma(e^-e^+ \to \mu^-\mu^+) = v_\mu(k_2)\gamma^\alpha\bar{u}_\mu(k_1)\,\frac{e^2}{s}\,u_e(q_1)\gamma_\alpha\bar{v}_e(q_2) \quad (3.111)$$

and (again, in the Feynman gauge)

$$\begin{aligned}&-i\mathcal{M}_Z(e^-e^+ \to \mu^-\mu^+)\\ &\quad= \frac{1}{4\sin^2\theta_W\cos^2\theta_W}\\ &\qquad\times v_\mu(k_2)\gamma^\alpha(g_V^\mu - g_A^\mu\gamma_5)\bar{u}_\mu(k_1)\,\frac{e^2}{s-m_Z^2}\,u_e(q_1)\gamma_\alpha(g_V^e - g_A^e\gamma_5)\bar{v}_e(q_2)\,.\end{aligned} \quad (3.112)$$

The momenta in these amplitudes are shown in Figure 3.22, and $(k_1+k_2)^2 = (q_1+q_2)^2 = s$.

The amplitude in Eq. (3.112) introduces a potential problem, namely that it blows up whenever the center of mass energy becomes close to the mass of the Z-boson and $s = m_Z^2$. I must resolve this problem before plunging into the computation. The solution comes from the discussion

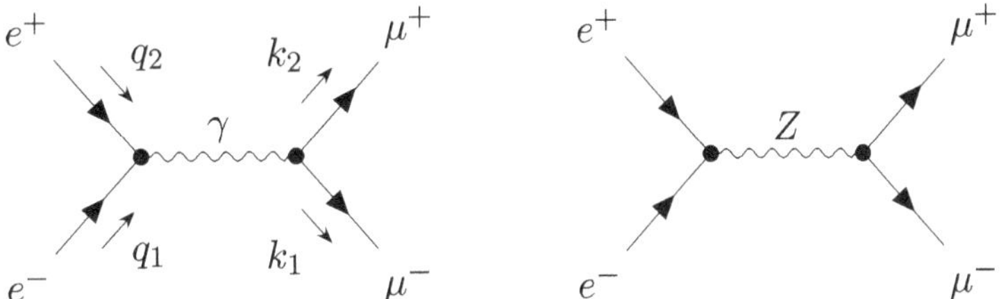

Figure 3.22 Feynman diagrams for electron–positron pair annihilation into a muon–antimuon pair. There are now two diagrams: one in which there is a photon in the s-channel and one in which there is a Z-boson.

about resonances in the first chapter. The Z-boson is like one of those resonances and the propagator must be written, as in Eq. (2.358),

$$D_Z^{\mu\nu}(s) = \frac{-ig^{\mu\nu}}{s - m_Z^2 + is\,\Gamma_Z/m_Z}, \tag{3.113}$$

in which the finite width Γ_Z has been introduced. Taking the product of the sum of the two amplitudes (3.111) and (3.112) times its conjugate, I find

$$|\mathcal{M}_\gamma|^2 + 2\,\mathrm{Re}\,\mathcal{M}_\gamma\mathcal{M}_Z + |\mathcal{M}_Z|^2. \tag{3.114}$$

Apart for the overall i's that are factorized out, the only imaginary part is in the denominator of the propagator in Eq. (3.113). The interference term, which is equal to twice the real part of the product of the two amplitudes, therefore receives a contribution proportional to

$$\mathrm{Re}\,\chi(s) = \frac{s - m_Z^2}{(s - m_Z^2)^2 + s^2\,\Gamma_Z^2/m_Z^2} \tag{3.115}$$

while the square of the Z-boson term contains a contribution proportional to

$$|\chi(s)|^2 = \frac{1}{(s - m_Z^2)^2 + s^2\,\Gamma_Z^2/m_Z^2}. \tag{3.116}$$

Averaging over the initial polarizations and summing over the final polarizations gives the three terms[3.39]

3.39 Remember:

$$g_V^\mu = g_V^e = -\frac{1}{2} + 2\sin^2\theta_W$$
$$g_A^\mu - g_A^e = -\frac{1}{2}.$$

$$|\overline{\mathcal{M}}_\gamma|^2 = \frac{c^4}{4s^2}\mathrm{Tr}\left[(\not{k}_1 - m_\mu)\gamma^\alpha(\not{k}_2 + m_m u)\gamma^\beta\right]\mathrm{Tr}\left[(\not{q}_1 - m_e)\gamma^u(\not{q}_2 + m_e)\gamma^\beta\right], \tag{3.117}$$

$$\begin{aligned}|\overline{\mathcal{M}}_Z|^2 = {} & \frac{e^4}{16\sin^4\theta_W\cos^4\theta_W}|\chi(s)|^2 \\ & \times \mathrm{Tr}\left[(\not{k}_1 - m_\mu)\gamma^\alpha(g_V^\mu - \gamma_5 g_A^\mu)(\not{k}_2 + m_m u)\gamma^\beta(g_V^\mu - \gamma_5 g_A^\mu)\right] \\ & \times \mathrm{Tr}\left[(\not{q}_1 - m_e)\gamma^\alpha(g_V^e - \gamma_5 g_A e)(\not{q}_2 + m_e)\gamma^\beta(g_V^e - \gamma_5 g_A^e)\right]\end{aligned} \tag{3.118}$$

```
In[50]:= ampSquared[0] = (amp[0] (ComplexConjugate[amp[0]]))//
             FeynAmpDenominatorExplicit//FermionSpinSum[#, ExtraFactor → 1/2^2]&//
             DiracSimplify//Simplify
```

$$\frac{2\,e^4\left(2\,m_e^2\left(2\,m_\mu^2+s-t-u\right)+2\,m_e^4+2\,m_\mu^4+2\,m_\mu^2\,(s-t-u)+t^2+u^2\right)}{s^2}$$

```
In[51]:= ampSquaredPolarized[0] =
             (ampPolarized[0] (ComplexConjugate[ampPolarized[0]]))//
             FeynAmpDenominatorExplicit//FermionSpinSum//DiracSimplify//Simplify

In[52]:= ampSquaredMassless1[0] = ampSquared[0]//ReplaceAll[#,{
             SMP["m_e"] → 0}]&
```

$$\frac{2\,e^4\left(2\,m_\mu^4+2\,m_\mu^2\,(s-t-u)+t^2+u^2\right)}{s^2}$$

Figure 3.23 FEYNCALC interface for the computation of the square of the amplitude.

and

$$2\,\mathrm{Re}\,\overline{\mathcal{M}_\gamma}\,\overline{\mathcal{M}_Z} = \frac{e^4}{8\,s\,\sin^2\theta_W\cos^2\theta_W}\mathrm{Re}\,\chi(s)$$
$$\times\ \mathrm{Tr}\left[(\not{k}_1-m_\mu)\gamma^\alpha(\not{k}_2+m_mu)\gamma^\beta(g_V^\mu-\gamma_5 g_A^\mu)\right]$$
$$\times\ \mathrm{Tr}\left[(\not{q}_1-m_e)\gamma^\alpha(\not{q}_2+m_e)\gamma^\beta(g_V^e-\gamma_5 g_A^e)\right]. \quad (3.119)$$

These Dirac traces can be done using FEYNCALC, the Mathematica package introduced in the previous section. For instance, $|\bar{\mathcal{M}}_\gamma|^2$ is manipulated as shown in Figure 3.23, and the other two terms can also be computed in the same way – the perfect exercise to review the Feynman rules and Diracology of the weak interactions. Take all the outputs and insert them into the cross section formula and, after some algebra, the cross section can be written as the sum of three terms:[3.40]

3.40 $\left(\frac{d\sigma}{d\Omega}\right)=\frac{1}{64\pi^2 s}\frac{p_f}{p_i}|\overline{\mathcal{M}}_{fi}|^2$.

$$\frac{d\sigma}{d\Omega}=\frac{\alpha^2}{8s}\left\{\underbrace{(1+\cos^2\theta)}_{\text{photon}}+\underbrace{\frac{2\,s\,\mathrm{Re}\,\chi(s)}{\cos\theta_W^2\sin\theta_W^2}\left[g_V^e g_V^\mu(1+\cos^2\theta)+2g_A^e g_A^\mu\cos\theta\right]}_{\text{interference}}\right.$$
$$\left.+\underbrace{\frac{s^2\,|\chi(s)|^2}{\cos\theta_W^4\sin\theta_W^4}\left\{\left[(g_V^e)^2+(g_A^e)^2\right]\left[(g_V^\mu)^2+(g_A^\mu)^2\right](1+\cos^2\theta)+8\,g_V^e g_V^\mu g_A^e g_A^\mu\cos\theta\right\}}_{Z\text{-boson}}\right\}. \quad (3.120)$$

While the original term coming from the photon propagation is symmetric in the scattering angle θ, the new terms arising because of the presence of the Z-boson depend on $\cos\theta$ and are not symmetric in going from the forward to the backward direction. The integrated cross section is shown in Figure 3.24.

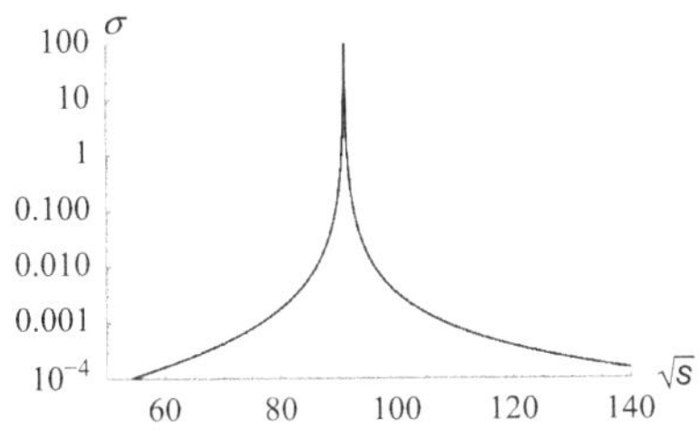

Figure 3.24 The resonant peak of the Z-boson. The peak is centered at $\sqrt{s} = m_Z$; a measure of the width of the peak provides a value for Γ_Z.

3.8 The Use of Unitarity, I

After invoking unitarity in the first part of this chapter to check Fermi's four-fermion β-decay interaction, I can use the same line of reasoning here for "discovering" the neutral currents in a way that does not proceed through the algebra of the $SU(2)$ group.

Let us look into the scattering of two neutrinos and the production of two gauge bosons, or, better, that part of the amplitude in which there are longitudinally polarized gauge bosons in the final state.[4] The process is (see Figure 3.25)

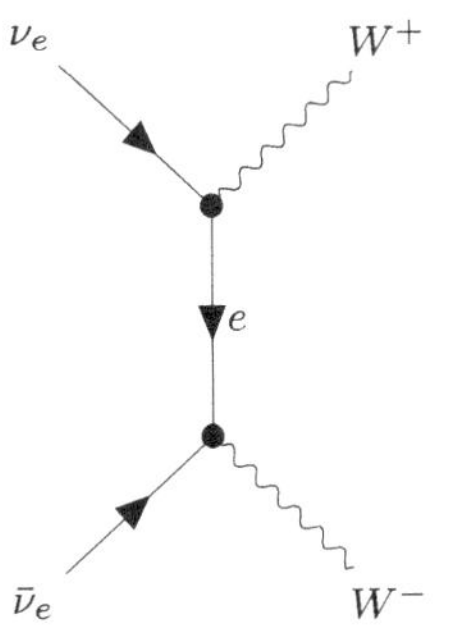

Figure 3.25 Feynman diagram for the process in Eq. (3.121).

$$\nu_e(p_1) + \bar{\nu}_e(p_2) \to W_L^+(q_1) + W_L^-(q_2). \tag{3.121}$$

The amplitude is obtained using the Feynman rules for the charged current (neglecting the electron mass):

$$\begin{aligned} \mathcal{M}_t(\nu_e \bar{\nu}_e \to W^+ W^-) \\ = -\frac{g^2}{2} \bar{v}(p_2)\gamma_\nu \frac{(\not{p}_1 - \not{q}_1)}{t} \frac{1-\gamma_5}{2} \gamma_\mu u(p_1) \varepsilon_L^\nu(q_1) \varepsilon_L^\mu(q_2), \end{aligned} \tag{3.122}$$

with $t = (p_1 - q_1)^2$.

In the center of mass frame, the momenta of the four particles are

$$\begin{aligned} p_1 &= \frac{\sqrt{s}}{2}(1,0,0,1) \\ p_2 &= \frac{\sqrt{s}}{2}(1,0,0,-1) \\ q_1 &= \frac{\sqrt{s}}{2}(1,\beta\,\sin\theta,0,\beta\,\cos\theta) \\ q_2 &= \frac{\sqrt{s}}{2}(1,\beta\,\sin\theta,0,-\beta\,\cos\theta)\,, \end{aligned} \tag{3.123}$$

where $\beta = \sqrt{1 - 4m_W^2/s}$. The longitudinal polarization of the gauge bosons can be written as

$$\begin{aligned} \varepsilon_L^\mu(q_1) &= \frac{\sqrt{s}}{2m_W}(\beta\,, \sin\theta, 0, \cos\theta) \\ \varepsilon_L^\mu(q_2) &= \frac{\sqrt{s}}{2m_W}(\beta\,, -\sin\theta, 0, -\cos\theta)\,. \end{aligned} \tag{3.124}$$

[4] I am following the nice presentation in J.C. Romão, arXiv:1603.04251.

The products of the fermion wave functions and the boson wave function can be approximated at high energy as follows:

$$\bar{v}(p_2)u(p_1) \propto \sqrt{s} \quad \text{and} \quad \varepsilon_L^\mu(q_2) \propto \frac{\sqrt{s}}{m_W}, \tag{3.125}$$

3.41 $(\not{p}_1 - \not{q}_1)u(p_1) = \not{q}_2 u(p_1)$.

for p_i, $q_i \propto \sqrt{s}$, so that the amplitude is approximated by[3.41]

$$\mathcal{M}_t(\nu_e\bar{\nu}_e \to W^+W^-) \simeq \frac{1}{2m_W^2}\bar{v}(p_2)\not{q}_2\frac{1-\gamma_5}{2}u(p_1) + O(1), \tag{3.126}$$

where $O(1)$ refers to parts of the amplitude that are sub-leading in powers of s. The amplitude in Eq. (3.126) grows as s for large center of mass energies. The cross section is going to be larger than the unitary bound for sufficiently large s. This spells disaster for the theory unless I can find another diagram that, combined with this one, leads to an amplitude with a better behavior. I need a diagram whose behavior for large s is something like $\mathcal{M}_t$ but with the opposite sign, so as to cancel the bad behavior of the first diagram. Where could such a diagram come from?

There are two possibilities: either an extra fermion giving a diagram like the one with the electron between the two neutrinos or an extra boson coupling to both the neutrinos and the W gauge bosons.

The extra fermion, let me call it E, must enter in a crossed channel diagram, Figure 3.26 (otherwise its contribution would add to the regular electron diagram) and gives

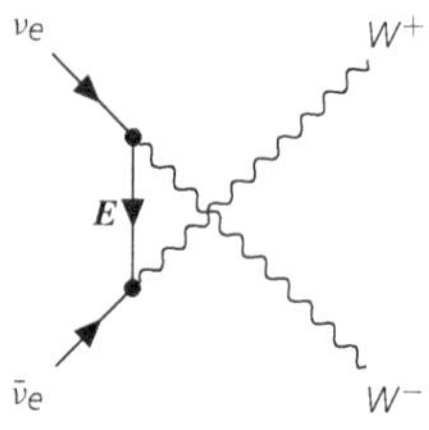

Figure 3.26 Feynman diagram for the amplitude $\mathcal{M}_u(\nu_e\bar{\nu}_e \to W^+W^-)$.

$$\mathcal{M}_u(\nu_e\bar{\nu}_e \to W^+W^-) = -\frac{g^2}{2}\bar{v}(p_2)\gamma_\nu\frac{(\not{p}_1 - \not{q}_2)}{u}\frac{(1-\gamma_5)}{2}\gamma_\mu u(p_1)\varepsilon_L^\nu(q_1)\varepsilon_L^\mu(q_2), \tag{3.127}$$

with $u = (p - q_2)^2$ and neglecting the mass of E. In the high-energy limit this gives

$$\mathcal{M}_u(\nu_e\bar{\nu}_e \to W^+W^-) \simeq -\frac{1}{2m_W^2}\bar{v}(p_2)\not{q}_2\frac{1-\gamma_5}{2}u(p_1) + O(1), \tag{3.128}$$

which nicely cancels the troublesome contribution from the electron diagram. So this is a possibility. The state E must be inserted into a $SU(2)_L$ multiplet, in fact a triplet, together with the electron and the neutrino. This extension makes the algebra close for the electric charge operator:

$$[T^+, T^-] = 2Q, \tag{3.129}$$

which was not possible before. All would be well except that no heavy fermion E with the required properties has ever been found.

Back to the drawing board.

Let me assume now that in addition to the charged gauge bosons I also have a neutral gauge boson coupled to the neutrinos and to the charged

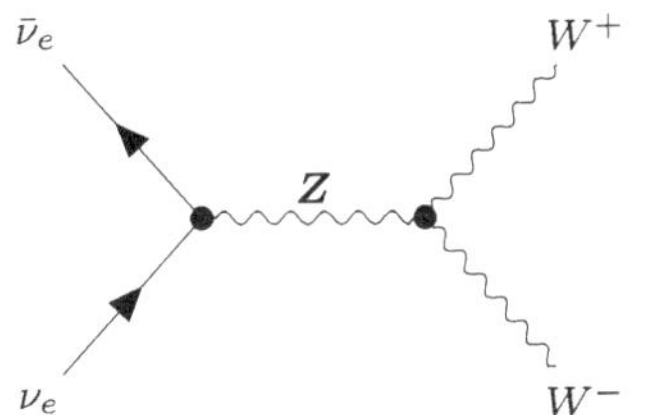

Figure 3.27 Feynman diagram for the amplitude $\mathcal{M}_s(\nu_e\bar{\nu}_e \to W^+W^-)$.

gauge bosons (see Figure 3.27). If this is the case, there is an additional diagram in which the two neutrinos go into the neutral boson, which afterwards decays into the two charged bosons. Assume that the coupling of this neutral boson to the other charged bosons is proportional to the same gauge coupling g as that of of the charged currents. Under these assumptions, the amplitude for this process is given by

$$\mathcal{M}_s(\nu_e\bar{\nu}_e \to W^+W^-) = -\frac{1}{2}\bar{v}(p_2)\gamma_\alpha\frac{1-\gamma_5}{2}u(p_1)\left[-g^{\alpha\beta}+\frac{(p_1+p_2)^\alpha(p_1+p_2)^\beta}{m_Z^2}\right]$$
$$\times\ \frac{\Gamma_{\mu\nu\beta}(q_1,q_2)}{s-m_Z^2}\varepsilon_L^\nu(q_1)\varepsilon_L^\mu(q_2), \qquad (3.130)$$

where $\Gamma_{\mu\nu\beta}(q_1,q_2)$ is the vertex between the neutral boson (of mass m_Z) and the W-bosons. I do not know the form of this vertex but I can try and fix it by requiring that $\mathcal{M}_s$ cancels the bad high-energy behavior of $\mathcal{M}_t$. Or I can just fast forward to the next chapter and take the form of this interaction to be

$$-ig\cos\theta_W\left[g_{\alpha\beta}(p-k)_\mu - g_{\beta\mu}(k-1)_\alpha + g_{\mu\alpha}(q-p)_\beta\right]. \qquad (3.131)$$

Using the same high-energy approximation as before I find (since $m_Z^2 = m_W^2\cos^2\theta_W$)

$$\mathcal{M}_s(\nu_e\bar{\nu}_e \to W^+W^-) \simeq -\frac{1}{2m_W^2}\bar{v}(p_2)\not{q}_2\frac{1-\gamma_5}{2}u(p_1) + O(1). \qquad (3.132)$$

The sum of the two amplitudes in Eq. (3.126) and Eq. (3.132) gives

$$\mathcal{M}_{s+t}(\nu_e\bar{\nu}_e \to W^+W^-) = O(1), \qquad (3.133)$$

which does not contains powers of s and leads to a well-behaved cross section.

I can see here the power of the unitarity constraint on the model: the neutral currents emerging from the requirement of having cross sections that are well behaved at large energies. Also, the intrinsic coherence of the model can be seen, where the high-energy behavior of the amplitudes is determined by the mechanism giving mass (and therefore their longitudinal components) to the gauge bosons. This behavior is ultimately connected to the nonlinear nature of the non-Abelian gauge theory and the coupling of the gauge bosons among themselves – a coupling not present in QED, which is an Abelian gauge theory.

To accomplish the unitarity of the amplitude, it had to be assumed that the neutral boson, let me call it by its name, the Z-boson of the electroweak theory, also couples to the charged gauge bosons, and that coupling is of a very precise form. But this is what is found, as I discuss in the next chapter, in writing the gauge theory of the group $SU(2)_L \times U(1)_Y$. It is this kind of internal consistency that gives us confidence in the model. Some people call this consistency beauty.[3.42]

3.42 Beauty is a dangerous concept. For the many physicists who claim, with Keats, that "Beauty is Truth, Truth Beauty" many more will tell you, with Boltzmann, that "If you are out to describe the truth, leave elegance to the tailor."

3.9 Exercises

1. **Hypercharge.** Find the assigned values of hypercharge Y for the quarks if the convention linking T^3 and Q is changed to $Q = T^3 + Y$ (which is often used).
2. **Polarized fermions.** Derive the differential cross section in Eq. (3.56) for the decay width of a polarized muon. Extend the result in the case in which the polarization of the final electron is also measured.
   [Hint: Use Eq. (3.59).]
3. **Extra fermion** E**.** Elaborate on the idea of having an extra fermion E in order to construct a the isospin charge $T^3 = Q$. Write explicitly the component of the isospin triplet and the charge Q. Why cannot E be the third-family lepton τ?
4. **High-energy behavior.** Verify explicitly that the cross section of the process $e^+e^- \to \gamma\gamma$ satisfies the unitary bound.
   [Hint: The photon has no longitudinal polarization.]
5. **Machine assisted computation.** Use your favorite software to do the Diracology and algebra to obtain Eq. (3.96).

4 Gauge Fields

Contents

The Lagrangian for the charge and neutral currents of the Standard Model contains fermion fields, what we call matter – leptons and quarks – and their interaction with spin-1 fields. These spin-1 fields are the W-, Z- and γ-bosons. As will become clear as I proceed, these fields are at the very center of the definition of the Standard Model as a gauge theory.

I had better start with the spin-1 field that should be the most familiar: the γ particle, that is, the photon.

Photons are the excitations of electric and magnetic fields. These fields obey (and are defined by) Maxwell's equations. To keep with a manifestly covariant notation, I write these equations in terms of the field strength tensor $F^{\mu\nu}$, which is a tensor defined in terms of the components of the electric $\vec{E}$ and magnetic $\vec{B}$ fields as follows:

$$F^{\mu\nu} = \begin{pmatrix} 0 & -E^1 & -E^2 & -E^3 \\ E^1 & 0 & -B^3 & -B^2 \\ E^2 & B^3 & 0 & -B^1 \\ E^3 & B^2 & B^1 & 0 \end{pmatrix} . \tag{4.1}$$

In this notation, Maxwell's equations can be written as

$$\begin{cases} \partial_\mu F^{\mu\nu} = J^\nu \\ \partial^\mu F^{\sigma\lambda} + \partial^\sigma F^{\lambda\mu} + \partial^\lambda F^{\mu\sigma} = 0 . \end{cases} \tag{4.2}$$

In Eq. (4.2), J^μ is the electromagnetic current, built out of charged matter. As soon as Maxwell's equations are written as in Eq. (4.2), the unification of the electric and magnetic forces, which is only implicit in the vectorial form of Maxwell's equations, becomes self-evident – as plain as day. The magnificent architecture of these equations is made even more manifest by taking the derivative of both sides of the first equation: this gives zero on the left-hand side because of the antisymmetry[4.1] of the space-time indices of $F^{\mu\nu}$ and therefore the right-hand side $\partial_\nu J^\mu$ must also be zero and – hey presto! – we have the conservation of the electromagnetic current.

4.1 $F^{\nu\mu} = -F^{\mu\nu}$.

In many applications of classical electrodynamics it is helpful to switch from the electric and magnetic field $\vec{E}$ and $\vec{B}$ to the scalar and vector potentials φ and $\vec{A}$ defined by

$$\vec{E} = -\vec{\nabla}\varphi - \frac{\partial \vec{A}}{\partial t} \qquad\qquad (4.3)$$
$$\vec{B} = \vec{\nabla} \times \vec{A} .$$

After this change of variables, one finds that the electric and magnetic fields are left unchanged if the potentials are replaced by new potentials φ' and $\vec{A}'$ defined as

$$\vec{A}'(x) = \vec{A}(x) - \vec{\nabla}\alpha(x) \qquad\qquad (4.4)$$
$$\varphi'(x) = \varphi(x) + \partial_t \alpha(x) ,$$

with α an arbitrary scalar function of space-time. This is easily verified by substituting the transformed potentials in Eq. (4.4) back into Eq. (4.3) and using the identities

$$\vec{\nabla} \times \vec{\nabla}\alpha(x) = 0 \quad \text{and} \quad -\vec{\nabla}\partial_t\alpha(x) + \partial_t\vec{\nabla}\alpha(x) = 0 . \qquad (4.5)$$

At the time when we first encounter it in some class on electromagnetism, this property (called **gauge invariance**) seems little more than a curiosity, maybe useful in solving one of those exercises very common in textbooks and rarely encountered in the real world.

The potential A^μ, which is a four-vector with φ as time component and $\vec{A}$ as space components, acquires a life of its own in the Lagrangian formulation because, together with $\partial_\nu A_\mu$, it plays the role of a generalized variable and momentum. In this framework, Maxwell's equations follow from the Lagrangian

$$\mathcal{L}^{cl}_{e.m.} = -\frac{1}{4} F_{\mu\nu} F^{\mu\nu} + J_\mu A^\mu , \qquad (4.6)$$

which describes classical electromagnetism and is the starting point of the quantum field theory of electrodynamics, that is, QED. The gauge invariance of Eq. (4.6) under the transformation

$$A'_\mu = A_\mu - \partial_\mu \alpha \qquad (4.7)$$

is guaranteed by the conservation of the current J_μ, which makes the term

$$-J_\mu \partial^\mu \alpha = \partial^\mu (J_\mu \alpha) , \qquad (4.8)$$

left over in Eq. (4.6) after the transformation, a vanishing surface term.[4.2]

The emphasis here is more on the conservation of the current than on gauge invariance. Yet there is a link. The conservation of the electromagnetic current can be understood in a different manner by looking at it in a form that contains explicitly the matter fields.

4.2 This autonomous life of the gauge potential is confirmed in the quantum theory by the existence of physical effects depending on its value – most notably the Aharonov–Bohm effect – when space-time is not simply connected.

If I change the wave function by a constant phase, as in

$$\psi' = e^{-i\alpha}\psi\,, \tag{4.9}$$

nothing is going to change because only $|\psi|^2$ has physical meaning. It is just a change of units of measure or, to use Weyl's terminology, of **gauge**.[4.3] This symmetry is associated, like any global symmetry, to a conserved Noether current.[4.4]

[4.3] This is H. Weyl's *Eichinvarianz*.

[4.4] The same notation α as in Eq. (4.4) is not accidental! Here, for the moment, α is a constant independent of space-time.

Matter in the Standard Model is provided by the fermion fields; they obey the Dirac equation, whose Lagrangian

$$L_\psi = i\bar{\psi}\gamma^\mu\partial_\mu\psi - m\bar{\psi}\psi \tag{4.10}$$

is left unchanged by the phase transformation

$$\begin{cases} \psi' = e^{-i\alpha}\psi \\ \bar{\psi}' = e^{i\alpha}\bar{\psi}\,, \end{cases} \tag{4.11}$$

and accordingly the Noether current,

$$J^\mu \equiv -\frac{\partial L_\psi}{\partial(\partial_\mu\psi)}i\psi + \frac{\partial L_\psi}{\partial(\partial_\mu\bar{\psi})}i\bar{\psi} = \bar{\psi}\gamma^\mu\psi\,, \tag{4.12}$$

is expected to be conserved, that is, $\partial_\mu(\bar{\psi}\gamma^\mu\psi) = 0$.

The change described by the transformation in Eq. (4.9) can be given a group structure if I write the phase as

$$e^{i\hat{q}\alpha} \tag{4.13}$$

and $\hat{q}$ as a 1×1 matrix (I know, it seems silly but bear with me). Written in this form the change of phase is a unitary transformation $\hat{U}$ generated by the group of 1×1 unitary matrices, which is known as $U(1)$.

Taking the time component of this conservation law and integrating over space one finds that the charge

$$Q = \int \mathrm{d}^3x\, J^0(x) \tag{4.14}$$

is conserved, that is, $\mathrm{d}Q/\mathrm{d}t = 0$.

The electromagnetic current can be seen as the Noether current corresponding to the phase independence of the wave function. In this way, a symmetry of the Lagrangian is turned into the physical conservation of a current.

This point of view lends itself to a generalization to larger groups. For example, take the group $SU(2)$. The phase transformation is now

$$\hat{U} = e^{i\sum_i \hat{\sigma}_i\hat{\alpha}_i}\,, \tag{4.15}$$

where σ_i are the Pauli matrices, and the parameters $\hat{\alpha}_i$ are equal to $\alpha\hat{e}_i$ with $\hat{e}_i$ the unit vectors in the group directions.

Here, one is reminded of the effect on the wave function of (infinitesimal) transformations such as translations,

$$\psi(\vec{r} + \Delta\vec{r}) = e^{i\Delta\vec{r}\cdot\frac{\partial}{\partial\vec{r}}}\psi(\vec{r}), \tag{4.16}$$

or rotations,

$$\psi(\varphi + \Delta\varphi) = e^{i\Delta\varphi\cdot\frac{\partial}{\partial\varphi}}\psi(\varphi), \tag{4.17}$$

in which the operators $\vec{p} \equiv \partial/\partial\vec{r}$ and $L_z \equiv \partial/\partial\varphi$ generate their respective transformations.[4.5] The Noether currents associated with translations and rotations live in space-time, the conserved quantities being the energy–momentum and the angular momentum. The only difference between Eq. (4.11) and Eq. (4.16) or Eq. (4.17) is that translations and rotations generate transformations taking place in space-time whereas phase transformations take place in the internal space of the corresponding group.

[4.5] Mathematically speaking, the phases mark positions on a **fiber** and these internal spaces are a **fiber bundle** attached to space-time.

4.1 *Problem Session*: Sum over the Polarizations

The gauge field, the photon, has only two independent polarizations and only two degrees of freedom whereas A_μ is a four-vector with four components. The price I pay for working with fully covariant quantities like A_μ is of having unphysical degrees of freedom around that must be eventually eliminated.

Consider the gauge potential written in Fourier space in terms of creation and annihilation operators:

$$A_\mu(x) = \int \frac{d^2\vec{k}}{(2\pi)^3}\frac{1}{\sqrt{\omega_{\vec{k}}}}\sum_\lambda\left[\varepsilon^\lambda_\mu\, a_{\vec{k}}\, e^{-ik\cdot x} + \varepsilon^{\lambda *}_\mu\, a^\dagger_{\vec{k}}\, e^{ik\cdot x}\right]. \tag{4.18}$$

For a photon moving with momentum $\vec{k}$ in the z-direction, $k^\mu = (\omega, 0, 0, \omega)$ and the two physical polarizations are given (in the helicity basis, as already introduced in Chapter 2) by

$$\varepsilon^{(\lambda=1)}_\mu = \frac{1}{\sqrt{2}}\begin{pmatrix}0\\-1\\-i\\0\end{pmatrix} \quad\text{and}\quad \varepsilon^{(\lambda=2)}_\mu = \frac{1}{\sqrt{2}}\begin{pmatrix}0\\1\\-i\\0\end{pmatrix}, \tag{4.19}$$

which satisfy $\varepsilon^{(\lambda=1,2)}\cdot k = 0$ and $\varepsilon^{(\lambda=1)}\cdot\varepsilon^{(\lambda=2)*} = -1$. These polarization vectors are the eigenvectors of the spin operator and satisfy the equation

$$\left(\vec{J}\cdot\hat{z}\right)^\nu_\mu \varepsilon^{(\lambda)}_\mu = \lambda\varepsilon^{(\lambda)}_\mu, \tag{4.20}$$

where J^i is the Lorentz generator given in Eq. (2.59) of Chapter 2.

4.6 $\Lambda^0_0 = 1$, $\Lambda^i_0 = \Lambda^0_j = 0$, $\Lambda^i_j = R^{ij}(\theta, \phi)$ with $R^{ij}(\theta, \phi)$ the rotation matrix away from the z-direction.

These two polarizations are rotated to[4.6]

$$\varepsilon_\mu^{\lambda=1}(k) = \Lambda_\mu^{\ \nu}\, \varepsilon_\nu^{\lambda=1}(\hat{z}) = \frac{1}{\sqrt{2}} \begin{pmatrix} 0 \\ -\cos\theta\cos\phi + i\sin\phi \\ -\cos\theta\sin\phi - i\cos\phi \\ \sin\theta \end{pmatrix}, \tag{4.21}$$

$$\varepsilon_\mu^{\lambda=2}(k) = \Lambda_\mu^{\ \nu}\, \varepsilon_\nu^{\lambda=2}(\hat{z}) = \frac{1}{\sqrt{2}} \begin{pmatrix} 0 \\ \cos\theta\cos\phi + i\sin\phi \\ \cos\theta\sin\phi - i\cos\phi \\ -\sin\theta \end{pmatrix}, \tag{4.22}$$

for a photon moving in the generic direction defined by the angles θ and ϕ with respect to the z-direction.

The other two polarizations,

$$\varepsilon_\mu^{(\lambda=0)} = \begin{pmatrix} 1 \\ 0 \\ 0 \\ 0 \end{pmatrix} \quad \text{and} \quad \varepsilon_\mu^{(\lambda=3)} = \begin{pmatrix} 0 \\ 0 \\ 0 \\ 1 \end{pmatrix}, \tag{4.23}$$

associated with the 0 and 3 components are unphysical. The completeness relation is defined by a sum over all four polarizations and is given by

$$\sum_{\lambda=0,1,2,3} \varepsilon_\mu^{\lambda *}(k)\varepsilon_\nu^{\lambda}(k) = -g_{\mu\nu}\,. \tag{4.24}$$

Yet why, you may ask, do the unphysical degrees of freedom associated with the longitudinal and time-like polarizations not enter in the computation of the amplitudes?

The killing of unphysical components cannot be done by hand because they come back through a Lorentz transformation. Luckily I can rely on the conservation of the current. Write the amplitude for the process in Figure 4.1 with one photon as an external leg as

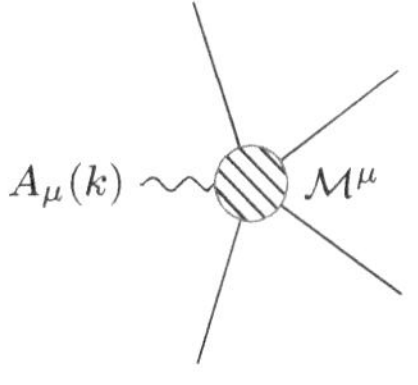

Figure 4.1 On-shell Ward identity: replace A_μ with its momentum k_μ.

$$\mathcal{M}(X \to Y\gamma) = \varepsilon_\mu^*(k,\lambda)\, \mathcal{M}^\mu\,, \tag{4.25}$$

where the photon is taken as going out. The amplitude $\mathcal{M}^\mu$ comprises the rest of the diagram and has the property of being orthogonal to the photon's momentum:

$$k_\mu \mathcal{M}^\mu = 0\,. \tag{4.26}$$

The condition in Eq. (4.26) – which is called a **Ward identity** – guarantees that the unphysical longitudinal and time-like polarizations do not contribute to the amplitude of the process. Equation (4.26) holds because of the conservation of the current and for photon fields that satisfy the equations of motion.

Because of this Ward identity, I can write the sum over the photon physical polarizations as follows:

$$\sum_{\lambda=0,1} \varepsilon_\mu^{\lambda *}(k)\varepsilon_\nu^\lambda(k) = \sum_{\lambda=0,1,2,3} \varepsilon_\mu^{\lambda *}(k)\varepsilon_\nu^\lambda(k) = -g_{\mu\nu} \tag{4.27}$$

as in Eq. (2.287) in the problem session on Compton scattering (in Chapter 2). Written in full,

$$\sum_{\lambda=1,2} \varepsilon_\mu^{\lambda *}(k)\varepsilon_\nu^\lambda(k) = -g_{\mu\nu} + \varepsilon_\mu^{0*}(k)\varepsilon_\nu^0(k) - \varepsilon_\mu^{3*}(k)\varepsilon_\nu^3(k) \tag{4.28}$$

$$\to \delta_{ij} - \frac{k_i k_j}{|\vec{k}|^2},$$

which is the result of taking $A_0 = 0$ and $\vec{k}\cdot\vec{A} = 0$ in the physical Coulomb gauge (see Section 4.5 on gauge fixing).

I can rewrite Eq. (4.28) by introducing a (non-covariant) fixed four-vector n^μ:

$$\boxed{\sum_{\lambda=1,2} \varepsilon_\mu^{\lambda *}(k)\varepsilon_\nu^\lambda(k) = -g_{\mu\nu} + \frac{k_\mu n_\nu + n_\mu k_\nu}{k\cdot n} - n^2\frac{k_\mu k_\nu}{(k\cdot n)^2}\,.} \tag{4.29}$$

The gauge invariance of the theory means that the last two terms in Eq. (4.29) do not contribute to the cross section because they are proportional to the momentum k^μ – which when contracted with the amplitude gives a vanishing contribution because of the Ward identity in Eq. (4.26). Explicitly, summing over the polarizations,

$$\begin{aligned}\sum_{\lambda=1,2} \mathcal{M}_\mu^* \mathcal{M}_\nu \varepsilon_\mu^{\lambda *}(k)\varepsilon_\nu^\lambda(k) &= \mathcal{M}_\mu^* \mathcal{M}_\nu\left(-g_{\mu\nu}\right) \\ &\quad + \mathcal{M}_\mu^* \mathcal{M}_\nu\left[\frac{k_\mu n_\nu + n_\mu k_\nu}{k\cdot n} - n^2\frac{k_\mu k_\nu}{(k\cdot n)^2}\right] \to 0 \\ &= \mathcal{M}_\mu^* \mathcal{M}_\nu\left(-g_{\mu\nu}\right).\end{aligned} \tag{4.30}$$

The second term goes to zero because of the Ward identity. Even though only the transverse terms $\mathcal{M}_1^*\mathcal{M}_1$ and $\mathcal{M}_2^*\mathcal{M}_2$ contribute, the sum over the polarizations can then be replaced with $-g_{\mu\nu}$.[4.7]

[4.7] Yet this is not guaranteed if I have more than just one external photon in the final or initial states. If this is the case I need to use the complete expression in Eq. (4.29).

For a massive spin-1 boson, the completeness relation becomes

$$\sum_{\lambda=1,2,3} \varepsilon_\mu^{\lambda *}(k)\varepsilon_\nu^\lambda(k) = -g_{\mu\nu} + \varepsilon_\mu^{0*}(k)\varepsilon_\nu^0(k), \tag{4.31}$$

where now $\varepsilon_\nu^0(k) = k^\mu/m$. Therefore

$$\boxed{\sum_{\lambda=1,2,3} \varepsilon_\mu^{\lambda *}(k)\varepsilon_\nu^\lambda(k) = -g_{\mu\nu} + \frac{k^\mu k^\nu}{m^2}\,.} \tag{4.32}$$

Because of the different numbers of physical polarizations, there is no way to go from the massive to the massless case in the limit $m \to 0$.

4.2 Gauge Freedom

What happens when the phase of a gauge transformation for a group symmetry is no longer a constant and depends instead on the position in space-time? This change is a momentous one.

If I now apply the transformations (writing out explicitly the space-time dependence in the phase)

$$\begin{cases} \psi' = e^{-iq\alpha(x)}\psi \\ \bar{\psi}' = e^{iq\alpha(x)}\bar{\psi} \end{cases} \tag{4.33}$$

to the Lagrangian in Eq. (4.10) I find that

$$\begin{aligned} \bar{\psi}'(i\gamma^\mu\partial_\mu - m)\psi' &= e^{iq\alpha(x)}\bar{\psi}i\gamma^\mu\partial_\mu e^{-iq\alpha(x)}\psi + e^{iq\alpha(x)}\bar{\psi}\,m\,e^{-iq\alpha(x)}\psi \\ &= \underbrace{\bar{\psi}q\gamma^\mu\partial_\mu\alpha(x)\psi}_{??} + \bar{\psi}(i\gamma^\mu\partial_\mu - m)\psi \end{aligned} \tag{4.34}$$

and the Lagrangian is no longer invariant! There is an extra term, the one highlighted by the question marks. This is bad. Yet there is hope because this extra term

$$\bar{\psi}q\gamma^\mu\partial_\mu\alpha(x)\psi \tag{4.35}$$

can be interpreted as the interaction of the fermion ψ with a vector field $\partial_\mu\alpha(x)$. This means that I can make the Dirac Lagrangian invariant if I write it as the Lagrangian for a Dirac fermion interacting with a vector field $A_\mu(x)$:

$$\bar{\psi}'(i\gamma^\mu\partial_\mu - m)\psi' + q\bar{\psi}'\gamma^\mu A'_\mu\psi', \tag{4.36}$$

which is equal to

$$\bar{\psi}(i\gamma^\mu\partial_\mu - m)\psi + q\bar{\psi}\gamma^\mu A_\mu\psi \tag{4.37}$$

if the vector field transforms as

$$A'_\mu(x) = A_\mu(x) - \partial_\mu\alpha(x)\,. \tag{4.38}$$

The transformation in Eq. (4.38) is precisely that of a gauge field. Therefore the Lagrangian for a Dirac fermion interacting with a gauge field,

$$\underbrace{D_\mu \equiv \partial_\mu + iqA_\mu}_{\text{covariant derivative}}$$

$$\mathcal{L}_\psi = \bar{\psi}(i\gamma^\mu \boldsymbol{D}_\mu - m)\psi\,, \tag{4.39}$$

where $D_\mu = \partial_\mu + iqA_\mu$ is the **covariant derivative**, which replaces the partial derivative ∂_μ of the free equation, is invariant for the combined gauge transformation in Eq. (4.33) and Eq. (4.38)

Local gauge invariance consists in the freedom of rotating the phase independently at different positions in space-time. It is **not** a symmetry and there are **no** conserved charges associated with it, as opposed to global gauge invariance, which is a symmetry and is related to the conservation of a charge.[4.8] Local gauge invariance implies the existence of constraints on the system – which is a consequence of Noether's second theorem, the first being related to current conservation as already discussed.

4.8 This is an important difference! You often hear "gauge symmetry", but think of it as "gauge freedom" in order to remember that there is no conserved quantity.

The freedom of rotating the phase by a different amount at each point in space-time is tantamount to having different coordinate systems for measuring the phase at each different space-time point. If I had a change of phase that is the same everywhere, I could set up a single coordinate system and measure every phase in terms of its coordinates. On the other hand, having the phase change from one point to the other amounts to having different coordinate systems at different points and the value of the local phases measured in terms of them.

The geometrical picture of a gauge theory follows from the freedom of setting local coordinate frames at each point of space-time. Because of this freedom, what we mean by a phase rotation, that is, the size of the phase, depends on the the local coordinate frame. To connect these multiple definitions I need a way to compare the value of my physical quantities at different space-time points, a way to transport these quantities. I need what is called a **connection**. For instance, for a Dirac fermion,

$$\psi(x + \mathrm{d}x) = (1 + \underbrace{qA_\mu \mathrm{d}x^\mu}_{\substack{\text{change due}\\ \text{to charge}}})\psi(x)\,. \tag{4.40}$$

This connection is the gauge potential. The covariant derivative contains this connection: the partial derivative keeps track of the intrinsic change of the wave function while the connection provides the change due to the change in the local coordinate frame.

This geometric picture is analogous to what happens if I take the Lorentz transformation of the energy–momentum and make it local. The result is general relativity! The change in a vector consists of its intrinsic change plus the change arising from the changing set of coordinates, that is, the change due to having to transport the vector from one point to another:

$$\mathrm{d}V^\mu = (\delta^\mu_\nu - \underbrace{\Gamma^\mu_{\nu\lambda}\mathrm{d}x^\lambda}_{\substack{\text{change due}\\ \text{to curvature}}})V^\nu\,. \tag{4.41}$$

Gauge freedom gives us the form of the Lagrangian of matter (the Dirac fermions) interacting through gauge fields. If I want my theory to be invariant under a local gauge transformation, the interaction must be given by the covariant derivative in Eq. (4.39). What works for the photon also works for the fields of the Standard Model. I only need to write the field strengths and covariant derivatives for the $SU(2)_L$ and $U(1)_Y$ groups. The field strengths for these groups are given by

$$F^A_{\mu\nu} = \partial_\mu W^A_\nu - \partial_\nu W^A_\mu + \underbrace{g\epsilon^{ABC} W^B_\mu W^C_\nu}_{\text{nonlinear term}} \tag{4.42}$$

$$B_{\mu\nu} = \partial_\mu B_\nu - \partial_\nu B_\mu \,. \tag{4.43}$$

The field strength of the non-Abelian gauge field W^A_μ contains, in addition to the linear terms, a term of self-interaction that the Abelian case of the hypercharge does not have.[4.9] The nonlinearity is controlled by the structure functions ϵ^{ABC} of the $SU(2)$ group, whose generators obey the Lie algebra given by

$$[T^A, T^B] = i\epsilon^{ABC} T^C \,. \tag{4.44}$$

4.9 The light sabers in Star Wars cannot work because of the linearity of the Maxwell equations. Two laser beams cross each other without interacting. To be picky: a small nonlinear interaction is generated by fermion loops but it is way too small an effect to be of any practical (let alone combat-grade) use.

Because of this term, the electroweak gauge fields can interact among themselves. The theory is nonlinear and there are terms providing for the interaction of three and four gauge bosons.

The classical Lagrangian is

$$\mathcal{L}_{\text{gauge}} = \underbrace{-\frac{1}{4}\sum_{A=1}^{3} F^A_{\mu\nu} F^{A\mu\nu}}_{SU(2)_L \text{ isospin}} - \underbrace{\frac{1}{4} B_{\mu\nu} B^{\mu\nu}}_{U(1)_Y \text{ hypercharge}} \,, \tag{4.45}$$

and the covariant derivative for the $SU(2)_L$ and $U(1)_Y$ gauge groups is defined as

$$D_\mu = \partial_\mu - \overbrace{ig\frac{1}{2}\,\vec{T}\cdot\vec{W}_\mu}^{SU(2)_L} - \overbrace{ig'\frac{1}{2}\,YB_\mu}^{U(1)_Y}. \tag{4.46}$$

The covariant derivative for the non-Abelian group $SU(2)_L$ acts on the corresponding group indices and contains the charged current in the Lagrangian in Eq. (3.7) of the previous chapter, with the interaction proportional to the $T^\pm$ matrices in isospin space coupled to the gauge bosons $W^\pm_\mu$. The neutral current interaction in the Lagrangian in Eq. (3.85) is a combination of the diagonal isospin matrix T^3 and the hypercharge Y. When acting on isospin doublets the derivative and the hypercharge parts must be understood as being multiplied by the identity matrix in isospin space.

The Lagrangian in Eq. (4.45) is not yet in the form utilized in the Standard Model. The correct expression in terms of the physical gauge bosons W, Z and γ is introduced in the next chapter after a detour to explain how these fields become massive.

4.3 *Problem Session*: Of Spinning Dancers and Falling Cats

To try to understand better what a gauge theory entails let us look for other systems, preferably in classical mechanics, with a behavior that is similar. It is often simpler to understand a new idea when it is applied in a context with which we are already familiar and which lends itself to a more concrete visualization.

If I look at a dancer like the one in Figure 4.2 spinning in a pirouette, I can see that she is going to spin faster as she folds her arms closer to her body.

Figure 4.2 Angular momentum of a spinning dancer [Dimitri Ctis/Getty images].

Rotations[4.10] like that of the dancer are described by the angular momentum

$$\vec{L} = \vec{r} \times \vec{p}, \tag{4.47}$$

which is linked to the angular velocity $\vec{\omega}$ of the body that is rotating. The relation can be quite complicated depending on how the mass of the body is distributed. If the body consists of a single piece and this piece is rotationally symmetric then the two quantities are proportional:

$$\vec{L} = I\,\vec{\omega}, \tag{4.48}$$

where I is the moment of inertia of the body. The moment of inertia quantifies both the total mass of the body and also the way this mass is distributed in the body geometry.

The angular momentum of the dancer is conserved. When she folds her arms the distance between them and the body decreases and $\vec{r}$ in Eq. (4.47) decreases – to maintain the angular momentum constant, the angular velocity must increase: the dancer spins faster. The change in angular velocity is only possible because her body is not rigid (as it would be if made of a single piece) and its geometry can be changed by moving the arms.

Here is a simple thought experiment. Get hold of a cat and, by holding his four legs, lean out of a low window, let him go (his back downwards)

[4.10] Rotations are fascinating: at first sight they appear to be no more complicated than motion in a straight line, yet they are another world altogether.

and watch what happens. The cat by rotating on himself would land on his four legs, as shown in the drawings in Figure 4.3. Once on the ground, the cat usually goes away outraged but unharmed.

But how does he do it? The cat starts out at rest, in your hands, and therefore with zero angular momentum. Angular momentum is conserved and should remain zero. Yet the cat in order to rotate must change his angular momentum locally, in such a way that the total is still zero. The cat can do this because he is not a rigid body – as you will become acutely aware if you try to hold him against his will.

A body that is not rigid can change shape. This change in shape can be thought of as a motion taking place in the space of all possible shapes of the body. For every given shape of his body, the cat is described by a point in this space. The cat (as shown in Figure 4.4) moves his tail, twists his neck and sends his front legs in the opposite direction with respect to his tail, and in this way moves in the internal space of his body shapes. This twisting and shape-shifting results in his final rotation, the rotation that brings him from being upside down to landing softly on his four paws. During these changes of shape the angular momentum as a whole remains constant and is conserved as it should be and is for the dancer spinning faster and faster as she folds her arms.

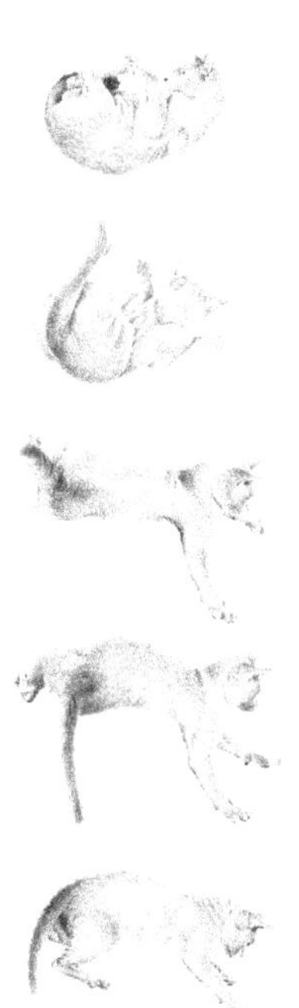

Figure 4.3 The falling cat: 1, 2, 3, 4, 5, and he is upright and ready to land on four legs [Dorling Kindersley/Getty images].

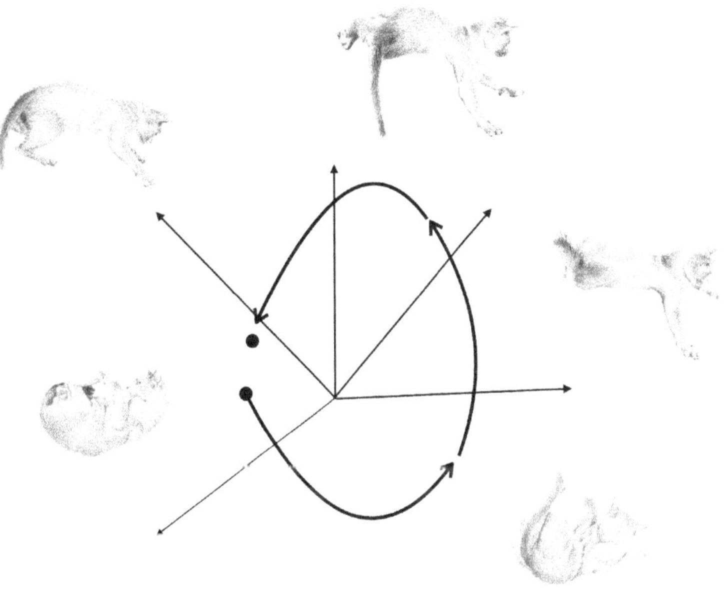

Figure 4.4 The motion undergone by the cat in the space of body shapes while falling. He starts out (lower dot) and ends up (upper dot) with the same body shape but rotated by 180°.

For a deformable body the angular momentum is not only proportional, as in Eq. (4.48), to the angular velocity $\vec{\omega}$ of the body as a whole but also contains a term that takes into account the motion of the various parts of which the body is composed, parts that can be labeled by an index i; in other words the angular momentum is given by[4.11]

[4.11] To simplify, let us assume that the overall inertia momentum I can still be used.

$$\vec{L} = I\vec{\omega} + I\sum_i \vec{A}_i v^i\,, \tag{4.49}$$

where $\vec{A}_i$ are some functions which make it possible to describe the rotation when the shape of the rotating body changes.

The necessity of such functions is best understood by considering the problem of defining the rotation for a deformable body. Whereas a rotation is usually defined with respect to the initial position of the body, if the body is not rigid then its deformations change its shape and it is no longer possible to distinguish what is a deformation and what is a rotation. Every rotation can always be redefined as a change in shape. For this reason, it is only the sum of the first term on the right-hand side of Eq. (4.49) (the rotation as a whole) and the second term (the change in shape) that remains constant – if the angular momentum is indeed conserved.[1]

For the particular case of the cat, for which the initial angular momentum vanishes ($\vec{L} = 0$), the total angular velocity is given by the sum of the angular velocities of the various parts of the body:

$$\vec{\omega} = -\sum_i \vec{A}_i v^i\,. \tag{4.50}$$

The equation relating the change in angular momentum in the frame of reference at rest in space to that fixed on the body,

$$\left.\frac{d\vec{L}}{dt}\right|_{space} = \left.\frac{d\vec{L}}{dt}\right|_{body} + \vec{\omega}\times\vec{L}, \tag{4.51}$$

tells me that the change in angular momentum measured at rest in space is equal to the change in momentum of the body plus the contribution due to the angular velocity of the rotating body.

The meaning of Eq. (4.51) is rather intuitive: in measuring the angular momentum of a rotating body, we must add the change due to the rotation of the body to the variation of the angular momentum if we are measuring the angular momentum of the body in a frame of reference which is at rest in space.

For the case of interest, the angular momentum as measured at rest is always conserved and therefore the term on the left-hand side of Eq. (4.51) vanishes, so that

[1] I first read about this interesting subject in A. Shapiro and F. Wilczeck, *Am. J. Phys.* **57** (1988) 514 and R. Montgomery, *Fields Institute Comm.* **1** (1993) 58.

$$\left.\frac{d\vec{L}}{dt}\right|_{body} = -\,\vec{\omega} \times \vec{L}, \tag{4.52}$$

and therefore, using Eq. (4.50),

$$\left.\frac{d\vec{L}}{dt}\right|_{body} = \sum_i \vec{A}_i \times \vec{L} v^i\,. \tag{4.53}$$

Equation (4.53) gives me the variation of the angular momentum as measured on the body and due to the relative motion of the various parts of the deformable body. The angular momentum is conserved (the derivative, as measured in the system at rest, is equal to zero) but if it is measured in the rotating system, it is the sum of the body deformations. This is the local form of the conservation of the angular momentum for which the change in the angular momentum in the rotating body is compensated at each point by the change in the velocities of the body parts so as to give a vanishing total angular momentum. The change of angular momentum in the rotating body is in part transferred to the deformation of the various body parts.

At the same time, the rotation of the body affects the motion of its parts, which are acted upon by a force, due to the rotation, with components

$$f_i = m_i \frac{dv_i}{dt} = \sum_j \vec{L} \cdot \vec{B}_{ij} v^j\,, \tag{4.54}$$

where $\vec{B}_{ij} = \partial_i A_j - \partial_j A_i$.

The force in Eq. (4.54) is called the **Coriolis force**, from the name of its discoverer who studied the effects on the Earth due to its rotation. On the Earth the effect of the daily rotation about its axis is rather modest (except for the large masses of air in the atmospheric system) but nevertheless it can be shown by recording the slow motion of rotation of the plane of oscillation of a pendulum. If the Earth did not rotate, this oscillation plane would be fixed. Instead, the Earth's rotation makes the plane rotate with a velocity which depends on the latitude: it takes 24 hours at the North Pole to complete a full circle.

Let me take stock and sum up what is going on.

The various parts of a deformable body that is rotating are accelerated because of the rotation. This acceleration is proportional to their relative velocities and given by the tensor B_{ij} in Eq. (4.54). The acceleration is due to the Coriolis force. The fields A_i connect the rotations in space with the deformations of the body and the change of angular momentum as measured on the reference frame fixed to the body. They connect, for the cat, the rotation in space with his body deformations. Since they are connections, I can call them gauge fields.

For the elementary particles, the gauge fields connect their motion in space with the motion in the internal spaces of the various groups under which they are charged. The internal space of the particles corresponds to the space of the bending and flexing of the body of the cat, shown in Figure 4.4. The motion in this internal space is seen as an acceleration, that is, a force in space-time.

The analogy between deformable bodies and elementary particles can be pushed further, to the equations of motion.[4.12] For an elementary particle, I can write:

[4.12] The equations for the elementary particles are in space-time with the indices running from 0 to 3 – while for the deformable bodies they live in space with all indices running from 1 to 3. This caveat makes the gauge force not quite like the Coriolis force.

$$\frac{dv_\mu}{dt} = \sum_\nu \vec{\sigma} \cdot \vec{F}_{\mu\nu}\, v^\nu\,, \tag{4.55}$$

to describe the motion in space-time (with the field strength $F_{\mu\nu}$ playing the role of B_{ij}). Equation (4.55) is the Lorentz force. The motion in the internal space is described by $\vec{\sigma}$ – the spin (or the isospin, or the color) variable here playing the role of $\vec{L}$ in the cat's case – which follows an equation similar to that for the angular momentum:

$$\frac{d\vec{\sigma}}{dt} = \sum_\mu \vec{A}_\mu \times \vec{\sigma}\, v^\mu\,. \tag{4.56}$$

4.4 *Problem Session*: A Story of Three Rods

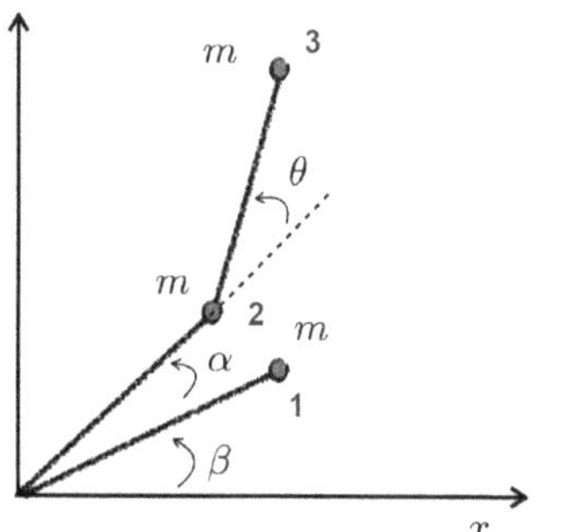

Figure 4.5 Three rods, two of which are hinged. The shape and position in space of the system is characterized by the three angles α, β and θ.

To show how the motion in the internal space of body shapes can result in a motion in space, a system can be used that, even though much simpler than a cat, has enough features to provide an example of deformable body. The system is shown in Figure 4.5.

It is made of three rods, two of which are joined with a hinge. Two of the rods are fixed at one end and can only rotate in the plane.[2]

All rods are of equal length R. There are three equal masses m, at their endpoints and at the hinge. The position in space of the body as a whole is fully specified by an angle, say, θ – the angle between the shorter rod and the horizontal axis. The shape of the body is given by two more angles: the one between the two rods (α) at the origin, and that at the hinge (β).

The position of the three rods in space corresponds to that of the cat as a whole – standing or, as imagined, with his paws in my hands and the head below; instead, the relative position of the rods and the angle at the hinge correspond to the various shapes the cat can assume as he flexes his furry body. The rods are a simplified model of the cat. They represent the proverbial spherical cow for the problem of the falling cat.

[2] From R.G. Littlejohn and M. Reinach, *Reviews of Modern Physics* **96** (1997) 217.

Here is a sequence of changes in the internal angles that produce – as for the falling cat – an overall net rotation of the system as a whole in space, yet leaving its final shape the same as at the beginning.

I will start with two rods along the direction x, one on top of the other (see Figure 4.6); the angles are

$$\theta = \alpha = \beta = 0\,. \tag{4.57}$$

For any change in the angles, the angular momentum $\vec{L}$ must remain the same (it is conserved). The three rods can only rotate on a plane, and their angular momentum is a vector with only one component, which is orthogonal to this plane and is given by

$$L = mR^2 \left[(4 + 2\cos\beta)\frac{d\theta}{dt} + (3 + 2\cos\beta)\frac{d\alpha}{dt} + (1 + \cos\beta)\frac{d\beta}{dt} \right]. \tag{4.58}$$

Taking $L = 0$ gives the equation

$$\frac{d\theta}{dt} = -\frac{3 + 2\cos\beta}{4 + 2\cos\beta}\frac{d\alpha}{dt} - \frac{1 + \cos\beta}{4 + 2\cos\beta}\frac{d\beta}{dt}\,. \tag{4.59}$$

The connection between the body shape (the angles α and β) and the rotation of the system as a whole (the angle θ) is given by what correspond to gauge fields, namely the coefficients of $\mathrm{d}\alpha/\mathrm{d}t$ and $\mathrm{d}\beta/\mathrm{d}t$ in Eq. (4.59):

$$A_\alpha = -\frac{3 + 2\cos\beta}{4 + 2\cos\beta} \quad \text{and} \quad A_\beta = -\frac{1 + \cos\beta}{4 + 2\cos\beta}\,. \tag{4.60}$$

Now I will go through the four steps indicated in Figure 4.6. First, I rotate θ by -75° clockwise ((b) in Figure 4.6). The two other rods must compensate (because of the conservation of the angular momentum) and rotate anticlockwise so that $\alpha = \pi/2$. Second, I rotate again θ clockwise to -102.7° ((c) in Figure 4.6) so that α goes back by 90° ((d) in Figure 4.6). Finally, I bring β back to its initial value.

The net result of this sequence of movements is the configuration (a) on the bottom of Figure 4.6. The system of the three rods has the same shape as at the start but the system as a whole is rotated by $\theta = -7.5^\circ$ with respect to the initial position in space.[4.13]

The simplified system of the three rods does the same trick as the cat who, when dropped upside down, turns on himself by flexing his body to land upright – the flexing being mimicked by the changes in internal angles of the rod system.

[4.13] Moving through points in the space of the shapes of their body to obtain a net overall rotation in space is common to cats but also to divers and astronauts. It is useful if you want to rotate without having to push against something.

4.5 Gauge Fixing

Now that we have got all excited by how great gauge fields are comes a rude awakening: there is no gauge freedom in the quantum theory. In order to quantize the theory the gauge freedom has to be restrained.

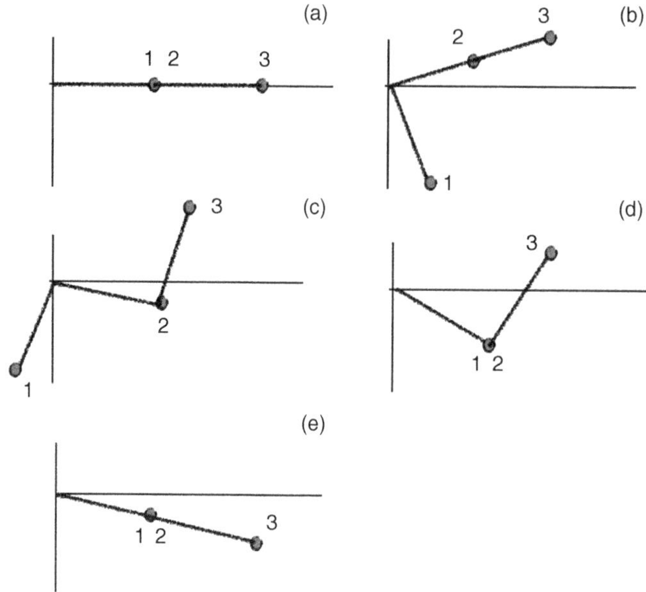

Figure 4.6 The five steps in the rotation: (a), (b), (c), (d) and (e). The rods end up back in the initial internal configuration but with an overall rotation of $\theta = -7.5°$ with respect to the surrounding space.

The need for somehow limiting the gauge freedom in the quantum theory is easy enough to understand. When I consider the equation of motion for the gauge field

$$\partial_\mu F^{\mu\nu} = \partial_\mu \partial^\mu A^\nu - \partial_\mu \partial^\nu A^\mu = J^\nu , \tag{4.61}$$

the kinetic part

$$A_\mu \left[g^{\mu\nu} \Box - \partial^\mu \partial^\nu \right] A_\nu \tag{4.62}$$

gives in momentum space the operator

$$- p^2 g_{\mu\nu} + p_\nu p_\mu . \tag{4.63}$$

I need to invert this operator in order to define the gauge boson propagator, but this is impossible because when I compute the determinant

$$\text{Det} \left(-p^2 g_{\mu\nu} + p_\nu p_\mu \right) \tag{4.64}$$

I find that the operator is singular with a zero eigenvalue.[4.14] Thus the operator in Eq. (4.63) cannot be inverted. The zero eigenvalue corresponds to an eigenvector consisting of gauge fields that are pure gauge.

This situation is common to all mechanical systems with a constraint, in this case, the conservation of the electromagnetic current.

4.14 If you are familiar with the path integral approach to quantum field theory, you will recognize here the problem of integrating over the many copies of gauge-equivalent orbits and the need to select a representative for each class of equivalent orbits by enforcing gauge fixing.

Since I need to invert the kinetic part of Eq. (4.63) in order to define the gauge boson propagator, something has to be done. If I modify the equations of motion by adding a term like

$$-\frac{1}{2\xi}\left(\partial_\mu A^\mu\right)^2 , \tag{4.65}$$

the operator in momentum space in Eq. (4.63) will be modified to

$$-p^2 g_{\mu\nu} + \left(1 - \frac{1}{\xi}\right) p_\nu p_\mu , \tag{4.66}$$

whose determinant has no zero eigenvalues.

So, I have "fixed" the gauge by adding the term in Eq. (4.65) to the Lagrangian of the gauge boson, which becomes

$$L = -\frac{1}{4}F^2_{\mu\nu} \underbrace{-\frac{1}{2\xi}\left(\partial_\mu A^\mu\right)^2}_{\text{gauge fixing}} + J_\mu A^\mu . \tag{4.67}$$

Because there are many gauge choices and they are often confusing, it is useful to include in Table 4.1 the most common gauges.

The most convenient gauge to work with depends on which particular computation you are doing.

The covariant gauge gives the propagator

$$D_{\mu\nu}(p) = \frac{-i}{p^2 + i\epsilon}\left(g_{\mu\nu} - (1-\xi)\frac{p_\mu p_\nu}{p^2}\right) . \tag{4.68}$$

Within this family, the choice $\xi = 1$ is often used to simplify computations. On the other hand, it is sometime advantageous to retain the explicit dependence of the propagator on ξ if you want to check your computation because in the end the dependence must drop out to give a gauge-invariant result.

The axial gauge is physical in the sense that only the two transverse polarizations are present. The propagator is given by

$$D_{\mu\nu}(p) = \frac{i}{p^2 + i\epsilon}\left(g_{\mu\nu} - \frac{p_\mu n_\nu + n_\mu p_\nu}{p \cdot n} + n^2 \frac{p_\mu p_\nu}{(p\cdot n)^2}\right) , \tag{4.69}$$

where n is an arbitrary vector n that can be taken to be light-like, that is, $n^2 = 0$. The propagator satisfies

$$D_{\mu\nu} p^\mu = D_{\mu\nu} n^\mu = 0 , \tag{4.70}$$

two conditions that kill the time-like and longitudinal polarizations.

Table 4.1 Some of the most common gauge-fixing conditions. These conditions are written for the Abelian case but are easily extended to the non-Abelian case. Those that may come up in actual computations are shown in bold type. The Lorenz gauge condition is named after the Danish physicist Ludvig Lorenz and not Hendrick Lorentz, who was also Danish and after whom the Lorentz transformations are named.

Gauge	Gauge-fixing condition	Use
Covariant: R_ξ-gauge	arbitrary ξ	Useful if you want to check gauge invariance or study the renormalizability
Covariant: Feynman	$\xi = 1$	Most useful to simplify computations
Covariant: Lorenz	$\partial_\mu A^\mu = 0$	Implemented from the start in the classical theory
Covariant: Landau	$\xi \to 0$	Classically like Lorenz's but implemented after the theory has been quantized
Covariant: Unitary	$\xi \to \infty$	Useful in spontaneously broken theories, it eliminates Goldstone modes
Non-covariant: Axial	$n \cdot A = 0$ $\xi \to \infty$	Utilized in QCD to keep only physical polarizations
Non-covariant: Coulomb	$\nabla \cdot A = 0$	Used in old quantization procedure (with transverse polarizations plus an instantaneous Coulomb potential)

4.6 *Problem Session*: The Nut and the Wrench

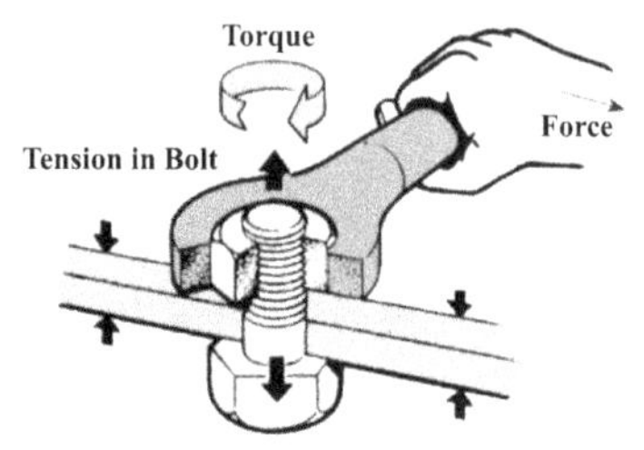

Figure 4.7 A wrench fastening a nut.

This is an exercise in classical mechanics that shows how the presence of a constraint prevents us from finding the solution and how gauge fixing can help.

Consider[3] fastening a nut with a wrench with a handle of length R by applying a force $\vec{F}$ (see Figure 4.7). The force $\vec{F}$ gives a torque $\vec{\tau}$ according to the equation

[3] This problem session follows notes by Hiren H. Patel.

$$\vec{R} \times \vec{F} = \vec{\tau}\,. \tag{4.71}$$

Whereas it is simple to compute the torque for a given $\vec{R}$ and $\vec{F}$, to determine, for a given $\vec{R}$, the value of the force $\vec{F}$ necessary to obtain a desired value of the torque $\vec{\tau}$ is perhaps less obvious. What is the magnitude and direction of the required force $\vec{F}$?

If I rewrite Eq. (4.71) in components as

$$\epsilon_{ijk} R_j F_k = \tau_i\,, \tag{4.72}$$

then this can be written as a product of matrices,

$$\begin{pmatrix} 0 & -R_z & R_y \\ R_z & 0 & -R_x \\ -R_y & R_x & 0 \end{pmatrix} \begin{pmatrix} F_x \\ F_y \\ F_z \end{pmatrix} = \begin{pmatrix} \tau_x \\ \tau_y \\ \tau_z \end{pmatrix}\,. \tag{4.73}$$

All I need to do to find $\vec{F}$ for given $\vec{R}$ and $\vec{\tau}$ is to invert the matrix in Eq. (4.73).[4.15] But there is a problem: the matrix in Eq. (4.73) is singular. It has a zero eigenvalue associated with the eigenvector

$$F_i = \alpha R_i\,. \tag{4.74}$$

This fact implies that:

- if a force $\vec{F}$ solves the problem, then the force $\vec{F}' = \vec{F} - \alpha\vec{R}$ also does;
- there is a constraint in the system, namely[4.16]

$$\vec{R} \cdot \vec{\tau} = 0\,, \tag{4.75}$$

because $R_i \epsilon_{ijk} R_j F_k = R_i \tau_i = 0$ because of the antisymmetry of the Levi–Civita tensor. This constraint tells me that the applied force must be orthogonal to the desired torque.

To solve this problem I add a gauge-fixing term to the matrix. For instance, I might only look for solutions for which $F_z = 0$. This condition can be recast as[4.17]

$$\vec{n} \cdot \vec{F} = 0\,, \tag{4.76}$$

with $\vec{n} = (0, 0, 1)$, so that the gauge-fixed equation is now given by

$$(\epsilon_{ijk} R_j + n_i n_k) F_k = \tau_i\,, \tag{4.77}$$

which replaces Eq. (4.72). Equation (4.77) is given in matrix form as

$$\begin{pmatrix} 0 & -R_z & R_y \\ R_z & 0 & -R_x \\ -R_y & R_x & n^2 \end{pmatrix} = \begin{pmatrix} F_x \\ F_y \\ F_z \end{pmatrix} = \begin{pmatrix} \tau_x \\ \tau_y \\ \tau_z \end{pmatrix}\,, \tag{4.78}$$

4.15 The analogy with the gauge fields case is made evident by writing Maxwell's equations

$$\left(\partial^2 \delta^\mu_\nu - \partial_\mu \partial_\nu\right) A^\nu = J^\mu$$

as

$$\begin{pmatrix} -\nabla^2 & -\partial_t \nabla_j \\ \partial_t \nabla_i & -\delta_{ij}\Box + \nabla_i \nabla_j \end{pmatrix} \begin{pmatrix} \phi \\ A_j \end{pmatrix} = \begin{pmatrix} \rho \\ J_i \end{pmatrix}.$$

Given $\vec{J}$ and ρ, solve with for ϕ and $\vec{A}$. There is a zero associated with the eigenvectors

$$\begin{pmatrix} \partial_t \alpha \\ -\vec{\nabla}\alpha \end{pmatrix}.$$

4.16 In the gauge fields case, the equation is

$$\partial_\mu \left(\partial^2 \delta^\mu_\nu - \partial_\mu \partial_\nu\right) A^\nu = \partial_\mu J^\mu = 0$$

and the constraint is the conservation of the current.

4.17 The analogy with the gauge fields case is here with the gauge fixing (Lorenz gauge)

$$\partial^\mu \partial_\nu.$$

The operator in Maxwell's equations is now

$$\begin{pmatrix} -\nabla^2 & -\partial_t \nabla_j \\ \partial_t \nabla_i & -\delta_{ij}\Box + \nabla_i \nabla_j \end{pmatrix} + \begin{pmatrix} \partial_t^2 & \partial_t \nabla_j \\ -\partial_t \nabla_j & -\nabla_i \nabla_j \end{pmatrix},$$

which gives

$$\begin{pmatrix} \Box & 0 \\ 0 & -\delta_{ij}\Box \end{pmatrix} \begin{pmatrix} \phi \\ A_j \end{pmatrix} = \begin{pmatrix} \rho \\ J_i \end{pmatrix};$$

this is now invertible.

which is now invertible. It follows that

$$\begin{pmatrix} F_x \\ F_y \\ F_z \end{pmatrix} = \frac{1}{R_z^2 n^2} \begin{pmatrix} R_x^2 & R_xR_y + n^2R_z & R_xR_z \\ R_xR_y + n^2R_z & R_y^2 & R_yR_z \\ R_xR_z & R_yR_z & R_z^2 \end{pmatrix} \begin{pmatrix} \tau_x \\ \tau_y \\ \tau_z \end{pmatrix}, \tag{4.79}$$

which gives the solution to the problem: the desired force is

$$\begin{pmatrix} F_x \\ F_y \\ F_z \end{pmatrix} = \frac{1}{R_z^2 n^2} \begin{pmatrix} n^2R_z\tau_z \\ -n^2R_z\tau_x \\ 0 \end{pmatrix} = \begin{pmatrix} \tau_y/R_z \\ -\tau_x/R_z \\ 0 \end{pmatrix}. \tag{4.80}$$

4.7 Ghosts

I want to look again at the process in which two fermions go into two gauge bosons and take the derivative with respect to the momenta of one of the gauge bosons. If the gauge bosons are on-shell, the amplitude vanishes in agreement with the Ward identity in Eq. (4.26). What happens if the gauge bosons do not obey the equations of motion and are off-shell? Let us check.

There are three diagrams, shown in Figure 4.8.

The amplitude for this process is made of three terms:

$$T^{AB}_{\mu\nu} = \underbrace{-ig^2\bar{v}(p_2)\frac{\tau^B}{2}\gamma_\nu \frac{1}{(\not{p}_1 - \not{k}_1) - m}\frac{\tau^A}{2}\gamma_\mu u(p_1)}_{(a)} \tag{4.81}$$

$$\underbrace{- ig^2\bar{v}(p_2)\frac{\tau^A}{2}\gamma_\mu \frac{1}{(\not{k}_1 - \not{p}_2) - m}\frac{\tau^B}{2}\gamma_\nu u(p_1)}_{(b)} \tag{4.82}$$

$$- ig^2\epsilon^{ABC}\Big[(k_1 - k_2)_\lambda g_{\mu\nu} + (k_1 + 2k_2)_\mu g_{\nu\lambda}$$
$$\underbrace{- (2k_1 + k_2)_\nu g_{\mu\lambda}\Big]\frac{1}{(k_1 + k_2)^2}\bar{v}(p_2)\frac{\tau^C}{2}\gamma_\lambda u(p_1)}_{(c)}. \tag{4.83}$$

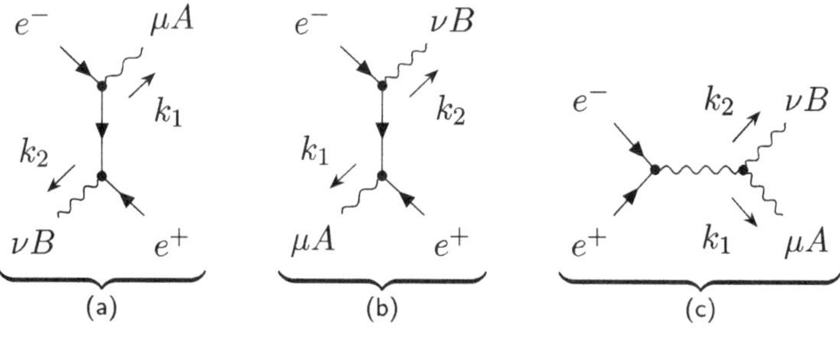

Figure 4.8 The three diagrams for two fermions going into two gauge bosons. Diagram (b) just corresponds to the cross-over of the final bosons and is required by gauge invariance. Diagram (c) is there because of the nonlinear coupling among gauge bosons.

The gauge-dependent part of the gauge boson in (c), proportional to $\xi(k_1+k_2)_\lambda(k_1+k_2)_\sigma$, ends up on the Dirac matrix of the fermion vertex and is therefore proportional to $\not{k}_1+\not{k}_2=\not{p}_1+\not{p}_2=0$, which vanishes because of the Dirac equation, so that the amplitude does not depend on the gauge choice.

Now looking at the Ward identity

$$k_1^\mu T_{\mu\nu}^{AB}, \tag{4.84}$$

it receives contributions from the three diagrams, namely

$$\begin{aligned}
k_1^\mu\,[(\mathrm{a})+(\mathrm{b})] &= -ig^2\bar{v}(p_2)\left[\frac{\tau^B}{2}\frac{\tau^A}{2}\gamma_\nu\frac{(\not{p}_1-\not{k}_1)+m}{(p_1-k_1)^2-m^2}\not{k}_1\right.\\
&\left.\quad+\frac{\tau^A}{2}\frac{\tau^B}{2}\not{k}_1\frac{(\not{k}_1-\not{p}_2)+m}{(k_1-p_2)^2-m^2}\gamma_\nu\right]u(p_1)\\
&= -ig^2\bar{v}(p_2)\left[\frac{\tau^A}{2},\frac{\tau^B}{2}\right]\gamma_\nu u(p_1)\\
&= g^2\epsilon^{ABC}\bar{v}(p_2)\frac{\tau^C}{2}\gamma_\nu u(p_1)
\end{aligned}\tag{4.85}$$

and

$$\begin{aligned}
k_1^\mu\,[(\mathrm{c})] =& -g^2\epsilon^{ABC}\Big[2k_1\cdot k_2 g_{\nu\lambda}+(k_1-2k_2)_\lambda k_{1\nu}\\
&-(2k_1+k_2)_\nu k_{1\lambda}\Big]\frac{1}{(k_1+k_2)^2)}\bar{v}(p_2)\frac{\tau^C}{2}\gamma_\lambda u(p_1)\\
=& \underbrace{-g^2\epsilon^{ABC}\bar{v}(p_2)\frac{\tau^C}{2}\gamma_\nu u(p_1)}_{=-\text{Eq. (4.85)}} && (4.86)\\
&\underbrace{-g^2\epsilon^{ABC}\frac{k_{1\nu}}{(k_1+k_2)^2}\bar{v}(p_2)\frac{\tau^C}{2}(\not{k}_1-\not{p}_2)u(p_1)}_{=0} && (4.87)\\
&-g^2\epsilon^{ABC}\frac{k_{2\nu}}{(k_1+k_2)^2}\bar{v}(p_2)\frac{\tau^C}{2}\not{k}_1 u(p_1)\,. && (4.88)
\end{aligned}$$

The term in Eq. (4.87) vanishes because of the Dirac equation. Next, the term in Eq. (4.85) exactly cancels that in Eq. (4.86). That is, the diagrams (a) and (b) are cancelled by these contributions to diagram (c). Notice that this is possible only because the triple gauge coupling has strength that is exactly g^2.

If the external boson is on-shell then I have to multiply Eq. (4.85) by the corresponding wave function $\varepsilon_2(k_2)$ and I get zero also for the term in Eq. (4.88), because $\varepsilon_2\cdot k_2=0$.

If the external boson is not on-shell, there is a non-vanishing term from Eq. (4.88). This is a problem because the Ward identity would be violated for off-shell external bosons.

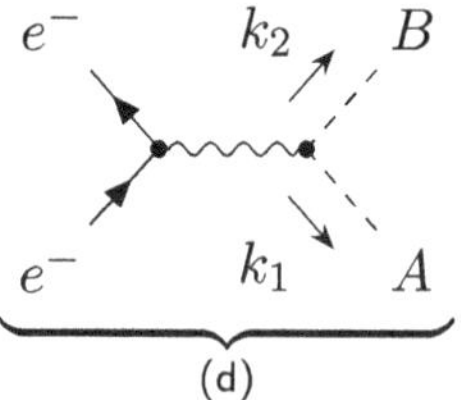

Figure 4.9 The extra diagram.

I can reason in a manner similar to the discussion of unitarity in Chapter 2 and look for an extra particle to make things right. What kind of particle must be added in order to make this unwanted contribution go away? I want to generate a diagram like that in Figure 4.9 by means of a scalar particle with a coupling to the gauge boson exactly like $-g\epsilon^{ABC}k_{1\mu}$. The amplitude would then have a new contribution:

$$T^{AB} = \underbrace{-ig^2\epsilon^{ABC}\frac{1}{(k_1+k_2)^2)}\bar{v}(p_2)\frac{\tau^C}{2}\not{k}_1 u(p_1)}_{(d)}\,, \tag{4.89}$$

and the Ward identity would have an extra term as well, given by

$$ik_2^\nu\,[(\mathrm{d})] = g^2\epsilon^{ABC}\frac{k_{2\nu}}{(k_1+k_2)^2}\bar{v}(p_2)\frac{\tau^C}{2}\not{k}_1 u(p_1)\,, \tag{4.90}$$

which would cancel the term in Eq. (4.88) – as desired.

The new scalar fields exactly cancel the contributions of the longitudinal polarizations of the gauge bosons, thus enforcing that only the physical polarizations contribute in the off-shell propagation of the gauge bosons. The presence of the new fields makes it possible to extend the Ward identities for quantum gauge fields that do not necessarily obey the classical equations of motion.

Inasmuch as the problem only arises for gauge fields off-shell, the new fields only enter in amplitudes containing closed loops. In this case, it is not enough to introduce the ghost fields with the coupling given in Eq. (4.89) because there are (in the case of a loop consisting of gauge boson propagators) two of them and their sign, whatever it may be, comes as a square and cannot cancel the propagation of the unphysical polarizations. The solution is to take the scalar states to be like fermions so that there is an overall (-1) multiplying the ghost diagram with respect to the original one.[4.18] The sum of the two diagrams is shown in Figure 4.10. Because the statistics are wrong, the new scalar is a **ghost field**. It does not exist as an asymptotic (physical) state.

Ghost fields only appear in loops of gauge bosons. I will not have to worry much about them in what follows but they are there, they are part of the quantum theory.

4.18 In the axial gauges the covariant derivative is

$$n_\mu\partial^\mu + gf^{ABC}n_\mu A^\mu_C \tag{4.91}$$

and because of the gauge the second term vanishes leaving the ghosts decoupled. Ghost fields are also decoupled in QED, where at the outset there are no nonlinear interactions of the gauge fields.

Figure 4.10 The gauge and ghost loops, the latter with sign (-1).

4.8 Quantum Gauge Theory*

Let me raise a slight quibble. The quantum theory of gauge fields requires gauge fixing. In the quantum theory there is no gauge freedom. In more than one way, the long previous discussion of gauge invariance is pointless and only important in the historical development of the subject. From the point of view of the definition of the Standard Model, gauge freedom only serves the purpose of introducing the concept of the covariant derivative. But it is confusing. As in the three-card trick, gauge freedom is first introduced, then explicitly limited by the gauge fixing.

What is interesting is that there is a residual symmetry. Gauge freedom is resurrected as **BRST symmetry**,[4.19] which is the global symmetry left over after fixing the gauge. When the subject is discussed in more advanced courses, only the BRST symmetry is employed in deriving the Ward identities. And perhaps that is indeed the best way of introducing and discussing gauge theories.

4.19 The weird acronym stands for C. Becchi, A. Rouet, R. Stora and I. V. Tyutin, who around the mid-1970s introduced this symmetry in the quantization of gauge fields.

While the BRST symmetry is itself a subject well worth a dedicated discussion, for me here it is sufficient to notice how the ghost field takes up the local part of the gauge transformations

$$\delta A^A_\mu(x) = \partial_\mu \alpha^A(x) - g\epsilon^{ABC} A^C_\mu(x)\alpha^B(x) \tag{4.92}$$

$$\delta\psi = -iT^A\alpha^A\psi, \tag{4.93}$$

which are left over after fixing the gauge, by writing the gauge function as

$$\alpha^A(x) = -g\omega c^A(x), \tag{4.94}$$

which turns it into a global symmetry, controlled by the space-time independent anticommuting variable ω.[4.20] The fields c^A are the ghost fields that have to be introduced to preserve the unitarity of the theory and are here made explicitly part of the Lagrangian.

4.20 The parameter ω is anticommuting because of the wrong statistics of the ghost fields.

The transformations for the gauge, fermion and ghost fields are

$$\begin{aligned} \delta A^A_\mu(x) &= \omega D_\mu c^A(x) \\ \delta\psi &= lg\omega T^A c^A(x)\psi \\ \delta c^A(x) &= -\frac{i}{\xi}\omega\partial^\mu A^A_\mu \\ \delta\bar{c}^A &= -\frac{g}{2}\omega\epsilon^{ABC}c^B(x)c^C(x). \end{aligned} \tag{4.95}$$

As it turns out, all is well. The residual symmetry of the gauge-fixed Lagrangian is enough to guarantee the Ward identities. In fact, the Ward identities are even more powerful than those that were introduced for the classical case, and they hold even for off-shell fields.

Finally, I arrive at the Lagrangian density for the gauge fields in the Standard Model. It is written in terms of the field strengths as (in the covariant gauge)

$$\mathcal{L}_{\text{gauge}} = \underbrace{-\frac{1}{4}\sum_{A=1}^{3} F^{A}_{\mu\nu} F^{A\mu\nu}}_{SU_L(2)\ \text{isospin}} - \underbrace{\frac{1}{4} B_{\mu\nu} B^{\mu\nu}}_{U_Y(1)\ \text{hypercharge}} \tag{4.96}$$

$$- \underbrace{\frac{1}{2\xi_2}(\partial^{\mu} W^{A}_{\mu})^2}_{\text{gauge fixing of } SU_L(2)} - \underbrace{\frac{1}{2\xi_1}(\partial^{\mu} B_{\mu})^2}_{\text{gauge fixing of } U_Y(1)} \tag{4.97}$$

$$- \underbrace{\sum_{a,b=1}^{3} \bar{c}^{a} \partial_{\mu} D^{\mu ab} c^{b}}_{\text{ghosts of } SU_L(2)} - \underbrace{\bar{c}\, \partial_{\mu} D^{\mu} c}_{\text{ghost of } U_Y(1)} \ . \tag{4.98}$$

The classical terms in Eq. (4.96) are augmented by the gauge fixing in Eq. (4.97) and the ghosts in Eq. (4.98).

For an infinitesimal BRST transformation the gauge-fixing term, for example, for the $SU(2)$ group

$$\delta_{\text{BRST}} L_{gf} = -\frac{1}{2\xi}(\partial^{\mu} W^{A}_{\mu})^2 , \tag{4.99}$$

and the corresponding ghost term

$$\delta_{\text{BRST}} L_{gh} = i \bar{c}^{A} \partial_{\mu} D^{\mu}_{AB} c^{B} \tag{4.100}$$

transform so as to cancel each other:

$$\delta_{\text{BRST}} (L_{gf} + L_{gh}) = 0 . \tag{4.101}$$

The quantum theory is a bit more complicated than the classical theory, which would have given a Lagrangian with only the first line, Eq. (4.96). There are also the gauge fixing and the ghost fields to take into account. They bring Feynman rules of their own and must be included in the computations wherever they are needed.

4.9 Exercises

1. **Total divergence.** Show that $\mathrm{Tr}\, F_{\mu\nu} \tilde{F}^{\mu\nu}$ is a total divergence. [Hint: It is the total divergence of the vector $K^{\rho} = \epsilon^{\rho\sigma\mu\nu} \mathrm{Tr}\, [F_{\sigma\mu} A_{\nu} + \frac{2}{3} A_{\sigma} A_{\mu} A_{\nu}]$.]

2. **Field strength.** Compute the commutator of the covariant derivatives $[D_\mu, D_\nu]$ for the Abelian case.
[Hint: It is $-ieF_{\mu\nu}$.]
3. **Gauge propagator.** Verify that the gauge propagator in the R_ξ gauge is orthogonal to the operator in Eq. (4.66).
4. **Feynman rule.** Find the Feynman rule for the interaction of the ghost fields and the gauge bosons from the Lagrangian term Eq. (4.98).
5. **Fermion loops.** Why does a fermion loop have a (-1) overall factor with respect to a loop made by a boson?
[Hint: Trace back this factor in the expansion of the exponential factor of the action, in the definition of the perturbative series in quantum field theory.]

5 Hidden Gauge Freedom

Contents

It is time to face the elephant in the room. Beautiful though they are, the gauge bosons of the weak interactions are massless and cannot be the mediators of the weak interactions. They do not reproduce Fermi's interaction at low energy. They make the weak interactions long range while they are most definitely not.

What if I were to add by hand a mass term to the Lagrangian of the gauge bosons? The term turning the W-bosons from massless to massive then looks like

$$\frac{m_W^2}{2} W_\mu^A W^{A\mu} , \tag{5.1}$$

and the resulting Lagrangian goes under Proca's name.

The problem is that this term **explicitly** breaks the gauge freedom since the Lagrangian will not remain the same if I replace the field W_μ^A with

$$W_\mu^A + \partial_\mu \alpha^A . \tag{5.2}$$

The term in Eq. (5.1) is indeed a gauge fixing, as are those I discuss in the previous chapter, but not a good one. It is too restrictive and does not preserve the BRST symmetry and the renormalizability of the theory.

As it turns out, there is a more subtle way to give mass to the gauge bosons than by just adding a symmetry-breaking term to the Lagrangian. The crucial insight comes from solid-state physics. There are systems described by Lagrangians that are symmetric under a given symmetry but having a ground state not sharing that symmetry, and, because of this feature, the gauge bosons of these systems behave as if they were massive particles.[5.1]

[5.1] Several classical examples are often mentioned: people sitting at a round dinner table with napkins uniformly distributed between the plates or a pencil standing on its tip. In these cases, it is the external perturbation – the first guest to choose one napkin, the pencil falling down – that moves the system to a non-symmetric state.

5.1 *Problem Session*: Spontaneous Symmetry Breaking

A phase transition occurs when a physical system undergoes a change in its physical properties.

Examples are the transitions between solid, liquid and gas as well as many others, such as the magnetization of a ferromagnet or the onset of superconductivity in a metal. Usually the various phases can be distinguished by the different values of an **order parameter**. Their classification is based on the order of the lowest discontinuous derivative of the free energy of the system with respect to some of the thermodynamical variables. In this scheme, a **first-order** phase transition has a discontinuity in the first derivative of the free energy. Familiar examples are the melting of ice or the boiling of water. In this kind of phase transition, heat is released (or absorbed) and the order parameter jumps from one value to another. In a **second-order** phase transition the first derivative of the free energy is continuous and the order parameter changes continuously during the transition. Ferromagnetic and superconducting transitions are examples of this second kind. Typically the phase at a higher temperature (for example, the gas) is more symmetric than that at the lower temperature (the frozen solid).

The study of phase transitions is an important chapter of thermodynamics and solid-state physics. It is also important in particle physics because a model used in the study of phase transitions turned out to play a role, indeed an essential one, in the Standard Model.

The relevant model is due to the physicist Lev Landau. It is based on a macroscopic description of the free energy of the system in terms of the dependence of the free energy on the order parameter.

In the Landau model, the free energy of the system is given by

$$F = F_0(T) + \frac{1}{2}\alpha_2(T)|\eta|^2 + \frac{1}{4}\alpha_4(T)|\eta|^4 + \cdots, \tag{5.3}$$

where η is the order parameter. This free energy is shown in Figure 5.1.

The free energy is symmetric under the change of sign of the order parameter

$$\eta \to -\eta\,. \tag{5.4}$$

Turning this feature around, the form of the free energy is such that this symmetry is preserved or, in other words, there are no odd powers of the order parameter in the free energy.

For

$$\alpha_2(T) = \alpha_2^0 \, \frac{T - T_c}{T_c} \tag{5.5}$$

and $\alpha_4(T) = \alpha_4^0 > 0$, there is a second-order phase transition, in which the order parameter changes in a continuous manner.

The potential of the free energy is shown in Figure 5.1.

The order parameter η in the broken-symmetry phase is found by minimizing the free energy; it is

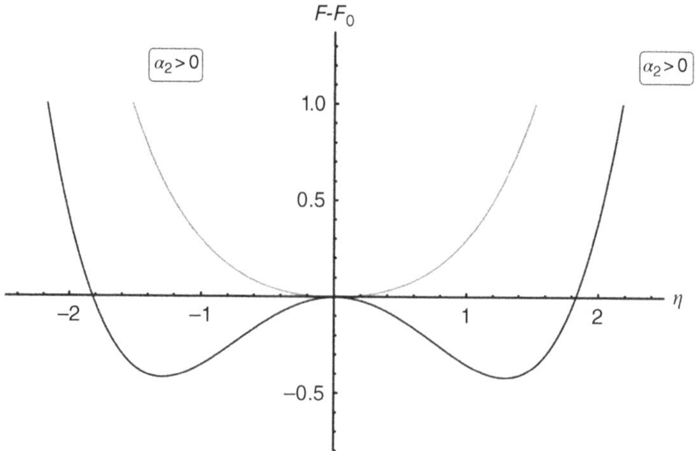

Figure 5.1 Free energy in the Landau model: the shape of the potential changes because the sign of the coefficient α_2 changes with temperature.

$|\eta|^2 = \frac{\alpha_2^0}{\alpha_4^0}(T_c - T)$, which can be written as

$$|\eta|^2 = -\frac{\tilde{\alpha}_2^0}{\alpha_4^0} \quad \text{for} \quad T < T_c, \tag{5.6}$$

for which the corresponding free energy is

$$F = F_0(T) - \frac{(\tilde{\alpha}_2^0)^2}{4\alpha_4^0}. \tag{5.7}$$

The system relaxes to the state with lower energy. In the ground state the invariance under a change of sign of η is lost. The symmetry is said to be **spontaneously broken**. It is broken spontaneously because there are no terms in the free energy that break the symmetry. It is the phase transition that puts the system into a ground state that is not symmetric.

Many physical systems can be described in terms of Landau's theory. A ferromagnet, for which the order parameter η is the magnetization of the spins, is often mentioned.

The Landau model can be implemented for a superconductor by taking the order parameter to be the wave function of the Cooper pairs[5.2] $\eta = |\Psi|$ to obtain the Ginzburg–Landau model of superconductivity. The free energy of the model is written as

$$F - F_0 = \frac{1}{2m}\left|\left(\nabla - 2\,ie\vec{A}\right)\Psi\right|^2 + \frac{1}{2}\alpha_2(T)|\Psi|^2 + \frac{1}{4}\alpha_4(T)|\Psi|^4 + \frac{1}{8\pi}(\nabla \times \vec{A})^2, \tag{5.8}$$

which is just as in Eq. (5.3) but for the addition of the electromagnetic potential[5.3] of the magnetic field $\vec{B}$ (and neglecting higher-order terms).

5.2 The Cooper pair is the loosely bound state of two electrons at the Fermi surface of the superconductor – it has charge $2e$.

5.3 $\vec{B} = \nabla \times \vec{A}$.

The free energy and the equations of motion are invariant under the gauge transformation

$$\begin{cases} \vec{A} \to \vec{A} + \nabla\phi \\ \Psi \to e^{ie\phi}\Psi \,. \end{cases} \tag{5.9}$$

As before, the second-order phase transition is accomplished by taking

$$\alpha_2(T) = \tilde{\alpha}_2^0 < 0 \quad \text{and} \quad \alpha_4(T) = \alpha_4^0 > 0 \,. \tag{5.10}$$

The order parameter in the broken-symmetry phase is given by the Cooper pair wave function being

$$\Psi = \sqrt{\frac{-\tilde{\alpha}_2^0}{\alpha_4^0}}\,\rho\, e^{i\phi}\,, \tag{5.11}$$

where ρ^2 is the density of Cooper pairs, which can be taken as constant inside the superconductor. The phase ϕ connects wave functions pointing in different directions in the group space $U(1)$.

What happens to the magnetic field (the gauge field in the problem) as the system relaxes to the lower energy?

The Lagrangian equation for the electromagnetic potential $\vec{A}$ is obtained from Eq. (5.8) to be

$$\frac{1}{4\pi}\left[\nabla \times \nabla \times \vec{A}\right] = \frac{2e}{m}\vec{j}, \tag{5.12}$$

where the current is

$$\vec{j} = \frac{1}{2i}\left(\Psi^\dagger \nabla\Psi - \Psi\nabla\Psi^\dagger\right) - 2e\vec{A}\Psi^\dagger\Psi\,. \tag{5.13}$$

The current can be rearranged by a gauge transformation

$$\vec{A}' = \vec{A} - \frac{1}{2e}\nabla\phi \tag{5.14}$$

to become

$$\vec{j} = \frac{-\tilde{\alpha}_2^0}{\alpha_4^0}\rho^2\left[2e\vec{A} + \nabla\phi\right] = \frac{-\tilde{\alpha}_2^0}{\alpha_4^0}\rho^2\left[2e\vec{A}'\right], \tag{5.15}$$

in which the phase ϕ of the Cooper pairs has been removed.

If I take the curl of both sides of Eq. (5.12) I find that

$$\nabla \times \nabla \times \vec{B} = \frac{8\pi e}{m}\nabla \times \vec{j} = \frac{16\pi e^2}{m}|\Psi|^2\vec{B}, \tag{5.16}$$

where the first term of the current in Eq. (5.13) has dropped out because the wave function in the broken-symmetry phase does not depend on the coordinates. Equation (5.16) can be written as[5.4]

5.4 Use $\nabla \times \nabla \times \vec{B} = \nabla(\nabla \cdot \vec{B}) - \nabla^2\vec{B}$.

$$\nabla^2\vec{B}(\vec{r}) = \frac{1}{\lambda^2}\vec{B}(\vec{r}) \tag{5.17}$$

with

$$\lambda = \sqrt{\frac{-m\tilde{\alpha}_2^0 \rho^2}{16\pi e^2 \alpha_4^0}} . \tag{5.18}$$

The solution of Eq. (5.17) is a magnetic field given by

$$B(\vec{r}) = B_0 e^{-|\vec{r}|/\lambda} , \tag{5.19}$$

which shows that the field has a finite penetration length, given by λ, into the superconductor, as if the photon had a mass.

This model of superconductivity teaches us that if the ground state of the theory is not symmetric under the gauge transformation, the gauge boson becomes massive even though the gauge freedom of the Lagrangian still holds. Whereas the free energy enjoys the symmetry in Eq. (5.9), the solution of the equation of motion of the magnetic field is not symmetric.

5.2 The Breaking of a Global Symmetry

In a relativistically invariant theory, the spontaneous symmetry breaking of a global symmetry is realized by a model for which the order parameter tracking the phase transition is the expectation value of a complex scalar field.

The Lagrangian for such a field ϕ is

$$L_\phi = (\partial_\mu \phi^*)(\partial^\mu \phi) - V(\phi) , \tag{5.20}$$

where the potential is given by

$$V(\phi) = \mu^2(\phi^*\phi) + \lambda(\phi^*\phi)^2 . \tag{5.21}$$

In Eq. (5.21) I am neglecting higher-order terms, which are not essential for the present discussion.

The Lagrangian in Eq. (5.20) is invariant under the field phase transformation

$$\phi \to \phi' = e^{i\alpha}\phi , \tag{5.22}$$

as can be checked by substituting the transformed field or, more simply, by noticing that only the absolute value of the field enters and therefore the phase is arbitrary.

To visualize the potential in Eq. (5.21) I can write out the real components of the field ϕ, as in

$$\phi = \frac{1}{\sqrt{2}} (\phi_1 + i\phi_2) , \tag{5.23}$$

to obtain

$$V(\phi_1, \phi_2) = \frac{\mu^2}{2}(\phi_1^2 + \phi_2^2) + \frac{\lambda}{4}(\phi_1^2 + \phi_2^2)^2 . \tag{5.24}$$

For $\mu^2 > 0$ the potential is a well, the lowest point of which has both the field components ϕ_1 and ϕ_2 vanishing. This point is the ground state of the model and is invariant under the phase transformation in Eq. (5.22). The ground state is called the **vacuum state** of the model.

The particle content of the theory is obtained by studying small oscillations about the lowest point of the well, that is, about the vacuum. I can see by inspection that there are two particles and their masses are the same and equal to μ.

No surprises so far. Just your garden-variety quantum field theory of a complex scalar field, but here is the twist. Consider now the case for which $\mu^2 < 0$. The shape of the potential is different, as shown in Figure 5.2. Instead of a unique minimum at $\phi_1 = \phi_2 = 0$, I have (taking the first derivative) all the values

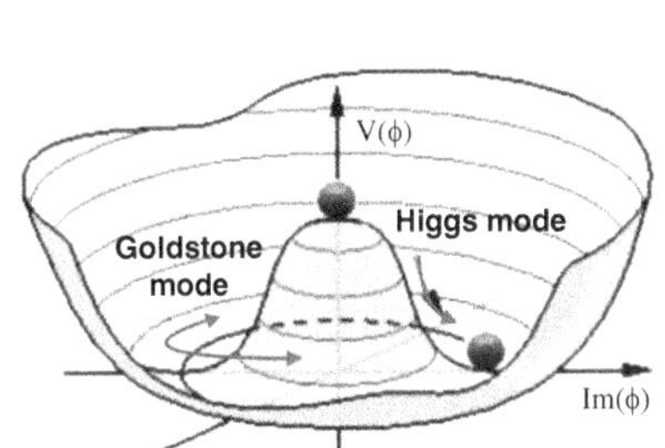

Figure 5.2 The "Mexican hat" potential when $\mu^2 < 0$ in Eq. (5.24).

$$\sqrt{\phi_1^2 + \phi_2^2} = \sqrt{\frac{-\mu^2}{\lambda}} \equiv v \tag{5.25}$$

as the lowest points of the potential. The minimum is degenerate. The transformation in Eq. (5.22) moves the system from one of these values to another.[5.5]

[5.5] The symmetry-breaking potential is at zero temperature. We must imagine, as the temperature is raised, the potential going back into one with a minimum around zero field. There is a transition (presumably around $T = 10^{15}\,K$, that is, after 10^{-12} s in the history of the Universe) to an earlier phase in which the symmetry is restored.

To find the particle content of the theory I need to study the small oscillations again, this time about a particular minimum among those that satisfy Eq. (5.25). Choose the minimum pointing in the ϕ_1-direction, write

$$\phi_1 = v + \eta \quad \text{and} \quad \phi_2 = \chi \tag{5.26}$$

and replace Eq. (5.26) in the kinetic part of the Lagrangian in Eq. (5.20) to obtain

$$\frac{1}{2}\partial_\mu(\eta + v - i\chi)\partial^\mu(\eta + v + i\chi) = \frac{1}{2}(\partial_\mu\eta)^2 + \frac{1}{2}(\partial_\mu\chi)^2 \tag{5.27}$$

since $\partial_\mu v = 0$ by definition. The potential becomes

$$V = \frac{\mu^2}{2}\left[(v+\eta)^2 + \chi^2\right] + \frac{1}{4}\lambda\left[(v+\eta)^2 + \chi^2\right]^2, \tag{5.28}$$

which (after multiplying out the square and using $\mu^2 = -\lambda v^2$) is

$$V = -\frac{1}{4}\lambda v^4 + \lambda v^2\eta^2 + \lambda v\eta(\eta^2 + \chi^2) + \frac{1}{4}\lambda(\eta^2 + \chi^2)^2. \tag{5.29}$$

By choosing the ground state to be the one in Eq. (5.26), I can pick one direction among those spanning the circle at the bottom of the Mexican hat potential. The symmetry in Eq. (5.22) is spontaneously broken.[5.6]

[5.6] As in the example mentioned in the beginning of this chapter: the first guest to choose a napkin on a round dinner table, with napkins uniformly distributed between the plates, brings the table to a non-symmetric configuration.

The particle content is obtained by considering only the terms quadratic in the fields η and χ: these are

$$\frac{1}{2}(\partial_\mu\eta)^2 - \lambda v^2\eta^2, \tag{5.30}$$

which describe a massive scalar particle with mass given by

$$m_\eta^2 = 2\lambda v^2\,, \tag{5.31}$$

and

$$\frac{1}{2}(\partial_\mu \chi)^2\,, \tag{5.32}$$

which is the kinetic term of a scalar particle with vanishing mass,

$$m_\chi = 0\,. \tag{5.33}$$

The massive mode η climbs the curvature of the well (and its mass is given by the second derivative of the curvature) while the massless mode χ goes around the minimum and therefore along a direction that is flat and gives a vanishing second derivative, and thus zero mass for the particle.

This result is actually rather general. It goes under the name of the **Goldstone theorem**. For any generator of the original group of the symmetry that is broken, there is a massless mode. These massless modes connect the various equivalent vacua, as the χ state in the example above runs over the circle made by the lowest points of the "Mexican hat" potential.

5.3 The Gauge Bosons Become Massive

What happens to the Lagrangian in Eq. (5.20) if I go through the same procedure after adding a gauge field? Partial derivatives are replaced by covariant ones in Eq. (5.20) and I have

$$L_\phi = (D_\mu \phi)^*(D^\mu \phi) - \frac{1}{4}F^{\mu\nu}F_{\mu\nu} - V(\phi)\,, \tag{5.34}$$

where the potential is the same as in Eq. (5.21). For $\mu^2 > 0$ there are, as before, two massive scalars and one massless photon. When $\mu^2 < 0$ the ground state is degenerate, with $\phi_1^2 + \phi_2^2 = -\mu^2/\lambda = v^2$. To find the spectrum in this case, expand the scalar field around this vacuum state as in Eq. (5.26) to obtain

$$\begin{aligned} L_\phi = &-\frac{1}{4}F^{\mu\nu}F_{\mu\nu} + \frac{1}{2}\partial_\mu \eta \partial^\mu \eta - \lambda v^2 \eta^2 \\ &\boxed{+\frac{1}{2}\partial_\mu \chi \partial^\mu \chi + \frac{1}{2}g^2 v^2 A_\mu A^\mu - gvA_\mu \partial^\mu \chi} \\ &+ gA_\mu \partial^\mu \eta \chi - gA_\mu \partial^\mu \chi \eta + \frac{1}{2}A_\mu A^\mu (\eta^2 + 2\eta v) \\ &+ \frac{1}{2}A_\mu A^\mu \chi^2 - \frac{1}{4}\lambda(\eta^2 + \chi^2)^2 - \lambda(\eta^2 + \chi^2)^2 \eta v\,, \end{aligned} \tag{5.35}$$

where the boxed terms can be rewritten as

$$\frac{g^2 v^2}{2}\left(A_\mu - \frac{1}{gv}\partial_\mu \chi\right)\left(A^\mu - \frac{1}{gv}\partial^\mu \chi\right) \tag{5.36}$$

or, after the gauge transformation $A'_\mu = A_\mu - \partial_\mu \chi / gv$, as

$$\frac{1}{2} g^2 v^2 A'_\mu A^{\mu'}, \tag{5.37}$$

which is a mass term for the gauge boson.[5.7]

5.7 The gauge boson has acquired a mass. The gauge freedom is hidden, not broken, because the Lagrangian remains gauge invariant – and the Ward identities are preserved.

The spectrum in this case is a massive spin-1 boson (with mass $m_A = gv$) and a massive spin-0 scalar (with mass $m_\eta = \sqrt{2\lambda} v$, as can be read off Eq. (5.35)). The Goldstone boson is gone! This is the **Anderson–Higgs mechanism**:[1] the gauge theory of a scalar field, the symmetry of which is spontaneously broken, has a gauge boson that acquires a mass. The same mechanism is at work for a photon propagating through a superconductor and becoming massive, as discussed in Problem Session 5.1.

It is not hard to extend this mechanism to the Standard Model gauge groups. A scalar field that lives in isospin space is needed, so as to break the Standard Model group generators. I take it to be a doublet of complex scalar fields,

$$\phi = \begin{pmatrix} \phi^+ \\ \phi^0 \end{pmatrix}, \tag{5.38}$$

with a vacuum expectation value given by[5.8]

5.8 It is the gauge-invariant quantity $|\langle\phi\rangle|$ that marks the spontaneous breaking of the global symmetry and the hiding of the gauge freedom. The vacuum expectation in Eq. (5.39) is not gauge invariant and cannot represent any physical property.

$$\langle 0|\phi|0\rangle \equiv \langle\phi\rangle = \frac{1}{\sqrt{2}} \begin{pmatrix} 0 \\ v \end{pmatrix}. \tag{5.39}$$

Before I proceed, let me check that this vacuum state is the one I actually want. Acting on it with the matrices of the generators $\vec{T}$ (for $SU(2)_L$) and Y (for $U(1)_Y$) of the Standard Model groups gives:

$$\begin{aligned} T_1 \langle\phi\rangle &= \frac{1}{2} \begin{pmatrix} 0 & 1 \\ 1 & 0 \end{pmatrix} \frac{1}{\sqrt{2}} \begin{pmatrix} 0 \\ v \end{pmatrix} = \frac{1}{2\sqrt{2}} \begin{pmatrix} v \\ 0 \end{pmatrix} \neq 0 \\ T_2 \langle\phi\rangle &= \frac{1}{2} \begin{pmatrix} 0 & -i \\ i & 0 \end{pmatrix} \frac{1}{\sqrt{2}} \begin{pmatrix} 0 \\ v \end{pmatrix} = -\frac{i}{2\sqrt{2}} \begin{pmatrix} v \\ 0 \end{pmatrix} \neq 0 \\ T_3 \langle\phi\rangle &= \frac{1}{2} \begin{pmatrix} 1 & 0 \\ 0 & -1 \end{pmatrix} \frac{1}{\sqrt{2}} \begin{pmatrix} 0 \\ v \end{pmatrix} = -\frac{1}{2\sqrt{2}} \begin{pmatrix} 0 \\ v \end{pmatrix} \neq 0 \\ Y \langle\phi\rangle &= (+1) \frac{1}{\sqrt{2}} \begin{pmatrix} 0 \\ v \end{pmatrix} = \frac{1}{\sqrt{2}} \begin{pmatrix} 0 \\ v \end{pmatrix} \neq 0 . \end{aligned} \tag{5.40}$$

[1] About the naming of this mechanism: The idea that the Goldstone bosons give mass to the gauge particles was first discussed in P.W. Anderson, *Physical Review* **130** (1963) 439. Work by Y. Nambu, *Physical Review Letters* **4** (1960) 380, should also be mentioned. The relativistic generalization was simultaneously and independently discussed in three papers: G. Guralnik, C.R. Hagen and T. Kibble, *Physical Review Letters* **13** (1964) 585; F. Englert and R. Brout, *Physical Review Letters* **13** (1964) 321; and P. Higgs, *Physical Review Letters* **13** (1964) 508. It is in Higgs' paper that the massive mode (the Higgs boson) is explicitly identified.

None of the generators annihilates the vacuum state $\langle\phi\rangle$. It means that all the group generators are broken. The vacuum is charged with $U(1)_Y$ and $SU(2)_L$ charges. The corresponding gauge bosons, as they propagate in this charged vacuum, acquire a mass.

Yet observe more closely. When acting on the vacuum state, the combination defining the electric charge gives

$$Q\langle\phi\rangle = \left(T_3 + \frac{Y}{2}\mathbb{1}\right)\begin{pmatrix}0\\ v\end{pmatrix} = \begin{pmatrix}1 & 0\\ 0 & 0\end{pmatrix}\frac{1}{\sqrt{2}}\begin{pmatrix}0\\ v\end{pmatrix} = 0 \tag{5.41}$$

and the generator corresponding to the electric charge is left unbroken: the photon remains massless.

From the point of view of the Standard Model gauge groups, what happens is that the symmetry is reduced:

$$SU(2)_L \otimes U(1)_Y \longrightarrow \underbrace{U(1)_{\text{e.m.}}}_{\text{unbroken}}, \tag{5.42}$$

which only goes to show that, at energies below the scale of symmetry breaking, the only symmetry is that related to the conservation of electric charge, as in QED.

The masses generated by the concealment of the gauge freedom come – as they do in Eq. (5.35) – from the covariant derivatives of the scalar field. Replacing the scalar field by its vacuum expectation value I obtain that the covariant derivative of the electroweak part of the Standard Model is given by

$$\begin{aligned}
D_\mu\phi &= \frac{1}{\sqrt{2}}\left(ig\frac{1}{2}\vec{T}\cdot\vec{W}_\mu + ig'\frac{1}{2}Y\mathbb{1}B_\mu\right)\begin{pmatrix}0\\ v\end{pmatrix}\\
&= \frac{i}{\sqrt{8}}\left[g\begin{pmatrix}0 & W^1_\mu\\ W^1_\mu & 0\end{pmatrix} + g\begin{pmatrix}0 & -iW^2_\mu\\ iW^2_\mu & 0\end{pmatrix}\right.\\
&\qquad \left. + g\begin{pmatrix}W^3_\mu & 0\\ 0 & -W^3_\mu\end{pmatrix} + g'\begin{pmatrix}YB_\mu & 0\\ 0 & YB_\mu\end{pmatrix}\right]\begin{pmatrix}0\\ v\end{pmatrix}\\
&= \frac{i}{\sqrt{8}}\begin{pmatrix}gW^3_\mu + g'YB_\mu & g(W^1_\mu - iW^2_\mu)\\ g(W^1_\mu + iW^2_\mu) & -gW^3_\mu + g'YB_\mu\end{pmatrix}\begin{pmatrix}0\\ v\end{pmatrix}\\
&= \frac{iv}{\sqrt{8}}\begin{pmatrix}g(W^1_\mu - iW^2_\mu)\\ -gW^3_\mu + g'YB_\mu\end{pmatrix},
\end{aligned} \tag{5.43}$$

which yields in the Lagrangian the term (now taking $Y = 1$ for the scalar doublet)

$$(D_\mu\phi)^\dagger(D^\mu\phi) \Rightarrow \frac{1}{8}v^2\left[\underbrace{g^2(W_1^2 + W_2^2)}_{2g^2W^+_\mu W^-_\mu} + \overbrace{(-gW^3_\mu + g'B_\mu)^2}^{(W^3_\mu\ B_\mu)\begin{pmatrix}g^2 & -gg'\\ -gg' & g'^2\end{pmatrix}\begin{pmatrix}W^3_\mu\\ B_\mu\end{pmatrix}}\right]. \tag{5.44}$$

The mass for the charged gauge bosons can be read directly from Eq. (5.44) to be

$$m_W = \frac{1}{2} g v, \tag{5.45}$$

while for the neutral gauge bosons I need to diagonalize the matrix in Eq. (5.44),

$$\begin{pmatrix} g^2 & -gg' \\ -gg' & g'^2 \end{pmatrix}. \tag{5.46}$$

The diagonalization gives two eigenvalues λ:

$$\boxed{\lambda = 0} \qquad A_\mu = \frac{1}{\sqrt{g^2 + g'^2}} \left(g' W_\mu^3 + g B_\mu \right) \qquad \text{(photon)}$$
$$\boxed{\lambda = g^2 + g'^2} \qquad Z_\mu = \frac{1}{\sqrt{g^2 + g'^2}} \left(g W_\mu^3 - g' B_\mu \right) \qquad \text{(Z-boson)} \tag{5.47}$$

and masses

$$m_Z = \frac{\sqrt{g^2 + g'^2}}{2} v \tag{5.48}$$

$$m_\gamma = 0. \tag{5.49}$$

The spectrum of the Standard Model gauge theory coupled to a scalar field that acquires the vacuum expectation value in Eq. (5.39) consists of massive spin-1 bosons.

Because of the rotation from the fields W_μ^3 and B_μ to A_μ and Z_μ the Feynman rules for the interactions among the gauge bosons have mixing coefficients which are controlled by the weak angle θ_W, defined by

$$\sin\theta_W = \frac{g'}{\sqrt{g^2 + g'^2}}$$
$$\cos\theta_W = \frac{g}{\sqrt{g^2 + g'^2}}, \tag{5.50}$$

and with $m_W = m_Z \cos\theta_W$.

I can write the Feynman rules for the electroweak bosons and their interactions. The interactions among the gauge bosons themselves have vertices with three and four gauge bosons. The Feynman rules for those with three particles are shown in Figure 5.3.

The Feynman rules for vertices with four particles are collected in appendix section A.6.

The excitation of the scalar field up the curved side of the potential shown in Figure 5.1 can be written as

$$\phi = \frac{1}{\sqrt{2}} \begin{pmatrix} 0 \\ v + h \end{pmatrix}, \tag{5.51}$$

vertex

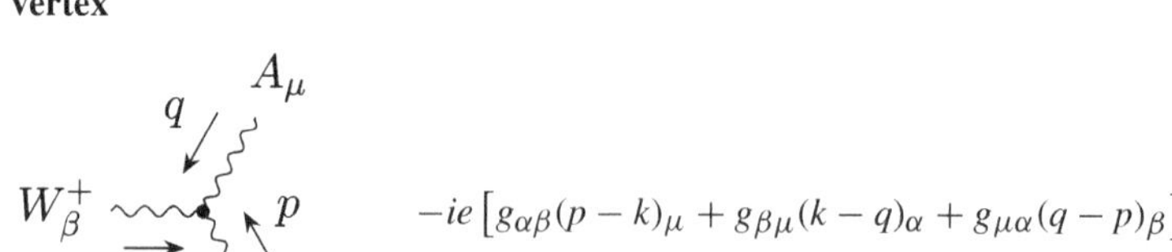

vertex

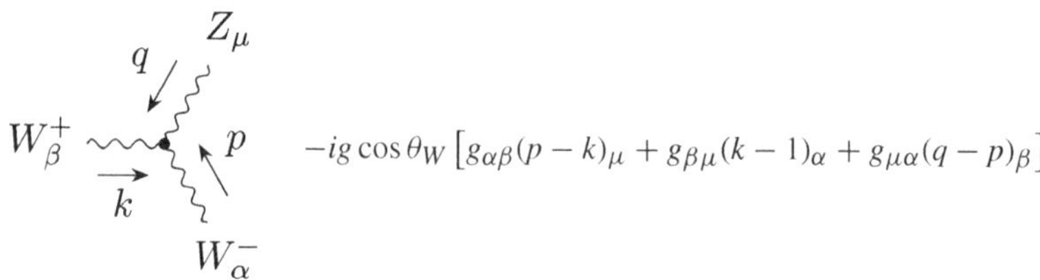

Figure 5.3 Three-particle vertices among gauge bosons.

which defines the Higgs boson h. Vertices for the interactions with one or two Higgs bosons are also given in appendix section A.6.

The photon, being neutral, does not couple to the Higgs boson.

The Higgs boson has interactions with itself from the potential in Eq. (5.21). Its Lagrangian is

$$\mathcal{L}_h = \frac{1}{2}(D_\mu h)(D^\mu h)^\dagger - \frac{m_h^2}{2}h^2 - \sqrt{\frac{\lambda}{2}}m_h h^3 + \frac{\lambda^4}{4}h^4 . \qquad (5.52)$$

Notice the cubic term generated by the vacuum expectation value. After substituting $\sqrt{\lambda} = m_h/\sqrt{2}v$ and $v = m_W/2g$, it gives rise to the Feynman rule in Figure 5.4. The rule for the quartic vertex is given in appendix section A.6.

I am not writing down the Feynman rules for the ghost and Goldstone fields because I do not need them. Finally, the propagators of the physical gauge bosons and the Higgs boson, which can be derived from the kinetic (bilinear) part of the Lagrangian, are shown in Figure 5.5.

5.4 *Problem Session*: Custodial Symmetry

The Lagrangian of the scalar boson in Eq. (5.20) can be written as

$$L_{\mathrm{h}} = -\frac{1}{2}\mathrm{Tr}\left[(D_\mu\Phi)^\dagger(D^\mu\Phi)\right] - \mu^2\mathrm{Tr}(\Phi^\dagger\Phi) - \lambda\mathrm{Tr}(\Phi^\dagger\Phi)^2, \qquad (5.53)$$

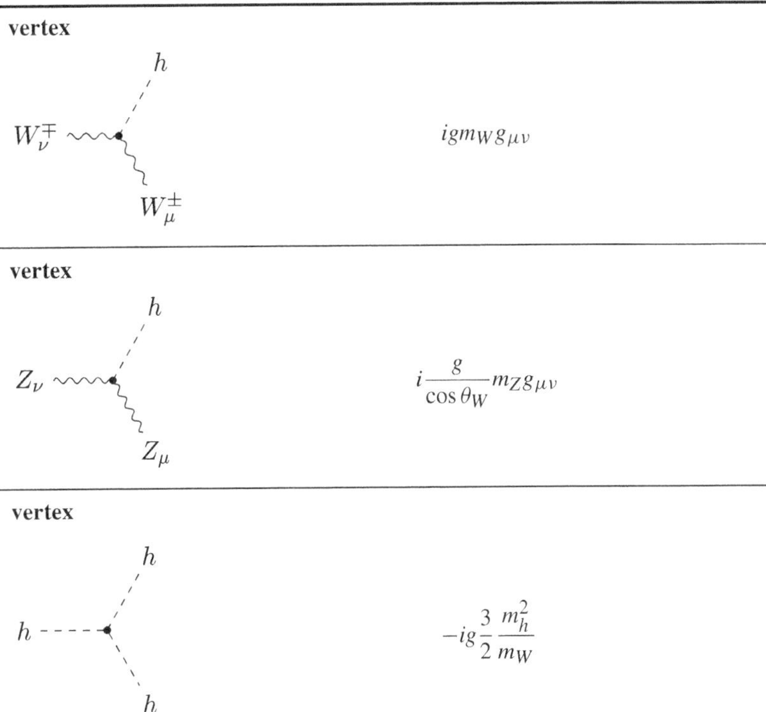

Figure 5.4 Vertices between the gauge bosons and the Higgs boson and between three Higgs bosons.

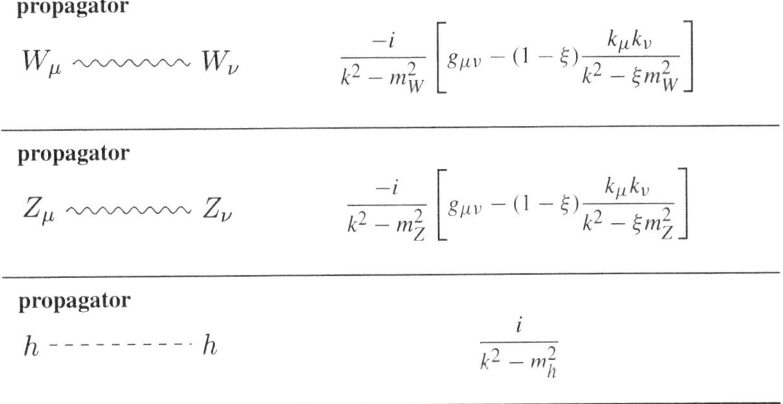

Figure 5.5 Propagators for the gauge and the Higgs bosons. The ξ are the gauge-fixing parameters.

collecting the four real components of the scalar boson isospin doublet in the matrix

$$\Phi = \begin{pmatrix} \phi^{0*} & \phi^{+} \\ \phi^{+*} & \phi^{0} \end{pmatrix} . \tag{5.54}$$

The covariant derivative is now defined by

$$D_\mu = \partial_\mu + \frac{1}{2\sqrt{2}}\left(ig\vec{T}\cdot\vec{W}_\mu + ig'Y\sigma_3 B_\mu\right), \tag{5.55}$$

where the Pauli matrix σ_3 is necessary in the last term because of the change of sign in the hypercharge in going from the first column of Φ to the second.

The Lagrangian in Eq. (5.53) for $g' = 0$ is invariant under the $SU(2)$ rotations described by the matrices

$$U_L = U_R = e^{i\alpha\sigma_3/2} \tag{5.56}$$

acting on the Higgs field as

$$\Phi \longrightarrow U_L \Phi U_R^\dagger . \tag{5.57}$$

Consider the symmetry[5.9]

5.9 The Lagrangian in Eq. (5.53) is invariant if I take $\Phi = \langle\Phi\rangle$ even with $g' \neq 0$.

$$SO(4) = SU(2)_L \otimes SU(2)_R . \tag{5.58}$$

The vacuum expectation value of the Higgs boson,

$$\langle\Phi\rangle = \frac{v}{\sqrt{2}}\mathbb{1}_{2\times 2}, \tag{5.59}$$

breaks the symmetry $SO(4)$ down to its diagonal part, a single $SU(2)$. This surviving symmetry is called **custodial**.

In first approximation, the electroweak physics of the Standard Model preserves this symmetry and traces of it can be found in the relationships among the Lagrangian parameters. For instance, the masses of the gauge bosons come from the term in the Lagrangian[5.10]

5.10 Assuming now that the two diagonal terms in Eq. (5.59) are different and are denoted by u and v.

$$M_{ab} = \langle\Phi\rangle\{g^A T^A, g^B T^B\}\langle\Phi\rangle = \begin{pmatrix} g^2 v^2 & 0 & 0 & 0 \\ 0 & g^2 v^2 & 0 & 0 \\ 0 & 0 & g^2 u^2 & -g' g\, u^2 \\ 0 & 0 & -g' g\, u^2 & -g'^2 u^2 \end{pmatrix}, \tag{5.60}$$

which must have $u = v$ if custodial symmetry holds. Because of the identity of the vacua, the masses of the gauge bosons satisfy the relation

$$m_W = m_Z \cos\theta_W . \tag{5.61}$$

On the other hand, the hypercharge interaction leads to an explicit breaking of the custodial symmetry because of the term proportional to the Pauli matrix σ_3 in the covariant derivative,

$$\mathrm{Tr}\,\partial_\mu \Phi^\dagger B^\mu \Phi\sigma_3 \to \mathrm{Tr}\,\partial_\mu \Phi^\dagger B^\mu \underbrace{U_R^\dagger \sigma_3 U_R}_{\neq \sigma_3} . \tag{5.62}$$

and any splitting of the masses of the matter isospin doublets in the currents also leads to this.

5.5 *Problem Session*: The Decay of the Gauge Boson *W*

The W-boson can decay through the charged current into leptons and neutrinos as well as pairs of quarks. Let me discuss the decay into leptons. The amplitude for the process can be read off the diagram in Figure 5.6 as

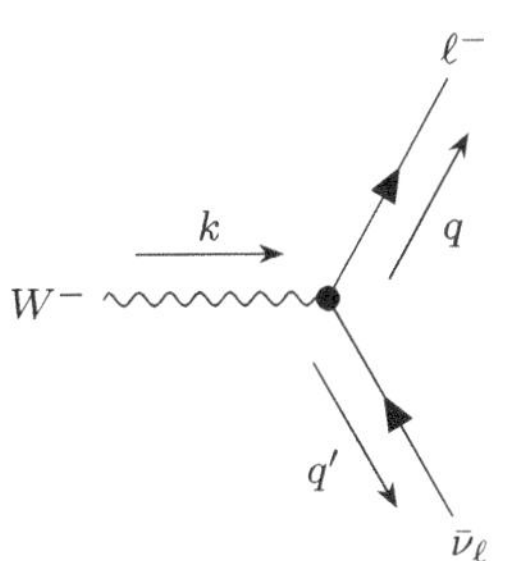

Figure 5.6 Feynman diagram for the decay of the W gauge boson.

$$\mathcal{M} = \frac{g}{2\sqrt{2}} \varepsilon_\mu(k,\lambda)\bar{u}_\ell(q,s)\gamma^\mu(1-\gamma_5)v_{\nu_\ell}(q',s')\,. \tag{5.63}$$

For a massive spin-1 boson there are three independent polarizations and therefore I have to average over these three the sum over the initial states. This sum over the three physical polarizations of the massive W gauge boson is given, as discussed in Chapter 3, by

$$\sum_{\lambda=1,2,3} \varepsilon_\mu^{\lambda *}(k)\varepsilon_\nu^{\lambda}(k) = -g_{\mu\nu} + \frac{k_\mu k_\nu}{m_W^2}\,. \tag{5.64}$$

From Eq. (5.63),[5.11]

5.11 $$k = q + q'$$ $$k^2 = m_W^2 = 2q\cdot q'\,.$$

$$\begin{aligned}\frac{1}{3}\sum_{\lambda,s,s'}|\mathcal{M}|^2 &= \frac{1}{3}\frac{g^2}{8}\sum_\lambda \varepsilon_\mu^{r\lambda *}(k)\varepsilon_\nu^{\lambda}(k)\\ &\quad\times\sum_{s,s'}\bar{u}_\ell(q,s)\gamma^\mu(1-\gamma_5)v_\ell(q',s')\times\bar{v}_\ell(q',s')\gamma^\nu(1-\gamma_5)u_\ell(q,s)\\ &= \frac{1}{3}\frac{g^2}{8}\left(-g_{\mu\nu}+\frac{k_\mu k_\nu}{m_W^2}\right)\underbrace{\mathrm{Tr}\left\{(\not{q}+m_e)\gamma^\mu(1-\gamma_5)\not{q}'\gamma^\nu(1-\gamma_5)\right\}}_{=8\left[q^\mu q'^\nu+q'^\mu q^\nu-(q\cdot q')g^{\mu\nu}+i\epsilon^{\alpha\mu\beta\nu}q_\alpha q'_\beta\right]}\\ &= \frac{g^2}{3m_W^2}\left[2(q\cdot k)(q'\cdot k)+m_W^2(q\cdot q')\right]\,.\end{aligned} \tag{5.65}$$

Equation 5.65 is ready to be put into the decay width formula, but it is somewhat obscure from the point of view of the contributions of the three polarizations of the W-boson. Before proceeding, it is useful to pause and write explicitly the three terms in the sum over the polarizations:

$$\sum_{\lambda=1,2,3} \varepsilon_\mu^{\lambda *}(k)\varepsilon_\nu^{\lambda}(k) = \underbrace{\varepsilon_L^\mu\varepsilon_L^\nu}_{W\text{-LONG}} + \underbrace{\varepsilon_+^\mu\varepsilon_+^\nu + \varepsilon_-^\mu\varepsilon_-^\nu}_{W\text{-TRAN}}\,. \tag{5.66}$$

The three polarizations are given, in the rest frame of the W-boson, by

$$\begin{aligned}\varepsilon_L^\mu &= (0,\,0,\,0\,1)\\ \varepsilon_+^\mu &= (0,\,-1,\,-i,\,1)/\sqrt{2}\\ \varepsilon_-^\mu &= (0,\,1,\,-i,\,1)/\sqrt{2}\,.\end{aligned} \tag{5.67}$$

Equation (5.65) can now be written as the sum of three contributions:[5.12]

$$|\mathcal{M}_L|^2 = \frac{g^2}{24}\varepsilon_L^{\mu}\,\varepsilon_L^{\nu}\mathrm{Tr}\left\{(\not{q}+m_e)\gamma_\mu(1-\gamma_5)\not{q}'\gamma_\nu(1-\gamma_5)\right\} = \frac{g^2E^2}{12}\sin^2\theta$$
$$|\mathcal{M}_+|^2 = \frac{g^2}{24}\varepsilon_+^{\mu}\,\varepsilon_+^{\nu}\mathrm{Tr}\left\{(\not{q}+m_e)\gamma_\mu(1-\gamma_5)\not{q}'\gamma_\nu(1-\gamma_5)\right\} = \frac{g^2E^2}{12}(1-\cos\theta)^2$$
$$|\mathcal{M}_-|^2 = \frac{g^2}{24}\varepsilon_-^{\mu}\,\varepsilon_-^{\nu}\mathrm{Tr}\left\{(\not{q}+m_e)\gamma_\mu(1-\gamma_5)\not{q}'\gamma_\nu(1-\gamma_5)\right\} = \frac{g^2E^2}{12}(1+\cos\theta)^2 .$$

5.12 $q = (E, E\sin\theta, 0, E\cos\theta)$ and $q' = (E, -E\sin\theta, 0, -E\cos\theta)$.

The angle θ is defined as between the initial direction of the decaying W and the direction of the final lepton. The angular dependence of the three amplitudes can be understood as the projection of the spin of the W-boson along the direction of the decay products: the lepton and its antineutrino. The W-boson acts, so to speak, as its own polarimeter.

Returning to Eq. (5.65), I can take the whole expression and compute the total width:

$$\begin{aligned}\Gamma(W^- \to e^-\bar{\nu}_e) &= \frac{1}{2k_0}\frac{1}{(2\pi)^2}\int\frac{\mathrm{d}^3\vec{q}}{2q_0}\frac{\mathrm{d}^3\vec{q}'}{2q_0'}\left(\frac{1}{3}\sum|\mathcal{M}|^2\right)\delta^{(4)}(q+q'-k)\\ &= \frac{g^2}{96\pi^2}\frac{1}{m_W^3}\int\frac{\mathrm{d}^3\vec{q}}{q_0}\frac{\mathrm{d}^3\vec{q}'}{q_0'}\delta^{(4)}(q+q'-k)\Big[2(q\cdot k)(q'\cdot k)+m_W^2(q\cdot q')\Big].\end{aligned} \tag{5.68}$$

I must now integrate over the phase space. I can use the phase-space integrals

$$k^\alpha k^\beta\underbrace{\int\frac{\mathrm{d}^3\vec{q}}{q_0}\frac{\mathrm{d}^3\vec{q}'}{q_0'}q_\alpha q_\beta'\delta^{(4)}(q+q'-k)}_{\frac{\pi}{6}(k^2g^{\alpha\beta}+2k^\alpha k^\beta)} = \frac{\pi}{6}(m_W^4+2m_W^4) = \frac{\pi}{2}m_W^4 \tag{5.69}$$

and

$$m_W^2\int\frac{\mathrm{d}^3\vec{q}}{q_0}\frac{\mathrm{d}^3\vec{q}'}{q_0'}q_\alpha q_\alpha'\delta^{(4)}(q+q'-k) = \frac{\pi}{6}(4m_W^4+2m_W^4) = \pi m_W^4 , \tag{5.70}$$

which have already been computed in a previous problem session, to obtain

$$\Gamma(W^- \to e^-\bar{\nu}_e) = \frac{g^2 m_W}{48\pi} = \frac{G_F m_W^3}{6\sqrt{2}\pi} \simeq 225\ \mathrm{MeV}. \tag{5.71}$$

Similar expressions can be found for the other channels in which the W can decay, namely $\mu^-\bar{\nu}_\mu$, $\tau^-\bar{\nu}_\tau$ and quarks. These partial widths together give a lifetime $\tau \simeq 10^{-24}$ s.

A bit of jargon is necessary here. If all the particles coming out of a decay are tracked, I can define the invariant mass of the decay

$$m_{\ell\ell}^2 = (p_{\ell^+}+p_{\ell^-})^2 . \tag{5.72}$$

This is the case for the decay of the Z-boson into charged pairs of leptons.

The same definition is not possible if some of the final states are invisible, as neutrinos are. The four components of the neutrino momentum are unknown. This is the case of the decay of the W-boson, for which the invariant mass reads[5.13]

5.13 In Eq. (5.73) the transverse momentum of the neutrino $\vec{p}_{\nu_T}$ is identified with the missing transverse momentum $\not{p}_T$, which is then identified with minus the sum of the lepton transverse momenta by using momentum conservation.

$$m_{\ell\nu_\ell}^2 = (E_\ell + E_\nu)^2 - \underbrace{(\vec{p}_{\ell_T} + \overset{\not{p}_T = -\sum \vec{p}_T}{\vec{p}_{\nu_T}})^2}_{\text{transverse}} - \underbrace{(p_{\ell_L} + p_{\nu_L})^2}_{\text{longitudinal}}; \qquad (5.73)$$

the quantities E_ν, $\vec{p}_{\nu_T}$ and p_{ν_L} in Eq. (5.73) cannot be observed.

Yet I can still define the **transverse mass**

$$\begin{aligned} m_{\ell\nu_{\ell_T}}^2 &= (E_\ell + \overset{\not{E}_T = -\sum \vec{E}_T}{E_\nu})^2 - \underbrace{(\vec{p}_{\ell_T} + \vec{p}_{\nu_T})^2}_{\text{transverse}} \\ &\simeq 2\vec{p}_{\ell_T} \cdot \vec{p}_{\nu_T} = 2E_{\ell_T}\not{E}_T(1 - \cos\phi_{\ell\nu}), \end{aligned} \qquad (5.74)$$

in which only the observed quantities (thanks to the determination of the missing energy and transverse momentum by way of the conservation laws) are included and the unknown longitudinal momentum of the neutrino is not. The vectorial transverse energies in the definition of the missing transverse energy $\not{E}_T$ are defined as vectors pointing from the interaction vertex to the center of the cell in the calorimeter (see Chapter 9) and of size proportional to the recorded energy deposited in the cell.

The transverse mass is bounded by the invariant mass:

$$0 \leq m_{\ell\nu_{\ell_T}}^2 \leq m_{\ell\nu_\ell}^2 . \qquad (5.75)$$

The different distributions for the two invariant masses are shown in Figure 5.7.

Why can I not determine the longitudinal momentum of the neutrino by using the same conservation law as used for the transverse components? The problem is that the total longitudinal momentum in a collision is not in general zero. In fact it is not well defined because is shared among the partons making up the protons in a hadronic collision.

But, then, why can I not find the longitudinal momentum of the neutrino in the W-boson decay by means of Eq. (5.73)? After all, if I use the conservation of momentum to determine the transverse momentum of the neutrino and conservation of energy to determine its energy, and identify the invariant mass with the actual mass of the W-boson, Eq. (5.73) becomes an equation in the only unknown quantity p_{ν_L}, namely

$$m_W^2 = (p_\nu + p_\ell)^2 = (E_\ell + E_\nu)^2 - (\vec{p}_{\ell_T} + \vec{p}_{\nu_T})^2 - (p_{\ell_L} + p_{\nu_L})^2, \qquad (5.76)$$

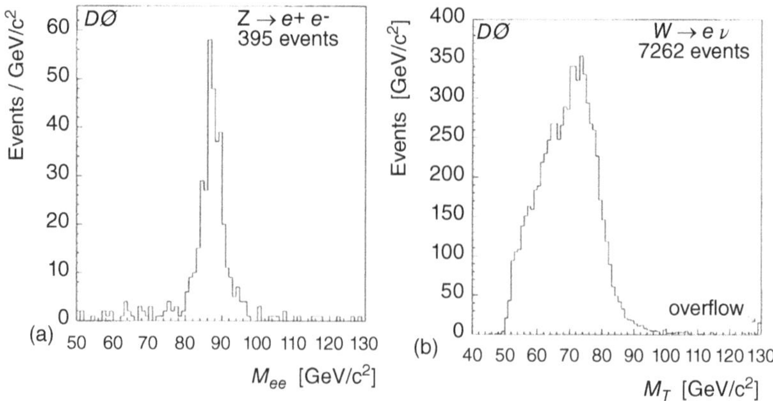

Figure 5.7 (a) The number of events as a function of the invariant mass in the decay $Z \to e^+e^-$. (b) The number of events as a function of the transverse invariant mass in the decay $W \to e\nu$; the curve would have a sharp drop but the finite efficiency of the detector and the center of mass boost make it into a smoother function [D0 Collaboration, *Physical Review* **D58** (2002) 012002].

which I can solve to find p_{ν_L}. Unfortunately, Eq. (5.76) is second order in the unknown quantity p_{ν_L}:

$$\underbrace{(p_{\ell_L}^2 - E_\ell^2)}_{a}\, p_{\nu_L}^2 + \underbrace{(m_W^2\, p_{\ell_L} + 2\, p_{\ell_L}\vec{p}_{\ell_T}\cdot\vec{p}_{\nu_T})}_{b}\, p_{\nu_L}$$
$$\underbrace{+\frac{m_W^4}{4} + (\vec{p}_{\ell_T}\cdot\vec{p}_{\nu_T})^2 + m_W^2\vec{p}_{\ell_T}\cdot\vec{p}_{\nu_T} - E_\ell^2\,\vec{p}_{\nu_T}^{\,2}}_{c} = 0 \qquad (5.77)$$

and I do not have a unique solution but rather two:

$$p_{\nu_L} = \frac{-b \pm \sqrt{b^2 - 4ac}}{2a}, \qquad (5.78)$$

and even no solution at all if $b^2 - 4ac < 0$. A hard fact of elementary algebra.

For this reason, in the decay of the W-boson I have to limit myself to the transverse mass $m_{\ell\nu_{\ell_T}}$. To find p_{ν_L} I need more sophisticated methods, which in any case come with their own uncertainty. Conclusion: only some approximated value of p_{ν_L} is available at the end.

5.6 Giving Mass to the Fermions

Matter fields in the Standard Model come as chiral fermions: the left-handed components are doublets under the gauge group $SU(2)_L$ while the right-handed components are singlets. This means that I cannot write a mass term such as $\bar{\psi}_L^D \psi_R$ because it would explicitly break the symmetry.[5.14]

5.14 It is a remarkable feature of the Standard Model that the scalar field gives mass not only to the gauge fields but also to all the fermions. Without the scalar boson neither the gauge fields nor the fermions can have a mass.

Indeed, to write a mass term I need another field that is a doublet of the group $SU(2)_L$, to make it possible to construct an invariant. This field is the scalar field ϕ already introduced to give mass to the gauge bosons. For such a field I can write a term like

$$-Y\left(\underbrace{\bar{\psi}_L}_{\text{doublet, } Y=-1}\overbrace{\boldsymbol{\phi}}^{\text{doublet, } Y=+1}\underbrace{\psi_R}_{\text{singlet, } Y=-2} + \bar{\psi}_R\boldsymbol{\phi}^\dagger\psi_L\right), \tag{5.79}$$

which is both invariant and of dimension four. Lagrangians like that in Eq. (5.79) are called after the physicist Hidaki Yukawa – a terminology that explains the Y in Eq. (5.79).[5.15]

5.15 The hypercharge of the Higgs field is determined by the invariance of this Lagrangian to be $Y = +1$.

This construction applied to the Standard Model proper gives, when the scalar boson takes its vacuum expectation value $\phi^0 = v/\sqrt{2}$, and for the first generation of leptons, a term in the Lagrangian

$$\begin{aligned}-Y_e\frac{1}{\sqrt{2}}&\left[(\bar{\nu}_e, \bar{e})_L\begin{pmatrix}0\\ v\end{pmatrix}e_R + \bar{e}_R(0, v)\begin{pmatrix}\nu_e\\ e\end{pmatrix}_L\right]\\ &= -\frac{Y_e v}{\sqrt{2}}(\bar{e}_L e_R + \bar{e}_R e_L) = -\underbrace{\frac{Y_e v}{\sqrt{2}}}_{m_e}\bar{e}e\,.\end{aligned} \tag{5.80}$$

The mass of the electron is given by

$$m_e = \frac{Y_e v}{\sqrt{2}}\,. \tag{5.81}$$

The neutrino remains massless. The same construction works for the down quarks.

But what to do with the "up" states in the quark sector? I need the rotated Higgs field

$$\tilde{\phi}^* = -i\tau_2\phi^* = \begin{pmatrix}0 & 1\\ -1 & 0\end{pmatrix}\begin{pmatrix}\phi^+\\ \phi^0\end{pmatrix} = \begin{pmatrix}\phi^0\\ -\phi^{+*}\end{pmatrix} \tag{5.82}$$

so that I have the Lagrangian terms

$$-Y_d\frac{1}{\sqrt{2}}\left[(\bar{u}, \bar{d})_L\begin{pmatrix}\phi^+\\ \phi^0\end{pmatrix}d_R + \bar{d}_R(\phi^+, \phi^0)\begin{pmatrix}u\\ d\end{pmatrix}_L\right] = -\underbrace{\frac{Y_d v}{\sqrt{2}}}_{m_d}\bar{d}d \tag{5.83}$$

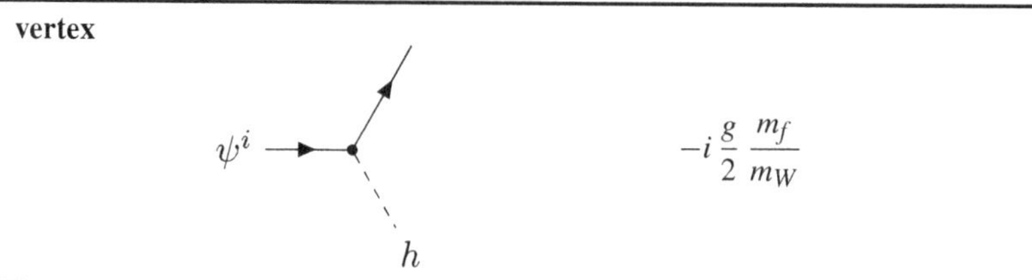

Figure 5.8 Vertex between fermions and the Higgs boson.

and

$$-Y_u\frac{1}{\sqrt{2}}\left[(\bar{u},\bar{d})_L\begin{pmatrix}\phi^0\\-\phi^{+*}\end{pmatrix}u_R+\bar{u}_R(\phi^0,\ -\phi^{+*})\begin{pmatrix}u\\d\end{pmatrix}_L\right]=-\underbrace{\frac{Y_u v}{\sqrt{2}}}_{m_u}\bar{u}u\,. \tag{5.84}$$

Altogether, the Yukawa Lagrangian for the Standard Model is given by

$$\begin{aligned} L_{\text{Yuk}} = &-Y_\ell^{ij}\frac{1}{\sqrt{2}}(\bar{\nu}_\ell^i,\bar{\ell}^i)_L\begin{pmatrix}\phi^+\\\phi^0\end{pmatrix}\ell_R^j \\ &-Y_d^{ij}\frac{1}{\sqrt{2}}(\bar{u}^i,\bar{d}^i)_L\begin{pmatrix}\phi^+\\\phi^0\end{pmatrix}d_R^j \\ &-Y_u^{ij}\frac{1}{\sqrt{2}}(\bar{u}^i,\bar{d}^i)_L\begin{pmatrix}\phi^{0*}\\-\phi^{+*}\end{pmatrix}u_R^j+\text{c.c.,}\end{aligned} \tag{5.85}$$

which means that the mass terms $Y_{ij}v/\sqrt{2}$ are not diagonal in flavor space. When I proceed and diagonalize these states defined by Eq. (5.85), they will enter as combinations of states of different generations in the charged electroweak currents. This mismatch between mass and interaction states gives rise to the rich physics of flavor in the Standard Model.

When the expansion of the scalar field

$$\phi^0=\frac{v+h}{\sqrt{2}} \tag{5.86}$$

is inserted in Eq. (5.85) I can extract the Feynman rule for the interaction between fermions and the Higgs boson h shown in Figure 5.8.

The vertex is proportional to the mass of the fermion.

5.7 *Problem Session*: The Fifth Force

If I were to count the number of forces in the Standard Model, what would I come up with? Three, right? There are the weak force, the strong force and the electromagnetic force. If I add gravity to these three I come up with the often repeated statement that there are only four fundamental forces in the Universe.

If this statement were true, I would expect that by taking the Standard Model forces to zero, that is, if I were to take the gaugeless limit $g = g' = 0$ with also the strong interaction (see the next chapter) $g_s = 0$ (and forgetting about gravity), no interaction would remain. All matter would be stable and free. In contrast with this expectation, in this limit the electron is unstable and decays in 10 ns![2]

To understand this fundamental property of the Standard Model, I need to consider the mass of the W-boson $m_W = gv/2$. This mass goes to zero when I take the limit $g \to 0$. At a certain point in this limiting process, the mass m_W becomes lighter than that of the electron m_e and the decay

$$e^- \to W^- + \nu_e \tag{5.87}$$

becomes possible with a width proportional to

$$g^2 \left(\varepsilon_T \cdot \varepsilon_T + \varepsilon_L \cdot \varepsilon_L \right) , \tag{5.88}$$

where ε_T and ε_L are the transverse and longitudinal polarizations of the gauge boson W.

Whereas $\varepsilon_T \propto 1$ and its contribution to the width vanishes as $g \to 0$, the longitudinal polarization ε_L is proportional $|p_W|/m_W$ and its contribution to the width goes as

$$\frac{g^2 |p_W|^2}{m_W^2} \simeq \frac{m_e^2}{v^2} \simeq G_F m_e^2, \tag{5.89}$$

to give

$$\Gamma_e = \frac{\sqrt{2}\, G_F m_e^3}{16\pi} , \tag{5.90}$$

which is independent of the coupling g. The naive expectation that $\Gamma_e \to 0$ when $g \to 0$ is frustrated by the presence of the longitudinal mode of the W-boson, which remains coupled.

Also, the scattering of the gauge bosons W and Z remains non-vanishing thanks to their longitudinal components, which are the Goldstone bosons connected with the hiding of the symmetry, the coupling of which is controlled by the Higgs potential and not the gauge interactions.

This is another way to see the Higgs mechanism: the transverse modes of the W-boson decouple for $g \to 0$ but the longitudinal mode remains interacting; this is the part associated with the Goldstone bosons. For that matter, the Yukawa coupling of the Higgs boson to a fermion goes like $\sqrt{G_F} m_f$ and also does not go to zero when g or $g' \to 0$.

The Higgs boson indeed carries an independent force, a fifth force that must be added to the usual four. Including gravity, there are **five** forces[5.16]

5.16 The five fundamental forces in the Universe.

[2] J.D. Bjorken, in *Proceedings of the 5th Moriond Workshop*, 1987.

(that we know of) in the Universe: the strong and weak forces, electromagnetism, gravity and the force due to exchange of the Higgs boson.

5.8 The Use of Unitarity, II

I now want to study the scattering of an electron–positron pair into longitudinal W-bosons:

$$e^-(p_1) + e^+(p_2) \to W_L^-(q_1) + W_L^+(q_2) . \tag{5.91}$$

I discussed in Chapter 2 a similar process, one with neutrinos in the initial state instead of electrons, and that process showed how the Z-boson is necessary in order to have a unitary S-matrix. In the case of electron pairs in the initial state, the full amplitude is given by three diagrams.[3] The first of these diagrams (Figure 5.9) is in the t-channel and is given by the amplitude

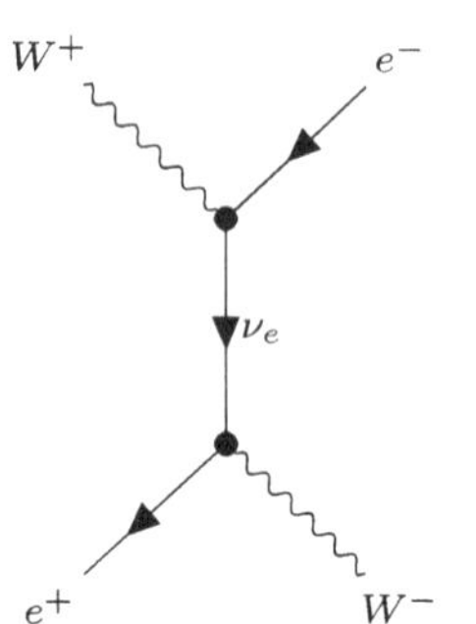

Figure 5.9 Feynman diagram for the amplitude $\mathcal{M}_t$.

$$\mathcal{M}_t = -\frac{g^2}{8\,t}\,\bar{v}(p_2)\gamma_\nu(1-\gamma_5)(\not{p}_1 - \not{q}_1)\gamma_\mu(1-\gamma_5)\gamma_\mu u(p_2)\varepsilon_L^\mu(q_1)\varepsilon_L^\nu(q_2) . \tag{5.92}$$

The second (Figure 5.10) has a photon in the s-channel with an amplitude given by

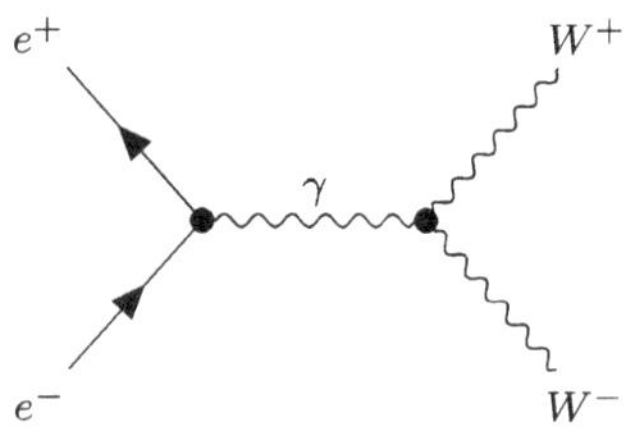

Figure 5.10 Feynman diagram for the amplitude $\mathcal{M}_{s1}$.

$$\mathcal{M}_{s1} = -\frac{g^2 \sin\theta_W}{s}\,\bar{v}(p_2)\gamma_\alpha u(p_2)\Gamma_{\nu\mu\alpha}\varepsilon_L^\mu(q_1)\varepsilon_L^\nu(q_2) , \tag{5.93}$$

and the third (Figure 5.11) has a Z-boson in the s-channel:

$$\mathcal{M}_{s2} = -\frac{g^2}{s - m_Z^2}\left[-g_{\alpha\beta} + \frac{(p_1+p_2)^\alpha (p_1+p_2)^\beta}{m_Z^2}\right] \tag{5.94}$$
$$\times\ \bar{v}(p_2)\Gamma_{\nu\mu\alpha}\gamma^\beta\left[g_L\frac{1-\gamma_5}{2} + g_R\frac{1+\gamma_5}{2}\right]u(p_2)\varepsilon_L^\mu(q_1)\varepsilon_L^\mu(q_2) .$$

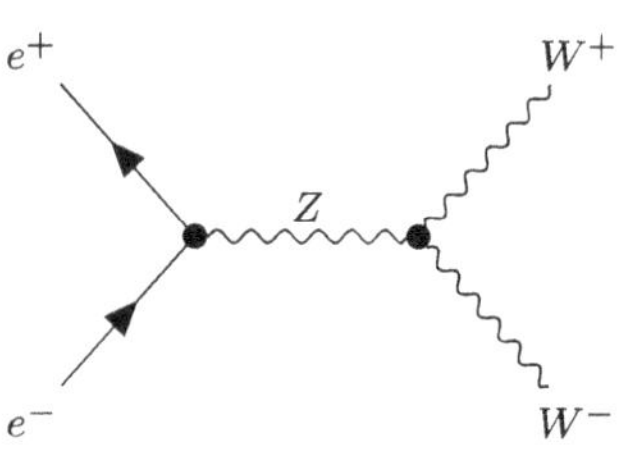

Figure 5.11 Feynman diagram for the amplitude $\mathcal{M}_{s2}$.

Take the same assignments of momenta and polarizations as those in Chapter 2, that is,

$$\begin{cases} p_1 = \frac{\sqrt{s}}{2}(1,0,0,1) \\ p_2 = \frac{\sqrt{s}}{2}(1,0,0,-1) \\ q_1 = \frac{\sqrt{s}}{2}(1,\beta\sin\theta,0,\beta\cos\theta) \\ q_2 = \frac{\sqrt{s}}{2}(1,-\beta\sin\theta,0,-\beta\cos\theta) \end{cases} \qquad \begin{cases} \varepsilon_L(q_1) = \frac{\sqrt{s}}{2m_W}(\beta,\sin\theta,0,\cos\theta) \\ \varepsilon_L(q_2) = \frac{\sqrt{s}}{2m_W}(\beta,-\sin\theta,0,-\cos\theta) , \end{cases}$$

where $\beta = \sqrt{1 - 4m_W^2/s}$; these expressions make it possible to single out the longitudinal part of the amplitudes.

[3] J.C. Romão, arXiv:1603.04251.

At high energy, where $s \gg m_W$, I have that

$$\bar{v}(p_2)u(p_1) \propto \sqrt{s} \quad \text{and} \quad \varepsilon_L^{\mu}(q_2) \propto \frac{\sqrt{s}}{m_W}, \tag{5.95}$$

and the sum of these three amplitudes is

$$\mathcal{M}_t + \mathcal{M}_{s1} + \mathcal{M}_{s2} = -\frac{g^2 m_e}{2}\frac{1}{s}\bar{v}(p_2)u(p_2)g^{\mu\nu}\varepsilon_L^{\mu}(q_1)\varepsilon_L^{\mu}(q_2) + O\left(\frac{4m_W^2}{s}\right), \tag{5.96}$$

which grows as $m_e\sqrt{s}/m_W^2$. To cancel this dangerous growth, I need a contribution of a new particle which exactly cancels the sum of the contributions of the others. This particle is the Higgs boson. I must remember this and add the diagram, depicted in Figure 5.12, in which the Higgs boson is exchanged:

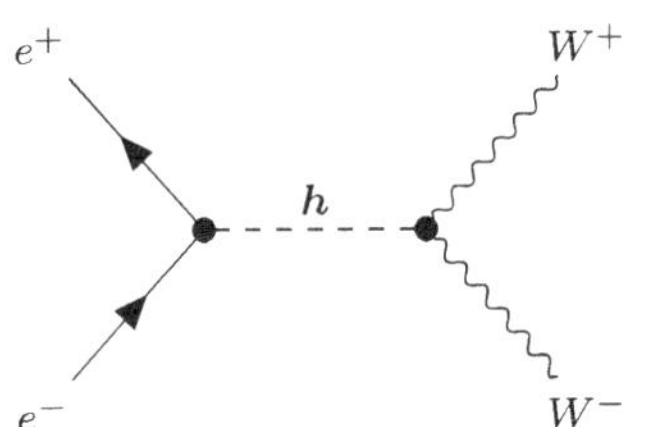

Figure 5.12 The extra diagram with a Higgs boson.

$$\mathcal{M}_h = \frac{g^2 m_e}{2}\frac{1}{s - m_h^2}\bar{v}(p_2)u(p_1)g_{\mu\nu}\varepsilon_L^{\mu}(q_1)\varepsilon_L^{\nu}(q_2). \tag{5.97}$$

The coupling to the electron is found from the Yukawa Lagrangian in Eq. (5.85), after the hiding of the gauge symmetry. It is precisely what is required to cancel the troublesome amplitude in Eq. (5.96).

If I had not already introduced the Higgs boson in order to give mass to the gauge bosons, and all the fermions as well, I would have to invent it in order to restore unitarity in the model. Turning this argument around, the couplings of the Higgs boson to gauge bosons and fermions are exactly those required to make the Standard Model a theory with preserved perturbative unitarity.

5.9 *Problem Session*: The Decay of the Top Quark

The decay of the top quark is computed by means of the Feynman rule for its interaction with the W-boson. The amplitude can be read off the Feynman diagram in Figure 5.13:

$$\mathcal{M} = \frac{g}{\sqrt{2}}V_{tb}\,\bar{u}(q)\gamma^{\mu}\left(\frac{1-\gamma_5}{2}\right)u(p)\varepsilon_{\mu}^{*}(k). \tag{5.98}$$

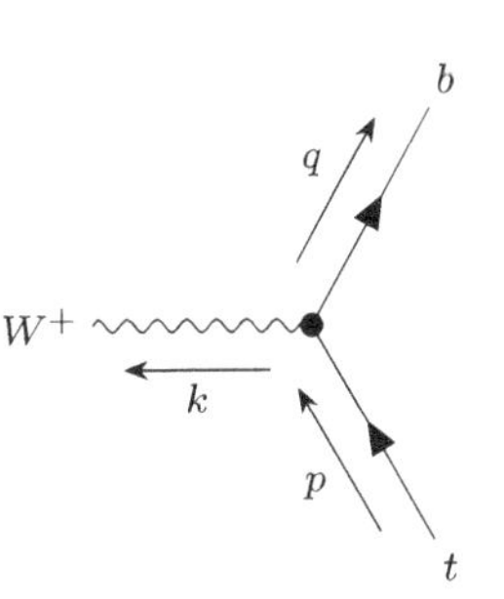

Figure 5.13 Feynman diagram for the decay of the t quark.

Taking the square of the modulus and summing over the polarizations yields

$$\frac{1}{2}\sum_{\text{pol}}|\mathcal{M}|^2 = \frac{g^2}{2}\left[q^{\mu}p^{\nu} + q^{\nu}p^{\mu} - g^{\mu\nu}(p\cdot q)\right]\underbrace{\sum_{\lambda}\varepsilon_{\mu}(k)\varepsilon_{\nu}^{*}(k)}_{-g_{\mu\nu}+\frac{k_{\mu}k_{\nu}}{m_W^2}}. \tag{5.99}$$

I can keep separate the longitudinal and transverse parts of the W-boson polarizations. For the transverse part I have

$$\underbrace{(-g_{\mu\nu})}\,\frac{g^2}{2}\left[q^\mu p^\nu + q^\nu p^\mu - g^{\mu\nu}(p\cdot q)\right] = \frac{g^2}{2}(2q\cdot p) = \frac{g^2}{2}(m_t^2 - m_W^2)\,. \tag{5.100}$$

This would be the contribution to the decay width even if the W-boson were massless.

There is, however, an extra contribution coming from the longitudinal polarization of the massive gauge boson. Because of the origin of the gauge boson mass from the Higgs mechanism, this contribution can be thought of as coming from the decay of the top quark into the Goldstone bosons connected with the hiding of the $SU(2)_L$ symmetry. The width from

5.17 $m_b = 0$

$q\cdot p = k\cdot p - m_W^2$

$2p\cdot k = m_t^2 + m_W^2$

$2p\cdot q = m_t^2 - m_W^2$

$2q\cdot p = m_t^2 - m_W^2\,.$

this part is[5.17]

$$\begin{aligned}
&\underbrace{\left(+\frac{k_\mu k_\nu}{m_W^2}\right)}\,\frac{g^2}{2}\left[q^\mu p^\nu + q^\nu p^\mu - g^{\mu\nu}(p\cdot q)\right]\\
&= \frac{g^2}{2m_W^2}\left[2(q\cdot p)(k\cdot p) - k^2(p\cdot q)\right]\\
&= \frac{g^2}{2m_W^2}\left[(m_t^2 - m_W^2)\frac{1}{2}(m_t^2 + m_W^2) - m_W^2\frac{(m_t^2 - m_W^2)}{2}\right]\\
&= \frac{g^2}{4m_W^2}(m_t^2 - m_W^2)\,m_t^2\,,
\end{aligned} \tag{5.101}$$

which is proportional to the top quark mass, as it should be, coming as it does from the decay into a Goldstone boson.

The sum of the two contributions is given by

$$\begin{aligned}
\frac{1}{2}\sum_{\text{pol}}|\mathcal{M}|^2 &= \frac{g^2}{4}\left(\frac{m_t^4}{m_W^2} - m_t^2 + 2m_t^2 - 2m_W^2\right)\\
&= \frac{g^2}{4}\frac{m_t^4}{m_W^2}\left(1 + \frac{m_W^2}{m_t^2} - 2\frac{m_W^4}{m_t^4}\right)\\
&= \frac{g^2}{4}\frac{m_t^4}{m_W^2}\left(1 - \frac{m_W^2}{m_t^2}\right)\left(1 + 2\frac{m_W^2}{m_t^2}\right),
\end{aligned} \tag{5.102}$$

5.18 $p^* = \frac{1}{2m_t}\sqrt{(m_t^2 - m_W^2)^2} = \frac{m_t}{2}\left(1 - \frac{m_W^2}{m_t^2}\right).$

which gives the width[5.18]

$$\begin{aligned}
\Gamma(t\to Wb) &= \underbrace{\frac{1}{32\pi^2 m_t^2}}_{=\frac{G_F m_t^5}{8\pi\sqrt{2}}}\,|p^*|\,\frac{1}{2}\sum_{\text{pol}}|\mathcal{M}|^2\int \mathrm{d}\Omega\\
&= \frac{g^2 m_t^3}{64\pi}\left(1 - \frac{m_W^2}{m_t^2}\right)^2\left(1 + 2\frac{m_W^2}{m_t^2}\right).
\end{aligned} \tag{5.103}$$

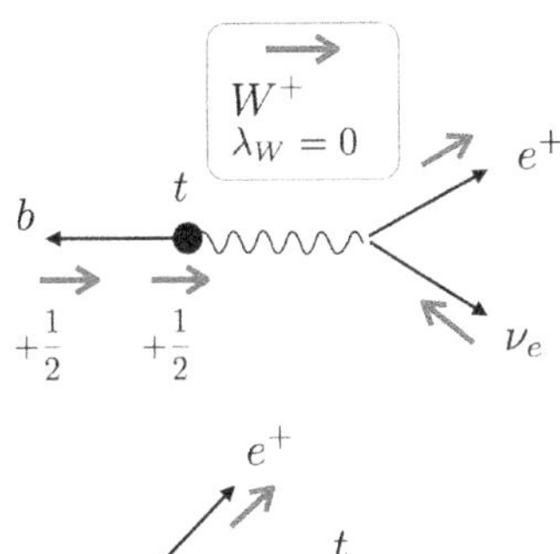

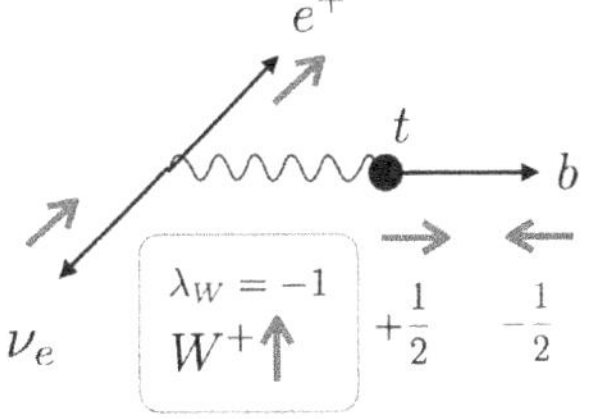

Figure 5.14 The system b–ν_e has zero spin and the electron's momentum is predominantly aligned with the spin of the top quark. These decays can be used to track the top quark polarization.

The composition of the polarizations is shown in Figure 5.14. I can identify in the decay the helicity fractions coming from the longitudinal ($\lambda_W = 0$) and the two transverse polarizations ($\lambda_W = \pm 1$) as I did in Eq. (5.65) for the first step of the process:

$$F_L \equiv \frac{\Gamma_{\text{W-LONG}}(t \to Wb)}{\Gamma(t \to Wb)} = \frac{m_t^2}{m_t^2 + 2\,m_W^2} \tag{5.104}$$

$$F_- \equiv \frac{\Gamma_{\text{W-TRAN}}(t \to Wb)}{\Gamma(t \to Wb)} = \frac{2m_W^2}{m_t^2 + 2\,m_W^2} \tag{5.105}$$

$$F_+ \equiv 0\,. \tag{5.106}$$

There is no contribution from the $\lambda_W = +1$ (right-handed) polarization because of angular momentum conservation, if one neglects the mass of the b quark as I have done.

The helicity fractions in Eq. (5.106) can be measured in the actual experiments to provide a precise test of the Standard Model interactions.

5.10 Exercises

1. **Ferromagnet.** Look up the properties of ferromagnetic materials and try to describe their property under a phase transition in terms of Landau's theory. What is the symmetry? And what is the configuration of the spins in the phase with spontaneously broken symmetry?
2. **Different vacua.** Choose for the Higgs-boson vacuum expectation value $\langle\phi^+\rangle = v/\sqrt{2}$ and $\langle\phi^0\rangle = 0$. Why is this choice not compatible with what we know of our world?
3. **Higgs boson decays.** What is the favorite channel for Higgs boson decay?
   [Hint: Consider the strength of its couplings to fermions and gauge bosons.]
4. **Vector boson scattering.** Write the amplitude for the scattering of the longitudinal components of the gauge boson W. Compute the cross section. How does it behave as a function of the center of mass energy? [Hint: Use the high energy momenta projections in Eq. (5.95).]
5. **Angular momentum.** Check that angular momentum is indeed conserved for the three possible polarizations $\lambda_W = 0, \pm 1$ of the W-boson in the decay of the top quark.

6 Strong Interactions

Contents

The strong interactions are described in the Standard Model by the gauge theory of the color group $SU(3)_c$. Matter is made of "colored" quarks interacting through the exchange of gluons, the gauge fields associated with the color group. The theory is called QCD by analogy to QED, with the electric charge replaced by the color (chromo) charges.

At first blush – and probably also after some consideration – this picture does not seem very plausible. We expect a theory of the strong interactions to be about hadrons and their interactions, indeed about the nuclear forces bounding nucleons inside the nuclei. Not a gauge theory with a small coupling, not the theory of invisible gluons and quarks.

For this reason it is harder to jump right in with the Lagrangian as I did for the weak interactions. We need first a suspension of disbelief, the willingness to suspend judgement concerning the implausibility of the narrative. The gauge theory of the strong interaction is the ultimate outcome of confronting the experimental results with our models. Somewhere during this process, what seemed central, the nuclear force, became the residual effect of a more fundamental force, and the hadrons the bound states of more fundamental degrees of freedom. Looking back, as I do in this chapter, everything falls magnificently into place as nuclear physics gives way to high-energy physics. Eventually it is also possible to go back and derive the properties of nuclear physics from those of QCD.

Having said that – it is easy to write down the Lagrangian. The QCD Lagrangian is that of a gauge theory based on the group $SU(3)_c$. The structure is like that of the QED Lagrangian, only the group structure is different:

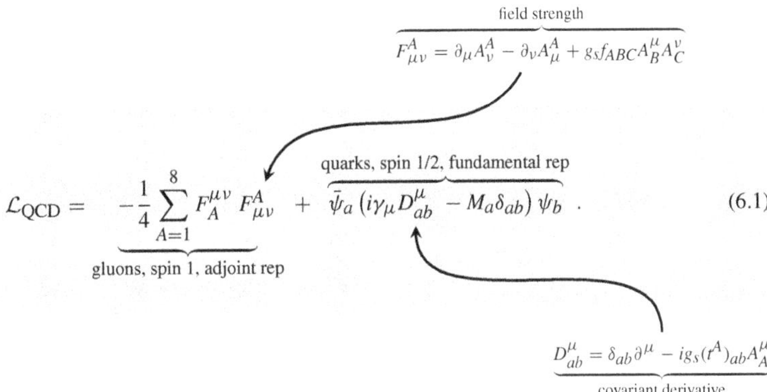

The color index a stands for the fundamental representation of the group (where the quarks ψ_a belong) whereas the index A is for the adjoint representation (for the gauge bosons, called **gluons**). The structure constants for the group $SU(3)$ are defined by their commutation relation:

$$[t_A, t_B] = if_{ABC}t_C, \tag{6.2}$$

where the matrices t_A are conventionally defined as $t_A = \lambda_A/2$, and the matrices λ_A are named after the physicist Murray Gell-Mann. They are for $SU(3)$ what the Pauli matrices T_i are for $SU(2)$. They are given by

$$\lambda_1 = \begin{pmatrix} 0 & 1 & 0 \\ 1 & 0 & 0 \\ 0 & 0 & 0 \end{pmatrix} \quad \lambda_2 = \begin{pmatrix} 0 & -i & 0 \\ i & 0 & 0 \\ 0 & 0 & 0 \end{pmatrix} \quad \lambda_3 = \begin{pmatrix} 1 & 0 & 0 \\ 0 & -1 & 0 \\ 0 & 0 & 0 \end{pmatrix}$$

$$\lambda_4 = \begin{pmatrix} 0 & 0 & 1 \\ 0 & 0 & 0 \\ 1 & 0 & 0 \end{pmatrix} \quad \lambda_5 = \begin{pmatrix} 0 & 0 & -i \\ 0 & 0 & 0 \\ i & 0 & 0 \end{pmatrix} \quad \lambda_6 = \begin{pmatrix} 0 & 0 & 0 \\ 0 & 0 & 1 \\ 0 & 1 & 0 \end{pmatrix}$$

$$\lambda_7 = \begin{pmatrix} 0 & 0 & 0 \\ 0 & 0 & -i \\ 0 & i & 0 \end{pmatrix} \quad \lambda_8 = \frac{1}{\sqrt{3}}\begin{pmatrix} 1 & 0 & 0 \\ 0 & 1 & 0 \\ 0 & 0 & -2 \end{pmatrix}. \tag{6.3}$$

It useful to remember the relations that these matrices satisfy. There are normalizations and sums:

$$\mathrm{Tr}\, t^A t^B = \frac{1}{2}\,\delta^{AB} \tag{6.4}$$

$$\mathrm{Tr}\, t^A_{ij} t^A_{jk} = \frac{4}{3}\,\delta_{ik} \tag{6.5}$$

$$\sum_{C,D} f^{ACD} f^{BCD} = 3\,\delta^{AB}\,. \tag{6.6}$$

In addition, the following relations,

$$\{t^A, t^B\} = \frac{1}{3}\delta^{AB} + d^{ABC}t^C \qquad (\mathrm{Tr}\, t^A = 0) \tag{6.7}$$

$$t^A t^B = T_R\left[\frac{1}{3}\delta^{AB} + (d^{ABC} + if^{ABC})t^C\right], \tag{6.8}$$

and the Fierz relationship

$$\sum_A t^A_{ab} t^A_{cd} = T_R \left[\delta_{ad}\delta_{bc} - \frac{1}{N_c}\delta_{ab}\delta_{cd} \right] \tag{6.9}$$

come in handy in the computations.

The choice of the group $SU(3)$ to describe the strong interactions is the final outcome of the gathering of pieces of various evidence. Granted – I want a gauge theory and I am used by now to see it as the theory of a gauge group. But why 3? Why $SU(3)$?

Traditionally, textbooks list four reasons that played an important role in closing in on the group $SU(3)$. I shall consider three.

Clue 1 The first reason comes from a problem with the symmetry properties of hadrons in the quark model. In the model, baryons are made of three quarks, each coming with its flavor and spin. Consider the spin-3/2 baryons and their quark content (written in terms of flavors u, d and s, and spin, up $\uparrow$ and down $\downarrow$):

$$\begin{array}{ll} \Delta^+ & (d^\uparrow d^\uparrow d^\uparrow) \\ \Delta^{++} & (u^\uparrow u^\uparrow u^\uparrow) \\ \Omega^- & (s^\uparrow s^\uparrow s^\uparrow)\,. \end{array} \tag{6.10}$$

You see the problem? The states at the quark level are three copies of the same wave function and as such are completely symmetric. But how can such a symmetric wave function possibly describe baryons that are fermions, since fermions have antisymmetric wave functions?

The solution is to add one degree of freedom, to wit, the color. This gives an additional index with which to play, so that baryon states, for example, the Δ^+, would be given by

$$\epsilon_{abc} d^\uparrow_a d^\uparrow_b d^\uparrow_c \tag{6.11}$$

in a completely antisymmetric form thanks to the Levi–Civita tensor and the new color indices.[6.1]

6.1 "...Three quarks for Muster Mark! Sure he hasn't got much of a bark And sure any he has it's all beside the mark." J. Joyce, *Finnegan's Wake*

The conclusion is that for the quark model to work and describe actual hadrons it needs to be augmented by the color degree of freedom. And I need three possible values of this color in order to antisymmetrize the baryon wave function.

Clue 2 The second reason for using the group $SU(3)$ comes from the study of the inclusive production of hadrons by electron pair annihilation. This process is best studied by factorizing out the leptonic channel and defining the ratio

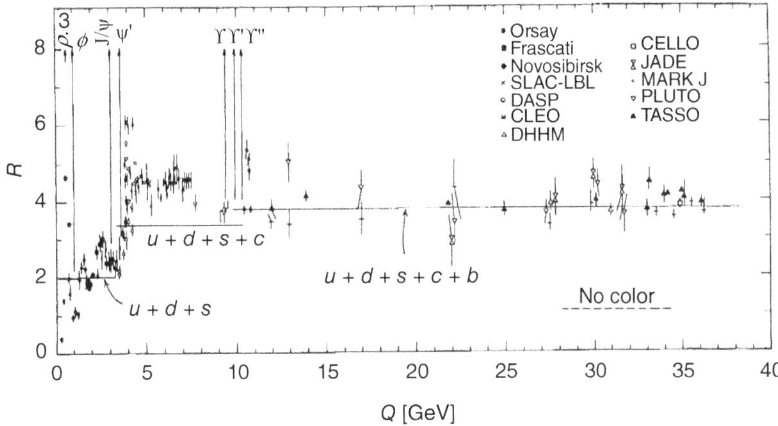

Figure 6.1 The ratio R in inclusive electron pair annihilation, as a function of the center of mass energy [R.L. Workman *et al.* (Particle Data Group), *Prog. Theor. Exp. Phys.* **2022** (2022) 083C01 http://pdg.lbl.gov].

$$
\begin{aligned}
R &= \left| \frac{\mathcal{M}(e^+e^- \to \text{ hadrons via } q\bar{q} \text{ pairs})}{\mathcal{M}(e^+e^- \to \mu^+\mu^-)} \right| \\
&= N_c \sum_j \sqrt{1 - \frac{4m_j^2}{s}} \left(1 + \frac{2m_j^2}{s}\right) Q_j^2 \,,
\end{aligned} \tag{6.12}
$$

where m_j are the quark masses and s is proportional to the square of the center of mass energy. The quarks are of various flavors: u, d, s, c and b (see Chapter 7).

The second identity in Eq. (6.12) follows from the coupling of quarks to the photon in the quark model, where they have charges Q_j and are free.[6.2]

6.2 Actually this computation does not seem right. We measure hadrons in the final states but compute as if they were free quarks. I come back to this in Chapter 7.

This ratio R has been studied for a wide range of energies – as shown in Figure 6.1. The narrow peaks correspond to the resonant production of bounded states of quarks. They are the very reason for the study of this process but this is not what is important here. What is important here is the values of R at the plateaus. For energies $E < 2m_c$ there is a first plateau around $R \simeq 2$. If I insert the charges of the quarks with a mass lighter than m_c, and take into account that each of them can take three possible colors, then the right-hand side of Eq. (6.12) is

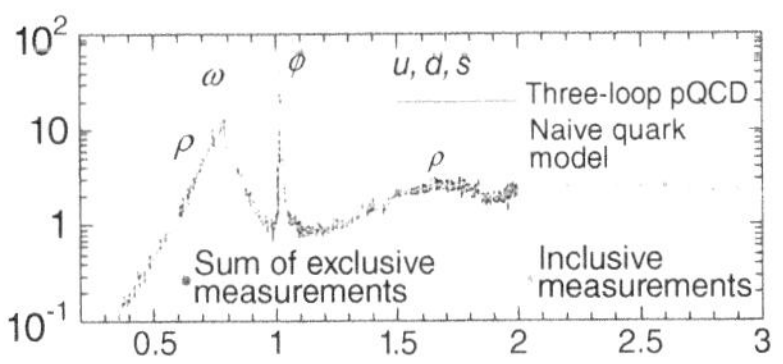

Figure 6.2 The ratio R up to 2 GeV [R.L. Workman *et al.* (Particle Data Group), *Prog. Theor. Exp. Phys.* **2022** (2022) 083C01 http://pdg.lbl.gov].

$N_c = 3$ for the colors

$$
\begin{aligned}
N_c \sum_j Q_j^2 &= 3 \times \left[\overbrace{\left(\frac{2}{3}\right)^2}^{u} + \overbrace{\left(-\frac{1}{3}\right)^2}^{d} + \overbrace{\left(-\frac{1}{3}\right)^2}^{s} \right] \\
&\simeq \boxed{2} \,,
\end{aligned} \tag{6.13}
$$

which agrees with the data in Figure 6.2 only because I have included a factor 3 for the number of colors.

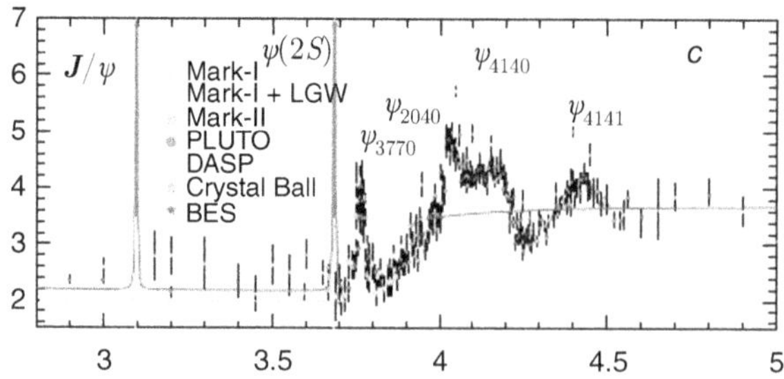

Figure 6.3 The ratio *R* up to 4 GeV. *c* refers to the *c* quark. [R.L. Workman *et al.* (Particle Data Group), *Prog. Theor. Exp. Phys.* **2022** (2022) 083C01 http://pdg.lbl.gov.]

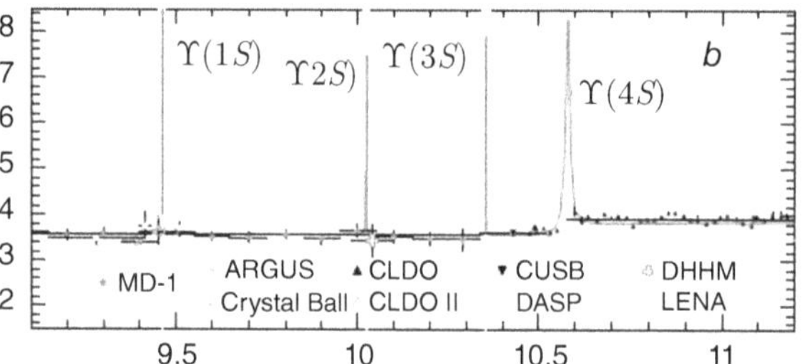

Figure 6.4 The ratio *R* up to 11 GeV. *b* refers to the *b* quark. [R.L. Workman *et al.* (Particle Data Group), *Prog. Theor. Exp. Phys.* **2022** (2022) 083C01 http://pdg.lbl.gov.]

The same is true as one moves towards higher energies $E > 2m_c$, where there is a second plateau (Figure 6.3) just above $R = 3$:

$$N_c \sum_j Q_j^2 = 3 \times \left[\overbrace{\left(\frac{2}{3}\right)^2}^{u} + \overbrace{\left(-\frac{1}{3}\right)^2}^{d} + \overbrace{\left(-\frac{1}{3}\right)^2}^{s} + \overbrace{\left(\frac{2}{3}\right)^2}^{c} \right] \simeq \boxed{3.33}. \tag{6.14}$$

Further up, for $E > 2m_b$, there is another plateau, as shown at the right in Figure 6.4, at a slightly higher level due to the possibility of producing the b quark:

$$N_c \sum_j Q_j^2 = 3 \times \left[\overbrace{\left(\frac{2}{3}\right)^2}^{u} + \overbrace{\left(-\frac{1}{3}\right)^2}^{d} + \overbrace{\left(-\frac{1}{3}\right)^2}^{s} + \overbrace{\left(\frac{2}{3}\right)^2}^{c} + \overbrace{\left(-\frac{1}{3}\right)^2}^{b} \right] \simeq \boxed{3.67}. \tag{6.15}$$

The agreement between the values of the various plateaus and the predictions of Eq. (6.12) is only possible because of the additional degree of freedom in the quark model and because of its three values.

Clue 3 The third, and more theoretical reason, for closing in on the group $SU(3)$ I leave it for a problem session.

The additional degree of freedom with three values seems to be solid. Yet why, you may ask, the group $SU(3)$? There are other groups of dimension 3:

$$\underbrace{SO(3), \quad SU(2), \quad Sp(1)}_{\simeq SO(3)} \quad \text{and} \quad SU(3)\,. \tag{6.16}$$

The first three are actually isomorphic so the choice boils down to $SO(3)$ versus $SU(3)$. Which one? The fundamental representation of $SO(3)$ is real and we need to distinguish quarks from antiquarks. Therefore I need complex representations. This leaves me with $SU(3)$.

Having justified the choice of $SU(3)_c$ as the gauge group of the strong interactions, I am ready to write down the Feynman rules as derived from the Lagrangian in Eq. (6.1): there is a vertex between a gluon and colored fermions (the quarks) and vertices with three and four gluons coming from the nonlinear structure of the non-Abelian field strengths. They are given in Figure 6.5.

The propagators for gluons and quarks are (no surprise here) given in Figure 6.6.

6.1 *Problem Session*: The Gauge Anomaly*

Symmetries are discussed in terms of classical Lagrangians. The fields transform and the Lagrangian remains the same. What happens when we take into account the quantum nature of our world? In most cases nothing, and the symmetries are carried over the quantum world.

Yet there are instances in which going to the quantum theory cannot be done without breaking some classical symmetries. This comes about because the contribution of the loop diagrams arising in the quantum theory can be divergent and require some form of regularization. It is the regularization that breaks the classical symmetry: it is not possible to find a regularization procedure that does not interfere with the symmetry.

This symmetry breaking can be a good or a bad thing. It could be good for a global symmetry that we would rather do without. It is a disaster if the breaking has to do with gauge freedom, where it would bring down the delicate mechanism of current conservation and Ward identities.

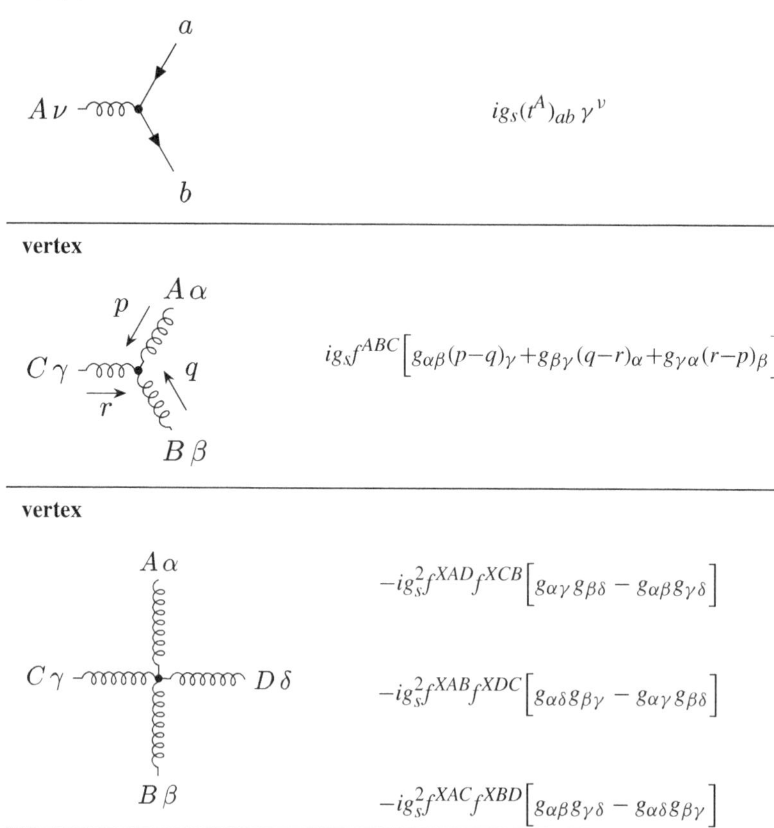

Figure 6.5 The basic vertices in QCD between quarks and gluons and among gluons.

gluon propagator

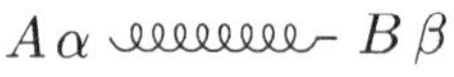

$A\,\alpha$ —— $B\,\beta$

$$\delta^{AB}\left[-g^{\alpha\beta} + (1-\xi)\frac{p^\alpha p^\beta}{p^2}\right]\frac{-i}{p^2}$$

quark propagator

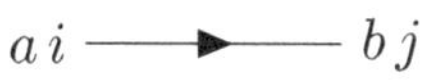

$a\,i$ —▶— $b\,j$

$$\delta^{ab}\frac{i}{(\not p - m)_{ji}}$$

Figure 6.6 The propagators of quarks and gluons.

In this section I discuss what is called the gauge anomaly. It arises by computing the one-loop integral in the triangle[6.3] diagrams shown in Figure 6.7. I attach at the vertices of the triangle an axial current and two vector currents of the Standard Model.

[6.3] It is a triangle in four space-time dimensions, a two-point blob in two dimensions and a square in six.

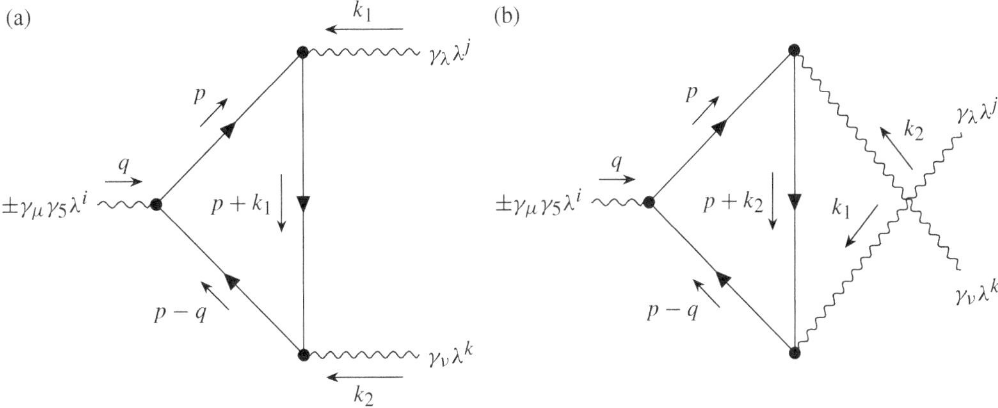

Figure 6.7 The loop diagrams giving rise to the gauge anomaly. The external wavy lines represent the insertion of the axial and vector currents.

The amplitude represented in Figure 6.7 is given by

$$
\begin{aligned}
&(2\pi)^4\delta^{(4)}(q+k_1+k_2)\Gamma^{ijk}_{\mu\nu\lambda}(k_1,k_2,q) \\
&\quad = -i\int \mathrm{d}^4x_1\mathrm{d}^4x_2\mathrm{d}^4x_3\, e^{ik_1\cdot x_1+ik_2\cdot x_2+iq\cdot x_3}\langle 0|A^i_\mu(x_1)V^j_\nu(x_2)V^k_\lambda(x_3)|0\rangle\,,
\end{aligned}
\tag{6.17}
$$

where the currents are

vector current

$$
A^i_\mu(x) = \bar{\psi}\frac{\hat{\lambda}^i}{2}\gamma_\mu\gamma_5\psi \quad \text{and} \quad V^j_\mu(x) = \bar{\psi}\frac{\hat{\lambda}^i}{2}\gamma_\mu\psi. \tag{6.18}
$$

axial current

The matrices λ^i stand for the various generators of the Standard Model groups.

Conservation of the vector and axial currents $\partial^\mu V^i_\mu - \partial^\mu A^i_\mu = 0$ can be written in Fourier space as

$$
q^\mu\Gamma^{ijk}_{\mu\nu\lambda}(k_1,k_2,q) = k_1^\lambda\Gamma^{ijk}_{\mu\nu\lambda}(k_1,k_2,q) = k_2^\nu\Gamma^{ijk}_{\mu\nu\lambda}(k_1,k_2,q) = 0\,. \tag{6.19}
$$

Equation (6.19) gives the Ward identities discussed in Chapter 3. Does the amplitude from the two diagrams in Figure 6.7 satisfy Eq. (6.19)? To find out I must – what else? – do a computation. Brace yourself because it is one of those that are a bit of a challenge.

The two loop diagrams in Figure 6.7 give (neglecting the fermion masses)

$$
\begin{aligned}
-i\Gamma^{ijk}_{\mu\nu\lambda}(k_1,k_2,q) = \overbrace{(-1)}^{\text{fermion loop}}\int \frac{\mathrm{d}^4p}{(2\pi)^4}\,&\overbrace{\mathrm{Tr}\left[\frac{i}{\not p}\gamma_\mu\gamma^5\hat\lambda^i\frac{i}{\not p-\not q}\gamma_\lambda\hat\lambda^j\frac{i}{\not p+\not k_1}\gamma_\nu\hat\lambda^k\right]}^{\text{diagram (a) in Figure 6.7}}\\
+(-1)\int\frac{\mathrm{d}^4p}{(2\pi)^4}\,&\underbrace{\mathrm{Tr}\left[\frac{i}{\not p}\gamma_\mu\gamma^5\hat\lambda^i\frac{i}{\not p-\not q}\gamma_\nu\hat\lambda^k\frac{i}{\not p+\not k_2}\gamma_\lambda\hat\lambda^j\right]}_{\text{diagram (b) in Figure 6.7}},
\end{aligned}
\tag{6.20}
$$

where the second integral (corresponding to diagram (b) in Figure 6.7) is obtained from the first by the substitution of $k_1 \to k_2$ and interchange of the indices λ and ν.

The gauge group matrices can be reorganized as follows:

$$
\mathrm{Tr}\left(\hat\lambda^i\hat\lambda^j\hat\lambda^k\right) = \frac{1}{2}\mathrm{Tr}\left[[\hat\lambda^i,\hat\lambda^j]\,\hat\lambda^k\right] + \frac{1}{2}\mathrm{Tr}\left[\{\hat\lambda^i,\hat\lambda^j\}\,\hat\lambda^k\right], \tag{6.21}
$$

where the first term is totally antisymmetric in the group indices and the second term is instead totally symmetric. The two terms in Eq. (6.20) multiplied by the antisymmetric combination give the difference between the two diagrams – which contributes to the renormalization of the three-gauge-boson vertex. The two terms multiplied by the symmetric combination give the sum of the two diagrams in which I am interested. Contracting the above vertex with momentum q gives[6.4]

$$
-iq^\mu\Gamma^{ijk}_{\mu\nu\lambda}(k_1,k_2,q) = \Xi^{1,ijk}_{\lambda\nu} + \Xi^{2,ijk}_{\lambda\nu}, \tag{6.22}
$$

where

$$
\begin{aligned}
\Xi^{1,ijk}_{\lambda\nu} =& \frac{i}{2}\,\mathrm{Tr}\left(\{\hat\lambda^i,\hat\lambda^j\}\,\hat\lambda^k\right)\int\frac{\mathrm{d}^4p}{(2\pi)^4}\\
&\times\mathrm{Tr}\left[\frac{1}{\not p}\gamma^5\gamma_\nu\frac{1}{\not p+\not k_1}\gamma_\lambda - \frac{1}{\not p+\not k_2}\gamma^5\gamma_\nu\frac{1}{\not p-\not q}\gamma_\lambda\right]
\end{aligned}
\tag{6.23}
$$

and

$$
\begin{aligned}
\Xi^{2,ijk}_{\lambda\nu} =& \frac{i}{2}\,\mathrm{Tr}\left(\{\hat\lambda^i,\hat\lambda^j\}\,\hat\lambda^k\right)\int\frac{\mathrm{d}^4p}{(2\pi)^4}\\
&\times\mathrm{Tr}\left[-\frac{1}{\not p+\not k_1}\gamma^5\gamma_\lambda\frac{1}{\not p-\not q}\gamma_\nu + \frac{1}{\not p}\gamma^5\gamma_\lambda\frac{1}{\not p+\not k_2}\gamma_\nu\right].
\end{aligned}
\tag{6.24}
$$

If I replace the momentum p (which is a dummy variable in any case) with $p+k_2$ in the first term of both integrals, the two terms in Eq. (6.23) and in Eq. (6.24) each become identical[6.5] but for their sign: $\Xi^{1,ijk}_{\lambda\nu}$ and $\Xi^{2,ijk}_{\lambda\nu}$ both vanish.[6.6]

[6.4] Use $\not q\gamma^5 = -\gamma^5\not q = \gamma^5(\not p-\not q)+\not p\gamma^5$ and the cyclicality of the trace. This substitution cancels some of the denominators.

[6.5] Use $-q = k_1 + k_2$.

[6.6] The naive expectation is not correct. The lesson is: do not manipulate divergent quantities as if they were finite.

I had better be careful because I am taking the difference between two divergent integrals. Here is where the frogs jump into the water.

I redo the subtraction by first introducing explicitly the shift s in the integration momentum of the second term in the integrals in Eq. (6.23) and Eq. (6.24), defining the two integrals

$$\tilde{\Xi}^{1,ijk}_{\lambda\nu} = \frac{i}{2}\,\mathrm{Tr}\left(\{\hat{\lambda}^i,\hat{\lambda}^j\}\,\hat{\lambda}^k\right)\int\frac{\mathrm{d}^4p}{(2\pi)^4} \times \mathrm{Tr}\left[\frac{1}{\not p}\gamma^5\gamma_\nu\frac{1}{\not p+\not k_1}\gamma_\lambda - \frac{1}{\not p+\not s+\not k_2}\gamma^5\gamma_\nu\frac{1}{\not p+\not s-\not q}\gamma_\lambda\right] \tag{6.25}$$

and

$$\tilde{\Xi}^{2,ijk}_{\lambda\nu} = \frac{i}{2}\,\mathrm{Tr}\left(\{\hat{\lambda}^i,\hat{\lambda}^j\}\,\hat{\lambda}^k\right)\int\frac{\mathrm{d}^4p}{(2\pi)^4} \times \mathrm{Tr}\left[-\frac{1}{\not p+\not k_1}\gamma^5\gamma_\lambda\frac{1}{\not p-\not q}\gamma_\nu + \frac{1}{\not p+\not s}\gamma^5\gamma_\lambda\frac{1}{\not p+\not s+\not k_2}\gamma_\nu\right]. \tag{6.26}$$

Each of these two integrals is in the form

$$i\int\frac{\mathrm{d}^4p}{(2\pi)^4}\Big[f_{\lambda\nu}(p) - f_{\lambda\nu}(p+\Delta)\Big], \tag{6.27}$$

and I need to compute this difference in the limit $\Delta\to 0$. This can be done by first expanding it in a Taylor series as

$$i\int\frac{\mathrm{d}^4p}{(2\pi)^4}\Big[f^{\lambda\nu}(p) - f^{\lambda\nu}(p+\Delta)\Big] = i\int\frac{\mathrm{d}^4p}{(2\pi)^4}\Big[\Delta^\mu\partial_\mu f_{\lambda\nu} + \cdots\Big], \tag{6.28}$$

where the integral on the right – which is finite – can be written as

$$i\int_{\partial S^4}\frac{\mathrm{d}^3S_\mu}{(2\pi)^4}\Delta^\mu f^{\lambda\nu}(p), \tag{6.29}$$

that is, an integral over the boundary of S^4 defined as $p\to\infty$.

The integration $\mathrm{d}^3S_\mu = |p|^2 p_\mu \mathrm{d}\Omega_3$, with $\Omega_3 = 2\pi^2$. In the case of interest,

$$f_{\lambda\nu}(p) = \mathrm{Tr}\left[\frac{1}{\not p}\gamma^5\gamma_\lambda\frac{1}{\not p+\not k_1}\gamma_\nu\right] = \frac{1}{p^2(p+k_1)^2}\mathrm{Tr}\left[\not p\gamma^5\gamma_\lambda(\not p+\not k_1)\gamma_\nu\right] = -4i\epsilon_{\lambda\alpha\nu\beta}\frac{k_1^\alpha p^\beta}{p^2(p+k_1)^2} \tag{6.30}$$

and I have, for $\Delta = s + k_2$, that[6.7]

6.7 $\int \mathrm{d}S_\mu p^\beta = \frac{\pi^2}{4}\delta^\beta_\mu$.

$$\tilde{\Xi}^{1,ijk}_{\lambda\nu} = -2\mathrm{Tr}\left[\{\hat{\lambda}^i,\hat{\lambda}^j\}\,\hat{\lambda}^k\right]\int_{\partial S^4}\frac{\mathrm{d}S_\mu}{(2\pi)^4}\epsilon_{\lambda\alpha\nu\beta}(s+k_2)^\mu k_1^\alpha p^\beta\frac{|p|^2}{p^2(p+k_1)^2} = -\frac{1}{16\pi^2}\mathrm{Tr}\left[\{\hat{\lambda}^i,\hat{\lambda}^j\}\,\hat{\lambda}^k\right]\epsilon_{\lambda\alpha\nu\beta}k_1^\alpha(k_2+s)^\beta. \tag{6.31}$$

Going through the same steps with $\Delta = k_1 - s$ gives

$$\tilde{\Xi}^{2,ijk}_{\lambda\nu} = +\frac{1}{16\pi^2}\mathrm{Tr}\left[\{\hat{\lambda}^i, \hat{\lambda}^j\}\,\hat{\lambda}^k\right]\epsilon_{\lambda\nu\alpha\beta}k_2^\alpha(k_1 - s)^\beta \tag{6.32}$$

and therefore

$$-iq^\mu\Gamma^{ijk}_{\mu\nu\lambda} = \frac{1}{16\pi^2}\mathrm{Tr}\left[\{\hat{\lambda}^i, \hat{\lambda}^j\}\,\hat{\lambda}^k\right]\epsilon_{\nu\lambda\alpha\beta}\left[2k_1^\alpha k_2^\beta + (k_1 + k_2)^\alpha s^\beta\right], \tag{6.33}$$

which vanishes for $s = -2k_2$. While it seems that there is indeed a cancellation between the two terms, before concluding anything I must check the conservation of the two vector currents. The same computation gives

$$-ik^{2\,\nu}\Gamma^{ijk}_{\mu\nu\lambda} = \frac{1}{16\pi^2}\mathrm{Tr}\left[\{\hat{\lambda}^i, \hat{\lambda}^j\}\,\hat{\lambda}^k\right]\epsilon_{\mu\lambda\alpha\beta}k_1^\alpha(s - k_2)^\beta \tag{6.34}$$

and

$$-ik^{1\,\lambda}\Gamma^{ijk}_{\mu\nu\lambda} = \frac{1}{16\pi^2}\mathrm{Tr}\left[\{\hat{\lambda}^i, \hat{\lambda}^j\}\,\hat{\lambda}^k\right]\epsilon_{\mu\nu\alpha\beta}k_2^\alpha(s + k_1)^\beta, \tag{6.35}$$

and it can be seen that for the choice $s = -2k_2$ these two currents are not conserved. Usually, however, we want the vector currents to be conserved and so we take $s = k_2 - k_1$. This choice leads to

$$-iq^\mu\Gamma^{ijk}_{\mu\nu\lambda} = -\frac{1}{16\pi^2}\mathrm{Tr}\left[\{\hat{\lambda}^i, \hat{\lambda}^j\}\,\hat{\lambda}^k\right]\epsilon_{\nu\lambda\alpha\beta}k_1^\alpha k_2^\beta \tag{6.36}$$

and the anomaly for the axial current.

The final result is[6.8]

$$\partial^\mu V^i_\mu = 0 \tag{6.37}$$

$$\partial^\mu A^i_\mu = \frac{g^2}{16\pi}\mathrm{Tr}\left[\{\hat{\lambda}^i, \hat{\lambda}^j\}\,\hat{\lambda}^k\right]\epsilon_{\mu\nu\alpha\beta}F_j^{\mu\nu}F_k^{\alpha\beta}, \tag{6.38}$$

and the axial current is no longer conserved. There is an anomaly because the symmetry has been broken. The Ward identity is violated and the gauge theory made inconsistent.

Maybe it is not as bad as it seems. The group factor, with its anticommutator, may actually give zero. It depends on the charges of the fermions running in the loop. I need to check.

It comes down to computing the group factor for all the four possible triangle diagrams, that is, the diagrams which contain all the possible electroweak currents. There are four possibilities, depending on which of the Standard Model groups are inserted at each vertex:[6.9]

- $[SU(2)_L]^3$: it is proportional to Tr $\left(T_i\,\{T_j, T_k\}\right) = \delta_{jk}\mathrm{Tr}\,T_i = 0.$ ✓
- $[U(1)_Y]^2 SU(2)_L$: Tr $\left(T_i\,\{\hat{Y}, \hat{Y}\}\right) = 2\,\hat{Y}^2\mathrm{Tr}\,T_i = 0.$ ✓
- $U(1)_Y[SU(2)_L]^2$: it is proportional to Tr $\left(\hat{Y}\,\{T_j, T_k\}\right) = \delta_{jk}\mathrm{Tr}\,\hat{Y}.$

6.8 As just shown, which current is conserved is a matter of definition. What is not a matter of definition is that at least one, or a combination of the two currents, is not conserved.

6.9 $\{T_j, T_k\} = \delta_{jk}\mathbb{1}$.

- $[U(1)_Y]^3$: it is proportional to $\mathrm{Tr}\left(\hat{Y}\{\hat{Y},\hat{Y}\}\right) = 2\mathrm{Tr}\,\hat{Y}^3$.

The first two group factors are identically zero and I do not have to worry about them. The other two can vanish only for a particular assignment of the fermion charges. It appears that there can be a vanishing group factor if[6.10]

6.10 Here, the charge assignments introduced in Chapter 2 are needed. Only the doublets enter in the sum of Eq. (6.39) because they must couple to the two $SU(2)$ vertices in the triangle. The factors of 2 keep track of the two degrees of freedom in the doublets.

$$\sum_{\text{doublets}} \hat{Y} = \sum_{L\text{ quarks}} \hat{Y} + \sum_{L\text{ leptons}} \hat{Y} = \overbrace{2N_c\left(\frac{1}{3}\right)}^{(u_L,\,d_L)} + \overbrace{2(-1)}^{(\nu_L,\,e_L)}$$
$$= \frac{2}{3}N_c - 2 \tag{6.39}$$

and

$$\sum \hat{Y}^3 = \sum_{L\text{ quarks}} \hat{Y}^3 - \sum_{R\text{ quarks}} \hat{Y}^3 + \sum_{L\text{ leptons}} \hat{Y}^3 - \sum_{R\text{ leptons}} \hat{Y}^3$$
$$= \overbrace{2N_c\left(\frac{1}{3}\right)^3}^{(u_L,\,d_L)} - \overbrace{N_c\left(\frac{4}{3}\right)^3}^{u_R} - \overbrace{N_c\left(-\frac{2}{3}\right)^3}^{d_R} + \overbrace{2(-1)^3}^{(\nu_L,\,e_L)} - \overbrace{(-2)^2}^{e_R}$$
$$= 6 - 2N_c \tag{6.40}$$

are both zero.[6.11]

6.11 Left- and right-handed chiral fermions contribute with opposite signs because of the axial coupling.

The requirement of the vanishing of the sum of the hypercharges as well as the cube of the same hypercharges gives two conditions and the solution:

$$\begin{cases} \frac{2}{3}N_c - 2 &= 0 \\ 6 - 2N_c &= 0 \end{cases} \quad \Longrightarrow \quad \boxed{N_c = 3}. \tag{6.41}$$

The anomaly vanishes but only if the number of colors is three! Actually, it vanishes only if the matter content in the Standard Model has exactly the charges that it does have.[6.12]

6.12 The cancellation works generation by generation. If it worked only for the three generations together then there would be an explanation for the presence of the three copies of fermions in the Standard Model.

Other two potentially dangerous triangle anomalies are those proportional to $U(1)_Y[SU(2)_c]^2$ (two gluon insertions in the triangle) and $U(1)_Y[G]^2$ (two graviton insertions in the triangle – and, yes, admittedly that does go beyond the Standard Model proper). They cancel for the charge assignments of the Standard Model since they are proportional to

$$\sum_{L\text{ quarks}} \hat{Y} - \sum_{R\text{ quarks}} \hat{Y} = 2N_c\left(\frac{1}{3}\right) - N_c\left(\frac{4}{3}\right) + N_c\left(\frac{2}{3}\right)$$
$$= 0 \quad (\text{independently of } N_c) \tag{6.42}$$

and[6.13]

$$\sum_{L\text{ quarks}} \hat{Y} - \sum_{R\text{ quarks}} \hat{Y} + \sum_{L\text{ leptons}} \hat{Y} - \sum_{R\text{ leptons}} \hat{Y}$$
$$= 2N_c\left(\frac{1}{3}\right) - N_c\left(\frac{4}{3}\right) + N_c\left(\frac{2}{3}\right) + 2(-1) - (-2)$$
$$= 0 \quad \text{(independently of } N_c\text{)}. \tag{6.43}$$

6.13 The coupling of the fermions to the gravitons is done by decomposing the metric in terms of vierbeins and thus having a covariant derivative $D_\mu \psi_\alpha = \partial_\mu + \frac{1}{2}\omega_\mu^{ab}(J_{ab})_\alpha^\beta \psi_\beta$ with ω^{ab} playing the role of gauge field. There are two diagrams in addition to the two in Figure 6.7 because of the two-graviton vertex.

The cancellation of the gravitational anomaly is particularly striking. Nowhere have gravitational interactions entered the Standard Model before; here they do and in such a graceful manner as to not invalidate the gauge structure.

I could have worked the other way around and, instead of starting with the known hypercharge assignments in the Standard Model, determined the possible charges by requiring the cancellation of the gauge anomalies. There are only two solutions: the Standard Model solution and another solution in which all doublets must have vanishing hypercharge.[6.14] The cancellation of the gauge anomaly uniquely assign the hypercharges to all fermions and provides a compelling explanation of why the hypercharge, and therefore the electric charge, is quantized.

6.14 This would no longer be true if there were a ν_R carrying hypercharge.

6.2 *Problem Session*: The Force Between Two Quarks

The best way to become acquainted with the Lagrangian in Eq. (6.1) is to use it to do a simple computation. If I take two colored quarks and put them near each other, I expect to be be able to see the force with which they interact. This computation can be done within the S-matrix framework by writing the amplitude for the scattering of the two quarks and going to the nonrelativistic limit where they move slowly and their momenta are little changed. This approximation in QED gives rise to the Coulomb potential between two electrons, and something similar should happen here.

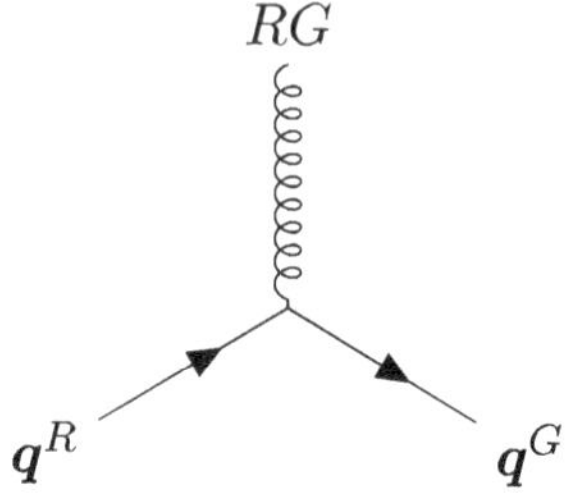

Figure 6.8 The colors in a vertex between quarks and gluons.

The vertex interaction $ig_s(t^A)_{ab}$ (Figure 6.8) in color space, for instance when the indices ab denote a red quark turning into a green quark, is given as (if $A = 1$)

$$\frac{ig_s}{2}\overbrace{\begin{pmatrix}1 & 0 & 0\end{pmatrix}}^{\textbf{red}\text{ quark}}\underbrace{\begin{pmatrix}0 & 1 & 0\\ 1 & 0 & 0\\ 0 & 0 & 0\end{pmatrix}}_{\text{Gell-Mann matrix }\lambda_1}\overbrace{\begin{pmatrix}0\\ 1\\ 0\end{pmatrix}}^{\textbf{green}\text{ quark}}. \tag{6.44}$$

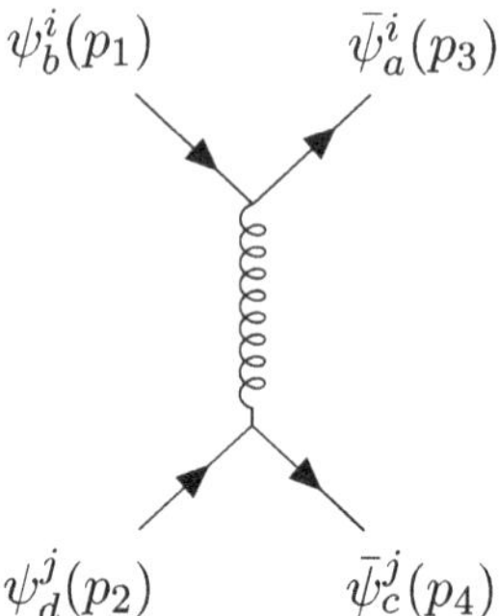

Figure 6.9 The colors in the interaction between two quarks.

I can take two of these vertices and construct the amplitude for the process in Figure 6.9, in which I am interested:

$$-i\mathcal{M} = \underbrace{\left[\bar{\psi}_c^j(p_4)(ig_s)(t^A)_{cd}\gamma^\mu\psi_d^j(p_2)\right]}_{\text{vertex}} \overbrace{\frac{-ig_{\mu\nu}}{t}\,\delta^{AB}}^{\text{gluon propagator } (\xi=1)}$$
$$\times \underbrace{\left[\bar{\psi}_a^i(p_3)(ig_s)(t^B)_{ab}\gamma^\nu\psi_b^i(p_1)\right]}_{\text{vertex}}, \tag{6.45}$$

where $t = -\vec{k}^2$ is the square of the exchanged momentum and i,j are flavor indices that I take to be different so as not to have to worry about the s-channel diagram. Keeping only the charge densities of the quarks (the 0-component of the vertices),

$$\rho = J^0 = \bar{\psi}_i^b\gamma^0\psi_i^a, \tag{6.46}$$

put all momenta on the mass shell and approximately equal ($p_i \simeq p$), and use the normalization of the wave functions to write $\psi_i^\dagger\psi_j = \delta_{ij}$. In this nonrelativistic approximation, the S-matrix can be written as

$$S = i\delta^{(4)}\left(\sum_i p_i\right) T_{\mathrm{NR}}, \tag{6.47}$$

where

$$T_{\mathrm{NR}} = \frac{g_s^2}{t}\sum_A (t^A)_{ab}(t^A)_{cd}, \tag{6.48}$$

which looks very simple; only the color factor remains to be evaluated.

At this point I could bring in the machinery of group theory, which tells me that

$$\sum_A (t^A)_{ab}(t^A)_{cd} = \frac{1}{2}\left[\delta_{da}\delta_{bc} - \frac{1}{3}\delta_{ab}\delta_{cd}\right], \tag{6.49}$$

but it is more instructive to do things by hand and check each possible contribution from each diagram.

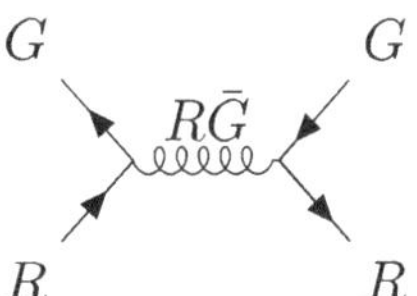

Figure 6.10 Quark colors changed by the interaction.

If one quark starts out with one color, say red (R), say, and converts to another, say green (G), the other quark must do the reverse (Figure 6.10). The process is mediated by a $R\bar{G}$ gluon, which is represented by the two matrices

$$\lambda_1 = \begin{pmatrix} 0 & 1 & 0 \\ 1 & 0 & 0 \\ 0 & 0 & 0 \end{pmatrix} \quad \text{and} \quad \lambda_2 = \begin{pmatrix} 0 & -i & 0 \\ i & 0 & 0 \\ 0 & 0 & 0 \end{pmatrix}. \tag{6.50}$$

The contribution to the color factor in Eq. (6.48) (remember that the matrices t are half of the Gell-Mann matrices λ) comes from inserting these matrices at the two vertices. If I take the first, the result in Eq. (6.44) is multiplied by itself, that is, a factor $1 \cdot 1$. Together, the two vertices give

$$\frac{1}{2}\left[1\cdot 1 + i\cdot(-i)\right] = 1. \tag{6.51}$$

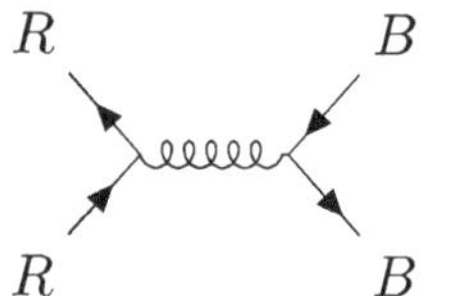

Figure 6.11 Quark colors unchanged by the interaction.

If now I take the case in Figure 6.11, in which the quarks instead remain with the same color, say R and B, respectively, the process must be mediated by gluons diagonal in the color, that is,

$$\lambda_3 = \begin{pmatrix} 1 & 0 & 0 \\ 0 & -1 & 0 \\ 0 & 0 & 0 \end{pmatrix} \quad \text{and} \quad \lambda_8 = \frac{1}{\sqrt{3}} \begin{pmatrix} 1 & 0 & 0 \\ 0 & 1 & 0 \\ 0 & 0 & -2 \end{pmatrix}, \tag{6.52}$$

with a contribution which is now

$$\frac{1}{2}\left[1 \cdot 0 + \frac{1}{\sqrt{3}} \cdot \frac{-2}{\sqrt{3}}\right] = -\frac{1}{3}. \tag{6.53}$$

I can see now where the two terms inside the brackets in Eq. (6.49) come from and proceed with more confidence. What should I substitute for the color sum in Eq. (6.49)? It depends on the pair of quarks I want to consider. For quarks in an antisymmetric color state $q\bar{q}$, the color-exchanging operator $\delta_{da}\delta_{bc}$ in Eq. (6.49) must have eigenvalue -1, and the color-conserving operator $\delta_{ab}\delta_{cd}$ has eigenvalue $+1$. In this case therefore the color sum is $-2/3$. For the symmetric state qq, both the color-exchanging and color-conserving operators must have eigenvalue $+1$ and the color-sum is $+1/3$. Thus the potential for $q\bar{q}$ is attractive and twice as strong as the repulsive potential for qq.

Consequently,

$$T_{\rm NR} = \begin{pmatrix} -2/3 \\ +1/3 \end{pmatrix} \frac{g_s^2}{t}. \tag{6.54}$$

($-2/3$ for $\bar{q}q$ (color antisymmetric); $+1/3$ for qq (color symmetric))

The quarks are bounded inside a meson state $|\pi\rangle$, the wave function of which is given by

$$|\pi\rangle = \frac{1}{\sqrt{3}}\left(R\bar{R} + B\bar{B} + G\bar{G}\right), \tag{6.55}$$

which gives me a factor $1/3$ in the amplitude, and I need to cover the 3! possible color assignments. The potential in space is obtained after Fourier transformation as[6.15]

$$V(\vec{r}) = \int \frac{d^3\vec{k}}{(2\pi)^3} e^{-i\vec{k}\cdot\vec{r}} T_{\rm NR}(\vec{k})$$
$$= \begin{pmatrix} -2/3 \\ +1/3 \end{pmatrix} \left(\frac{g_s^2}{4\pi}\right) \frac{1}{r}\; 3! \left(\frac{1}{3}\right) = \begin{pmatrix} -4/3 \\ +2/3 \end{pmatrix} \frac{\alpha_s}{r}, \tag{6.56}$$

(3!: color multiplicity; $-4/3$: $\bar{q}q$ (attractive); $+2/3$: qq (repulsive))

6.15 The most famous Fourier transformation:

$$\int \frac{d^3\vec{k}}{(2\pi)^3} e^{-i\vec{k}\cdot\vec{r}} \frac{1}{\vec{k}^2} = \frac{1}{4\pi}\frac{1}{r}.$$

where I have introduced $\alpha_s \equiv g_s^2/(4\pi)$, the fine-structure constant for the strong interactions.

The result of this computation shows that the nonrelativistic potential between a pair of quarks is very much the same as between a pair of electrically charged particles. It is attractive between a quark and its antiquark and repulsive between quarks (or antiquarks). Its dependence in space is that of a potential such as the Coulomb potential.[6.16]

[6.16] Had the gluon been a scalar particle, the force would have always been attractive no matter what the charge of the quarks is.

6.3 Symmetries of the Strong Interactions

The Lagrangian in Eq. (6.1) does not contain the hadrons we see in experiments. Yet its structure encodes the symmetries that we know are obeyed by these hadrons.

What are the symmetries of the QCD Lagrangian in Eq. (6.1)? Some are **exact** symmetries.[6.17] They are:

[6.17] First, I must convince myself that electroweak corrections are not going to disrupt these symmetries. In other words: does one expect corrections $O(\alpha)$? The answer is no. The generators of the electroweak group $SU(2)_L \times U(1)_Y$ commute with those of the group $SU(3)_C$. Because of this, electroweak corrections can enter only through operators of dimension larger than 4. I expect the first correction to come in with dimension 6 and be

$$O\left(\frac{\alpha}{m_W^2}\right) \approx O(G_F)\,. \qquad (6.57)$$

This is reassuring. Electroweak corrections are going to be small, of the order of Fermi's constant, and under control.

- the gauge symmetry of the group $SU(3)_c$, which is kind of obvious;
- the discrete symmetries C, P and T. Parity is manifestly conserved because of the vector-like nature of the interaction. Charge requires a bit of a discussion because this transformation acts on the Gell-Mann matrices in the current. I have that

  $$\hat{C}\bar{q}\gamma^\mu \frac{\lambda^A}{2} q\hat{C}^{-1} = -\bar{q}\gamma^\mu \left(\frac{\lambda^A}{2}\right)^T q\,. \qquad (6.58)$$

  Now λ_1, λ_3, λ_4, λ_6 and λ_8 are symmetric matrices while λ_2, λ_5 and λ_7 are antisymmetric. The current in the QCD Lagrangian therefore transforms as

  $$J_\mu^A \to -\eta^A J_\mu^A\,, \qquad (6.59)$$

  where

  $$\eta^A = \begin{cases} +1 & \text{for } a = 1,3,4,6,8 \\ -1 & \text{for } a = 2,5,7\,. \end{cases} \qquad (6.60)$$

  To have an invariant Lagrangian the gauge fields must transform like the current, that is,

  $$A_\mu^A \to -\eta^A A_\mu^A. \qquad (6.61)$$

  This is the right transformation, since it gives

  $$G_{\mu\nu}^A \to -\eta^A G_{\mu\nu}^A \qquad (6.62)$$

  thanks to the the property of f^{ABC} of being non-zero only for an odd number of indices which themselves are odd. Finally, T conservation follows from the conservation of P and C and the CPT theorem.

Other symmetries are only **approximate** because they hold only for particular values of the fermion masses. They are symmetries in the flavor space of the Standard Model:

$\boxed{m_u \simeq m_d}$ The first approximate symmetry is obtained if the quark masses in the isospin doublets are degenerate. In this case, the isospin doublet can be freely rotated,

$$\begin{pmatrix} u \\ d \end{pmatrix}' = e^{i\alpha_i \sigma_i} \begin{pmatrix} u \\ d \end{pmatrix}, \tag{6.63}$$

and there is a $U(2)_V$ symmetry in the isospin. This group can be decomposed[6.18] as follows:

6.18 More precisely, $U(2) \to U(1) \otimes SU(2)/\mathbb{Z}_2$.

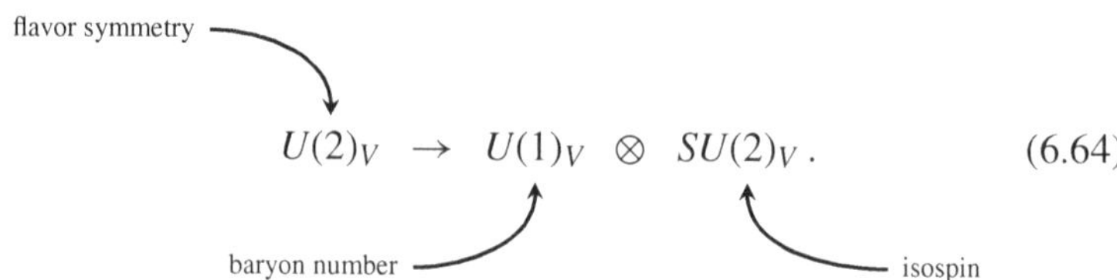

$$U(2)_V \;\rightarrow\; U(1)_V \;\otimes\; SU(2)_V\,. \tag{6.64}$$

The isospin symmetry is encoded in the hadron spectroscopy. Many hadrons have almost the same mass and the same spin and parity. For example:

$$\begin{matrix} n & 939.5\text{ MeV} \\ p & 938.9\text{ MeV} \end{matrix} \quad \text{and} \quad \begin{matrix} \pi^- & 139.6\text{ MeV} \\ \pi^0 & 135\text{ MeV} \\ \pi^+ & 139.6\text{ MeV}. \end{matrix} \tag{6.65}$$

The neutron and the proton, or the pions, form components of isospin doublets or triplets. This classification can be pushed a bit further if we include also the strangeness. The group becomes $SU(3)_V$ and the states organized (in the "eightfold way") in multiplets, as shown in the diagrams in Figure 6.12.

A lot was made of these symmetries in the dark ages of particle physics. They helped in classifying the bewildering abundance of particles that were being produced at the accelerators. Yet this classification has problems of its own, namely, why do we find decuplets in the baryon spectrum but none in that of the mesons? Why are there the representations of $SU(3)$ $\underline{1}$, $\underline{8}$, $\underline{10}$ but not $\underline{3}$, $\underline{6}$ and $\underline{27}$? The answer to these questions is rooted in the underlining symmetry of the QCD Lagrangian in Eq. (6.1). In the quark model mesons consist of two quarks and baryons of three. The representations of the $SU(3)_V$ group must be constructed by means of the fundamental representations carried by the quarks. In terms of group representations,

$$\begin{aligned} Q\bar{Q} &= \underline{3} \otimes \underline{\bar{3}} = \underline{1} \oplus \underline{8} \\ QQQ &= \underline{3} \otimes \underline{3} \otimes \underline{3} = \underline{1} \oplus \underline{8} \oplus \underline{8} \oplus \underline{10}\,. \end{aligned} \tag{6.66}$$

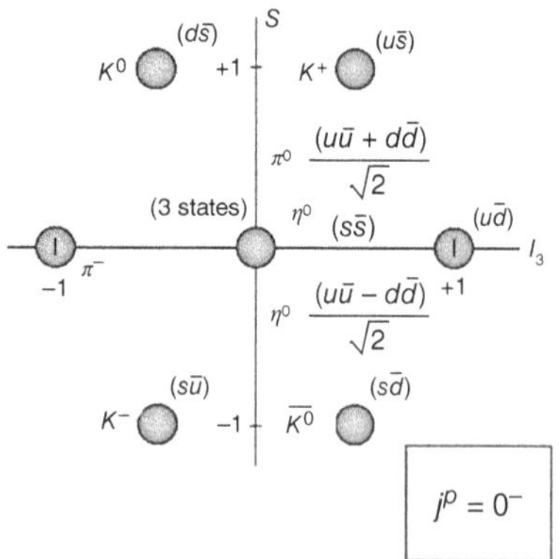

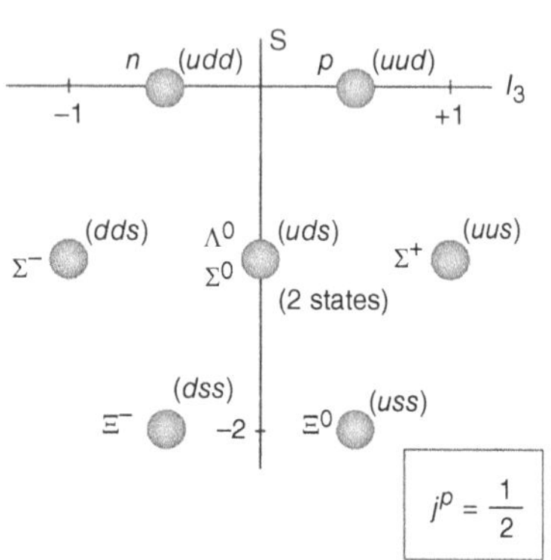

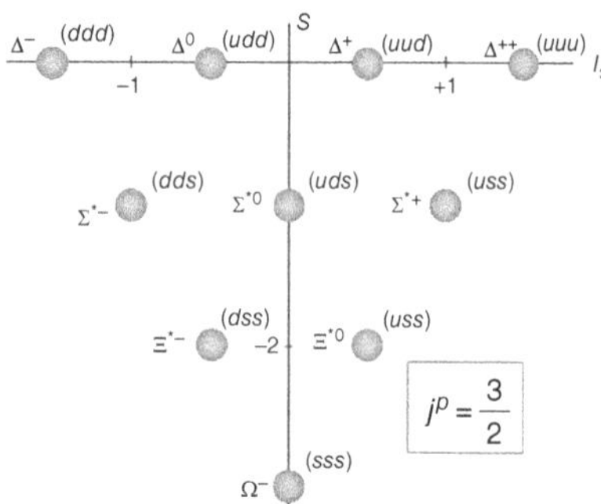

Figure 6.12 The classification of mesons ($J^P = 0$) and baryons ($J^P = 1/2$ and $3/2$) in the weight diagrams of $SU(3)$ in the celebrated eightfold way.

The fact that quarks are triplets of $SU(3)$ has the consequence that only octets and decuplets of $SU(3)_V$ are going to be present in the spectrum of baryons consisting of three quarks.

$\boxed{m_u \simeq m_d \simeq 0}$ There is a second approximate symmetry that is true only if the quark masses vanish. In this case there is a larger symmetry because I can rotate the left-handed fields independently of the right-handed fields:

$$\begin{cases} q'_L &= e^{i\alpha_i\sigma_i} q_L \\ q'_R &= e^{i\beta_i\sigma_i} q_R \,. \end{cases} \tag{6.67}$$

The symmetry consists of there being two copies of the isospin symmetry: $SU(2)_L \otimes SU(2)_R$. This symmetry is not seen in the spectrum of the hadrons. It is understood to be spontaneously broken – as in

$$SU(2)_L \otimes SU(2)_R \xrightarrow[\text{spontaneous breaking}]{} SU(2)_V \otimes U(1)_V \,. \tag{6.68}$$

The axial $SU(N_f)_A$ symmetry is realized in the Nambu–Goldstone way. Because of the Goldstone theorem, I expect to have massless bosons in the spectrum, one for each of the broken generators of the group. They are, for $N_f = 2$, the pions, which in fact are the lightest mesons. They should be massless particles. Their masses, which explicitly break the $SU(2)_A$ symmetry, come from the finite masses of the quarks u and d. They are called pseudo-Goldstone bosons.

Summing up: the QCD Lagrangian in Eq. (6.1) has four symmetries represented by the groups in the following product:

realized in the spectrum → $SU(N_f)_V$; broken by anomaly (see Section 6.5) → $U(1)_A$

$$SU(N_f)_V \otimes U(1)_V \otimes U(1)_A \otimes SU(N_f)_A \,. \tag{6.69}$$

baryon number → $U(1)_V$; Goldstone bosons → $SU(N_f)_A$

As appears from this discussion on the symmetries, an amazing amount of information is packed in the Lagrangian in Eq. (6.1). The slow work of uncovering and classifying the hadrons out of the intricacies of the strong interactions performed in the 1960s is brought to full fruition.

6.4 The θ-Term*

It is possible to add to the Lagrangian in Eq. (6.1) a gauge- (and Lorentz-) invariant term[6.19]

$$\theta \, \frac{g_s^2}{16\pi^2} \epsilon^{\mu\nu\alpha\beta} F^A_{\mu\nu} F^A_{\alpha\beta} = \theta \, \frac{g_s^2}{8\pi^2} F^A_{\mu\nu} \tilde{F}^A_{\mu\nu} \,. \tag{6.70}$$

6.19 The factors of 2π are there to normalize the result. The dual field strength is defined by

$$\tilde{F}^A_{\mu\nu} = \epsilon^{\mu\nu\alpha\beta} F^A_{\alpha\beta}/2.$$

The term in Eq. (6.70) is called a θ-term – from the arbitrary constant θ in front.[6.20] At first sight, such a term seems irrelevant. I can always write

$$\frac{g_s^2}{16\pi^2}\epsilon^{\mu\nu\alpha\beta}F^A_{\mu\nu}F^A_{\alpha\beta} = \partial_\mu K^\mu\,, \tag{6.71}$$

where

$$K^\mu = \frac{g_s^2}{4\pi^2}\epsilon^{\mu\alpha\beta\gamma}\left[A^A_\alpha F^A_{\beta\gamma} - \frac{ig_s}{3}f_{ABC}A^A_\alpha A^B_\beta A^C_\gamma\right] \tag{6.72}$$

is the Chern–Simon current, and I find that (by using the Green theorem connecting integrals over a surface to those over the volume inside it)

$$\frac{g_s^2}{8\pi^2}\int \mathrm{d}^4x\, F^A_{\mu\nu}\tilde{F}^A_{\mu\nu} = \oint \mathrm{d}^3S_\mu K^\mu, \tag{6.73}$$

which gives a boundary contribution

$$\int \mathrm{d}^3rK^\mu\Big|_{t\to-\infty}^{t\to+\infty} = \frac{g_s^3}{12\pi^2}\epsilon^{\mu\alpha\beta\gamma}f_{ABC}\int \mathrm{d}^3rA^A_\alpha(r)A^B_\beta(r)A^C_\gamma(r)\,. \tag{6.74}$$

The fields at $t \to \pm\infty$ must be pure gauge fields and $F^A_{\mu\nu} = 0$ (which explains why there is no contribution from the first term in Eq. (6.72)). For an Abelian gauge field, I could have ignored the total divergence. This is the case for QED. In the general case of non-Abelian fields, the question is whether the term in Eq. (6.74) vanishes or not. The vanishing of the θ-term can be rephrased in terms of the vanishing of the gauge fields at infinitely large distances: are they pure gauge fields? Yes, they are. However, their asymptotic space is non-trivial and pure gauge fields can end up into two different asymptotic vacua, giving rise to a non-vanishing contribution.

To understand this contribution, first write the pure gauge field as

$$\sum_A T^A A^\mu_A \to \frac{i}{g_s}\Omega^\dagger\partial^\mu\Omega, \tag{6.75}$$

where Ω is an element of the gauge group. Actually, it only needs to be an element of $SU(2)$, which is the sphere S^3. This subgroup of $SU(3)$ already contains all the relevant topological features needed.

Also, the asymptotical R^3 space is a sphere when completed by the two points at infinity. Therefore

$$\Omega : R^3 \simeq S^3 \to SU(2) \simeq S^3\,, \tag{6.76}$$

which is a mapping between Euclidean space-time and the gauge group, which I can parametrize as

$$\Omega(\vec{\Theta}) = \frac{\Theta_0\mathbb{1} + i\vec{\Theta}\cdot\vec{\sigma}}{\sqrt{\Theta_0^2+\vec{\Theta}^2}}\,. \tag{6.77}$$

[6.20] How about a θ-term for the gauge group $SU(2)_L$? Or, for that matter, $U(1)_Y$? The θ-term for the gauge groups $SU(2)_L$ can be rotated away by means of the baryonic current, and $U(1)_Y$ is Abelian and with a boundary term that is a total derivative and can be set to zero.

Equation (6.74) can be written as

$$\frac{1}{12\pi^2}\int_{S_3} \mathrm{d}^3r\,\epsilon^{ijk}\mathrm{Tr}\,[(\Omega^\dagger\partial_i\Omega)(\Omega^\dagger\partial_j\Omega)(\Omega^\dagger\partial_k\Omega)]\,. \tag{6.78}$$

The integration on the S_3 border can be done after inserting (for elements Ω close to the identity)

$$\partial_k\Omega\,\Omega^{-1} = i\sigma_k \tag{6.79}$$

6.21 $\mathrm{Tr}\,\sigma_i\sigma_j\sigma_k = 2i\epsilon_{ijk}$ and $\epsilon^{ijk}\epsilon_{ijk} = 6$.

and[6.21]

$$\mathrm{Tr}\,[(\Omega^\dagger\partial_i\Omega)(\Omega^\dagger\partial_j\Omega)(\Omega^\dagger\partial_k\Omega)] = (i)^3\mathrm{Tr}\,\sigma_i\sigma_j\sigma_k = 2\epsilon_{ijk}, \tag{6.80}$$

6.22 $\int_{S^3}\mathrm{d}^3r = 2\pi^2$.

which yields[6.22]

$$\frac{1}{12\pi^2}\int_{S^3}\mathrm{d}^3r\,2\,\epsilon^{ijk}\epsilon_{ijk} = 1. \tag{6.81}$$

Thus

$$\frac{g_s^2}{16\pi^2}\int \mathrm{d}^4x\,\epsilon^{\mu\nu\alpha\beta}F^A_{\mu\nu}F^A_{\alpha\beta} = \frac{2\times 12\pi^2}{24\pi^2} = 1\,. \tag{6.82}$$

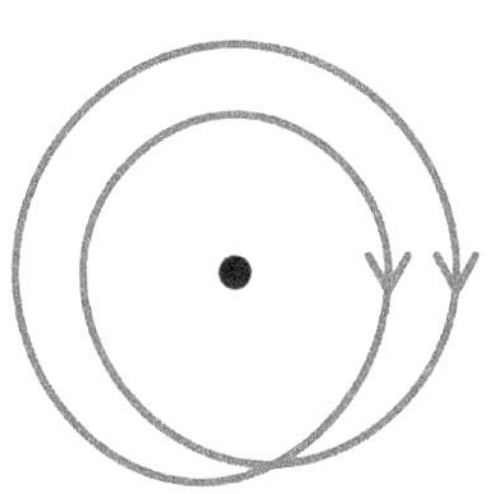

Figure 6.13 Curve with winding number = −2.

The number 1 on the right-hand side is actually the **winding number**, the number of times the gauge configuration in $SU(2)$ wraps around the non-trivial vacuum of $SU(3)$, which has the homotopy structure

$$\pi_3(S^3) = \mathbb{Z}, \tag{6.83}$$

so that in general the θ-term is $\theta\nu$ with $\nu \in \mathbb{Z}$. This explains why the constant θ is periodic: $\theta + 2\pi = \theta$. Figure 6.13 shows a curve winding twice around a singularity.

The non-perturbative gauge transformation leading from one sector to another (characterized by their winding numbers) is called the **instanton**. It represents the quantum tunneling between two distinct vacua of QCD. The probability of such a tunneling can be estimated (again in Euclidean space where it is bounded from below) by first considering that[6.23]

6.23 Use the identity $\epsilon_{\mu\nu\lambda\sigma}\epsilon^{\mu\nu}_{\ \ \rho\xi} = 2(\delta_{\lambda\rho}\delta_{\sigma\xi} - \delta_{\lambda\xi}\delta_{\sigma\rho})$.

$$0 \le (F^A_{\mu\nu} \pm \tilde{F}^A_{\mu\nu})^2 = 2\Big(F^A_{\mu\nu}F^{A\,\mu\nu} \pm F^A_{\mu\nu}F^{A\,\mu\nu}\Big) \tag{6.84}$$

and therefore

$$\frac{1}{4}F^A_{\mu\nu}F^{A\,\mu\nu} \ge \pm\frac{8\pi}{g^2}\frac{g_s^2}{32\pi^2}F^A_{\mu\nu}\tilde{F}^{A\,\mu\nu}\,. \tag{6.85}$$

Since

$$\nu = \frac{g_s^2}{32\pi^2}F^A_{\mu\nu}\tilde{F}^{A\,\mu\nu}, \tag{6.86}$$

it follows that

$$S \ge \frac{8\pi}{g_s^2}|\nu|\,. \tag{6.87}$$

Equation (6.87) becomes an identity if I take the self-dual field strength $F = \pm\tilde{F}$, which is the case for the instanton solution. The probability of the transition between these two vacua is proportional to

$$e^{-8\pi^2\nu/g_s^2}\,. \tag{6.88}$$

One last thing. By the chiral rotations

$$\begin{cases} \psi_L & \to e^{-i\text{Arg Det } M}\psi_L \\ \psi_R & \to e^{i\text{Arg Det } M}\psi_R\,, \end{cases} \tag{6.89}$$

which remove the imaginary parts from the quark mass term, the anomaly of the axial current,

$$\partial_\mu J_A^\mu = \frac{g^2}{16\pi^2}\text{Tr}\,\tilde{F}^A_{\mu\nu}F_A^{\mu\nu}, \tag{6.90}$$

is shifted. The effect of these rotations on the action S is given by Noether's theorem

$$\delta S = \int \mathrm{d}^4x\delta L, \tag{6.91}$$

where

$$\delta L = \sum \delta\phi\partial_\mu J_A^\mu \tag{6.92}$$

and in this case $\delta\phi = \text{Arg Det}\,M$. If the rotation is not a symmetry, there will be a shift in the Lagrangian density proportional to the divergence of the corresponding current. The invariant quantity is therefore

$$\bar{\theta} = \theta + \text{Arg Det}\,M. \tag{6.93}$$

6.5 Global Chiral Anomalies

In going from the discussion on the θ-term in QCD to the chiral anomalies I must remind myself that in the first case the term with the dual field strength is in the Lagrangian while in the anomaly computation it is in an amplitude. In the first case I can and do reduce it to a boundary term.

The only thing I can do in the second case is to redefine the anomalous current by adding the Chern–Simon current to restore conservation. But the Chern–Simon current is not gauge invariant and therefore this only makes the theory worse. This distinction is important in order to comprehend why even a $U(1)$ anomaly – which has no boundary term – cannot be eliminated in the amplitude computation.

Going back now to the global symmetries of the strong interactions, some of them show the presence of anomalies. These are all global symmetries and the presence of anomalies leads to no inconsistencies.

- **Chiral anomaly and** $U(1)_A$ The divergence of the axial current is

$$\partial_\mu J_5^{a\,\mu} = n_f \frac{g_s^2}{32\pi^2} \mathrm{Tr}\,[T^a \mathrm{Tr}\,\lambda_A \lambda_B \tilde{F}_A^{\mu\nu} F_{\mu\nu}^B]\,, \tag{6.94}$$

where the field strengths are those of the gauge field of $SU(3)_c$. Because of the trace over the isospin matrix T^a, only the isospin singlet η' (for which T^a is replaced by the identity matrix $\mathbb{1}$) receives a non-vanishing contribution. This anomaly breaks the $U(1)_A$ symmetry (as advertised in the previous section) and no Goldstone boson is present in the spectrum, as confirmed by the large η–η' mass splitting. The anomalous term is non-negligible and is given by the instanton amplitude as

$$\exp\left[-\frac{2\pi\nu}{\alpha_s(1/\rho)}\right]\,, \tag{6.95}$$

with ρ the instanton size. The symmetry is badly broken.

- **Baryon number non-conservation** The classical Lagrangian of the Standard Model enjoys two additional $U(1)$ symmetries, linked to the baryon and lepton numbers. These numbers are assigned as $B = 1/3$, $L = 0$ for quarks and $B = 0$, $L = 1$ for leptons. While these symmetries, and the conservation they bring, are not required, they come for free, they are **accidental**. The two charges B and L are the time components of the corresponding currents:

$$J_B^\mu = \sum_i \bar{q}_i \gamma^\mu q_i \quad \text{and} \quad J_L^\mu = \sum_i \bar{\ell}_i \gamma^\mu \ell_i\,. \tag{6.96}$$

The conservation of these two currents is broken by the anomaly. The triangle diagram in this case consists of one baryon, or lepton, current and two $SU(2)_L$ currents. By going through the same steps as before I find that (for the first generation of quarks and leptons)

$$\sum_{L \text{ quarks}} \hat{B} = 2N_c \left(\frac{1}{3}\right) = 2\,, \tag{6.97}$$

and similarly

$$\sum_{L \text{ leptons}} \hat{L} = 2\,, \tag{6.98}$$

which tells me that both baryon and lepton number symmetries are broken. This result could spell disaster for the lifetime of the proton. We know that

$$\tau(p^+ \to \pi^0 e^+) > 5.5 \times 10^{32} \text{ years.} \tag{6.99}$$

The non-conservation of the baryon current is (for one fermion generation)

$$\Delta B = \int \mathrm{d}^4x\, \partial_\mu J_B^\mu = -\frac{g^2}{16\pi^2} \int \mathrm{d}^4x\, F_A^{\mu\nu} \tilde{F}_{\mu\nu}^A = -2\nu, \tag{6.100}$$

proportional to the instanton transition, which in turn is proportional to

$$\exp\left[-\frac{2\pi\nu}{\alpha_2(m_Z)}\right] \simeq 10^{-80\nu} \tag{6.101}$$

and is therefore very small unless the temperature is large enough – as it is in the early universe where violation of the baryon number can be used, in principle, to generate an asymmetry between matter and antimatter via baryogenesis, by the mechanism of having a **sphaleron** transition.[6.24]

[6.24] The sphaleron configuration is made possible because the temperature bath brings the system close to the energy at the top of the potential between θ vacua.

The combination $B - L$ remains unbroken and the associated combination of currents reflects a good accidental global symmetry of the Standard Model which could, if you are so inclined, be gauged.

- π^0 **and** η **decays** Because the anomaly in $\pi^0 \to \gamma\gamma$ and $\eta^0 \to \gamma\gamma$ is Abelian, it contributes only to the amplitudes with no θ-term in the Lagrangian. The pion (and the η) couples to the axial current with strength F_π:

$$\langle 0|J_5^{a=3\,\mu}|\pi(p)\rangle = -i\sqrt{2}F_\pi p^\mu\,, \tag{6.102}$$

and, replacing the anomalous contribution to the divergence of the current,

$$\partial_\mu J_5^{a=3\,\mu} = -\frac{e^2}{8\pi^2}\mathrm{Tr}\,[T^3 Q^2 \tilde{F}^{\mu\nu}F_{\mu\nu}]\,, \tag{6.103}$$

compute the width can be computed. This term is actually the largest contribution to the width of these two particles. It was, historically, the first instance of the anomaly in high-energy physics, the place where the whole story began.[6.25]

[6.25] There is no amplitude similar to Eq. (6.102) mediated by the baryonic or leptonic anomalies because the corresponding anomalous currents are not associated with any known particle, as is the case for the axial current and the pion.

The trace over the group matrices T^3 and

$$Q = \frac{1}{3}\begin{pmatrix} 2 & 0 \\ 0 & -1 \end{pmatrix}, \tag{6.104}$$

which comes from the electric charge assignments for the u and d quarks, gives

$$N_c\left(\frac{2}{3}\right)^2\left(\frac{1}{2}\right) + N_c\left(-\frac{1}{3}\right)^2\left(-\frac{1}{2}\right) = \frac{1}{2}\frac{N_c}{3}. \tag{6.105}$$

Thus (after computing the width from the triangle one-loop diagram – which is left as an exercise) I find that

$$\Gamma(\pi^0 \to 2\gamma) = \frac{N_c\alpha^2 m_\pi^2}{144\pi^3 f_\pi^2} = \left(\frac{N_c}{3}\right)^2 \times 1.11 \times 10^{16}\ \mathrm{s}^{-1}\,, \tag{6.106}$$

in agreement with the experimental value

$$\Gamma(\pi^0 \to 2\gamma) = (1.119 \pm 0.08) \times 10^{16}\ \mathrm{s}^{-1} \tag{6.107}$$

only if $N_c = 3$!

6.6 *Problem Session*: Debye Screening and Pauli Paramagnetism

The vacuum in quantum field theory is far from being empty and is a medium with many characteristic properties. These properties modify the way in which the classical fields behave. To help in giving a feeling of what happens, in this problem session I discuss electromagnetism in a medium. It is an interesting subject of in its own right, well worth studying.

Debye Screening

For a negative charge $Q = Ze$ in a ionic solution such as water, with average density c_0, I expect the local density of charge ρ to be the sum of the contributions of the negative and positive ions, the densities of which are controlled by the Maxwell distribution at the temperature T:

$$\rho = Zec_0 \left[e^{-ZeV/k_BT} - e^{+ZeV/k_BT} \right] \simeq -2Z^2e^2c_0 \frac{V}{k_BT}, \qquad (6.108)$$

where k_B is the Boltzmann constant.

The electric potential at a distance λ_D from the charge is given by Coulomb's law:

$$V = \frac{Q}{4\pi\varepsilon\lambda_D}, \qquad (6.109)$$

where ε is the permittivity of the ionic solution, so that

$$\rho \simeq -2Z^2e^2c_0 \frac{Q}{4\pi\varepsilon\lambda_D k_BT}. \qquad (6.110)$$

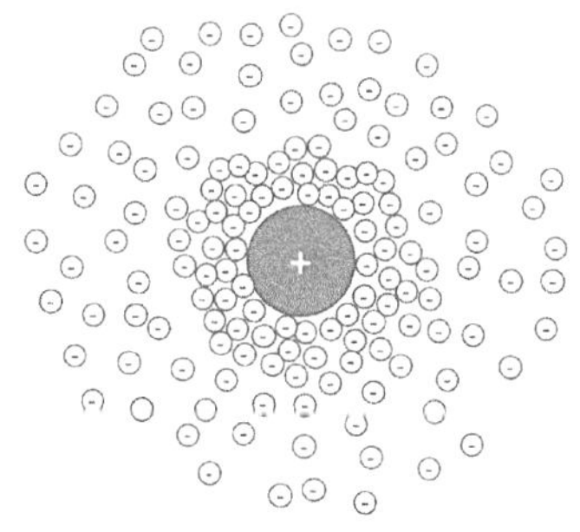

Figure 6.14 Debye screening of charge in an ionic solution such as water.

The ions within the solution arrange themselves as in Figure 6.14 so as to be on average closer to the charge Q and thus partially screen it. I can find the charge Q_D within a sphere of radius λ_D is given by

$$Q_D = \frac{4\pi}{3}\lambda_D^3\rho. \qquad (6.111)$$

by replacing the charge density ρ from Eq. (6.110) to obtain

$$Q_D = -\frac{8\pi}{3}\lambda_D^2 Z^2e^2c_0 \frac{Q}{4\pi\varepsilon k_BT}. \qquad (6.112)$$

Requiring that the induced charge Q_D compensates the original charge Q, I find

$$\lambda_D = \frac{3\varepsilon k_BT}{8\pi Z^2e^2c_0}. \qquad (6.113)$$

The Coulomb potential of the charge Q is modified and looks like that in Figure 6.15. The effective charge Qe^{-r/λ_D} is smaller at distances

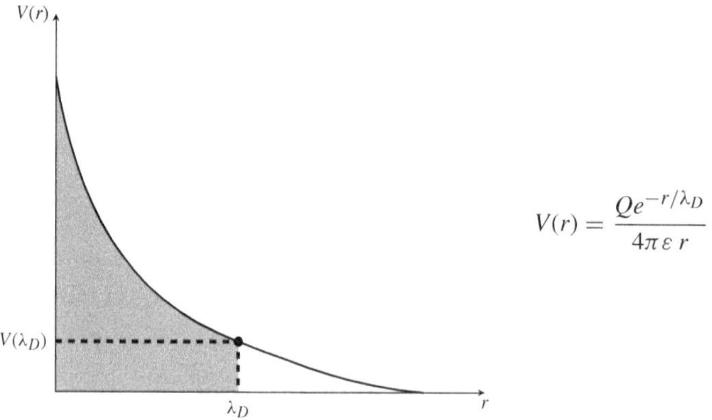

Figure 6.15 The Coulomb potential screened by the ions.

larger than λ_D and goes to zero for very large distances. On the contrary, for distances below λ_D the full charge Q is recovered. This behavior is described by taking a new permittivity

$$\varepsilon' = \varepsilon\, e^{r/\lambda_D}\,, \tag{6.114}$$

which for $\varepsilon = 1$ (such as for water) will always be bigger than 1 for $r > \lambda_D$. The ionic solution is diamagnetic.

Pauli Paramagnetism

A Fermi gas of N free electrons in a volume V fills all levels up to the Fermi energy

$$E_F = \frac{1}{2m}\left(\frac{3\pi^2 N}{V}\right)^{2/3}. \tag{6.115}$$

From Eq. (6.115), the number of states is given by

$$N = \frac{V}{3\pi^2}\,(2mE_F)^{3/2}\,, \tag{6.116}$$

with density per unit energy

$$D(E_F) = \frac{\mathrm{d}N}{\mathrm{d}E_F} = \frac{V}{2\pi^2}(2m)^{3/2}\sqrt{E_F}\,. \tag{6.117}$$

The electrons have an intrinsic magnetic moment equal to one Bohr magneton μ_B. The energy of these electrons in an external magnetic field $\vec{B}$ is

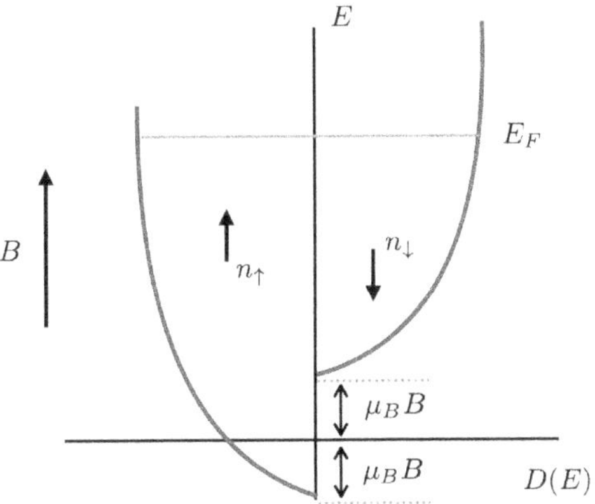

Figure 6.16 Pauli paramagnetism in a gas of free electrons: the energy distribution in a weak external magnetic field.

$$E = -\vec{\mu} \cdot \vec{B}\,, \tag{6.118}$$

where the vector $\vec{\mu}$ can be aligned ($E = -\mu_B B$) or anti-aligned ($E = \mu_B B$) to the external magnetic field.

For a weak magnetic field, for which $\mu_B B \ll kT_F$, the energy levels are as in Figure 6.16: the numbers of electrons with spins parallel or antiparallel to the external magnetic field are given by

$$\begin{aligned} n_\uparrow &= \frac{1}{2}\int_{-\mu_B B}^{E_F} \mathrm{d}E D(E_F + \mu_B B) \\ &\simeq \frac{1}{2}\int_{-\mu_B B}^{E_F} \mathrm{d}E D(E_F) + \frac{1}{2}\mu_B B D(E_F) \\ n_\downarrow &= \frac{1}{2}\int_{-\mu_B B}^{E_F} \mathrm{d}E D(E_F - \mu_B B) \\ &\simeq \frac{1}{2}\int_{-\mu_B B}^{E_F} \mathrm{d}E D(E_F) - \frac{1}{2}\mu_B B D(E_F)\,, \end{aligned} \tag{6.119}$$

which gives rise to a net magnetization

$$M = \mu_B (n_\uparrow - n_\downarrow) = D(E_F)\mu_B^2 B\,. \tag{6.120}$$

The magnetic susceptibility χ is obtained as[6.26]

6.26 The vacuum permeability $\mu_0 = 1$ in natural units.

$$\chi = \frac{\partial M}{\partial B} = \mu_B^2 D(E_F) \simeq \frac{3N\mu_B^2}{2E_F}\,, \tag{6.121}$$

which is a positive number. In other words, the magnetic field is enhanced by the magnetization that it has itself induced in the medium. Since the magnetic permeability μ is related to the susceptibility by

$$\mu = 1 + \chi\,, \tag{6.122}$$

I have that μ is always larger than 1 for Pauli paramagnetism. Relativistic invariance requires ($c = 1$)

$$\mu\,\varepsilon = 1 \tag{6.123}$$

and therefore the paramagnetic medium must have electric permittivity $\varepsilon < 1$.

These two examples show that the value of the charge, and therefore the properties of the Coulomb potential, depend on those of the medium in which the charge is to be found:

- If the medium is diamagnetic and $\varepsilon > 1$, the charge is screened, as happens for an electric charge in water. The charge grows as you get closer to the source.
- If the medium is paramagnetic and $\varepsilon < 1$, as in the case of Pauli paramagnetism, the charge is anti-screened. The charge decreases as you get closer to the source.

6.7 The Running of the QCD Coupling

To try to understand how the perturbative theory of quarks and gluons ends up being the theory of the strong interactions of hadrons I need to look into how the color charge is modified by quantum effects.

In QED the electric charge is small and well defined at low energy and can be matched directly to the coupling parameter in the classical Lagrangian. In QCD things are more complicated because the matching between charge and coupling parameter is done at high energy.

I will look first at QED. The charge is defined by the interaction vertex between matter (the fermion fields) and the photon, as shown in Figure 6.17.

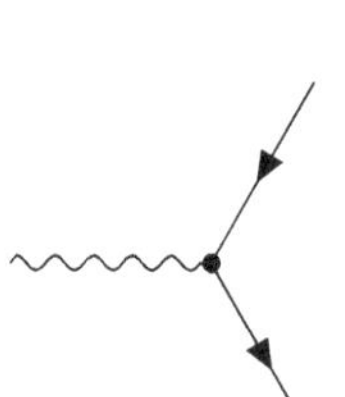

Figure 6.17 Interaction vertex in QED.

The key ingredient is quantum loops. When loops are taken into account in the classical Lagrangian, the photon propagator (Feynman gauge, Figure 6.18) attached to the vertex interaction with the fermions,

$$\frac{-ig_{\mu\nu}}{q^2}, \tag{6.124}$$

Figure 6.18 Propagation of the photon to the vertex at tree level.

is modified to

$$\frac{-ig_{\mu\nu}}{q^2} - \frac{iI_{\mu\nu}(q^2)}{q^4}, \tag{6.125}$$

where

$$\begin{aligned} I_{\mu\nu}(q^2) &= -e^2 \int \frac{\mathrm{d}^4k}{(2\pi)^4} \frac{\mathrm{Tr}\left[\gamma_\mu(\not{k} + m_e)\gamma_\nu(\not{k} - \not{q} + m_e)\right]}{(k^2 - m_e^2)[(k-q)^2 - m_e^2]} \\ &= -ig_{\mu\nu}q^2\, I(q^2) + q_\mu q_\nu\, J(q^2) \end{aligned} \tag{6.126}$$

it does not contribute, because of current conservation

is the contribution of the Feynman diagram in Figure 6.19 with the loop. The function $I(q^2)$ is given (after a computation I do not need to dwell on here, leaving it as an exercise to the reader) by[6.27]

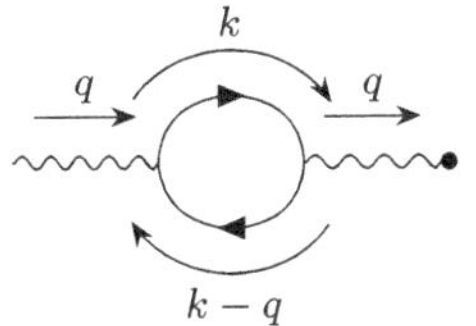

Figure 6.19 Propagation of the photon to the vertex at one-loop level.

6.27 $Q^2 = -q^2$.

$$I(q^2) = \frac{e^2}{12\pi^2}\left\{\int_{m_e^2}^{\Lambda^2}\frac{dz}{z} - 6\int_0^1 dz(1-z)\log\left[1-\frac{q^2}{m_e^2}z(1-z)\right]\right\} \tag{6.127}$$

$$= \frac{e^2}{12\pi^2}\left[\overbrace{\int_{m_e^2}^{\Lambda^2}}^{\text{cutoff}}\frac{dz}{z} - f\left(\frac{Q^2}{m_e^2}\right)\right] = \frac{e^2}{12\pi^2}\left[\log\frac{\Lambda^2}{m_e^2} - f\left(\frac{Q^2}{m_e^2}\right)\right],$$

where the cutoff has to be introduced because the integral is divergent, and

$$f\left(\frac{Q^2}{m_e^2}\right) = -6\int_0^1 dz(1-z)\log\left[1+\frac{Q^2}{m_e^2}z(1-z)\right]. \tag{6.128}$$

Insertion of this correction into the photon propagator modifies the charge with which the photon couples to fermions. The charge factor is effectively changed from e^2 to

$$e^2\underbrace{\left\{1-\frac{e^2}{12\pi^2}\left[\log\frac{\Lambda^2}{m_e^2} - f\left(\frac{Q^2}{m_e^2}\right)\right]\right\}}_{e_R^2(0)/e^2}. \tag{6.129}$$

There are two important features in Eq. (6.129).

The first is that there is a divergent part that has been tamed by a cutoff. This part, as indicated, goes into the definition of the parameter of the classical Lagrangian at this order in the perturbative expansion. In other words, what I identify with the electric charge is no longer e but is now e_R. In this way, the cutoff-dependent part is absorbed into the definition of what I call the electric charge.

The second important feature is that I can write

$$\begin{aligned} e^2 &= e_R^2(0) + \frac{e_R^4(0)}{12\pi^2}f\left(\frac{Q^2}{m_e^2}\right) + \cdots \\ &= e_R^2(0)\left[1+\frac{e_R^2(0)}{12\pi^2}f\left(\frac{Q^2}{m_e^2}\right)\right] + O(e_R^4), \end{aligned} \tag{6.130}$$

which shows how the value of the renormalized charge comes to depend on the momentum with which it is probed.[6.28]

6.28 The momentum dependence corresponds in space to a distance. Large momenta correspond to short distances and, *vice versa*, small momenta to large distances.

The last line in Eq. (6.130) can be rewritten (for $\alpha = e^2/4\pi$) as

$$\underbrace{\alpha(Q^2)}_{\text{running coupling}} = \underbrace{\alpha(0)}_{=\frac{1}{137}\text{(as measured in Thomson scattering)}}\left[1 + \frac{\alpha(0)}{3\pi} f\left(\frac{Q^2}{m_e^2}\right) + O(\alpha^2)\right]. \tag{6.131}$$

After reintroducing the function $f(Q^2/m_e^2)$, in the limit of large Q^2 it follows that

$$f\left(\frac{Q^2}{m_e^2}\right) \xrightarrow[Q^2 \gg m_e^2]{} \log\frac{Q^2}{m_e^2}. \tag{6.132}$$

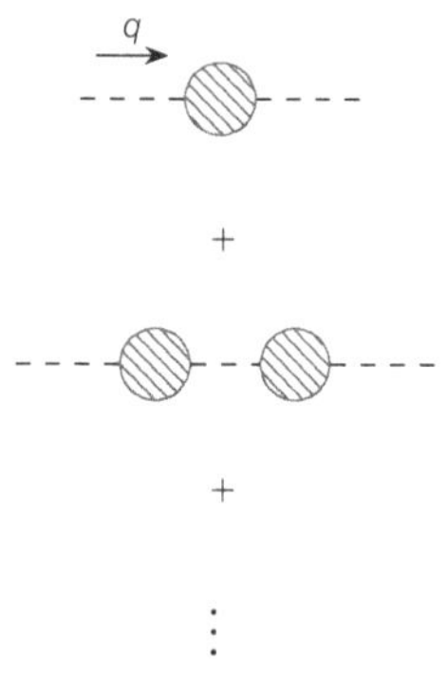

Figure 6.20 The geometrical series of blobs.

I can think of the first term in Eq. (6.131) as the first term (Figure 6.20) in a geometrical series – the following terms being represented by blobs containing the higher-order corrections – and re-sum it to obtain

$$\alpha(Q^2) = \frac{\alpha(\mu^2)}{1 - \dfrac{\alpha(\mu^2)}{3\pi}\log\dfrac{Q^2}{m_e^2}}. \tag{6.133}$$

The result in Eq. (6.133) is expressed in a compact manner by defining the so-called β-**function**,

$$\beta(\alpha) \equiv \frac{\mathrm{d}\alpha(Q^2)}{\mathrm{d}\log Q^2} = -\underbrace{\beta_0}_{-\frac{1}{3\pi}\text{ in QED}}\,\alpha^2 + \cdots, \tag{6.134}$$

which is the logarithmic derivative of the charge with respect to the squared momentum. In QED the sign of β_0 is negative. The charge accordingly is screened, becoming weaker as we move away.

Things are different in QCD because of the gluons and their self-interactions. What takes place can be understood by thinking of Gauss's law[1] for a color charge density ρ^A:

$$\vec{\nabla}\cdot\vec{E}^A = \rho^A - g_s f^A_{BC}\vec{A}^B\cdot\vec{E}^C, \tag{6.135}$$

in which the contribution of the self-interaction of the gluons is represented by the second term on the right-hand side. If I have, for instance, a red (R) charge surrounded by (virtual) gluons carrying a green (G) charge, the nonlinear term generates a blue (B) electric field, the gradient of which is given by

$$\vec{\nabla}\cdot\underbrace{\vec{E}^B}_{\text{induced electric field}} = -g_s f^B_{GR}\underbrace{\vec{A}^G}_{\text{virtual gluon}}\cdot\underbrace{\vec{E}^R}_{\text{original electric field}}. \tag{6.136}$$

[1] The Gauss's law example is borrowed from M. Peskin and D. Schroeder, *Quantum Field Theory* (Perseus Books, 1995).

At the same time, the field $\vec{E}^B$ will generate a red electric field, the size of which is given by Gauss's law:

$$\vec{\nabla} \cdot \vec{E}^R = -g_s f^R_{GB} \vec{A}^G \cdot \vec{E}^B \,. \tag{6.137}$$

The induced extra field causes an increase (since $f^R_{GB} = f^B_{GR} = -1$) in the initial red electric field $\vec{E}^R$ generated by the red charge ρ^R. The effective red charge is larger than the one measured from the charge density ρ^R. The QCD vacuum is populated by virtual gluons and behaves like a Pauli paramagnetic substance.

The intuitive picture provided by Gauss's law is made quantitative by a long computation (see the problem session), which gives the coefficient β_0 of the β-function in QCD:

$$\beta_0 = \frac{\overbrace{11\, N_c}^{N_c = 3\text{ (gluons)}} - \overbrace{2\, N_f}^{N_f = 6\text{ (quarks)}}}{12\pi} = \frac{21}{12\pi} > 0 \,. \tag{6.138}$$

In Eq. (6.138) I can recognize two terms. The first depends on the number of colors and comes from gluon loops. The second term is proportional to the number of flavors and comes from fermion loops. Whereas the sign of this second terms is like that in QED (and in fact the physics is very much the same), the sign of the first term is positive and corresponds to the self-interacting Gauss's law of QCD. As long as the first term is larger than the second, the overall sign of β_0 is positive and the effective charge of QCD grows as we move away from its source. In the opposite direction, for very high energies, at very short distances, the charge becomes smaller and smaller. This phenomenon is called **asymptotic freedom**.

To discuss the physics of the running coupling $\alpha_s(Q^2)$, first replace m_e in Eq. (6.133) with a generic mass scale μ and the coefficient in front of the logarithm with β_0. The running charge (see Figures 6.21, 6.22) can thus be written as

$$\alpha_s(Q^2) = \frac{\alpha_s(\mu^2)}{1 + \beta_0\, \alpha_s(\mu^2) \log \dfrac{Q^2}{\mu^2}} \,, \tag{6.139}$$

which can be rewritten as

$$\alpha_s(Q^2) = \frac{1}{\beta_0 \log \dfrac{Q^2}{\Lambda^2_{\text{QCD}}}} \,, \tag{6.140}$$

where $\beta_0 = (11N_c - 2N_f)/12\pi$ for QCD and

$$\Lambda^2_{\text{QCD}} = \mu^2 \exp\left[\frac{-1}{\beta_0\, \alpha_s(\mu^2)}\right] \tag{6.141}$$

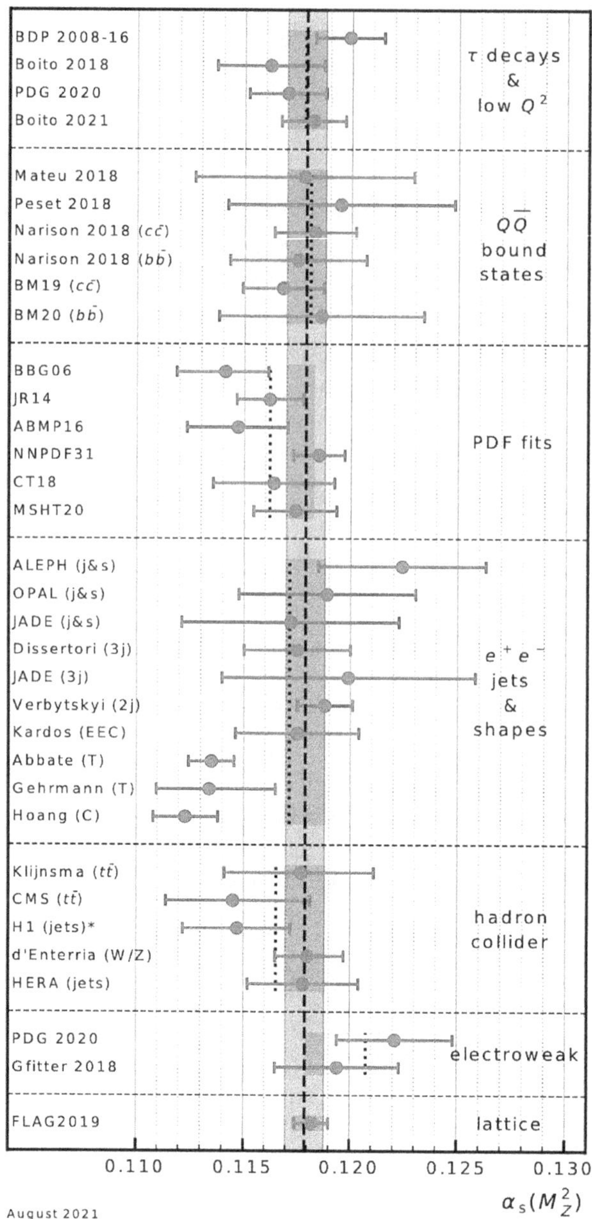

Figure 6.21 Experimental determination of the value of the QCD charge at energy equal to m_Z [R.L. Workman *et al.* (Particle Data Group), *Prog. Theor. Exp. Phys.* **2022** (2022) 083C01 http://pdg.lbl.gov].

is a **dimensional** parameter which replaces the dimensionless coupling α_s. As for α_s, Λ_{QCD} is a free parameter of the model that has to be fixed by experiment.

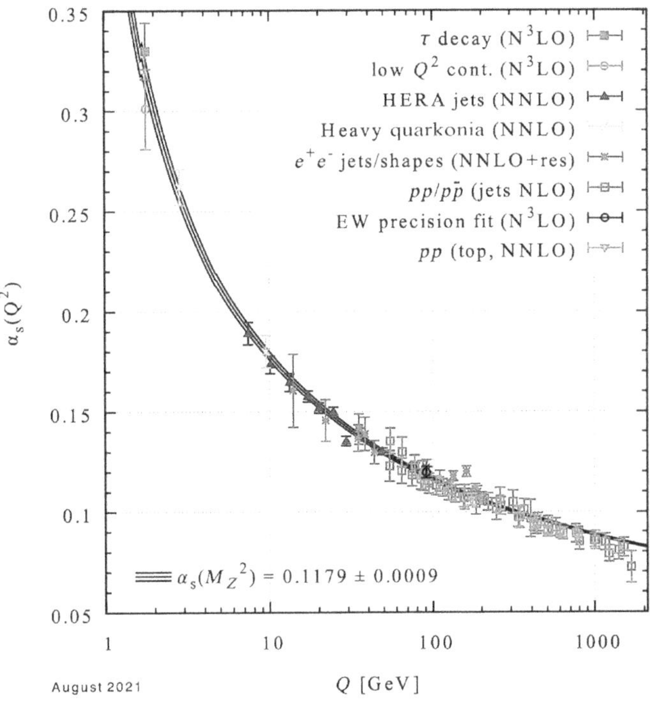

Figure 6.22 Experimental determination of the value of the QCD charge at different energies, showing the running of the coupling [R.L. Workman *et al.* (Particle Data Group), *Prog. Theor. Exp. Phys.* **2022** (2022) 083C01 http://pdg.lbl.gov].

This apparently trivial step is actually a momentous one.[6.29] Something very peculiar is going on here. I started with a classical Lagrangian with no dimensionful parameters (neglecting the fermion masses) and I end up with a dimensionful parameter. This phenomenon is called **dimensional transmutation**. The quantum corrections must be regularized and this process gives rise to a dimensional parameter in a Lagrangian that had none to start with.

[6.29] The classical QCD Lagrangian (with all masses taken to be zero) is scale invariant. This invariance is broken by the regularization procedure, and the scale symmetry is anomalous. The scale Λ_{QCD} is born out of the scale symmetry anomally. It sets the scale of all possible bound states of quarks in hadrons. The vacuum expectation value of the Higgs field sets the mass parameters in the Standard Model Lagrangian but the masses of protons and neutrons – and therefore the masses of what we usually refer to ordinary things – are set by Λ_{QCD}.

The scale Λ_{QCD} defines the energy regime in which the perturbative expansion is valid because

$$\alpha_s \leq 1 \tag{6.142}$$

when the square of the energy Q in the process satisfies

$$Q^2 \geq \Lambda_{\text{QCD}}^2 . \tag{6.143}$$

Because of the dimensionful parameter Λ_{QCD}, the theory has a natural scale at which the running coupling becomes strong and bound states are formed. Quarks and gluons cannot propagate for distances larger than $1/\Lambda_{\text{QCD}}$; they are confined. This feature provides a first explanation of how the Lagrangian in Eq. (6.1) can describe the hadrons even if the degrees of freedom are quarks and gluons.

6.8 *Problem Session*: The β-Function of QCD

The computation of the β-function of QCD is a sort of rite of passage for all practicing quantum field theorists. Like sailors first crossing the equator, you are put through a challenging ritual which is somewhat tedious but that at the end makes you a better person.

It is a computation that can be done by hand (highly recommended once in your lifetime) but here I prefer to point out how to use some software to help and make sure that no trivial mistakes are made. I am going to use FeynCalc again.

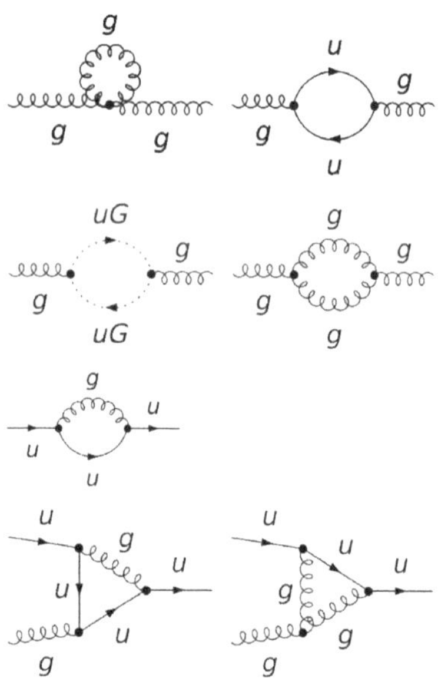

Figure 6.23 Feynman diagrams entering in the one-loop computation of the β-function in QCD: g stands for a gluon, u for a quark and uG for the ghost.

The program automatically generates the diagrams required for the computation. As shown in Figure 6.23, there are loop corrections to the propagators of both the quarks and the gluons, and the vertices among quarks and gluon. These corrections must include loop diagrams with ghosts in addition to quarks and gluons.

Counter-terms must also be added to the Lagrangian because of the divergencies in the loop momentum integrals. The value of the one-loop-order β-function is obtained by identifying the coefficient of one of these counter-terms in each of the diagrams. This coefficient is the residue of the pole $1/\varepsilon$ in **dimensional regularization**, where all integrals are done in $D = 4 - \varepsilon$ space-time dimensions to make them finite (instead of using a cut off, as I did before).

The diligent reader should read about the dimensional regularization of the loop integrals in a textbook on quantum field theory. The appendix includes the formulas necessary to follow what FeynCalc is actually doing.

First, it is necessary to set up the program: it runs within the Mathematica™ software. As it happens, the computation of the β-function is among the prewritten examples included in the software. Familiarization with the definitions and command lines takes a couple of hours – a time shorter than that necessary to actually compute the β-function from scratch. The final output consists of the lines shown in Figure 6.24, which nicely provides the result I need:

$$\beta_0 = -\frac{2N_f - 11N_c}{24\pi}, \tag{6.144}$$

which is the value I used in Eq. (6.138).

```
sol[5] = Simplify[Flatten[Solve[ampQGlVertexDiv[5] == 0,
         SMP["d_g"]]]]
solMS4 = sol[5] /. {SMP["d_g"] -> SMP["d_g^MS"]}
solMSbar4 = sol[5] /. {SMP["d_g"] -> SMP["d_g^MSbar"],
        1/Epsilon -> SMP["Delta"]}
```

$$\left\{\delta_g \to \frac{\alpha_s\,(2\,N_f - 11\,C_A)}{24\,\pi\,\varepsilon}\right\}$$

Figure 6.24 FEYNCALC compact computation of the β_0-coefficient.

6.9 Renormalization Group Equations*

All vertices in quantum field theory need to be regularized because, in general, they contain divergent quantities. This is done in a variety of manners. The most brutal of which is just to cut off the integrals over the internal momenta. The momentum integration becomes convergent and we can work with finite quantities. At the end of the computation, the cut off Λ is taken to infinity, and the singular contributions are absorbed in the definition of the parameters of the theory.

When this procedure is applied to a generic vertex with n_A external gauge bosons and n_q fermions, its regularized value is related to the bare vertex (computed at the classical level) as follows:

$$\Gamma_{n_A\,n_q}(k_i, p_j; g; \Lambda) = \left[Z_A(\Lambda)\right]^{n_A/2}\left[Z_q(\Lambda)\right]^{n_q/2}\overset{\text{bare}}{\Gamma^{(0)}_{n_A\,n_q}}(k_i, p_j; g_0)\,, \tag{6.145}$$

where the coupling

$$g = g(\overset{\text{bare coupling}}{g_0}, \Lambda) \tag{6.146}$$

is a function of the bare (classical) coupling parameter and the cut off.[6.30] The factors $Z_i(\Lambda)$ contain the regularization of the external particles wave functions. The cut off Λ is introduced by the regularization; it is the ultimate source of dimensional transmutation in QCD – by which a classical theory without an explicit mass scale ends up with having a scale.

6.30 I am working in a theory with only massless particles.

The classical (bare) vertex depends neither on the regularization nor on the scale Λ. The independence of the bare vertex from the renormalization scale gives the equation

$$\Lambda\frac{\mathrm{d}}{\mathrm{d}\Lambda}\Gamma^{(0)}_{n_A\,n_q}(k_i, p_j; g_0) = 0\,, \tag{6.147}$$

which is called the **renormalization group equation**. By replacing the bare vertex with its regularized expression I can write Eq. (6.147) as

$$\left[\Lambda\frac{\partial}{\partial\Lambda} + \beta(g)\frac{\partial}{\partial g} - n_A\gamma_A(g) - n_q\gamma_q(g)\right]\Gamma_{n_A\,n_q}(k_i, p_j;\, g;\, \Lambda) = 0\,, \tag{6.148}$$

with[6.31]

6.31 $\beta(g, \lambda)$ does not depend on the gauge choice only, in dimensional regularization.

$$\beta(g) = \lim_{\Lambda\to\infty}\Lambda\frac{\partial}{\partial\Lambda}g(g_0, \Lambda) \tag{6.149}$$

$$\gamma_i(g) = \frac{1}{2}\lim_{\Lambda\to\infty}\Lambda\, Z_i^{-1}\frac{\partial}{\partial\Lambda}Z_i(g_0, \Lambda)\,. \tag{6.150}$$

The terminology here is that $\beta(g)$ is called the **β-function** (as already introduced in Eq. (6.134)) and the $\gamma_i(g)$ are the **anomalous dimensions** because, as will shortly become clear, they modify (by their quantum corrections) the naive dimensions of the vertex.

By construction, the vertex $\Gamma_{n_A,\,n_q}$ has dimension[6.32]

6.32 With M the signpost for a generic scale.

$$\dim\Gamma_{n_A,\,n_q} = [M]^{4-n_A-\frac{3}{2}n_q}\,, \tag{6.151}$$

and I can use Euler's differential equation for homogenous functions to write

$$\left[\lambda\frac{\partial}{\partial\lambda} + \Lambda\frac{\partial}{\partial\Lambda} + \frac{3}{2}n_q + n_A - 4\right]\Gamma_{n_A\,n_q}(\lambda k_i, \lambda p_j;\, g;\, \Lambda) = 0\,, \tag{6.152}$$

which I can then use to replace the logarithmic derivative with respect to Λ and rewrite Eq. (6.148) as

$$\Bigg[\lambda\frac{\partial}{\partial\lambda} - \beta(g)\frac{\partial}{\partial g} + \underbrace{\left(\frac{3}{2} + \gamma_q\right)n_q + (1 + \gamma_A)n_A}_{\text{deviation from naive scaling}} - 4\Bigg]\Gamma_{n_A\,n_q}(\underbrace{\lambda k_i, \lambda p_j}_{\text{change of scale}};\, g;\, \Lambda) = 0\,. \tag{6.153}$$

In the form above, the renormalization group equation tells me how the vertex function is changed by rescaling all the external momenta by a factor λ.

Equation (6.153) is a differential equation that can be solved by the method of the characteristic curve. First introduce the curve $\bar{g}(\lambda, g)$ defined by the equation

$$\frac{\mathrm{d}\bar{g}(\lambda, g)}{\mathrm{d}\log\lambda} = \beta(\bar{g}) \tag{6.154}$$

with initial condition $\bar{g}(1, g) = g$. The function $\bar{g}(\lambda, g)$ is called the **running** coupling constant. The solution of Eq. (6.154) is given by

$$\int_1^{\lambda}\frac{\mathrm{d}\lambda'}{\lambda'} = \log\lambda = \int_g^{\bar{g}}\frac{\mathrm{d}g'}{\beta(g')} \tag{6.155}$$

which can be inserted back into Eq. (6.153) to obtain the solution

$$\Gamma_{n_A n_q}(\lambda k_i, \lambda p_j; g) = \Gamma_{n_A n_q}(k_i, p_j; \bar{g}(\lambda, g)) \underbrace{\lambda^{4-n_A-\frac{3}{2}n_q}}_{\text{naive scaling}}$$
$$\times \exp\left\{-\int_g^{\bar{g}} dg' \left[n_A \gamma_A(g') + n_q \gamma_q(g')\right] \beta^{-1}(g')\right\} . \tag{6.156}$$

The meaning of the solution in Eq. (6.156) is that a change in the momenta by a factor λ can be expressed by the dependence of the vertex function on the running coupling constant, while being modified by a multiplicative factor. The multiplicative factor contains the naive scaling of the vertex function plus the quantum corrections, as given by the anomalous dimensions.

The characteristic curve in the neighborhood of a fixed point g^* for which

$$\beta(g^*) = 0 \tag{6.157}$$

can be studied by expanding the β-function around the fixed point as

$$\beta(g) = \beta_0 \, (g - g^*)^r \qquad r \geq 1. \tag{6.158}$$

The solution is

$$\begin{cases} \bar{g} - g^* = (g - g^*)\lambda^{\beta_0} & r = 1 \quad \text{(simple zero)} \\ (\bar{g} - g^*)^{r-1} = \dfrac{(g - g^*)^{r-1}}{1 - \beta_0(r-1)(g - g^*)^{r-1} \log \lambda} & r > 1 . \end{cases} \tag{6.159}$$

The coupling runs toward the fixed point as the momenta are rescaled to larger values (Figure 6.25) in the ultraviolet limit or smaller values in the infrared limit. The running depends on the sign of β_0:

$$\bar{g}(\lambda, g) \to g^* \qquad \text{for} \qquad \begin{cases} \lambda \to \infty & \beta_0 < 0 \quad \text{UV} \\ \lambda \to 0 & \beta_0 > 0 \quad \text{IR} . \end{cases} \tag{6.160}$$

For $\beta_0 < 0$, the running coupling goes to the ultraviolet fixed point; it goes toward the infrared fixed point for $\beta_0 > 0$.

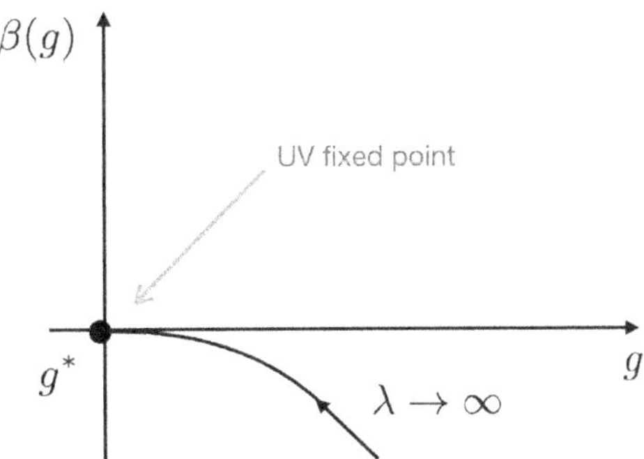

Figure 6.25 The running toward g^* of the coupling constant in the plane $(g, \beta(g))$.

In QCD, it is found that

$$\beta(g) = \beta_0 \frac{g^3}{16\pi^2} \qquad (r = 3) \tag{6.161}$$

with $\beta_0 = (2N_f - 11N_c)/3$ and $g^* = 0$ an ultraviolet stable fixed point (as long as $N_f < 33/2$). Since $\beta_0 < 0$ is negative, the theory is asymptotically free with the running coupling going to zero as $\lambda \to \infty$. The fixed point is not a simple zero and the running is logarithmic. The logarithmic dependence on the momentum is a characteristic signature of the QCD Lagrangian.

6.10 *Problem Session*: Λ_{QCD} vs. Mass Scale

To get some feeling about the running of the QCD coupling is useful to estimate Eq. (6.140) in a few representative cases.

First rewrite Eq. (6.140) as

$$\frac{Q^2}{\Lambda^2_{\text{QCD}}} = \exp \frac{2\pi}{3\,\alpha_s(Q^2)}, \tag{6.162}$$

taking $\beta_0 = 3/2\pi$ (for $N_c = 3$ and $N_f = 2$). Suppose now that the momentum Q^2 is taken equal to the proton mass m_p^2 so that

$$\frac{m_p^2}{\Lambda^2_{\text{QCD}}} = \exp \frac{2\pi}{3\,\alpha_s(m_p^2)}. \tag{6.163}$$

Now take the coupling $\alpha_s(m_p^2)$ to be of the same order as the weak interactions, say $\alpha_s(m_p^2) \approx 0.01$. This is not its value but it is what it would be if the strong interactions were weaker.

After inserting these numbers, I obtain that

$$\Lambda_{\text{QCD}} \approx 10^{-30}\ \text{GeV}, \tag{6.164}$$

that is, a length scale[6.33]

6.33 10^{12} km is approximately 0.1 light year.

$$1/\Lambda_{\text{QCD}} \approx 10^{16}\ \text{cm} \approx 10^{12}\,\text{km!} \tag{6.165}$$

To get a feeling for this immense size, consider the Compton length of the proton, which is about 1 fm – thus the confinement scale is 10^{30} times larger. In plain words, the quarks inside the proton would have to travel 0.1 light year before realizing they are confined.

A more realistic case is the top quark. For this quark, $\alpha_s(m_t)$ is indeed weak (close to 0.1) and I have that

$$\frac{m_t}{\Lambda_{\text{QCD}}} \approx 10^3, \tag{6.166}$$

that is, $\Lambda_{\rm QCD} \approx 200$ MeV and the associated length scale is 1 fm. The Compton length of the top quark is about 10^{-3} fm. Clearly the confinement scale is about 1000 times bigger than the Compton length of the top quark. The top quark behaves as if it were not confined.

Finally, if I make the same estimate for the actual proton, for which $\alpha_s(m_p^2) \approx 1$, I find that

$$\frac{m_p}{\Lambda_{\rm QCD}} \approx 2\,, \tag{6.167}$$

and $\Lambda_{\rm QCD} \approx 500$ MeV, that is, 1 fm, which is also the Compton wavelength of the proton. This time the two scales are about the same and I do expect quarks inside the proton to be restrained by confinement.

6.11 What about the Nuclear Force?

The question in the heading of this section brings me back to the initial problem of describing the strong interactions. I did a detour into perturbative QCD and tried to convince you that QCD is indeed the theory of the strong interactions. Yet a problem remains inasmuch as the degrees of freedom of QCD, quarks and gluons, are not those of the hadrons we see in nuclear interactions. The existence of different degrees of freedom is explained by confinement – the other face of QCD asymptotic freedom – but how to go from one set of states to the other remains veiled by the non-perturbative nature of the mapping.

How can then nuclear forces be described?

The only systematic way to compute such a force is by pushing QCD into the non-perturbative regime. This is done by lattice QCD.[6.34] The result is a nucleon–nucleon potential like that in Figure 6.26. This result confirms that the long-distance behavior of QCD does indeed give rise to nuclear physics. Nuclear forces are the residual interactions of quarks and gluons when taken at a distance comparable to those in the nuclei.

6.34 Lattice QCD is the theory of the QCD Lagrangian discretized on a lattice. It is a renormalizable theory that is well-defined in the strong coupling limit. The correlations among the quarks are computed by simulating the QCD Lagrangian by Monte Carlo methods on a super-computer. The results are eventually extrapolated to the continuum limit of zero lattice spacing.

A more phenomenological approach relies on the symmetries of the QCD Lagrangian. The nuclear forces must appear following the spontaneous breaking of chiral symmetry, which is the symmetry governing the low-energy physics of the strong interactions. We know that the degrees of freedom are the Goldstone bosons of the breaking of chiral symmetry, in this case pions (and kaons). In this picture, the pions are the degrees of freedom into which the high-energy degrees of freedom, quarks and gluons, turn, in going to low-energy processes. We do not have a detailed description of the transformation – because of its non-perturbative nature – but we can describe the pions by means of a Lagrangian in which all terms respecting chiral symmetry and Lorentz invariance are written and ordered in terms of the momentum they require in order to be relevant. I have

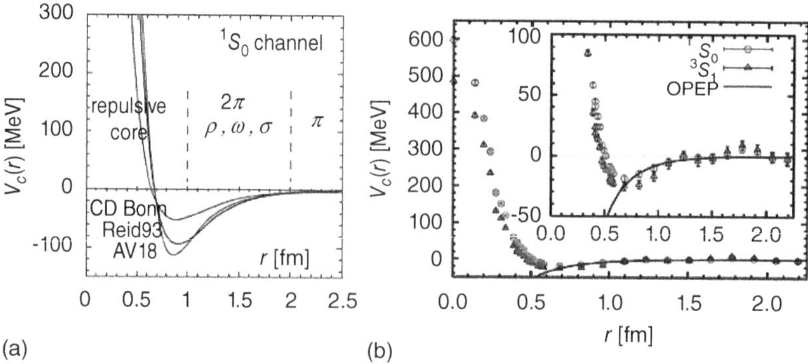

Figure 6.26 Lattice-QCD-generated nuclear potential. [From N. Ishii, A. Aoki. and T. Hatsuda, *Physical Review Letters* **99** (2007) 022001.] (a) The potential, with the identification of the regions where the contributions of the exchange of the various mesons dominate. (b) The actual points generated by the lattice simulation of the interactions among the quarks.

$$L_{\chi \mathrm{PT}} = -\frac{1}{2}\partial_\mu \vec{\pi} \cdot \partial^\mu \vec{\pi} - \frac{1}{2} m_\pi^2 \vec{\pi}^{\,2} + \cdots , \tag{6.168}$$

where $\vec{\pi}$ represents the isospin triplet state of the pions. The first two terms of the Lagrangian in Eq. (6.168) are determined by the symmetry requirement. The other, higher-order, terms have the operator structure determined but arbitrary coefficients weighting their contribution. These coefficients need to be determined from experiment.

Still missing are the nucleons. Again following the Lorentz and chiral symmetry I introduce them by writing the Lagrangian

$$\begin{aligned} L_{\chi N} = -\bar{N}\Big[\not{\partial} + m_N + \frac{2i}{F_\pi}\gamma_5 g_A \vec{t} \cdot \not{\partial}\vec{\pi} \\ + \frac{2i}{F_\pi^2}\vec{t} \cdot (\vec{\pi} \times \not{\partial}\vec{\pi}) \Big] N - (\bar{N}\Gamma_\alpha N)(\bar{N}\Gamma^\alpha N), \end{aligned} \tag{6.169}$$

which now includes both the Goldstone boson, in this case the pions, and the nucleons – the latter written as isospin doublets. The constant $F_\pi \simeq 195$ MeV is taken from the decay of the pion, $g_A \simeq 1.25$, from the nucleon axial current. The matrices Γ^α are constrained by Lorentz and isospin invariance and take into account direct interactions between the nucleons. The matrices $\vec{t}$ are the usual Pauli matrices and act in isospin space.

The effective Lagrangian in Eq. (6.169) – in a way dual to that of QCD – is valid at long distances. It contains the degrees of freedom that are active in this regime: pions and nucleons.

The nuclear potential describing the interaction of two nucleons in nuclei can be derived from Eq. (6.169) by taking the nonrelativistic limit and making a Fourier transformation, to obtain the two-nucleon potential

$$V_{\text{two-nucleon}} = 2\,(C_S + C_T\,,\vec{\sigma}_1 \cdot \vec{\sigma}_2)\delta^{(3)}(\vec{x}_1 - \vec{x}_2)$$
$$- \left(\frac{2g_A}{F_\pi}\right)^2 (\vec{t}_1 \cdot \vec{t}_2)(\vec{\sigma}_1 \cdot \vec{\nabla}_1)(\vec{\sigma}_2 \cdot \vec{\nabla}_2) Y(|\vec{x}_1 - \vec{x}_2|), \quad (6.170)$$

where $Y(r) = \exp(-m_\pi r)/4\pi r$ is the Yukawa potential arising through the exchange of the pions, and $\vec{\sigma}_i$ are the nucleon spins. The two constant C_S and C_T cannot be determined by the formalism alone and need to be fixed by the data.

This potential gives a reasonably good description of the nuclear potential among nucleons in nuclei. It is derived by exploiting the symmetry structure of the QCD Lagrangian and in particular the consequences of the spontaneously broken chiral symmetry, identifying pions and nucleons as the long-distance degrees of freedom propagating after quarks and gluons have been confined inside the hadrons.

6.12 *Problem Session*: The Strong-*CP* Problem*

The θ-term in the QCD Lagrangian violates *CP* symmetry. This can be understood by rewriting it in terms of the electric and magnetic fields,

$$F^A_{\mu\nu}\tilde{F}^A_{\mu\nu} = 4\vec{E}^A \cdot \vec{B}^A, \quad (6.171)$$

and recalling the opposite properties of these fields under C and P transformations.

The existence of this term leads to *CP* violation in the strong sector of the Standard Model.

The neutron, for instance, acquires a *CP*-violating electric dipole momentum. To compute it, first rotate the θ-term away by performing a $U(1)_A$ transformation of the quark fields, which leads to a term

$$\sum_k \bar{q}_{L\,k} q_R m_k e^{i\bar{\theta}/n_F} + \bar{q}_R q_{L\,k} m_k e^{-i\bar{\theta}/n_F}$$
$$= \sum_k m_k \bar{q}_k e^{i\gamma^5\bar{\theta}/n_F} q_k \simeq \sum_k m_k \bar{q}_k \left(1 + i\gamma^5 \frac{\bar{\theta}}{n_F}\right) q_k \quad (6.172)$$

in their mass Lagrangian, which can be rearranged (to have the same $\bar{\theta}$ for all the masses) as

$$m_k + m_k i\gamma^5 \frac{\bar{\theta}}{n_F} \to m_k + iA\gamma^5\bar{\theta}, \quad (6.173)$$

where (writing the determinant in flavor space)

$$(1 + i\bar{\theta}\gamma^5)\prod_k m_k = \prod_k (m_k + iA\gamma^5\bar{\theta}) + O(\bar{\theta}^2)$$
$$= \prod_k m_k + iA\bar{\theta}\gamma^5 \sum_l \prod_{k\neq l} m_k + O(\bar{\theta}^2), \quad (6.174)$$

which gives (for $n_F = 3$)

$$A = \frac{\prod_k m_k}{\sum_l \prod_{k\neq l} m_k} = \frac{m_u m_d m_s}{m_u m_d + m_u m_s + m_d m_s} \simeq \frac{m_u m_d}{m_u + m_d}\,. \tag{6.175}$$

This way, I end up with a term in the quark Lagrangian given by

$$\bar\theta A(\bar u i\gamma^5 u + \bar d i\gamma^5 d)\,. \tag{6.176}$$

To compute the electric dipole moment of the neutron I must go from the quark-level Lagrangian in Eq. (6.176) to one in terms of the hadrons. In the spirit of the effective Lagrangian in Eq. (6.169), this quark-level interaction can be modeled as an extra coupling between nucleons and pions that is thus given by two terms:[6.35]

6.35 The coefficient c_+ is obtained by means of the identity $\langle \pi^a N|\bar q i\gamma^5 q|N\rangle = \langle N|\bar q i\gamma^5 \tau^a q|N\rangle/F_\pi$ and is based on the splitting of the baryon octet masses, giving $c_+ = (m_\Xi - m_N)/(2m_s - m_u - m_d)$. There is a bit of black magic here that I have taken from the original work, R.J. Crewther *et al.*, *Physics Letters* **B 88** (1979) 123.

$$-\frac{\bar\theta A c_+}{F_\pi}\pi^a \bar N \tau^a N + i\frac{g_A m_N}{F_\pi}\pi^a \bar N \tau^a N\,. \tag{6.177}$$

These terms can be written in terms of the proton, neutron and pion fields by expanding the isospin matrices and fields (a triplet for the pions, a doublet for the nucleons) as

$$-\sqrt{2}\frac{\bar\theta c_+ A}{F_\pi}\left(\pi^+ \bar p n + \pi^+ \bar n p\right) - i\sqrt{2}\frac{g_A m_N}{F_\pi}\left(\pi^+ \bar p \gamma^5 n + \pi^+ \bar n \gamma^5 p\right). \tag{6.178}$$

The relevant diagram is depicted in Figure 6.27 where at the vertices I can insert either one or the other term in Eq. (6.178). The amplitude for the process is given by

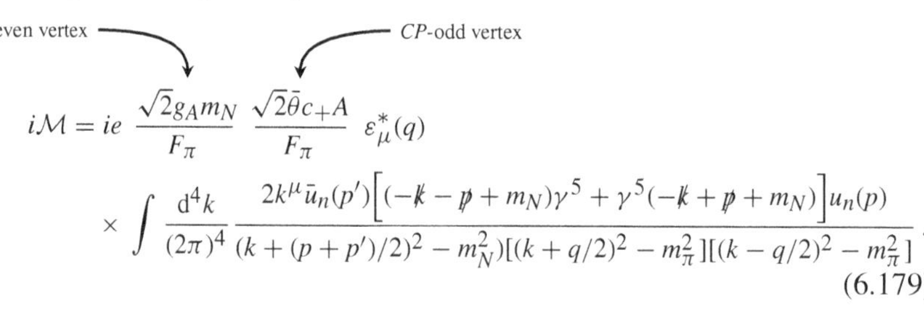

$$i\mathcal{M} = ie\,\frac{\sqrt{2} g_A m_N}{F_\pi}\,\frac{\sqrt{2}\bar\theta c_+ A}{F_\pi}\,\varepsilon^*_\mu(q)$$
$$\times \int \frac{\mathrm{d}^4 k}{(2\pi)^4}\frac{2k^\mu \bar u_n(p')\Big[(-\not k - \not p + m_N)\gamma^5 + \gamma^5(-\not k + \not p + m_N)\Big]u_n(p)}{(k+(p+p')/2)^2 - m_N^2)[(k+q/2)^2 - m_\pi^2][(k-q/2)^2 - m_\pi^2]}\,. \tag{6.179}$$

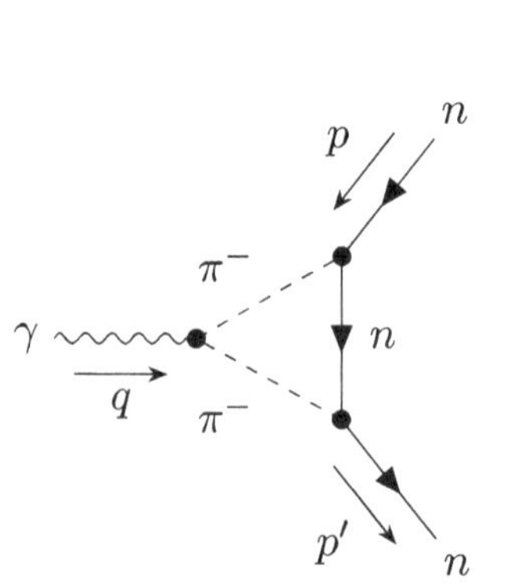

Figure 6.27 The Feynman diagram for the *CP*-violating electric dipole moment of the neutron.

The Diracology of the numerator gives just a factor $2m_N \bar u_N(p')\gamma^5 u(p)$. Setting $q = 0$ gives

$$\mathcal{M} = 4e\frac{g_A m_N^2 \bar\theta c_+ A}{F_\pi^2}\,\varepsilon^*_\mu(q)\int \frac{\mathrm{d}^4 k}{(2\pi)^4}\frac{2k^\mu \bar u_n(p)\gamma^5 u(p)}{(2p\cdot k)(k^2 - m_\pi^2)^2}u_n(p)\,. \tag{6.180}$$

The term proportional to the electric dipole momentum is identified by using the Gordon identity, which is given in this case by

$$(p+p')^\mu \bar u_n \gamma^5 u_n = \bar u_n \sigma^{\mu\nu} q_\nu i\gamma^5 u_n\,. \tag{6.181}$$

The loop integral is logarithmically divergent but the divergence is regularized by the mass of the pion in the IR and that of the nucleon in the UV, so that I obtain

$$\mathcal{M} = \frac{e\bar{\theta} g_A c_+ A}{8\pi^2 F_\pi^2} \log \frac{m_N^2}{m_\pi^2} \, \varepsilon_\mu^*(q) \bar{u}_n(p') \sigma^{\mu\nu} q_\nu \gamma^5 u_n(p) , \tag{6.182}$$

which gives for the electric dipole momentum of the neutron[6.36]

$$d_n^e = \frac{e\bar{\theta} c_+ g_A A}{8\pi^2 F_\pi^2} \log \frac{m_N^2}{m_\pi^2} \simeq \bar{\theta} \times 3.2 \times 10^{-16} \, e \, \mathrm{cm} \tag{6.183}$$

6.36 In quantum field theory, the electric dipole moment is given by the coefficient of the operator $\bar{u}_N(p')\sigma^{\mu\nu} q_\nu \gamma^5 u_N(p)$.

after taking $c_+ = 1.7$.

The estimate in Eq. (6.183) must be compared with the experimental limit, which is

$$d_n^e < 10^{-26} \, e \, \mathrm{cm} , \tag{6.184}$$

which shows that $\bar{\theta}$ must be very small indeed, less than 10^{-10}. The presence of the θ-term in the QCD Lagrangian indeed gives a *CP* violation which can be measured in the non-vanishing electric dipole moment of the neutron. The result is very much too large to be compatible with what we know. This is a problem if you do not like to assign very small numbers to the coupling constants of your model.

6.13 Exercises

1. **$\pi^0 \to \gamma\gamma$.** Compute the width for this decay from the one-loop triangle diagram and check that Eq. (6.106) is correct.
   [Hint: Use the integrals in Eq. (6.20) after substituting the appropriated group matrices.]
2. **Group reps.** Verify the group decompositions in Eq. (6.66) for the quarks in mesons $\underline{3} \otimes \underline{\bar{3}} = \underline{1} \oplus \underline{8}$ and hadrons $\underline{3} \otimes \underline{3} \otimes \underline{3} = \underline{1} \oplus \underline{8} \oplus \underline{8} \oplus \underline{10}$.
   [Hint: Use the Young tableaux to find the irreducible representations.]
3. **Glueballs.** Find the possible angular momentum J of the lowest bound states of gluons. Can these states have electric charge? What is the order of magnitude of their mass? Into what states do they decay?
4. **Λ_{QCD}.** I can think of many models of strong interactions, each defined by the same QCD Lagrangian but with a different value of Λ_{QCD}. What would be the first consequence for the hadron spectrum of having $\Lambda_{\mathrm{QCD}} = 1$ TeV instead of its actual value, around 500 MeV?
5. **N_f.** Plot the running of the QCD coupling for different values of N_f. Notice the variation of the slope as N_f becomes larger. What happens to the QCD β-function if there are more flavors of fermions and $N_f = 16$?
   [Hint: Just insert this number for N_f into Eq. (6.138) and notice the change in the running of α_s.]

Part II

Where the Model is Put to Use

7 The Electroweak Theory at Work

Contents

Now that I know about the Standard Model, what am I going to do with it? Wear it on a T-shirt, perhaps, like that in Figure 7.1.

What else? The Standard Model answers all our questions about high-energy physics. At the time of writing of this book, there was no evidence of physics beyond the Standard Model. The predicted values of all observables are in reasonable agreement with the experimental data; only few of them show some tension (that is, a discrepancy of more than three standard deviations) in the global fit.

There exists an open, and long-standing, question about bringing together gravity and the Standard Model. The energies at which these two theories may overlap are so high that we have no experimental data to guide us but for those in cosmology. There is also a problem with missing matter in cosmic evolution and galaxy dynamics which has been addressed by conjecturing the existence of a new form of matter not

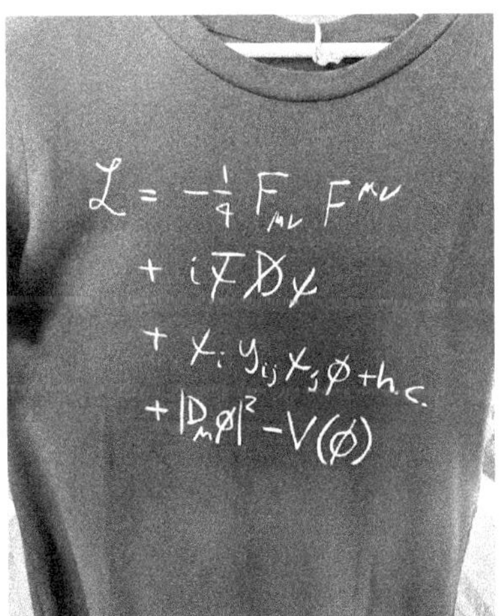

Figure 7.1 T-shirt showing the Lagrangian of the Standard Model.

electromagnetically charged and therefore "dark". Though interesting, these are questions for the theory of gravity and not for the Standard Model proper.

All others possible shortcomings of the Standard Models that you might hear about are of the "Why" variety of questions[7.1] ("Why so many particles?", "Why the hierarchy among the values of these masses?", "Why are the neutrino masses so small?", "Why is the electroweak scale so much smaller than the Planck mass?"), whereas good physics is mostly done by addressing the "How" kind of questions. The "Why" kind of physics sometime helps to prompt us to new directions but most often leads us astray. "Why" questions have a poor record of leading to progress in our understanding of the fundamental interactions.

Consider the same kind of enquiry into the origin of the specific orbits of the various planets in the solar system. One can speculate – and for instance Euler did try to put the Platonic solids into some sort of correspondence with these orbits – but there is no real point in doing this. The values of the orbits are just given by the initial conditions and these are beyond our dynamical explanation.[7.2]

The Standard Model is a theory whose structure is determined by the choice of the gauge groups ($SU(3)_c \otimes SU(2)_L \otimes U(1)_Y$) and the representations for the fermions ($SU(2)$ doublets for left-handed fields, singlets for right-handed fields and so on). Once these two choices have been made, the model is determined. Or, is it?

The operators entering the Lagrangian are dictated by the symmetries and also by the order in energy we want to keep.[7.3] One starts with operators of dimension four and then goes on to those of dimensions five, six and so on. For energies sufficiently low, one only need worry about the leading operators, that is, those of dimension 4. It is a peculiarity of Lagrangians consisting of these operators to require the knowledge of only a finite number of parameters – the theory is said to be renormalizable.[7.4] It can be concluded that the Lagrangian of the Standard Model is completely fixed by its symmetries (gauge groups and Lorentz) and by the requirement of renormalizability.

The classical Lagrangian of the Standard Model contains **26 parameters**. They are listed in Table 7.1. By parameters I mean quantities that are free, not fixed by the model and require experimental data in order to be determined.

The number of parameters would go down to **19** if the neutrinos were massless and up to **27** if they were Majorana particles. The parameters would be only **8** (three masses, three couplings and the two in the Higgs potential) if I restricted myself to the first generation of fermions – to those, that is, of everyday life.

7.1 "[...] Theirs not to reason why," as in Alfred Lord Tennyson's ill-fated light brigade.

7.2 The same is true for total solar eclipses. While the Moon and the Sun have very different actual sizes, their apparent diameters are very close when seen from the Earth. This feature makes total eclipses possible. Is the fine tuning of the apparent diameters a coincidence or should we take solar eclipses as a crucial clue to some hidden dynamics waiting to be discovered?

7.3 This point of view is the endpoint of our thinking about quantum field theory and the Standard Model: from renormalizability as a necessary requirement to a feature of the low-energy approximation.

7.4 Non-renormalizable operators of dimension larger than 4 exist (the Standard Model being an effective theory) but can be probed only at energies that are indeed very high – unless they break some of the Standard Model symmetries. A noteworthy example of the latter is provided by operators that break baryon number symmetry and leading to the decay of the proton.

Table 7.1 The parameters of the Standard Model Lagrangian.

No. of parameters	Parameter names	Symbols
12	Fermion masses	$m_u, m_d, m_c, m_s, m_t, m_b$ $m_e, m_\mu, m_\tau, m_1, m_2, m_3$
4 + 4	Flavor mixing matrices	$V_{\rm CKM}$ $V_{\rm PMNS}$
3	Gauge couplings	g_s, g, g'
2	Parameters in the Higgs potential	μ, λ
1	θ-Term in QCD	$\bar{\theta}$

Whether 26 (or 8) parameters are too many or too few for a model depends on how many different observables I can compute in terms of these parameters.[7.5]

[7.5] Just to give an example: A model with two parameters that reproduces one observable is not very good. It is mostly useless. You would be surprised by how many models with more parameters than predicted observables are nonetheless discussed and used.

The procedure (it is the same in all fields of physics and even beyond physics) is to use as many observables as the number of free parameters I have in my model to fix these parameters. Once these parameters have been fixed, I compute what the model predicts for other observables and compare these computed observables with the data. The more observables that are correctly predicted by the model, the better the model will be.

7.1 Precision Tests

To understand the procedure for testing the Standard Model, let us consider again the electroweak theory. The Lagrangian for the interactions of the matter fields is (as at the beginning of Chapter 2):

$$\begin{aligned}\mathcal{L}_\psi = &-\sum_i Q^i \bar{\psi}_i \gamma^\mu \psi_i A_\mu \\ &- \frac{g}{2\sqrt{2}} \sum_i \bar{\psi}_i^{\rm D} \gamma^\mu (1-\gamma_5) \left(T^+ W_\mu^+ + T^- W_\mu^-\right) \psi_i^{\rm D} \\ &- \frac{g}{2\cos\theta_W} \sum_i \psi_i \gamma^\mu \left(g_V^l - g_A^l \gamma_5\right) \psi_i Z_\mu .\end{aligned} \tag{7.1}$$

If I limit myself to the first-generation leptons and neglect all fermion masses, the only parameters of this Lagrangian are g, e and θ_W. Because $e = g \sin\theta_W$, I actually have only two parameters:

$$e \quad \text{and} \quad \sin\theta_W , \tag{7.2}$$

that is, the electromagnetic coupling and the weak angle. Because the gauge bosons are massive, I must add their masses as two additional parameters. They are both related to the vacuum expectation value of the Higgs boson v, which is the third free parameter since

$$m_W = \frac{1}{2} g v = \frac{e v}{2 \sin\theta_W} \tag{7.3}$$

and

$$m_Z = \frac{v}{2}\sqrt{g^2 + g'^2} = \frac{e v}{2 \sin\theta_W \cos\theta_W} = \frac{m_W}{\cos\theta_W}\,. \tag{7.4}$$

The values of the three free parameters e, $\sin\theta_W$ and v can be determined by looking at three observables: $\hat{\alpha}$, $\hat{G}_F$ and $\hat{m}_Z$.[1]

— The Thomson limit of $\gamma^* \to e^+e^-$, which gives me $\hat{\alpha}$. I take its value to be

$$\hat{\alpha} = 1/137.035999084\ldots$$

— The decay of the muon, which gives me $\hat{G}_F$. I take its value to be

$$\hat{G}_F = 1.1663787(6) \times 10^{-5}\ \text{GeV}^{-2}\,.$$

— The mass of the Z-boson, which gives $\hat{m}_Z$ directly. I take its value to be

$$\hat{m}_Z = 91.1876 \pm 0.0021\ \text{GeV}.$$

These three observables I call the input values, adding a hat to indicate measured quantities.

These three experimentally determined quantities together fix the three parameters of the Lagrangian[7.6] because

$$e = \sqrt{4\pi\hat{\alpha}} \tag{7.5}$$

$$v = \frac{1}{\sqrt{\sqrt{2}\hat{G}_F}} \tag{7.6}$$

$$\sin^2\theta_W \cos^2\theta_W = \frac{\pi\hat{\alpha}}{\sqrt{2}\hat{G}_F\hat{m}_Z^2}\,. \tag{7.7}$$

The last equation can be solved to obtain[7.7]

$$\sin^2\theta_W = \frac{1}{2} - \frac{1}{2}\sqrt{1 - 4\hat{\xi}}\,, \tag{7.8}$$

where $\hat{\xi} = \pi\hat{\alpha}/\sqrt{2}\hat{G}_F\hat{m}_Z^2$.

Does the Standard Model predict the mass m_Z? Does it predict the value of the electric charge? These are not the right questions. I can let the mass of the Z-boson have its measured value $\hat{m}_Z$ if I take it as one of the input values. That can always be done if m_Z is the only observable I am going to measure. What is really needed is the reproduction of all the observables that have been (and are going to be) measured.[7.8]

7.6 The first two come from their respective definitions. The third one comes from the definitions $m_W = gv/2$ and $m_W = m_Z \cos\theta_W$ together with $e = g\sin\theta_W$. Combining them I find that $\sin^2\theta_W \cos^2\theta_W = e^2v^2/4m_Z^2$, which can be solved to give the desired expression.

7.7 The uncertainties in the input values must be propagated through these relations; for a parameter $Q = f(x_1, \ldots, x_n)$ I have that

$$\delta Q = \sqrt{\left(\frac{\partial f}{\partial x_1}\delta x_1\right)^2 + \cdots + \left(\frac{\partial f}{\partial x_n}\delta x_n\right)^2}\,.$$

7.8 In this way – which is the actual way in which the Standard Model is put to the test – there is no room for the divergencies of quantum field theory. They all cancel in the procedure as long as the inputs are taken from the physical observables.

[1] PDG (2020) https://pdg.lbl.gov.

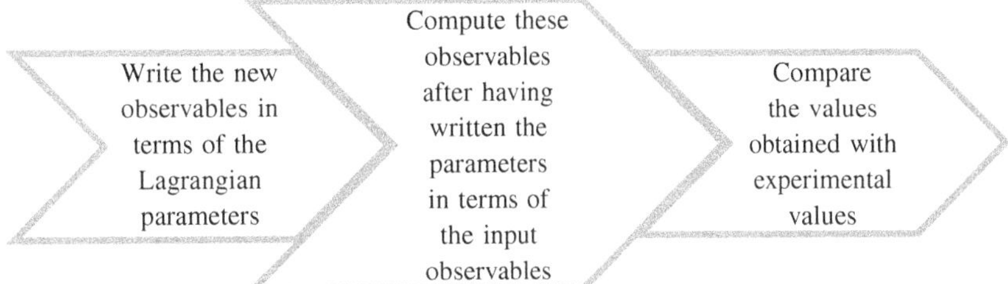

Figure 7.2 The arrow boxes sum up the three steps of the testing procedure.

If I want to test how well the Standard Model reproduces some observables I have to write these observables in terms of the parameters of the Lagrangian (e, $\sin\theta_W$ and v) and compare their measured values with what I obtain after rewriting the parameters of the Lagrangian in terms of the input observables $\hat{\alpha}$, $\hat{G}_F$ and $\hat{m}_Z$. The arrow boxes in Figure 7.2 sum up the procedure.

In this section I take as the observables I want to estimate in the Standard Model the mass of the gauge boson W, the left–right asymmetry and the width of the gauge boson Z for decay into lepton pairs, that is,

$$\hat{m}_W \qquad \hat{A}_{LR} \qquad \hat{\Gamma}_{\ell^+\ell^-}, \tag{7.9}$$

and compare their experimental values with what I obtain by writing them in terms of the three parameters of the Lagrangian – after having fixed the latter by means of the three input observables.

It is a minimal implementation in which for three inputs I derive only three observables – it serves as an illustration of how to proceed to the more general case in which more observables are derived.

The mass of the gauge boson W is readily written in terms of the parameters in the Lagrangian:

$$m_W^2 = m_Z^2 \cos^2\theta_W = \frac{e^2 v^2}{4\sin^2\theta}. \tag{7.10}$$

The left–right asymmetry is defined as the difference over the sum of the cross sections for the production of left- and right-handed electrons by the gauge boson Z:

$$A_{LR} = \frac{\sigma_L(Z \to e_L^+ e_L^-) - \sigma_R(Z \to e_R^+ e_R^-)}{\sigma_L(Z \to e_L^+ e_L^-) + \sigma_R(Z \to e_R^+ e_R^-)}. \tag{7.11}$$

It can be computed by simply replacing the couplings in the interaction Lagrangian for the left- and right-haded electrons:

$$-\frac{g}{\cos\theta_W} Z_\mu \Big[\overbrace{g_L^e}^{-\frac{1}{2}+\sin^2\theta_W} \bar{e}_L \gamma^\mu e_L + \underbrace{g_R^e}_{\sin^2\theta_W}\, e_R \gamma^\mu e_R \Big]. \tag{7.12}$$

The asymmetry contains the cross section, and therefore the squares of the coefficients, so that

$$A_{LR} = \frac{g_L^{e2} - g_R^{e2}}{g_L^{e2} + g_R^{e2}} = \frac{\left(\frac{1}{2} - \sin^2\theta_W\right)^2 - \sin^4\theta_W}{\left(\frac{1}{2} - \sin^2\theta_W\right)^2 + \sin^4\theta_W} . \tag{7.13}$$

Finally, for the decay width of the Z-boson I have

$$\Gamma_{\ell^+\ell^-} = \frac{1}{64\pi}\frac{|\overline{\mathcal{M}}|^2}{m_Z}\sqrt{1 - 4\frac{m_\ell^2}{m_Z^2}} , \tag{7.14}$$

where[7.9]

7.9 Now neglecting m_e, gives $\mathrm{Tr}\,\not{p}_e\gamma^\mu\not{p}_{\bar{e}}\gamma_\mu[(g_V^e)^2 + (g_A^e)^2] = 4m_Z^2[(g_V^e)^2 + (g_A^e)^2]$. The factor 1/3 in Eq. (7.15) comes from the average over the initial polarizations.

$$\begin{aligned} |\overline{\mathcal{M}}|^2 &= \frac{g^2}{4\cos^2\theta}\,\frac{1}{3}\,\mathrm{Tr}\left[(\not{p}_e + m_e)\gamma^\mu(g_V^e - g_A^e\gamma^5)(\not{p}_{\bar{e}} - m_{\bar{e}})\gamma_\mu(g_V^e + g_A^e\gamma^5)\right] \\ &= \frac{g^2}{3\cos^2\theta}m_Z^2\left[(g_V^e)^2 + (g_A^e)^2\right] , \end{aligned} \tag{7.15}$$

and therefore[7.10]

7.10 $g^2 = \dfrac{e^2}{\sin^2\theta_W}$.

$$\begin{aligned} \Gamma_{\ell^+\ell^-} &= \frac{g^2}{192\pi\cos^2\theta_W}m_Z\Big[\underbrace{(g_V^e)^2}_{-\frac{1}{2}+2\sin^2\theta_W} + \underbrace{(g_A^e)^2}_{\frac{1}{2}}\Big] \\ &= \frac{e^2}{192\pi\sin^2\theta_W\cos^2\theta_W}m_Z\left[(-\frac{1}{2} + 2\sin^2\theta_W)^2 + \frac{1}{4}\right] . \end{aligned} \tag{7.16}$$

The three observables can now be written in terms of the parameters and the latter in terms of their input observables to obtain:

$$\begin{cases} m_W^2 & = \dfrac{e^2v^2}{4\sin^2\theta_W} = \dfrac{\pi\sqrt{2}\hat{\alpha}}{\hat{G}_F\left(1 - \sqrt{1 - 4\hat{\xi}}\right)} \\ \Gamma_{\ell^+\ell^-} & = \dfrac{\sqrt{2}\hat{G}_F\hat{m}_Z^3}{48\pi}\left[\left(\dfrac{1}{2} - \sqrt{1 - 4\hat{\xi}}\right)^2 + \dfrac{1}{4}\right] \\ A_{LR} & = \dfrac{-1 + 2\sqrt{1 - 4\hat{\xi}}}{3 - 8\hat{x} - 2\sqrt{1 - 4\hat{\xi}}} . \end{cases} \tag{7.17}$$

After propagating the errors in the input parameters, the computed numbers can be compared with the experimental values:

$$A_{LR} = \underbrace{0.2960 \pm 0.00956}_{\text{computed}} \qquad \hat{A}_{LR} = \underbrace{0.1544 \pm 0.0060}_{\text{exp. value}}$$

$$\Gamma_{\ell^+\ell^-} = \underbrace{84.843 \pm 0.0012 \text{ MeV}}_{\text{computed}} \qquad \hat{\Gamma}_{\ell^+\ell^-} = \underbrace{83.984 \pm 0.086 \text{ MeV}}_{\text{exp. value}}$$

$$m_W = \underbrace{80.939 \pm 0.003 \text{ GeV}}_{\text{computed}} \qquad \hat{m}_W = \underbrace{80.379 \pm 0.012 \text{ GeV}}_{\text{exp. value}}.$$

These numbers seem pretty good but I must pay attention to their relative errors. The errors, given as standard deviations σ, are very small and the differences between computed and experimental numbers are large when measured in terms of these errors. Errors and therefore the discrepancies between two numbers are usually written down in terms of their significance, that is, the number of standard deviations σ. It can be seen that

$$\begin{aligned} A_{LR} \neq \hat{A}_{LR} \quad &\text{at} \quad 24\,\sigma \\ \Gamma_{\ell^+\ell^-} \neq \hat{\Gamma}_{\ell^+\ell^-} \quad &\text{at} \quad 10\,\sigma \\ m_W \neq \hat{m}_W \quad &\text{at} \quad 47\,\sigma, \end{aligned}$$

which are huge discrepancies.[7.11]

[7.11] What did I learn from this exercise? The Standard Model is ruled out by experiments! Maybe I should have checked before going through with the first six chapters!

Before pushing the panic button I must remember that the numbers I obtained where derived from tree level relations. They fail because I have not included the quantum corrections. Maybe victory can still be snatched from the jaws of defeat.

7.2 *Problem Session*: Enter the Quantum Corrections*

To introduce the matching between the parameters of the Lagrangian and the physical observables at one loop I need the self-energy of the gauge bosons. These are not the only one-loop corrections but are the dominant ones (the vertex correction in $Z \to \bar{b}b$ is the exception).

These corrections are all in the form of the modification of the gauge boson propagator,[2] which I depict as in Figure 7.3 and write, by using Lorentz covariance, as

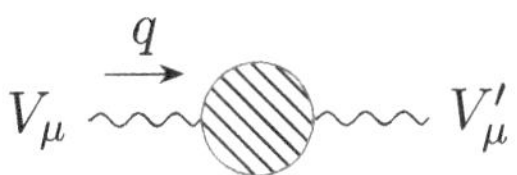

Figure 7.3 The full propagator for the gauge bosons. The blob represents all the possible loop corrections.

$$i\Pi_{VV'}(q^2)g^{\mu\nu} + i\Delta_{VV'}(q^2)q^\mu q^\nu, \tag{7.18}$$

since it is a rank-2 tensor that can only depend on the metric and the incoming momentum. In the basic case, in which $V = V'$, $\Pi_{VV}(q^2)$ modifies the gauge boson propagator, which becomes

[2] Here, I mostly follow J.D. Wells, *TASI Lecture Notes*, arXiv:hep-ph/05112342.

$$iD_V^{\mu\nu}(q^2) = \frac{-ig^{\mu\nu}}{q^2 - m_V^2}\left[1 + i\Pi_{VV}(q^2)\frac{-i}{q^2 - m_V^2} + i\Delta_{VV}(q^2)q^\mu q^\nu \frac{-i}{q^2 - m_V^2} + \cdots\right], \tag{7.19}$$

where the terms proportional to $q^\mu q^\nu$ can be safely neglected as long as the propagator is between fermions with negligible masses.

If I keep going and replace the ellipses in Eq. (7.19). I obtain a geometrical series[7.12] that can be re-summed, giving

7.12 $1 + x + x^2 + \cdots = \frac{1}{1-x}$

$$iD_V^{\mu\nu}(q^2) = \frac{-ig^{\mu\nu}}{q^2 - m_V^2 - \Pi_{VV}(q^2)} + \cdots. \tag{7.20}$$

It seems that the effect of this correction is just a shift in the value of the gauge boson mass:

$$m_V^2 \longrightarrow m_V^2 + \Pi_{VV}(m_V^2). \tag{7.21}$$

It is not necessary to actually compute this shift; I will just take it as a number.[7.13]

7.13 $\Pi_{VV}(q^2)$ is a divergent quantity and its computation is part of the renormalization of the parameters of the classical Lagrangian, as discussed for the charge in QCD. As long as one only works with physical quantities that are observables, these divergences must cancel out, leaving finite quantities.

The first step in including the one-loop corrections seems rather simple: in connecting the Lagrangian mass parameters I must shift the relation by an amount given by the self-energies, namely

$$m_Z^2 = \frac{e^2v^2}{4\sin^2\theta_W\cos^2\theta_W} + \Pi_{ZZ}(m_Z^2) \tag{7.22}$$

$$m_W^2 = \underbrace{\frac{e^2v^2}{4\sin^2\theta_W}}_{\text{as before, in Eq. (7.10)}} + \underbrace{\Pi_{WW}(m_W^2)}_{\text{the shift}}, \tag{7.23}$$

which will replace Eq. (7.10).

A similar insertion must go into the diagram that defines the Fermi constant, as shown in Figure 7.4, which is now given by

$$\frac{G_F}{\sqrt{2}} = \frac{g^2}{8m_W^2}\left[1 - i\Pi_{WW}(q^2)\frac{i}{q^2 - m_W^2}\right]_{q^2\to 0} = \overbrace{\frac{1}{2v}}^{\text{as before, in Eq. (7.8)}}\left[1 - \overbrace{\frac{\Pi_{WW}(0)}{m_W^2}}^{\text{the shift}}\right]. \tag{7.24}$$

Equation (7.24) replaces the relation in Eq. (7.8).

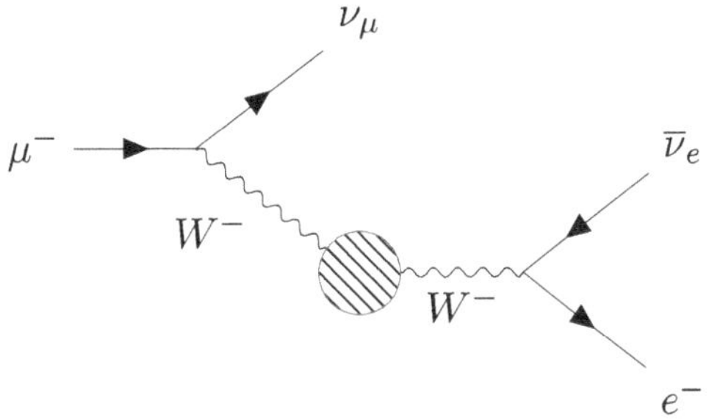

Figure 7.4 The renormalization of the gauge boson propagator in β-decay, on which the definition of the Fermi constant G_F is based.

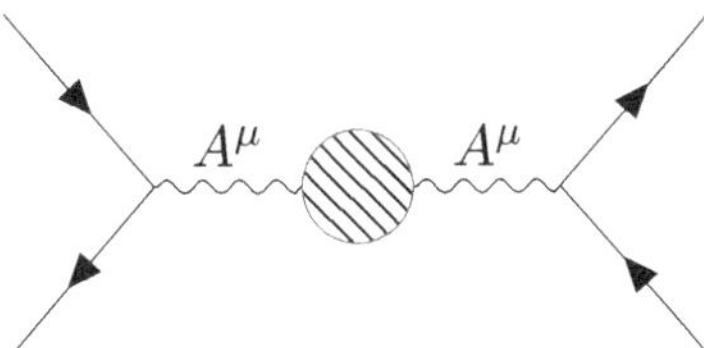

Figure 7.5 The modification of the single-photon exchange and the definition of the constant α.

Also, the relation connecting the parameter e of the Lagrangian to the fine-structure constant α is changed by a correction represented by the diagram in Figure 7.5, which implies that

$$-i\frac{4\pi\alpha}{q^2}\bigg|_{q^2\to 0} = -i\frac{e^2}{q^2}\left[1+\frac{\Pi_{AA}(q^2)}{q^2}\right]_{q^2\to 0} \tag{7.25}$$

and therefore

$$\alpha = \frac{e^2}{4\pi}\left[1+\Pi'_{AA}(0)\right], \tag{7.26}$$

where

$$\Pi'_{AA}(0) = \lim_{q^2\to 0}\frac{\Pi_{AA}(q^2)}{q^2}\,. \tag{7.27}$$

Here I encounter a difficulty. I don't know how to compute $\Pi'_{AA}(0)$. The problem is that in the blob there are also contributions from the insertion of hadrons, which I do not know how to compute in perturbation theory. Never mind. Let us just decide that the observable to be used is not $\hat{\alpha}(0)$ but rather $\hat{\alpha}(m_Z)$ – which is the fine-structure constant computed at an energy sufficiently high not to have to worry about hadrons. I now have

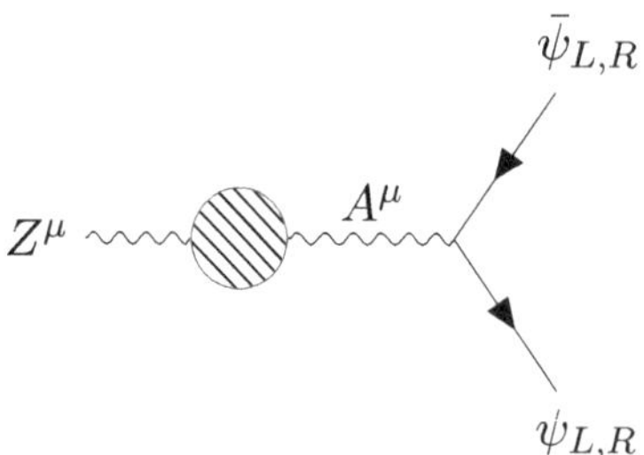

Figure 7.6 The Feynman diagram for the renormalized decay width for the Z gauge boson.

$$\alpha(m_Z) = \frac{e^2}{4\pi}\left[1 + \frac{\overbrace{\Pi'_{AA}(m_Z^2)}^{\text{the shift}}}{m_Z^2}\right]. \tag{7.28}$$

The three parameters of the Lagrangian are now linked to the input parameters $\hat{G}_F$, $\hat{\alpha}$ and $\hat{m}_Z$ by

$$v^2 = \frac{1}{\sqrt{2}\hat{G}_F}\left[1 - \frac{\Pi_{WW}(0)}{\hat{m}_W^2}\right] \tag{7.29}$$

$$e^2 = 4\pi\hat{\alpha}(m_Z^2)\left[1 - \frac{\Pi_{\gamma\gamma}(\hat{m}_Z^2)}{\hat{m}_Z^2}\right] \tag{7.30}$$

$$\sin^2\theta_W\cos^2\theta_W = \frac{\pi\hat{\alpha}(\hat{m}_Z^2)}{\sqrt{2}\hat{G}_F\hat{m}_Z^2}\left[1 + \Delta_\Pi\right], \tag{7.31}$$

where

$$\Delta_\Pi = \frac{\Pi_{ZZ}(\hat{m}_Z^2)}{\hat{m}_Z^2} - \frac{\Pi_{WW}(0)}{\hat{m}_W^2} - \frac{\Pi_{\gamma\gamma}(\hat{m}_Z^2)}{\hat{m}_Z^2}. \tag{7.32}$$

The above relations replace those at tree level.

The observables written in terms of the parameters of the Lagrangian are modified as well. The decay width of the Z-boson is changed when I include the diagram in Figure 7.6. The blob insertion gives a common factor shifting the couplings in the Z-boson vertex:

$$g_L^e = \left(-\frac{1}{2} + \sin^2\theta_W\right) + i\Pi_{Z\gamma}(m_Z^2)\frac{-i}{m_Z^2}\sin\theta_W\cos\theta_W \tag{7.33}$$

$$g_R^e = \sin^2\theta_W + i\Pi_{Z\gamma}(m_Z^2)\frac{-i}{m_Z^2}\sin\theta_W\cos\theta_W\,, \tag{7.34}$$

which amounts to just an overall shift in the mixing parameter:

$$\sin^2\theta_W \to \sin^2\theta_W - \sin\theta_W\cos\theta_W\frac{\Pi_{Z\gamma}(m_Z^2)}{m_Z^2}. \tag{7.35}$$

The insertion of the blobs in the decay width of the Z-boson is already included in Eq. (7.35) and the mass shift, except for the effect of a higher-order term in the expansion of the polarization tensor:

$$\Pi_{ZZ}(q^2) = \Pi_{ZZ}(m_Z^2) + \Pi'_{ZZ}(m_Z^2)(q^2 - m_Z^2) + \cdots, \tag{7.36}$$

where the prime stands for the derivative in the Taylor expansion. Insertion of this term modifies the propagator of the Z-boson, which becomes

$$\frac{-ig^{\mu\nu}}{q^2 - m_Z^2 - \Pi_{ZZ}(q^2)} = \frac{-iZg^{\mu\nu}}{q^2 - m_Z^2}, \tag{7.37}$$

where

$$Z = 1 + \Pi'_{ZZ}(m_Z^2) + \cdots \tag{7.38}$$

and

$$\Pi'_{ZZ}(m_Z^2) \simeq \frac{\Pi_{ZZ}(m_Z^2) - \Pi_{ZZ}(0)}{m_Z^2}. \tag{7.39}$$

Accordingly, it follows that

$$\Gamma_{\ell^+\ell^-} = \frac{e^2}{192\pi \sin^2\theta_W \cos^2\theta_W} m_Z \left[\left(-\frac{1}{2} + 2\sin^2\theta_W\right)^2 + \frac{1}{4}\right]^2 \times \left[1 + \frac{\Pi_{ZZ}(m_Z^2)}{m_Z^2} - \frac{\Pi_{ZZ}(0)}{m_Z^2}\right], \tag{7.40}$$

which replaces Eq. (7.16).

It takes some algebraic manipulation to write the observables in terms of the inputs. At the end I find that

$$\begin{aligned}
m_W^2 &= \hat{x}\left[1 - \frac{\cos^2\theta_W}{\cos^2\theta_W - \sin^2\theta_W}\Delta_\Pi - \frac{\Pi_{\gamma\gamma}(m_Z^2)}{m_Z^2} - \frac{\Pi_{WW}(0)}{m_W^2} + \frac{\Pi_{WW}(m_W^2)}{m_W^2}\right] \\
\sin^2\theta_W &= \sin^2\theta_W + \frac{\sin^2\theta_W\cos^2\theta_W}{\cos^2\theta_W - \sin^2\theta_W}\Delta_\Pi - \frac{\Pi_{\gamma Z}(m_Z^2)}{m_Z^2} \\
\Gamma_{\ell^+\ell^-} &= \Gamma^{\text{tree}}_{\ell^+\ell^-}\left[1 - a\frac{\cos^2\theta_W \sin^2\theta_W}{\cos^2\theta_W - \sin^2\theta_W}\Delta_\Pi + \xi\cos\theta_W\sin\theta_W\frac{\Pi_{\gamma Z}(m_Z^2)}{m_Z^2}\right. \\
&\quad \left. + \frac{\Pi_{WW}(0)}{m_W^2} - \frac{\Pi_{ZZ}(0)}{m_Z^2}\right],
\end{aligned} \tag{7.41}$$

where $\xi = -8(-1+4\sin^2\theta_W)/[(-1+4\sin^2\theta_W)^2+1]$, $\Gamma^{\text{tree}}_{\ell^+\ell^-}$ is given by Eq. (7.16) and all parameters on the right-hand side must be taken at their tree values.

Quantity	Value	Fit	Pull
M_Z [GeV]	91.1876 ± 0.0021	91.1882 ± 0.0020	0.3
Γ_Z [GeV]	2.4952 ± 0.0023	2.4947 ± 0.0014	-0.2
σ_{had} [nb]	41.540 ± 0.037	41.484 ± 0.015	-1.5
M_h[GeV]	125.1 ± 0.2	125.1 ± 0.2	0.0
M_W[GeV]	80.379 ± 0.013	80.359 ± 0.006	-1.5
R^0_ℓ	20.767 ± 0.025	20.742 ± 0.017	-1.0
A_{LR}	0.1499 ± 0.0018	0.1470 ± 0.0005	0.1
$\sin^2\theta_W$	0.23148 ± 0.00033	0.23153 ± 0.00006	0.1
$\alpha_s(M_Z^2)$	0.1194 ± 0.0029	0.1203 ± 0.0030	1.3

Figure 7.7 The global fit of the electroweak observables in the Standard Model. As an example, I wrote down the result for nine observables which include masses, widths, asymmetries and couplings. The pull represents the difference between the value from the fit and the experimental value divided by the uncertainty (1σ).

If I now insert the input values, values for the observables are obtained that are in agreement (within 2σ or less) with the data:

$$A_{LR} = \underbrace{0.1469 \pm 0.0003}_{\text{computed}} \qquad \hat{A}_{LR} = \underbrace{0.1544 \pm 0.0060}_{\text{exp. value}}$$

$$\Gamma_{\ell^+\ell^-} = \underbrace{83.942 \pm 0.009 \text{ MeV}}_{\text{computed}} \qquad \hat{\Gamma}_{\ell^+\ell^-} = \underbrace{83.984 \pm 0.086 \text{ MeV}}_{\text{exp. value}}$$

$$m_W = \underbrace{80.361 \pm 0.006 \text{ GeV}}_{\text{computed}} \qquad \hat{m}_W = \underbrace{80.379 \pm 0.012 \text{ GeV}}_{\text{exp. value}}.$$

A more efficient way is to just write all observables in terms of the parameters of the Lagrangian and then produce a global fit by the least squares method. The result is shown in Figure 7.7 for nine of them. The fit reproduces a large number of observables – the values of which are linked by their dependence on the parameters of the Lagrangian of the Standard Model. The fit works well – not at the classical level, however, but only after the quantum corrections have been included. Its success is not only for the model but also for the quantum mechanics on which it is based (together with relativity).

7.3 Atomic Parity Violation

Low-energy observables are those measured at energies well below the characteristic scale of electroweak symmetry breaking.

A number of small corrections to low-energy observables are expected from the Standard Model. They can be measured best in those settings where the dominant interactions from electrodynamics are protected by some symmetry, thus giving no contribution.

I can look for these effects in atomic systems. Parity is conserved by electromagnetism but violated by the electroweak interactions. I expect to find energy transitions due to the parity-violating interactions.

A correction can be computed by considering the interaction energy of the electrons orbiting around a nucleus made of Z protons and N neutrons. While the bulk of the energy of the system comes from the electromagnetic interaction, there is a small contribution originating in the exchange of a Z-boson (Figure 7.8), which I can write as

$$\mathcal{L}^{\text{int}} = -\frac{G_F}{2\sqrt{2}} j^e_\mu(x) \Big[\sum_i^Z \underbrace{j_i^{p\,\mu}(x)}_{\text{proton}} + \sum_i^N \underbrace{j_i^{n\,\mu}(x)}_{\text{neutron}} \Big] Z_\mu(x) , \tag{7.42}$$

where the currents for the electrons, protons and neutrons are given by

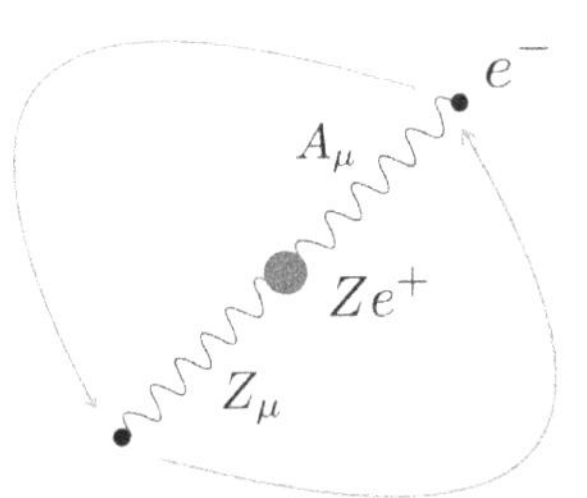

Figure 7.8 Cartoon of an atom with an electron bound by both photon and Z-boson exchanges.

$$\begin{aligned} j_i^{e\,\mu} &= \bar{\psi}_e \gamma^\mu (c_V^e - \gamma^5) \psi_e \\ j_i^{p\,\mu} &= \bar{\psi}_p \gamma^\mu (c_V^p - \gamma^5) \psi_p \\ j_i^{n\,\mu} &= \bar{\psi}_p \gamma^\mu (c_V^n - \gamma^5) \psi_p , \end{aligned} \tag{7.43}$$

with the effective couplings defined as

$$c_V^x = \frac{T_3^x - 2Q^x \sin^2\theta_W}{T_3^x} , \tag{7.44}$$

as in the electroweak Lagrangian except for the normalization factor T_3^x. The effective couplings for the nucleons are given by the quark model (*uud* quarks in the proton and *udd* in the neutron):

$$c_V^p = 2c_V^u - c_V^d = 1 - 4\sin^2\theta_W \quad \text{and} \quad c_V^n = c_V^u - 2c_V^d = -1, \tag{7.45}$$

while that for the electrons is

$$c_V^e = 1 - 4\sin^2\theta_W . \tag{7.46}$$

The atomic exchange of the Z-boson can be approximated by the Fermi effective Lagrangian

$$\mathcal{L}^{\text{int}}_{\text{eff}} = -\frac{G_F}{2\sqrt{2}} j^e_\mu(x) \left[\sum_{i=1}^Z j_i^{p\,\mu}(x) + \sum_{i=1}^N j_i^{n\,\mu}(x) \right] , \tag{7.47}$$

where the sums run over the proton and neutron content in the atomic nucleus.

Since I am interested in an atomic system, I may as well take the nonrelativistic limit for the wave function of the nucleons. In this limit, only the component $\bar{u}\gamma^0 u$ survives as equal to the nuclear density $\rho(x)$. Momentum- and spin-dependent terms can be neglected as long I require just the leading effect. The nuclear current in Eq. (7.47) thus becomes[7.14]

7.14 $\rho_p \simeq \rho_n \simeq \rho$.

$$\frac{G_F}{2\sqrt{2}}\left[\sum_{i=1}^{Z} j_i^{p\,\mu}(x)+\sum_{i=1}^{N} j_i^{n\,\mu}(x)\right] \simeq \delta_{\mu 0}\underbrace{\left[c_V^p Z\rho_p(x)-N\rho_n(x)\right]}_{(Zc_V^p-N)\rho(x)=Q_W\rho(x)}, \qquad (7.48)$$

and the interaction Lagrangian is, in this approximation, equal to

$$\mathcal{L}_{\text{eff}}^{\text{int}} = -\frac{G_F}{2\sqrt{2}} Q_W \rho(x)\left[c_V^e \psi_e^\dagger \psi_e - \psi_e^\dagger \gamma_5 \psi_e\right]. \qquad (7.49)$$

The parity-violating term is the second term in Eq. (7.49), the term arising from the axial part of the electron current, which gives

$$-\frac{G_F}{2\sqrt{2}} Q_W \rho(x) \psi_e^\dagger \gamma_5 \psi_e , \qquad (7.50)$$

for the extra energy interaction I am looking for in atomic transitions. The coefficient

$$Q_W = -N + Z(1 - 4\sin^2\theta_W) \qquad (7.51)$$

characterizes the strength of the atomic transition that violates parity.

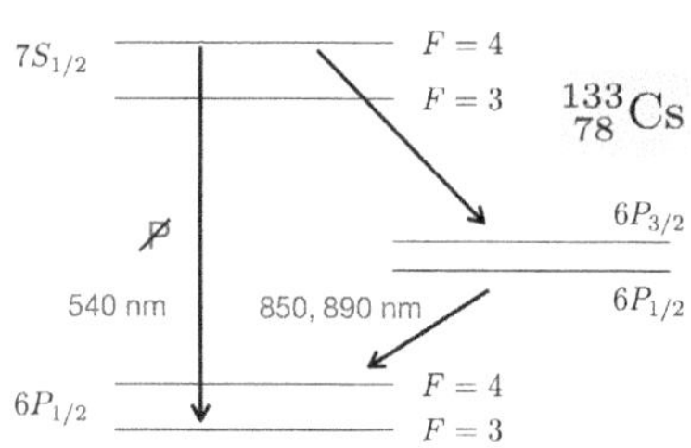

Figure 7.9 Atomic transitions in cesium.

Whereas electric dipole transitions between states with the same parity cannot arise in QED (without external fields), they can be induced by the weak interaction energy in Eq. (7.50), which does not conserve parity. These transitions have been searched for in cesium atoms (Figure 7.9), where it is found that the forbidden transition $6S$–$7S$ can be excited by the mixing of the opposite-parity states by the electron–nucleon weak interaction in Eq. (7.50). The coefficient Q_W has been obtained by first exciting the atoms by a laser beam to the higher state $7S$ and then measuring the amount of light emitted in the decay sequence. It was found to be[3]

$$Q_W\left(^{133}_{78}\text{Cs}\right) = -73.16(29)_{exp}(20)_{th}, \qquad (7.52)$$

a number to be compared with Eq. (7.51), which, for the values for $N = 78$ and $Z = 55$ for cesium, gives (taking $\sin^2\theta_W = 0.23$)

$$Q_W = -78 + 55\,(1 - 4\sin^2\theta_W) = -73.6, \qquad (7.53)$$

in reasonably good agreement. Better agreement is reached after including QED and other corrections to Eq. (7.50).

3 J. Guéna *et al.*, *Physical Review Letters* **90** (2003) 143001.

7.4 Flavor Physics

The kinetic and gauge interaction part of the fermion Lagrangian in the Standard Model,

$$\sum_k \bar{\psi}_k i \not{D} \psi_k, \tag{7.54}$$

has a very large global symmetry group because all the fermion fields can be rotated freely. There are five fermion multiplets for each generation and three of these generations. I therefore have the symmetry

$$SU(3)_Q \otimes SU(3)_u \otimes SU(3)_d \otimes SU(3)_L \otimes SU(3)_\ell \tag{7.55}$$

after removing the $U(1)$ parts which are anomalous (except for the combination $B - L$). The groups $SU(3)$ in Eq. (7.55) act by rotating the quark fields through the three generations.

This huge symmetry is explicitly broken by the Yukawa Lagrangian

$$\begin{aligned} -\mathcal{L}_{\text{Yuk}} = & -Y_\ell^{ij} \frac{1}{\sqrt{2}} (\bar{\nu}_\ell^i, \bar{\ell}^i)_L \begin{pmatrix} \phi^+ \\ \phi^0 \end{pmatrix} \ell_R^j \\ & - Y_d^{ij} \frac{1}{\sqrt{2}} (\bar{u}^i, \bar{d}^i)_L \begin{pmatrix} \phi^+ \\ \phi^0 \end{pmatrix} d_R^j \\ & - Y_u^{ij} \frac{1}{\sqrt{2}} (\bar{u}^i, \bar{d}^i)_L \begin{pmatrix} \phi^{0*} \\ -\phi^{+*} \end{pmatrix} u_R^j + \text{h.c.}, \end{aligned} \tag{7.56}$$

giving mass to all the fermions.

The lepton terms can be diagonalized right away because of the extra freedom arising because the neutrino had no mass. I can try and diagonalize the quark terms as well by rotating each quark field,[7.15]

$$\textbf{mass}\ \text{eigenstates} \implies \left\{ \begin{array}{l} u_R \to V_{u_R} u_R \\ d_R \to V_{d_R} d_R \\ u_L \to V_{u_L} u_L \\ d_L \to V_{d_L} d_L \end{array} \right\} \Longleftarrow \textbf{interaction}\ \text{eigenstates}$$

in such a way that[7.16]

$$V_{u_L}^\dagger Y_u V_{u_R} = Y'_u \quad \text{and} \quad V_{d_L}^\dagger Y_d V_{d_R} = Y'_d, \tag{7.57}$$

where the primed matrices are now diagonal and give the masses of the quarks as follows:

$$\begin{pmatrix} m_u & 0 & 0 \\ 0 & m_c & 0 \\ 0 & 0 & m_t \end{pmatrix} = \frac{v}{\sqrt{2}} \begin{pmatrix} (Y'_u)_{11} & 0 & 0 \\ 0 & (Y'_u)_{22} & 0 \\ 0 & 0 & (Y'_u)_{33} \end{pmatrix} \tag{7.58}$$

[7.15] Since there have been already many rotations of the quark fields, let me stress here that the V_k matrices are 3×3 and represent rotations of the quark fields through the three generations.

[7.16] Notice that the Yukawa matrices are not necessarily Hermitian. An arbitrary matrix M can always be transformed in a matrix that is real, positive and diagonal, $M' = P^\dagger M Q$, by means of two unitary matrices P and Q.

and

$$\begin{pmatrix} m_d & 0 & 0 \\ 0 & m_s & 0 \\ 0 & 0 & m_b \end{pmatrix} = \frac{v}{\sqrt{2}} \begin{pmatrix} (Y'_d)_{11} & 0 & 0 \\ 0 & (Y'_d)_{22} & 0 \\ 0 & 0 & (Y'_d)_{33} \end{pmatrix} . \tag{7.59}$$

In this form, in terms of the mass eigenstates, the Lagrangian $\mathcal{L}_{\text{Yuk}}$ correctly defines the quark propagators.

These rotations leave the kinetic terms unchanged. What happens to the gauge interaction terms? The neutral currents are left unchanged. The right-handed fields transform as follows[7.17]

7.17 $[V_k, T^3] = 0$.

$$\begin{aligned} &\bar{u}_R(g\not{W}_3 + g'\not{B})u_R + \bar{d}_R(g\not{W}_3 + g'\not{B})d_R \\ &\quad \Rightarrow \bar{u}_R V^\dagger_{u_R}(g\not{W}_3 + g'\not{B})V_{u_R}u_R + \bar{d}_R V^\dagger_{d_R}(g\not{W}_3 + g'\not{B})V_{d_R}d_R \\ &\quad \Rightarrow \bar{u}_R \underbrace{V^\dagger_{u_R} V_{u_R}}_{=1}(g\not{W}_3 + g'\not{B})u_R + \bar{d}_R \underbrace{V^\dagger_{d_R} V_{d_R}}_{=1}(g\not{W}_3 + g'\not{B})d_R, \end{aligned} \tag{7.60}$$

and therefore remain the same. The same holds for the left-handed fields.

The Higgs boson interaction is left unchanged by these rotations, as is the neutral current. There are no flavor-changing transitions induced by the Higgs boson in the Standard Model. This property is true because there is just one scalar doublet. It would not be true if there were more than one scalar doublet.

On the contrary, the charged currents transform as

$$\begin{aligned} &\frac{g}{\sqrt{2}}\bar{u}_L \not{W}^+ d_L + \frac{g}{\sqrt{2}}\bar{d}_L \not{W}^- u_L \\ &\quad \Rightarrow \frac{g}{\sqrt{2}}\Big(\bar{u}_L \underbrace{V^\dagger_{u_L} V_{d_L}}_{V_{\text{CKM}}} \not{W}^+ d_L + \bar{d}_L \underbrace{V^\dagger_{d_L} V_{u_L}}_{V^\dagger_{\text{CKM}}} \not{W}^- u_L\Big), \end{aligned} \tag{7.61}$$

and the rotation matrices do not simplify – since they always come from different rotations, one from the u-component of the field and the other from the d-component. I am left with the vertices of the charged current modulated by the CKM matrix:

$$V_{\text{CKM}} = \begin{pmatrix} V_{ud} & V_{us} & V_{ub} \\ V_{cd} & V_{cs} & V_{cb} \\ V_{td} & V_{ts} & V_{tb} \end{pmatrix} . \tag{7.62}$$

The V_{CKM} matrix is a 3×3 unitary matrix and therefore has nine parameters. The quark fields can still be rotated, leaving the masses diagonal, to reduce these to four parameters: three angles θ_{12}, θ_{23} and θ_{13} and one phase δ. The matrix can be written in terms of Euler angles as

$$V_{\text{CKM}} = \hat{C}\hat{B}\hat{A}, \tag{7.63}$$

where

$$\hat{A} = \begin{pmatrix} \cos\theta_{12} & \sin\theta_{12} & 0 \\ -\sin\theta_{12} & \cos\theta_{12} & 0 \\ 0 & 0 & 1 \end{pmatrix} \tag{7.64}$$

$$\hat{B} = \begin{pmatrix} \cos\theta_{13} & 0 & \sin\theta_{13}\, e^{-i\delta} \\ 0 & 1 & 0 \\ -\sin\theta_{13}\, e^{i\delta} & 0 & \cos\theta_{13} \end{pmatrix} \tag{7.65}$$

$$\hat{C} = \begin{pmatrix} 1 & 0 & 0 \\ 0 & \cos\theta_{23} & \sin\theta_{23} \\ 0 & -\sin\theta_{23} & \cos\theta_{23} \end{pmatrix}. \tag{7.66}$$

The Wolfenstein form of $V_{\rm CKM}$ exchanges the angles and phase of the standard-form matrix with the four real parameters λ, A, ρ and η. Of course, one can parametrize the matrix in any way one prefers, but the Wolfenstein choice is made here to emphasize the hierarchy found in the data as an expansion in the parameter λ. In this form one has

$$V_{\rm CKM} = \begin{pmatrix} 1-\lambda^2/2 & \lambda & A\lambda^3(\rho - i\eta) \\ -\lambda & 1-\lambda^2/2 & A\lambda^2 \\ A\lambda^3(1-\rho-i\eta) & -A\lambda^2 & 1 \end{pmatrix} + O(\lambda^4), \tag{7.67}$$

where $\sin\theta_{12} = \lambda$, $\sin\theta_{23} = A\lambda^2$ and $\sin\theta_{13} = A\lambda^3(\rho + i\eta)$.

Going back to Chapter 2, where charged currents were discussed, the Feynman rule for the charged currents allows fermions to jump across different generations. The probability of these jumps is controlled by the square of the elements of the CKM matrix.

The familiar possibilities in the β-decay of a u quark of going not only into a d quark but also into a strange quark (with a probability given by the Cabibbo angle, the cosine of which is numerically equal to the parameter λ) is here understood within the greater context of the full CKM mixing matrix.

The unitarity of the CKM matrix gives the relations

$$\sum_k V^{ik}_{\rm CKM} V^{jk*}_{\rm CKM} = \delta^{ij}. \tag{7.68}$$

Looking at the off-diagonal sums: three complex numbers that sum to zero can be represented by a triangle in the complex plane. There are six of them but they are mostly very thin. The more substantial among these triangles is given by the relation

$$\boxed{V_{ud}V^*_{ub} + V_{cd}V^*_{cb} + V_{td}V^*_{tb} = 0,} \tag{7.69}$$

which is usually rewritten by dividing by $V_{cd}V^*_{cb}$ so as to obtain

$$\frac{V_{ud}V^*_{ub}}{V_{cd}V^*_{cb}} + 1 + \frac{V_{td}V^*_{tb}}{V_{cd}V^*_{cb}} = 0; \tag{7.70}$$

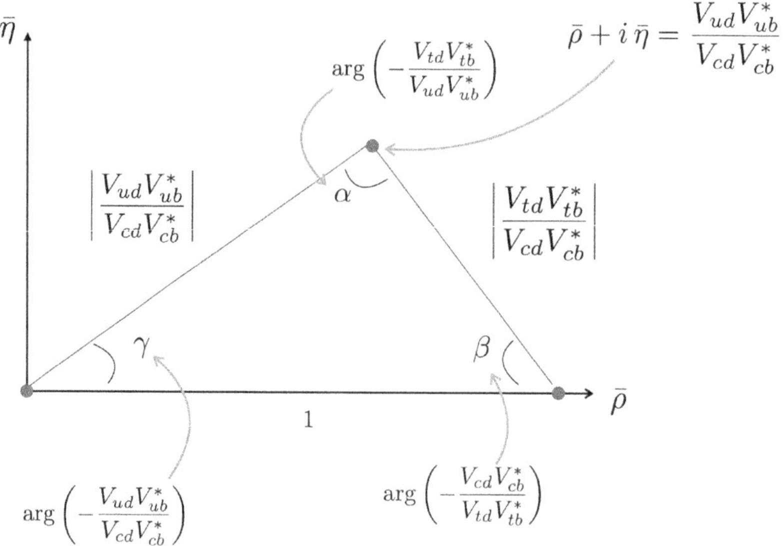

Figure 7.10 The unitary triangle.

this can be represented in the complex plane of the variable

$$\bar{\rho} + i\bar{\eta} = \frac{V_{ud}V_{ub}^*}{V_{cd}V_{cb}^*}, \tag{7.71}$$

as shown in Figure 7.10.

How well does the unitarity of the CKM matrix work? As a quick check, take the row

$$|V_{ud}|^2 + |V_{us}|^2 + |V_{ub}|^2 = 1 \tag{7.72}$$

and plug in the numbers for the three matrix elements, which can be found in the particle data booklet. I can neglect the last term, which is very small ($|V_{ub}| = (3.82 \pm 0.24) \times 10^{-3}$; its square is $O(10^{-6})$). The other two terms are given by[4]

$$|V_{ud}| = 0.97370 \pm 0.00014 \tag{7.73}$$

and

$$|V_{us}| = 0.2245 \pm 0.0008\,, \tag{7.74}$$

which gives (keeping three significant digits)[7.18]

7.18 The error σ of a number x propagates in x^2 as $\delta x^2 = 2x\sigma$, the two errors are added in quadrature.

$$\begin{aligned} |V_{ud}|^2 + |V_{us}|^2 &= (0.948 \pm 0.000273) + (0.050 \pm 0.000359) \\ &= \underbrace{0.998 \pm 0.000451}_{\text{it should be 1}}\,, \end{aligned} \tag{7.75}$$

[4] PDG (2020) https://pdg.lbl.gov.

which is off by 0.2% from unity. There is a violation of unitarity of two parts in a thousand. This represents a deviation of 4.4σ. Not a disaster but still ... The problem is that these values do not come directly but they must be extracted from the observables. As before for the electroweak precision tests, I must include quantum effects in the process of extraction. On top of that, I have to deal with long-distance effects brought in by the hadronic matrix elements. The following problem session contains the observables from which the the CKM matrix elements are extracted.

7.5 *Problem Session*: The Elements of the CKM Matrix

This may easily seem to be the most boring problem session ever. But it is not. The processes to be examined are the basis of the flavor physics that provides so much of the current experimental interest in the Standard Model. The Standard Model blossoms in a variety of possible processes and brings to fruition the description of the many different hadrons and leptons and their intricate decays.

Where to search for the various CKM matrix elements? What is needed is to single out processes where the CKM matrix elements enter and have as few uncertainties (from long-distance physics, mainly) as possible. The precision in the determination of the parameters depends on how well these long-distance corrections are known or, if not under control, can be factorized out.

Weak decays of quarks are classified by the presence of hadrons or leptons in the final states: lepton decays are those with only leptons, non-leptonic decays only have hadrons and semi-leptonic decays may involve both. It is the semi-leptonic decays that provide the best determination of many of the CKM matrix elements.

This quick survey of the determination of the values of the CKM matrix element cannot do justice to the work that goes into each determination of each parameter.[5] There are radiative corrections to be taken into account. On top of these, the form factors, connecting the quarks to the mesons, must be estimated. The largest uncertainty originates in these form factors since they require long-distance physics which is known within only a limited degree of precision.

- The most precise measurement of $|V_{ud}|$ comes from the decay rates of nuclei experiencing super-allowed β-decays, like that in Figure 7.11. In these nuclear transitions, the wave function of the entire nucleus is left

[5] All experimental values are taken, again, from PDG (2020) https://pdg.lbl.gov.

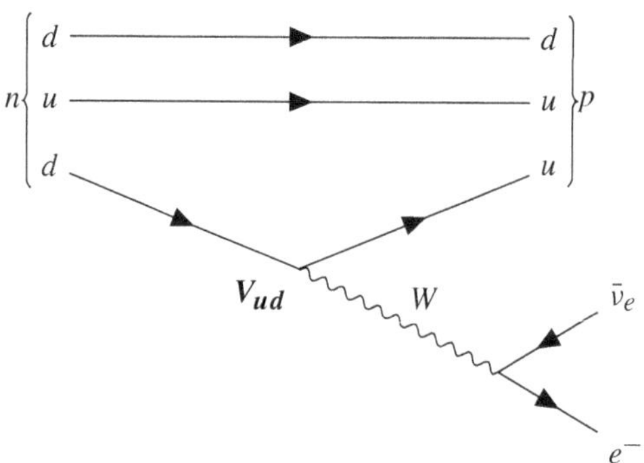

Figure 7.11 Feynman diagram for β-decay.

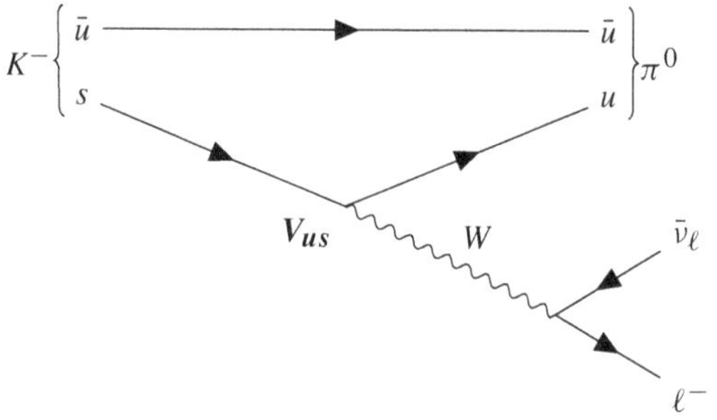

Figure 7.12 Feynman diagram for the semi-leptonic decay of K^-.

unchanged since the decay involves no change in angular momentum nor in parity.

These processes provide clean theoretical predictions and can be evaluated by precise calculations that do not require strong assumptions or approximations, allowing for a very precise determination:

$$|V_{ud}| = 0.97420 \pm 0.00021\,. \tag{7.76}$$

The error is dominated by theoretical uncertainties from nuclear Coulomb distortions and radiative corrections.

- The determination of $|V_{us}|$ comes from semi-leptonic and leptonic kaon decays, including charged, $K^- \to \pi^0\ell^-\nu_\ell$ (Figure 7.12), and neutral decays, $K_L \to \pi^0\ell\nu_\ell$, decays and the $K^+ \to \mu^+\nu_\mu$ leptonic decay, with the main limitation arising from the uncertainties on the form factors and decay constant. The average from these measurements provides

$$|V_{us}| = 0.2243 \pm 0.0005\,. \tag{7.77}$$

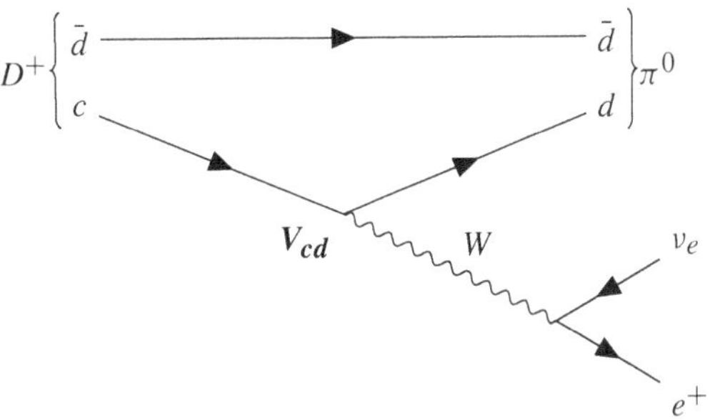

Figure 7.13 Feynman diagram for the determination of V_{cd}.

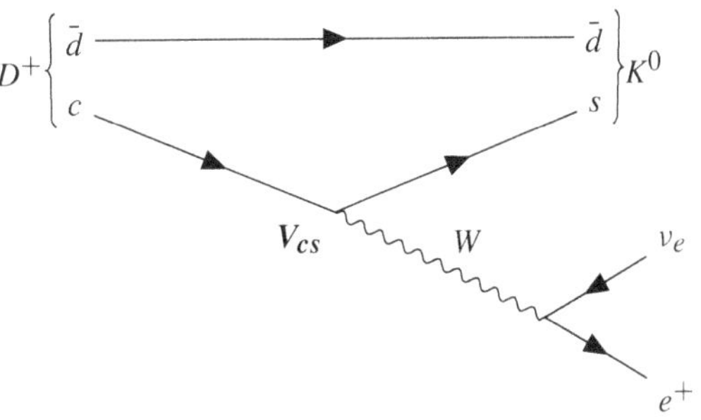

Figure 7.14 Feynman diagram for the determination of V_{cs}.

- The determination of $|V_{cd}|$ is currently based on leptonic and semi-leptonic charm decays, namely $D^+ \to \pi e^+ \bar{\nu}_e$ (Figure 7.13) and $D^+ \to \mu^+ \nu_m u$. Also, $|V_{cs}|$ can be obtained from semi-leptonic decays like $D \to K \ell \nu_\ell$ (Figure 7.14) and leptonic decays like $D_s^+ \to \mu \nu_\mu$ or $D_s^+ \to \tau \nu_\tau$. The uncertainty comes from the estimate of the relevant form factors.

  It is found that

$$|V_{cs}| = 0.997 \pm 0.017 \quad \text{and} \quad |V_{cd}| = 0.218 \pm 0.004 \,. \tag{7.78}$$

- Similarly to the matrix elements for the first two generations, the matrix elements $|V_{ub}|$ and $|V_{cb}|$ can be accessed through semi-leptonic decays.

  In the case of $|V_{ub}|$ one can use either exclusive or inclusive measurements to extract the CKM matrix element. The exclusive determination benefits from a better known form factor, that of the decay $B \to \pi \ell \nu_\ell$ (Figure 7.15), which can be combined with measurements of the differential decay rate.

  The determination of $|V_{cb}|$ is based on various decays. The exclusive decay $B \to D \ell \nu_\ell$ (Figure 7.16) provides the most precise value because there exists a satisfactory estimate of the corresponding form factor.

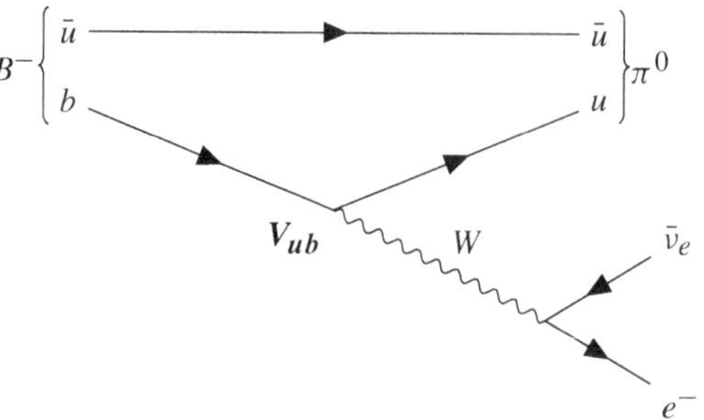

Figure 7.15 Feynman diagram for the determination of V_{ub}.

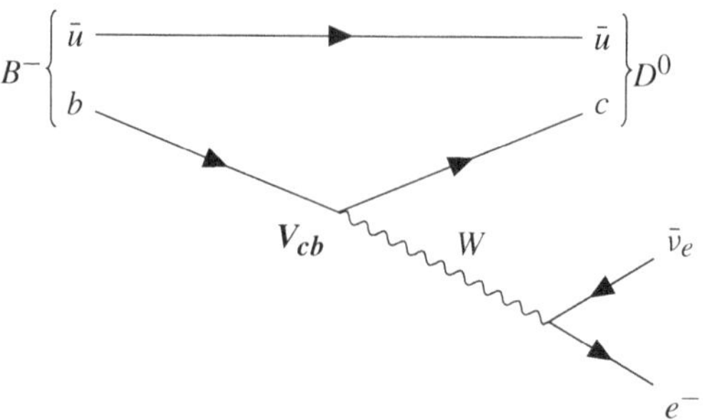

Figure 7.16 Feynman diagram for the determination of V_{cb}.

The various determinations together provide the values

$$|V_{ub}| = 0.0039 \pm 0.00036 \quad \text{and} \quad |V_{cb}| = 0.0422 \pm 0.0008 \,. \tag{7.79}$$

Figure 7.17 shows these and other determinations of $|V_{ub}|$ and $|V_{cb}|$ as well as the result of the overall fit.

- $|V_{td}|$ and $|V_{ts}|$ cannot be determined from the decays of mesons since there are no mesons made up of top quarks. Their value can only come from loop diagram corrections containing the matrix element. Two of them are shown in Figure 7.18. They enter the box diagrams that define the mass difference between the mesons B^0 and $\bar{B}^0$. There are two of these mass differences, Δm_d and Δm_s, depending on whether the B^0_d- or the B^0_s-meson is considered:

  The mass differences are measured as inverse frequencies in the oscillation between the neutral B-mesons. It is found that

  $$\Delta m_d = (0.5064 \pm 0.0019)\ \text{ps}^{-1} \quad \text{and} \quad \Delta m_d = (17.757 \pm 0.021)\ \text{ps}^{-1} \,.$$

  Once these parameters are known, the CKM matrix elements can be extracted:

  $$|V_{ts}| = 0.0394 \pm 0.0023 \quad \text{and} \quad |V_{td}| = 0.0081 \pm 0.0005 \,. \tag{7.80}$$

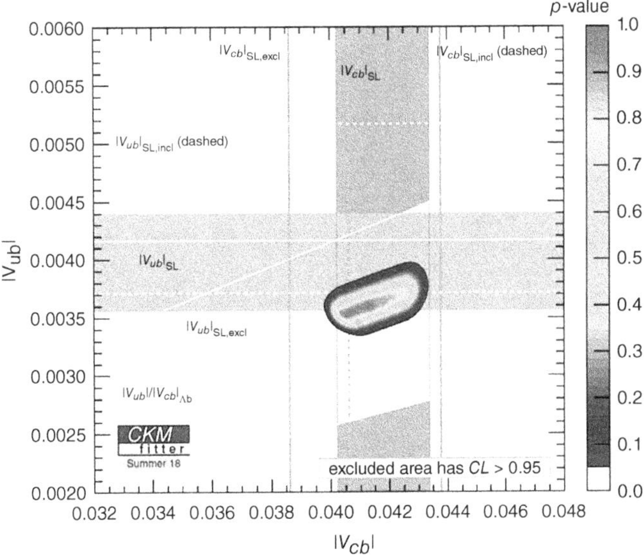

Figure 7.17 Determination of the CKM matrix elements $|V_{ub}|$ and $|V_{cb}|$. The ovals show the result of an overall fit of all CKM matrix elements [CKMfitter Group (J. Charles *et al.*), *European Physical Journal* **C41** (2005) 1–31 [hep-ph/0406184]].

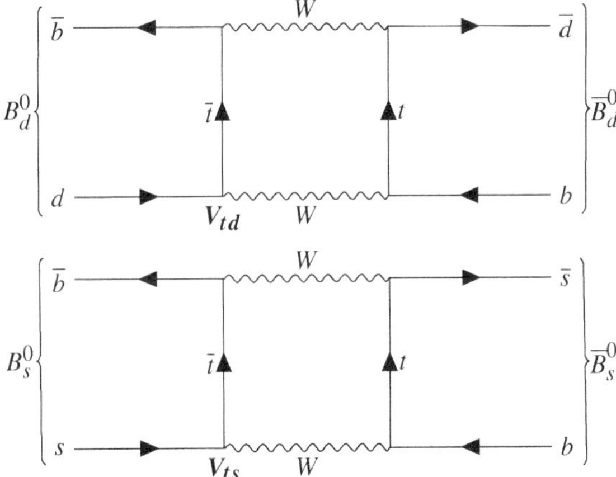

Figure 7.18 Feynman box diagrams for the mass difference between the mesons B^0 and $\bar{B}^0$.

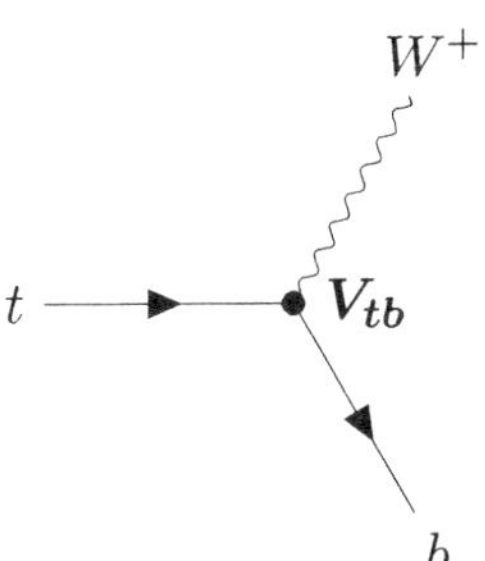

Figure 7.19 The decay of the top quark and the determination of V_{tb}.

- Finally $|V_{tb}|$ comes from the ratio $BR(t \to Wb)/BR(t \to Wq)$ of the decays of the top quark; one such decay is shown in Figure 7.19. It is found to be

$$|V_{tb}| = 1.019 \pm 0.025 \,. \tag{7.81}$$

That discussion was for the size of the matrix elements. The angles of the unitary triangle bring in the CP-violating phase. It is the B-meson system that has provided the measurements of these angles. More precisely:

- the value of the angle α comes from the charmless decays of the B-meson: $B^0 \to \pi^+\pi^-$, $B^0 \to \rho\pi$ and $B^0 \to \rho\rho$. It is found that
$$\alpha = \left(84.5^{+5.9}_{-5.2}\right)^\circ ; \tag{7.82}$$
- the value of the angle β comes from the measurement of the golden channel $B^0 \to J/\psi\, K^0$, which gives
$$\sin 2\beta = 0.691 \pm 0.017 ; \tag{7.83}$$
- and the value of the angle γ comes from many different techniques, most of them relaying on the interference in B-meson decays via $b \to c$ and $b \to u$ transitions, for instance in $B \to DK$ followed by a $D^{(*)}$ decay chosen to allow for interference. The combination of these measurements gives
$$\gamma = \left(73.5^{+4.2}_{-5.1}\right)^\circ . \tag{7.84}$$

I have gone through this exercise and written down the actual values of all these elements because, when all of them are brought together, the sides and the angles have values just right to close the unitary triangle. It is awe inspiring: everything works out, the various pieces combine together and out of the sizes of nine matrix elements and three angles we have the sides of the unitarity triangle precisely matched!

Figure 7.20 makes clear how difficult it is to change even a small detail of the electroweak Lagrangian of the Standard Model. If you start tampering with it, you do this at your own risk. If what is circulating in a loop, or the size of a certain coupling, is changed, the sides of the unitary triangle do not close any longer and the amazing fit is lost.

7.6 Flavor-Changing Neutral Currents

The rotation necessary to diagonalize the Yukawa Lagrangian leaves the neutral currents unchanged. These currents do not change flavor within a generation.

That is why they are called neutral. They also do not change flavor across generations. Changes of flavor only come together with a change of charge, as in the case of charged currents.

This feature is supported by looking at the branching ratio of a flavor-changing transition, for example
$$\mathrm{Br}(K^+ \to \mu^+\nu_\mu) = 0.64 \tag{7.85}$$

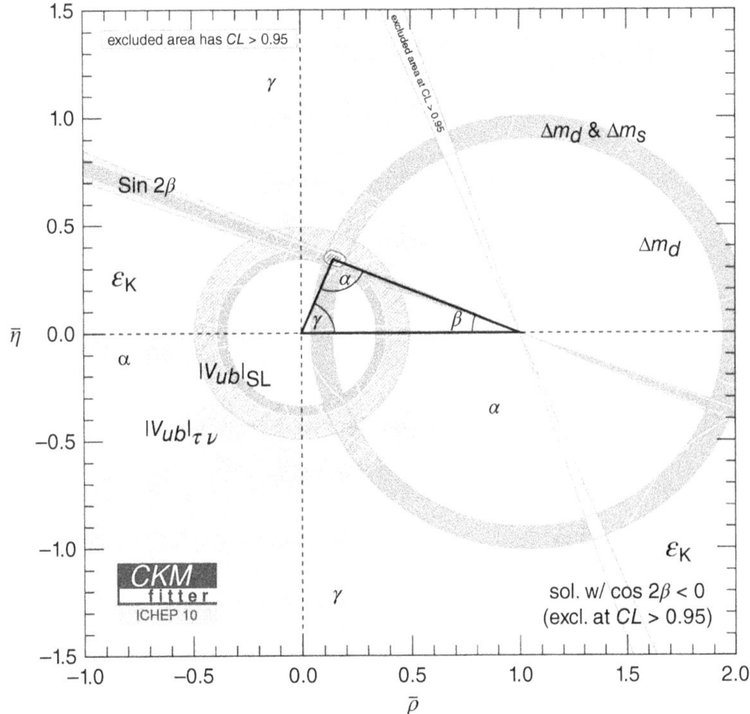

Figure 7.20 The most amazing test of the Standard Model. The matrix elements and the phases of the CKM matrix combine to close the unitary triangle [CKMfitter Group (J. Charles *et al.*), *European Physical Journal* **C41** (2005) 1–31 [hep-ph/0406184], updated results and plots available at: http://ckmfitter.in2p3.fr].

(which proceeds through the charged current) and comparing it with the similar flavor-changing transition to be carried by a neutral current,

$$\mathrm{Br}\,(K_L \to \mu^+\mu^-) = 7 \times 10^{-9}, \tag{7.86}$$

which is indeed very suppressed.

In the Standard Model this result is explained in a natural way by the neutral currents not changing flavor at the tree level. The small branching ratios of flavor-changing neutral currents can be explained in a natural manner by the process having to go through at least one loop. I can fashion such a transition by sewing together two charged currents so that the flavor is changed at the two vertices.

As an example, consider the transition $b \to s\gamma$. Whereas at the level of the classical Standard Model Lagrangian there is no such transition, it is generated by taking two charged currents and making a loop out of them as in Figure 7.21.

The amplitude for the the $b \to s\gamma$ transition is obtained by inserting propagators and vertices and doing the loop integration. This amplitude turns out to be

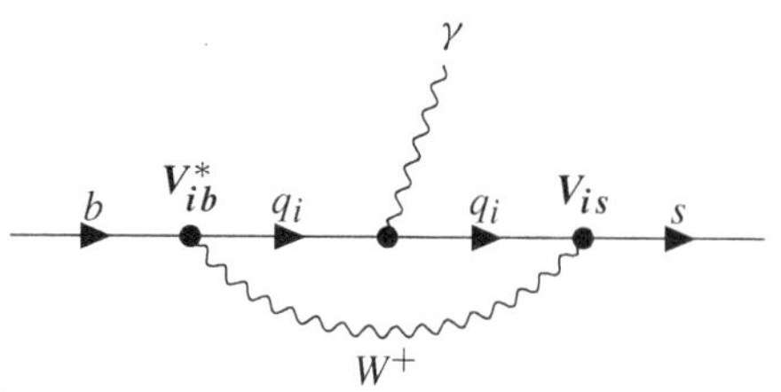

Figure 7.21 The Feynman diagram for the transition $b \to s\gamma$.

$$\mathcal{M} = \sum_i V^*_{ib} V_{is} f(m_i) , \tag{7.87}$$

where the function $f(m_i)$ – the Inami–Lim function – comes from the loop integration where I have to sum over the quarks u, c and t which can be inserted in the loop. If I neglect the masses of the quarks in the loop, the Inami–Lim function becomes an overall function that can be factorized out, leaving the amplitude in Eq. (7.87) to be proportional to

$$\mathcal{M} \propto \sum_i V^*_{ib} V_{is} , \tag{7.88}$$

which vanishes because of the unitarity of the CKM matrix.

The amplitude is therefore proportional to the masses of the quarks running in the loop. The Inami–Lim function goes as m_i^2/m_W^2 for small fermion masses times a factor $g^2/16\pi^2$ from the loop integration, and I can write

$$\mathcal{M} \propto e \frac{g^2}{16\pi^2} \frac{m_i^2}{m_W^2} , \tag{7.89}$$

which shows that I have a double suppression: one from the loop and the other from the masses in the internal propagator that I have to retain. The suppression of the flavor-changing neutral current due to the masses is called the **GIM mechanism** after the initials of the physicists who first pointed out this feature.

All in all, I expect the amplitude for the $b \to s\gamma$ transition to be of order

$$\frac{e}{16\pi^2} V^*_{cb} V_{cs} G_F m_c^2 , \tag{7.90}$$

where the factor e comes from the photon coupling, $1/16\pi^2$ from the loop integration[7.19] and $G_F m_c^2$ from the GIM mechanism. It is the charm quark that gives the dominant contribution even though its mass is smaller than that of the top quark because the CKM matrix elements are larger.

7.19 More about this loop factor in the next problem session.

These results explain why flavor-changing neutral currents in the Standard Model are possible but very suppressed – arising, as they do, at the one-loop order and because of the GIM mechanism.

7.7 *Problem Session*: Dipole Moments

Two other observables that can potentially show small corrections due to the Standard Model are the electric and magnetic dipole moments of particles.

The interaction energy of a dipole is given by the classical Hamiltonian

$$\mathcal{H}_{\text{int}} = \underbrace{-\mu \frac{\vec{S}}{|\vec{S}|} \cdot \vec{B}}_{\text{MDM}} \underbrace{- d \frac{\vec{S}}{|\vec{S}|} \cdot \vec{E}}_{\text{EDM}} \tag{7.91}$$

for the electromagnetic interaction of a particle with spin $\vec{s}$. These interactions can be written in quantum field theory as

$$\mu \frac{\vec{S}}{|\vec{S}|} \cdot \vec{B} \quad \rightarrowtail \quad e\bar{\psi}\gamma_\mu \psi A^\mu + \frac{a}{4m}\bar{\psi}\sigma_{\mu\nu}\psi F^{\mu\nu} \tag{7.92}$$

$$d \frac{\vec{S}}{|\vec{S}|} \cdot \vec{E} \quad \rightarrowtail \quad i\frac{d}{2}\bar{\psi}\sigma_{\mu\nu}\gamma_5\psi F^{\mu\nu}, \tag{7.93}$$

where $\mu = ge/2m$ and $a = g - 2$. These two operators are of dimension five. I have already discussed the anomalous magnetic moment of the electron in Chapter 1.

Recalling the properties under T and P transformations of the electromagnetic fields and of an axial vector such as spin,

$$\begin{aligned} \hat{T}: &\quad \vec{E} \to \vec{E}, \quad \vec{B} \to -\vec{B}, \quad \vec{S} \to -\vec{S} \\ \hat{P}: &\quad \vec{E} \to -\vec{E}, \quad \vec{B} \to \vec{B}, \quad \vec{S} \to \vec{S}, \end{aligned} \tag{7.94}$$

shows that

$$\begin{aligned} \text{MDM} &\quad \rightarrowtail \quad \hat{P}\text{ even} \quad \hat{T}\text{ even} \quad (CP\text{ even}) \\ \text{EDM} &\quad \rightarrowtail \quad \hat{P}\text{ odd} \quad \hat{T}\text{ odd} \quad (CP\text{ odd}). \end{aligned} \tag{7.95}$$

The electric dipole moment is therefore of particular interest inasmuch as it violates CP invariance.[7.20] I can estimate the order of its size as expected from the Standard Model. This estimation is a useful exercise in judging the size of a given amplitude before doing any computation, and also of why computations are sometime necessary.

7.20 Here I consider the *CP* violation arising from the phase in the CKM matrix. In Chapter 5, I discussed *CP* violation from the θ-term in the strong interaction Lagrangian.

So here goes.

The electric dipole moment operator is not in the Lagrangian. It must come from loop corrections. One-loop diagrams, however, do not carry the phase necessary to give CP violation because the CKM matrix elements always come with their conjugate since the external states are the same, as in the diagram in Figure 7.22, in which $\text{Im}\, V_{id}V^*_{id} = \text{Im}\,|V_{id}|^2 = 0$. I therefore need at least two-loop diagrams where there are four CKM matrices and the phase $\text{Im}\,(V^*_{td}V^*_{tb}V_{bc}V_{cd}) \neq 0$. Such a diagram is shown

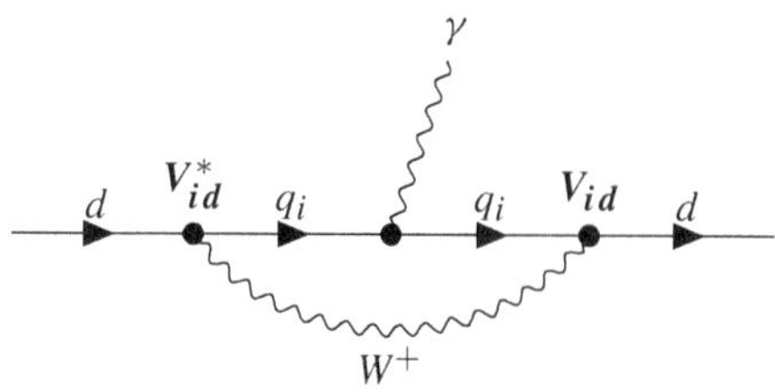

Figure 7.22 Feynman diagram that could give rise to an electric dipole moment but in fact it does not.

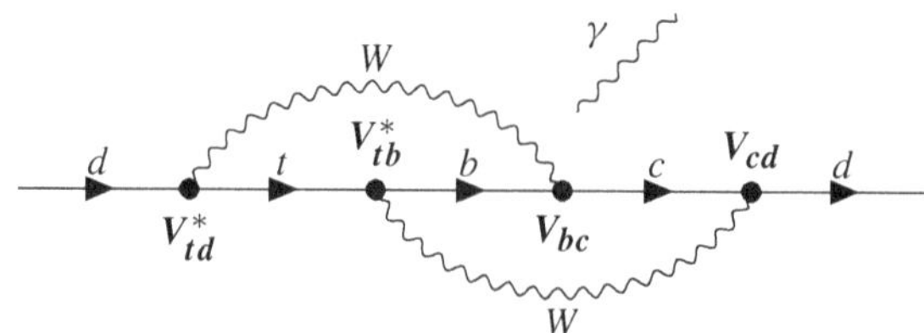

Figure 7.23 Two-loop Feynman diagram for the electric dipole moment. The hovering photon can be attached to any charged line.

in Figure 7.23, in which I can attach the photon line to any of the charged propagators.

What is the suppression factor due to loop integrals? For finite contributions it comes from the various factors of 2 and π from the normalization of the loop integrals. This is not a minor point since the expansion in the perturbative theory is based on the fact that the gauge coupling always comes in the combination $e^2/4\pi$ rather than just e^2. Since e in natural units is about 0.3 while $e^2/4\pi$ is the fine-structure constant and equal to the much smaller number $1/137$, it is only because of these factors of 2 and π that the perturbative series is well-behaved.[7.21] This feature gives the rule that there is a suppression factor $g^2/16\pi^2$ for each loop.

7.21 Remember that the series in quantum field theory are asymptotic series. They do not converge. Only the first terms in the expansion sum up to give a reliable approximation to the solution, after which adding further terms takes you away from the solution.

Altogether, the order of the dipole moment correction is suppressed by several factors,

1st loop factor → ; ← GIM plus 2nd loop factor

$$d \simeq \frac{e}{16\pi^2}\, m_c^2\, G_F\, m_d\,, \tag{7.96}$$

← chirality

and is going to be a rather small number.

The conclusion of this analysis is that a first contribution to the electric dipole moment is expected to been at two loops and small. But, lest it

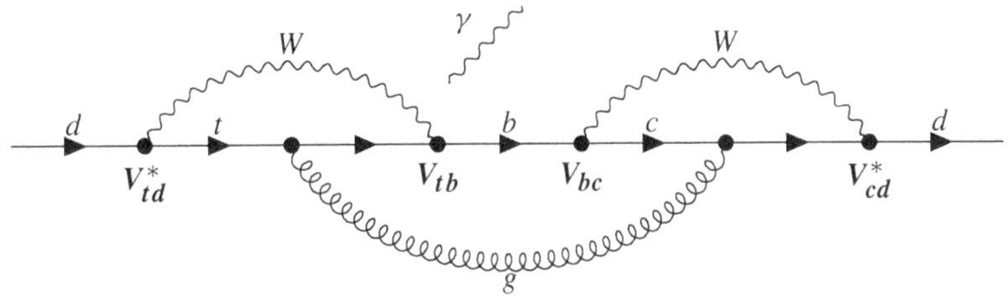

Figure 7.24 The three-loop Feynman diagram. The hovering photon line, as before, can be attached to any charged line.

be thought that computations are redundant and we can get away with just order-of-magnitude estimates, here is the twist: the first non-vanishing contribution comes only at the three-loop order! See Figure 7.24

The two-loop diagrams conspire to cancel each other out. This result could not have been predicted by the order-of-magnitude analysis. It requires a full computation. And a very careful one since we are looking for a cancellation among different terms.

In the end, the electric dipole moment is expected to be really small, $d \simeq 10^{-34} e$ cm. It has not been probed yet by the experiments looking for it, for instance, using neutrons.

7.8 Exercises

1. **Divergences.** Compute the blob corrections (in some regularization scheme) and retain the divergent terms. Show that these divergent terms cancel out in the corrections to the observables.
2. **Matrix diagonalization.** Prove that an arbitrary matrix M can always be transformed into a real, positive and diagonal matrix $M' = P^\dagger M Q$ by means of two unitary matrices P and Q.
   [Hint: $M^\dagger M$ and $MM^\dagger$ are Hermitian even if M is not, and they can be diagonalized by a unitary transformation.]
3. **Form factor.** The decay of a pion is given by the matrix element $\langle 0|A_\mu|\pi\rangle$. Use symmetry to determine the form of the current A_μ. How can you determine the size of this matrix element?
4. **Flavor-changing neutral currents.** Can an interaction with photons or gluons give flavor-changing neutral currents? How about such an interaction with Higgs bosons?
   [Hint: Use symmetry arguments.]

5. **Jarlskog invariant.** Write the so-called Jarlskog invariant $J = \sum \mathrm{Im}\left[V_{ij}V_{kl}V_{il}^{*}V_{kj}^{*}\right]$ in terms of the Wolfenstein parameters for the CKM matrix. Show that J measures, in a parametrization-independent manner, the amount of CP violation.
[Hint: J depends on all the mixing angles.]

8 Quantum Chromodynamics at Work

Contents

Strong interactions are, well, strong. You have hadrons interacting and breaking up into a huge number of other hadrons. It is hard to understand what is going on. Physicists stumbled along during the 1960s trying out many ideas (the key words here are current algebra, analytic S-matrix, Regge trajectories and string theory) without much real understanding.

The study of the strong interactions changed when the detectors were moved from being in front of the collision to 90° with respect of the line of collision. It was a breakthrough. Very few events were recorded in this geometry in comparison with the former one but these few were telling about the structure deep inside hadrons.

In a way, it is the old story of Rutherford's electron–atom scattering all over again but reversed this time, the data showing a cross section smaller than expected for large scattering angles. This softening is shown in Figure 8.1 and it can be understood by assuming that the target, the proton, is not a point charge but rather has some internal structure.

A similar picture of a target with some internal structure is shown by a plot of the cross section as a function of the energy (for a fixed scattering angle), with peaks at the energy values where the substructure behaves as a coherent target and the scattering is elastic. The internal substructures can be more than one, as in the cross section depicted in Figure 8.2 where you can first see the peak for elastic scattering between the electrons and the nucleus as a whole, the region where nuclear resonances are excited, and then, increasing the energy, you can see the same pattern repeated as the electrons interact with the nucleons inside the atom (giving rise to a semi-elastic peak).

As the energy is further increased, the target breaks apart and the scattering is no longer elastic. This is the regime of deep inelastic scattering. The proton itself behaves as if it is made of some constituents, the **partons**.

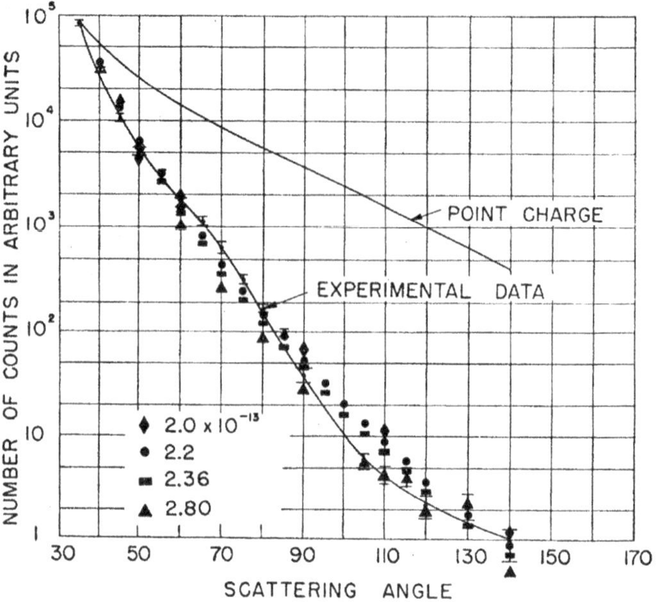

Figure 8.1 Cross section of electrons scattering off a proton as a function of the scattering angle [From R. Hofstadler *et al.*, *Physical Review* **92** (1953) 978].

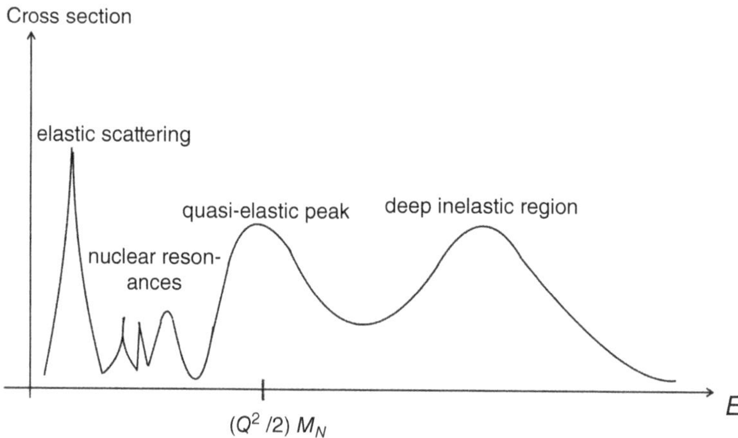

Figure 8.2 Schematic of the cross section for the scattering of electrons on a target nucleus.

8.1 $e^+e^- \to$ Hadrons

I want to circle back to Chapter 5 on the strong interactions and the computation of the ratio R in the production of hadrons by electron pair annihilation. At the time the result did not look quite right. How can a computation using quarks describe a process with hadrons? Here I want to try and understand why such a computation works, the shaky framework notwithstanding.

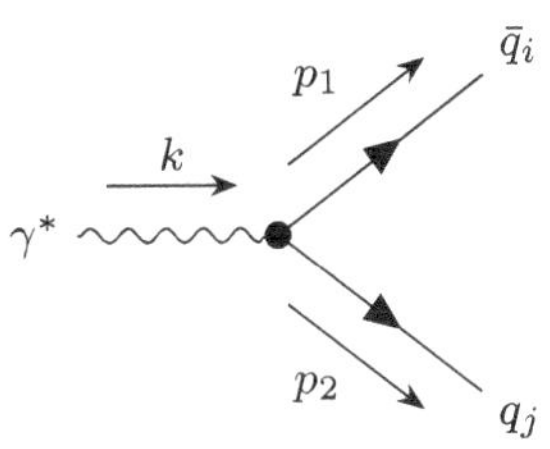

Figure 8.3 Off-shell photon decaying into a quark–antiquark pair.

What I need to consider is the leading correction to the diagram in Figure 8.3, that is, the process $e^+ + e^- \to \gamma^* \to q + \bar{q}$ in QCD:

$$e^+ + e^- \to \gamma^* \to q + \bar{q} + g\,, \tag{8.1}$$

where there is a gluon g in the final state. The electron-pair annihilation into the virtual photon γ^* can be factorized and I only need to estimate the process $\sigma_{gq\bar{q}}$ shown in Figure 8.3, the final decay of the virtual photon.

There are two diagrams, as shown in Figure 8.4, to be taken into account. They give rise to the two amplitudes[8.1]

8.1 Take $k = p_1 + p_2 + p_3$ and $k_1 = p_2 + p_3$, $k_2 = p_1 + p_3$, $E_{CM} = \sqrt{k^2} = k$.

$$\mathcal{M}_1 = \bar{u}(p_2)(-ig_s\gamma^\nu t^A_{ij})\left(\frac{i\not{k}_1}{k_1^2}\right)(-iee_q\gamma^\mu)v(p_1)\varepsilon_\mu\varepsilon'^*_\nu \tag{8.2}$$

$$\mathcal{M}_2 = \bar{u}(p_2)(-iee_q\gamma^\mu)\left(\frac{i\not{k}_2}{k_2^2}\right)(-ig_s\gamma^\nu t^A_{ij})v(p_1)\varepsilon_\mu\varepsilon'^*_\nu\,, \tag{8.3}$$

where e_q are the quark charges, $u(p_2)$ and $v(p_1)$ are the quark and antiquark spinors and ε_μ and ε'^*_ν are the gluon polarization vectors.

The cross section is factorized into the initial electron-pair vertex and the decay of the virtual photon. The latter part is obtained by means of the usual formula

$$\mathrm{d}\sigma_{gq\bar{q}} = \frac{1}{2E_{\mathrm{CM}}}|\overline{\mathcal{M}}|^2\mathrm{d}\Phi_{\mathrm{LIPS}}, \tag{8.4}$$

where the phase space is given by

$$\mathrm{d}\Phi_{\mathrm{LIPS}} = \frac{\mathrm{d}^3\vec{p}_1}{(2\pi)^3 2E_1}\frac{\mathrm{d}^3\vec{p}_2}{(2\pi)^3 2E_2}\frac{\mathrm{d}^3\vec{p}_3}{(2\pi)^3 2E_3}(2\pi)^4\delta^{(4)}(k - p_1 - p_2 - p_3)\,. \tag{8.5}$$

I can define the new variables $x_i = 2E_i/K$ so that (neglecting the quark masses)[8.2]

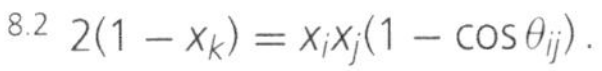

8.2 $2(1 - x_k) = x_i x_j(1 - \cos\theta_{ij})$.

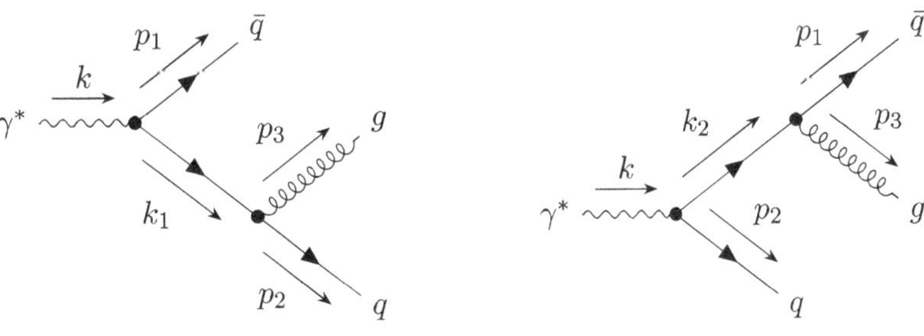

Figure 8.4 The diagrams for the two leading corrections to $\gamma^* \to q\bar{q}$.

$$\frac{p_i \cdot p_j}{K^2} = \frac{E_i E_j}{K^2}(1 - \cos\theta_{ij})$$
$$= \frac{x_i x_j}{4}(1 - \cos\theta_{ij}) = \frac{1}{2}(1 - x_k) \qquad (8.6)$$

and

$$s = K^2(1 - x_2)$$
$$t = K^2(1 - x_1) \qquad (8.7)$$
$$u = K^2(1 - x_3)$$

for the Mandelstam variables.

The variables x_i vary over the region shown in Figure 8.5,

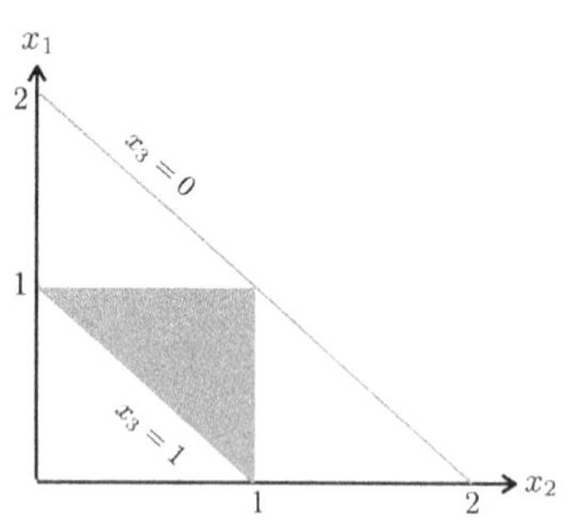

Figure 8.5 Dalitz plot for the variables x_1 and x_2.

$$0 \le x_1 \le 1$$
$$1 - x_1 \le x_2 \le 1, \qquad (8.8)$$

which spans the kinematic space of the process.

The phase space element can be rewritten in terms of the new variables. First, integrate over the momentum $\vec{p}_3$ and obtain

$$\int \mathrm{d}^3\vec{p}_3\, \delta^{(4)}(k - p_1 - p_2 - p_3) = \underbrace{\delta(K - E_1 - E_2 - E_3)}_{x_1+x_2+x_3=2}. \qquad (8.9)$$

In the remaining integrations over

$$\frac{\mathrm{d}^3\vec{p}_1}{(2\pi)^3 2E_1}\frac{\mathrm{d}^3\vec{p}_2}{(2\pi)^3 2E_2} = \frac{1}{2(2\pi)^4}E_1\mathrm{d}E_1 E_2\mathrm{d}E_2\,\mathrm{d}\cos\theta_{12}, \qquad (8.10)$$

I can do the angular part (with θ_{12} the angle between p_1 and p_2) and write

$$\int_{-1}^{+1}\mathrm{d}\cos\theta_{12}\,\frac{\delta(K - E_1 - E_2 - E_3)}{2E_3} = \frac{2}{x_1 x_2 K^2}. \qquad (8.11)$$

Gathering the various terms, I find that

$$\mathrm{d}\Phi_{\mathrm{LIPS}} = \frac{K^2}{16(2\pi)^3}\mathrm{d}x_1\mathrm{d}x_2. \qquad (8.12)$$

To proceed, first take $\mathcal{M}_1$ and compute the square of its modulus:[8.3]

8.3 $k_1^2 = t$, $p \cdot p_1 = (u + s)/2$ and $k_1 \cdot p_2 = t/2$.

$$|\overline{\mathcal{M}}_1|^2 = \frac{g_s^2 e^2 e_q^2}{p_a^4}\underbrace{\mathrm{Tr}\,(t^A t^A)}_{=4}\underbrace{\mathrm{Tr}\,(\not{p}_2\gamma_\nu\not{p}_a\gamma_\mu\not{p}_1\gamma_\mu\not{p}_a\gamma_\nu)}_{=16(2p_1\cdot p_a p_2\cdot p_a - p_1\cdot p_2 p_a\cdot p_a)}$$
$$= 32\,e^2 e_q^2 g_s^2\left(\frac{st}{t^2}\right) = 32\,e^2 e_q^2 g_s^2\frac{1 - x_2}{1 - x_1}. \qquad (8.13)$$

I can do the same with $\mathcal{M}_2$ and obtain

$$|\overline{\mathcal{M}}_2|^2 = 32\,e^2 e_q^2 g_s^2\frac{1 - x_1}{1 - x_2}. \qquad (8.14)$$

There is also an interference term between the two amplitudes, which gives

$$2\,\overline{\mathcal{M}_1\mathcal{M}_2^*} = 32\,e^2e_q^2\,g_s^2\frac{2uK^2}{st} \tag{8.15}$$

$$= 32\,e^2e_q^2\,g_s^2\left[\frac{2}{(1-x_1)(1-x_2)} - \frac{2}{(1-x_1)} - \frac{2}{(1-x_2)}\right].$$

Altogether I have

$$\mathrm{d}\sigma_{gq\bar{q}} = \frac{4}{3}\frac{\alpha_s}{2\pi}\,\sigma_0\,\frac{x_1^2+x_2^2}{(1-x_1)(1-x_2)}\mathrm{d}x_1\mathrm{d}x_2\,, \tag{8.16}$$

where $\sigma_0 = 4\pi\alpha^2e_q^2/s$ is the Born cross section and $\alpha_s = g_s^2/4\pi$ as usual. Equation (8.16) is integrated,

$$\boxed{\sigma_{gq\bar{q}} = \frac{4}{3}\frac{\alpha_s}{2\pi}\,\sigma_0\int_0^1\mathrm{d}x_1\int_{1-x_1}^1\mathrm{d}x_2\frac{x_1^2+x_2^2}{(1-x_1)(1-x_2)}\,,} \tag{8.17}$$

to give the total cross section. Recall the notation: the two quarks are labeled as 1 and 2, the gluon as 3. These are the three partons taking part in the process.

Just by looking at the cross section in Eq. (8.17) and at Figure 8.6, I can see that there are singularities when the gluon (parton number 3) has momentum p_3 collinear to that of either quark (partons number 1 and 2), that is,

$$\left.\begin{array}{ll} x_1\to 1 & \theta_{32}\to 0 \\ x_2\to 1 & \theta_{31}\to 0\end{array}\right\}\quad \text{parton 3 is } \textbf{collinear} \text{ with parton 1 or 2,} \tag{8.18}$$

or when its momentum becomes soft ($p_3 \to 0$), that is, when

$$(1-x_1)\to 0 \quad \text{and} \quad (1-x_2)\to 0 \quad \text{parton 3 is } \textbf{soft}. \tag{8.19}$$

This singular structure is quite interesting. It comes from the particles being massless or being produced along directions that are close together.[8.4] It takes place at small momenta; it is not an ultraviolet

[8.4] If the particles are massive then $(p_1+p_2)^2 = 2E_1E_2(1-\cos\theta)$ becomes $(p_1+_2)^2 - m^2 = 2E_1E_2(1-\beta\cos\theta)$ and the singularities go away (so they are also called **mass singularities**).

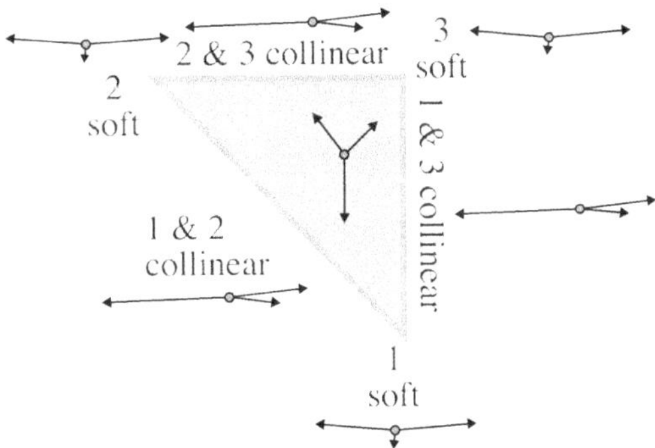

Figure 8.6 Momentum configurations giving rise to singularities.

divergence as are those found in estimating loop diagrams. It is an infrared divergence. The gluon here could have been a photon (no non-Abelian feature has entered the computation) and the same singularities would have been present.

In order to understand this result, I must regularize Eq. (8.17). The simplest way consists in giving a mass m_G to the gluons. Redoing the cross section computation,[1] I now obtain

extra terms from the gluon mass

$$\sigma_{gq\bar{q}} = \frac{2}{3}\frac{\alpha_s}{\pi}\sigma_0 \times \int \mathrm{d}x_1 \int \mathrm{d}x_2 \frac{1}{(1-x_1)(1-x_2)} \left\{ x_1^2 + x_2^2 + \mu\left[2\,(x_1+x_2) - \frac{(1-x_1)^2 + (1-x_2)^2}{(1-x_1)(1-x_2)} \right] + 2\mu^2 \right\}, \tag{8.20}$$

where $\mu^2 = m_G^2/Q^2$.

The integration region (Figure 8.7) is modified by the finite gluon mass to

$$0 \leq x_1 \leq 1-\mu$$
$$1-\mu-x_1 \leq x_2 \leq \frac{1-x_1-\mu}{1-x_1}. \tag{8.21}$$

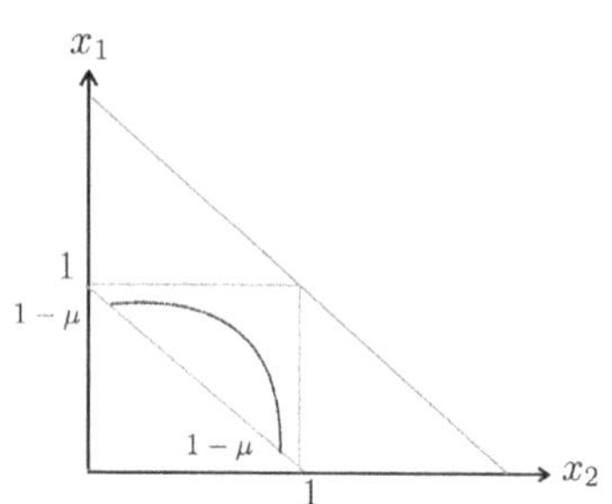

Figure 8.7 Dalitz plot for the variables x_1 and x_2 in the presence of a massive gluon.

8.5
$$\int_0^{1-\mu} \frac{\mathrm{d}x_1}{1-x_1} = -\log\mu$$
$$\int_0^{1-\mu} \frac{\mathrm{d}x_1}{1-x_1}\log x_1 = -\frac{1}{2}\log^2\mu$$
$$\int_0^{1-\mu} \frac{\mathrm{d}x_1\,(1+x_1^2)}{1-x_1}\log x_1 = -\frac{\pi^2}{3}+\frac{5}{4}$$
$$\int_0^{1-\mu} \frac{\mathrm{d}x_1\,(1+x_1^2)}{1-x_1}\log(1-x_1) = -\log^2\mu+\frac{7}{4}$$
$$\int_0^{1-\mu} \frac{\mathrm{d}x_1\,\mu}{(1-x_1)^2} = 1$$
$$\int_0^{1-\mu} \frac{\mathrm{d}x_1\,\mu^2}{(1-x_1)^3} = \frac{1}{2}.$$

The integration can now be performed to yield[8.5]

$$\sigma_{gq\bar{q}} = \frac{2\alpha}{3\pi}\sigma_0\left[\log^2\mu^2 + 3\log\mu - \frac{\pi^2}{3} + 5\right]. \tag{8.22}$$

To obtain Eq. (8.22) I drop all terms proportional to μ whenever they are non-singular and use the integrals given in note 8.5. Notice that some of the integrals look as though they should go to zero as $\mu \to 0$ but they do not.

Before proceeding it must be noted – and it is the crucial observation – that the soft divergence also appears in the interference between the Born term and the one-loop vertex correction. It is because of this interference term that a higher-order contribution gives an $O(\alpha_s)$ term in the cross section, that is, a contribution of the same order of those coming from the leading-order diagrams. The diagram is shown in Figure 8.8. Computing the loop diagram gives (after being regularized with a massive gluon)

$$\sigma_{\text{virtual}} = \frac{2\alpha}{3\pi}\sigma_0\left[-\log^2\mu^2 - 3\log\mu - \frac{2\pi^2}{3} - \frac{7}{2}\right]. \tag{8.23}$$

[1] Here, I am following R.D. Field, *Applications of Perturbative QCD* (Addison-Wesley, 1986), which is one of the few textbooks containing the often challenging computations of QCD.

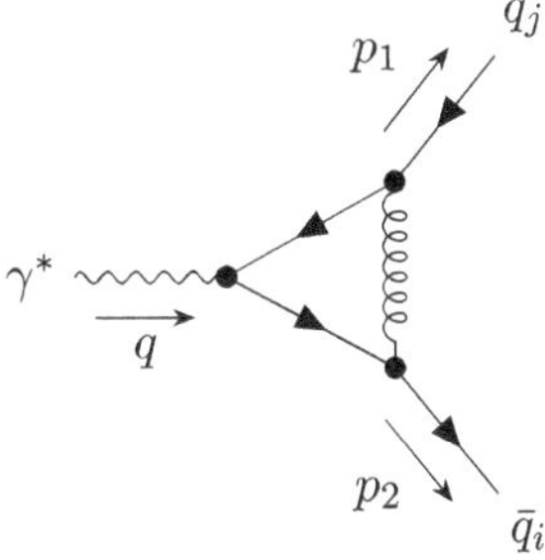

Figure 8.8 The Feynman diagram for the one-loop vertex correction.

If now I sum over all final states, I obtain the inclusive cross section, in which both $\sigma_{gq\bar{q}}$ and σ_{virtual} are included.[8.6] The total cross section is finite[8.7] and equal to

$$\sigma_{gq\bar{q}} + \sigma_{\text{virtual}} = \frac{2\alpha}{3\pi}\sigma_0\left[\frac{3}{2}\right] = \frac{\alpha_s}{\pi}\sigma_0 \,. \tag{8.24}$$

8.6 The contributions of the self-energy of the external quark legs are zero in the Landau gauge.

8.7 The finiteness of the total cross section is an example of the Kinoshita–Lee–Nauenberg theorem, which guarantees that divergencies like the one I have encountered, and that are due to the quasi-degeneracy of the final states, are eliminated when summing over all degenerate states.

The ratio R is therefore given by

QCD corrections

$$R = 3\Big(1 + \frac{\alpha_s}{\pi}\Big) \sum_j \sqrt{1 - \frac{4m_j^2}{s}} \left(1 + \frac{2m_j^2}{s}\right) Q_j^2, \tag{8.25}$$

where now the leading term is as before, that is, the term proportional to 1 in Eq. (8.25), and the first QCD correction.

Equation (8.25) shows that the result obtained by taking only quarks in the final states is the dominant one. A small correction of order $O(\alpha_s)$ comes from QCD. To obtain this result I had to put together final states without gluons and with one gluon in addition to the two produced quarks. These are different final states and can only be combined when computing an inclusive cross section. The case of the ratio R is an example of the more general result: the **inclusive** total cross sections in QCD do not have infrared singularities; they are infrared safe.

But the physics of inclusive cross sections is a bit tedious. I really want to look at more exclusive processes. Because these processes are going to be in general infrared divergent, I need to define infrared-safe observables. The simplest example (at least conceptually if not computationally) is provided by the **Sterman–Weinberg jets**. These observables are obtained by singling out, in the exclusive cross section for the production of hadrons in electron-pair annihilation, finite contributions obtained by cutting off the collinear and soft divergences.

In practice, I retain the entire Born and the virtual cross sections but, out of the cross section with one real gluon emission, I only retain the part for which the energy of the emitted gluon ℓ^0 is limited by $\ell^0 < \epsilon E$ or,

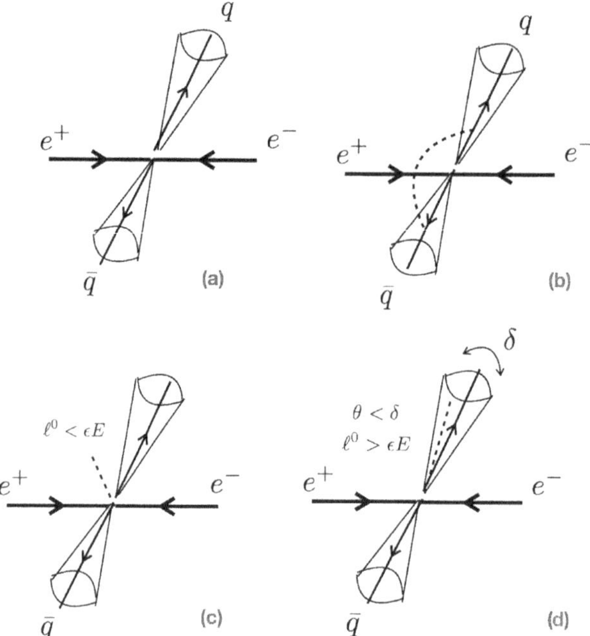

Figure 8.9 The four processes included in the definition of the Sterman–Weinberg jets. The dashed lines represent either virtual (b) or real (c), (d) gluons, the real gluons being constrained by their energy and angle, as indicated.

when $\ell^0 > \epsilon E$, only angles $\theta < \delta$. The two quantities ϵ and δ cut off the collinear and soft infrared-divergent contributions. The four contributions to the Sterman–Weinberg jets are shown in Figure 8.9.

The regularized integrals are

$$\begin{aligned} \sigma^{(c)}_{\text{real}} &= \frac{3\alpha_s}{2\pi}\sigma_0 \int_0^{\epsilon E} \frac{\mathrm{d}E}{E} \int_{\theta=0}^{\pi} \frac{4\mathrm{d}\cos\theta}{(1-\cos\theta)(1+\cos\theta)} \\ \sigma^{(d)}_{\text{real}} &= \frac{3\alpha_s}{2\pi}\sigma_0 \int_{\epsilon E}^{E} \frac{\mathrm{d}E}{E} \left[\int_{\theta=0}^{\delta} \frac{4\mathrm{d}\cos\theta}{(1-\cos\theta)(1+\cos\theta)} \right. \\ &\qquad \left. + \int_{\theta=\pi-\delta}^{\pi} \frac{4\mathrm{d}\cos\theta}{(1-\cos\theta)(1+\cos\theta)} \right], \end{aligned} \tag{8.26}$$

where $\sigma_0 = \sigma^{(a)}_{\text{Born}}$ is the Born cross section, to which I must add the virtual contribution

$$\sigma^{(b)}_{\text{virtual}} = -\frac{2\alpha}{3\pi}\sigma_0 \int_0^{\epsilon E} \frac{\mathrm{d}E}{E} \int_{\theta=0}^{\pi} \frac{4\,\mathrm{d}\cos\theta}{(1-\cos\theta)(1+\cos\theta)}. \tag{8.27}$$

The sum of the terms provides a finite result:

$$\begin{aligned}\sigma &= \sigma^{(a)}_{\text{Born}} + \sigma^{(c)}_{\text{real}} + \sigma^{(d)}_{\text{real}} + \sigma^{(b)}_{\text{virtual}} \\ &= \sigma_0 - \frac{2\alpha_s}{3\pi}\sigma_0 \int_{\epsilon E}^{E} \frac{\mathrm{d}E}{E} \int_{\theta=\delta}^{\pi-\delta} \frac{4\,\mathrm{d}\cos\theta}{(1-\cos\theta)(1+\cos\theta)} \\ &= \sigma_0 \left[1 - \frac{3\alpha}{2\pi} \log\epsilon \, \log\delta^2\right]. \end{aligned} \tag{8.28}$$

The cross section for the jets in Eq. (8.28) predicts that the leading correction to the Born cross section consists of events with two back-to-back jets. The dominance of these jets is a prediction of QCD. The relative weight of two jets over three or more jets in the events can be estimated in powers of the coupling constant α_s, three jets being $O(\alpha_s)$ less frequent than two.

8.2 *Problem Session*: The Soft and the Collinear

This problem session is devoted to making explicit that the infrared singularities encountered in the QCD computations do indeed come from long-distance physics.

To do this, I have to switch to space-time and translate the momenta of the scattering process into space-time intervals.[2] This is done by doing a Fourier transformation of the amplitude

$$\mathcal{M}(x) = \int \mathrm{d}^4 k\, e^{-ik\cdot x} \mathcal{M}(k), \tag{8.29}$$

where $k = p_2 + p_3$ in the first diagram, for $\gamma^* \to g q_j q_i$, and $k = p_1 + p_3$ in the second diagram.

Before inserting in this expression the momenta of the scattering process, I will change to **light-cone variables**, where the components of all momenta are written as

$$a = (a^+, a^-, \vec{a}_T) \quad \text{where} \quad a^\pm = \frac{a^0 + a^3}{\sqrt{2}}, \tag{8.30}$$

so that I have

$$a \cdot b = a^+ b^- + a^- b^+ - \vec{a}_T \cdot \vec{b}_T \quad \text{and} \quad a^2 = 2a^+ a^- - \vec{a}_T^{\,2}. \tag{8.31}$$

In these coordinates, if I take the z-axis along the direction of the momentum k of a parton that is either soft or collinear to another parton then

$$k^+ \gg 1 \quad \text{and} \quad \vec{k}_T \to 0, \tag{8.32}$$

[2] D.E. Soper, arXiv:hep-ph/9702203.

while

$$k^- = \frac{\vec{p}_{3T}^{\,2}}{2q_j^+} + \frac{\vec{p}_{3T}^{\,2}}{2p_3^+}\,. \tag{8.33}$$

The component k^- is therefore small either when $\vec{p}_{3T}^{\,2}$ is small (which corresponds to a collinear divergence) or when $\vec{p}_{3T}^{\,2}$ and p_3^+ are both small and $p_3^+ \propto |\vec{p}_{3T}^{\,2}|$ (which corresponds to a soft divergence). Taking the Fourier transform of the amplitude, I have

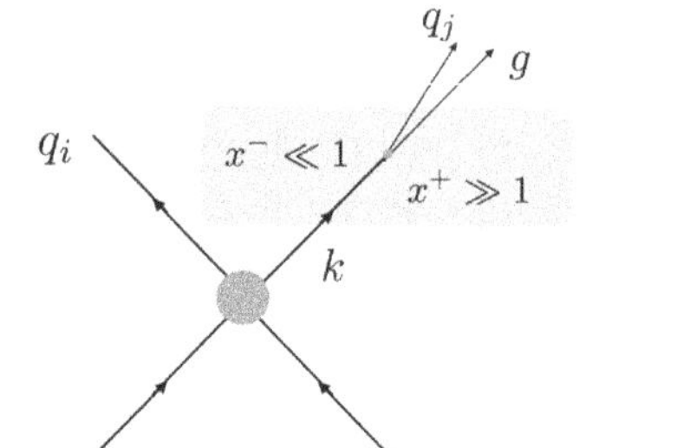

Figure 8.10 The dominant kinematic configuration and its space-time representation.

$$M(x) = \int \mathrm{d}^4 k\, e^{-ik^+x^- - ik^-x^+ + i\vec{k}_T\cdot\vec{x}_T}\, \mathcal{M}(k)\,, \tag{8.34}$$

from which it can be seen that the integral is dominated by the region where the phase of the exponential function is of order 1 and therefore where, as in Figure 8.10,

$$x^- \ll 1 \quad \text{and} \quad x^+ \gg 1\,, \tag{8.35}$$

for $k^+ \gg 1$ and $k^- \ll 1$ respectively. This region of space-time corresponds to events taking place at a large distance from the scattering along the x^+ direction.

8.3 The Parton Model

The **parton model** was introduced to describe the deep inelastic scattering (DIS) of electrons on nucleons. The model works in an energy regime where the scattering takes place with a large momentum transfer and the process is inelastic. The target nucleon breaks apart after the interaction.

A few kinematic preliminaries are necessary. The electron imparts to the target (a nucleon) a momentum r, which is time-like so that it is convenient to set $Q^2 = -r^2$.

Whereas in an elastic process there is only one independent variable, for an inelastic scattering the variables are two. I can take (ν, Q^2), where ν is the energy transferred by the electron,

$$\nu = \frac{r\cdot P}{M} = \omega' - \omega\,, \tag{8.36}$$

or (x, Q^2), where

$$x = \frac{Q^2}{2M\nu}\,, \tag{8.37}$$

for P the momentum of the nucleon and M its rest mass.

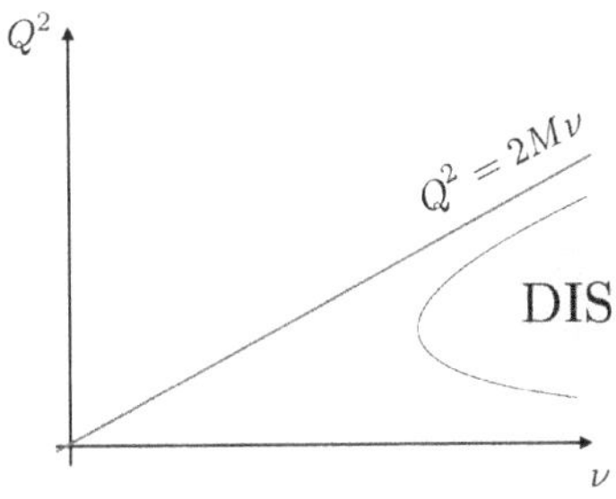

Figure 8.11 The DIS region in the energy plane.

The deep inelastic scattering regime is defined by the region shown in Figure 8.11, where both ν and Q^2 are large.

The cross section can be written by means of the general expression

$$d\sigma = \frac{1}{4\omega E|\vec{v}|}\frac{1}{4}\sum_{\text{pol}}|\mathcal{M}|^2(2\pi)^4\delta^{(4)}(k'+P_X-k-P)\frac{d^3\vec{k}'}{(2\pi)^3 2\omega'}\frac{d^3\vec{P}_X}{(2\pi)^3 2E_X}, \tag{8.38}$$

where the final momentum $\vec{P}_X$ and energy E_X take into account the fragments of the broken-up nucleon (as in Figure 8.12).

The amplitude is given by

$$\mathcal{M} = e^2\bar{u}_e^{\sigma'}(k')\gamma_\alpha u_e^\sigma(k)\,\frac{1}{q^2}\,\langle X|J_\alpha|P\rangle, \tag{8.39}$$

where u_e^σ are the electron spinors and I still have not committed myself to an explicit choice for the electromagnetic hadron current J_α.

The average of the squared amplitude is given by

$$\frac{1}{4}\sum_{\text{pol}}|\mathcal{M}|^2 = \left(\frac{e^2}{Q^2}\right)^2\left\{\frac{1}{2}\sum_{\sigma,\sigma'}\bar{u}_e^{\sigma'}(k')\gamma_\alpha u_e^\sigma(k)\,\bar{u}_e^\sigma(k)\gamma_\beta u_e^{\sigma'}(k')\right\}$$
$$\times\left\{\frac{1}{2}\sum_{\text{pol}}\langle P|J^\alpha|X\rangle\langle X|J^\beta|P\rangle\right\}. \tag{8.40}$$

Equation (8.40) can be rewritten as[8.8]

8.8 $d^3\vec{k}' = (\vec{k}')^2 d|\vec{k}'|d\Omega' = (\omega')^2 d\omega' d\Omega'$ and $|\vec{v}| \to 1$ for $m_e \to 0$ and the proton at rest.

$$d\sigma = \frac{d^3\vec{k}'}{4\omega\omega'|\vec{v}|}\frac{1}{(2\pi)^2}\left(\frac{e^2}{Q^2}\right)^2 L_{\alpha\beta}W^{\alpha\beta} = \frac{\omega'}{\omega}\frac{\alpha^2}{Q^4}L_{\alpha\beta}W^{\alpha\beta}d\omega'\,d\Omega', \tag{8.41}$$

where

$$L_{\alpha\beta} = \frac{1}{2}\sum_{\sigma,\sigma'}\bar{u}_e^{\sigma'}(k')\gamma_\alpha u_e^\sigma(k)\,\bar{u}_e^\sigma(k)\gamma_\beta u_e^{\sigma'}(k') \tag{8.42}$$

is given by (neglecting the electron mass)

$$L_{\alpha\beta} = 2\,(k_\alpha k'_\beta + k'_\alpha k_\beta - g_{\alpha\beta}k'\cdot k), \tag{8.43}$$

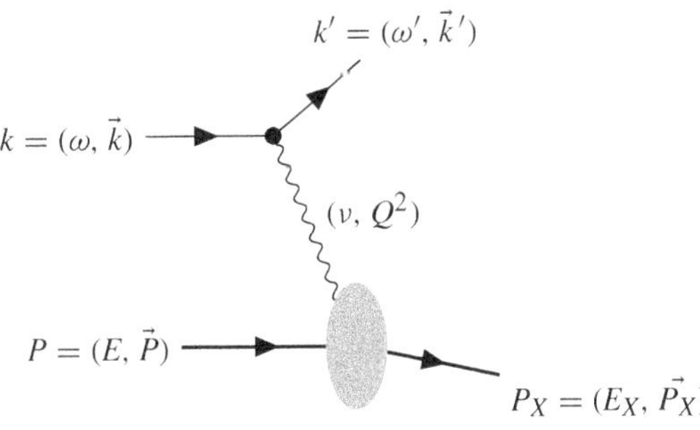

Figure 8.12 Kinematics of the deep inelastic scattering of an electron on a target nucleon.

and

$$W^{\alpha\beta} = \frac{1}{4\pi M}\int \frac{d^3\vec{P}_X}{(2\pi)^3 2E_X}\frac{1}{2}\sum_{\text{pol}} \varkappa^\alpha \varkappa^{\beta *}\delta^{(4)}(k' + P_X - k - P), \quad (8.44)$$

where

$$\varkappa^\alpha = \langle P|J^\alpha|X\rangle \quad (8.45)$$

is the matrix element of the electromagnetic current of the hadron (the symbol on the left-hand side is a form of kappa). Thus I have a name for what I do not know how to compute yet.

The only thing I know for sure about $W^{\alpha\beta}$ is that it has two Lorentz indices and depends only on the four-vectors r and P. This is sufficient to write it in terms of structure functions:

$$\begin{aligned} W^{\alpha\beta} &= g^{\alpha\beta}F_1(r,P) + P^\alpha P^\beta F_2(r,P) + (r^\alpha P^\beta + r^\beta P^\alpha)F_3(r,P) \\ &+ r^\alpha r^\beta F_4(r,P) + \underbrace{i\epsilon^{\alpha\beta\mu\nu}P^\mu r^\nu F_5(r,P)}_{=0 \text{ for parity-conserving EM}} . \end{aligned} \quad (8.46)$$

There is another piece of information I can use: the electromagnetic hadron current, whatever that might be, is conserved and therefore

$$r^\alpha W_{\alpha\beta} = r^\beta W_{\alpha\beta} = 0, \quad (8.47)$$

which tells me that

$$\begin{cases} r_\beta(F_1 + P\cdot rF_3 + r^2F_4) = 0 \\ P_\beta(r\cdot PF_2 + r^2F_3) = 0 . \end{cases} \quad (8.48)$$

Solving Eq. (8.48) gives

$$F_4 = -\frac{1}{r^2}\left(F_1 - \frac{(r\cdot P)^2}{r^2}F_2\right) \quad \text{and} \quad F_3 = -\frac{r\cdot P}{r^2}F_2 . \quad (8.49)$$

The Lorentz structure of the hadron tensor is therefore given by

$$\begin{aligned} W^{\alpha\beta} &= \left(-g_{\alpha\beta} + \frac{r_\alpha r_\beta}{r^2}\right)\frac{F_1(x,Q)}{2M} \\ &+ \left(P^\alpha - \frac{P\cdot r}{r^2}r^\alpha\right)\left(P^\beta - \frac{P\cdot r}{r^2}r^\beta\right)\frac{F_2(x,Q)}{M^2\nu} . \end{aligned} \quad (8.50)$$

It is just a matter of algebra to find the longitudinal projection[8.9]

$$\frac{4x^2}{\nu}P^\alpha P^\beta W_{\alpha\beta} = F_2 - xF_1 \quad (8.51)$$

and the trace

$$2Mg^{\alpha\beta}W_{\alpha\beta} = -3F_1 + \frac{1}{x}F_2 . \quad (8.52)$$

Both relations will come in handy in a moment.

[8.9] Neglecting terms of order the quark masses, such as $w^2P^2 = m_{q'}^2$, and using the definition $-r^2 = 2M\nu x$.

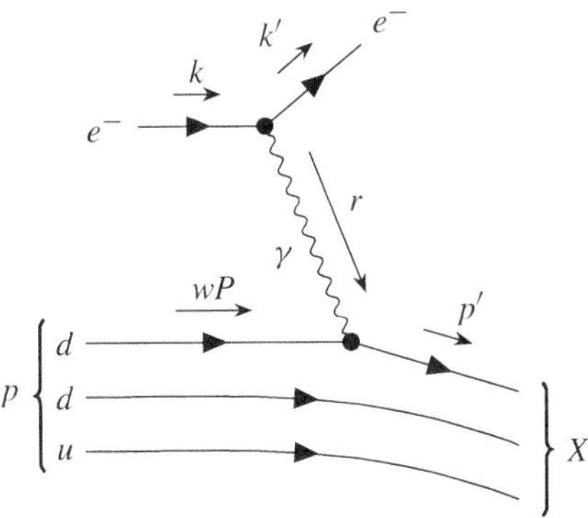

Figure 8.13 Kinematics of the deep inelastic scattering of an electron on a target nucleon in the parton model. The struck parton carries a fraction w of the nucleon momentum P.

To proceed I have to write an explicit expression for $\varkappa^\alpha = \langle P|J^\alpha|X\rangle$. The parton model assumes that the hadron current $\varkappa^\alpha$ in Eq. (8.45) is just the same as if the photon were to strike a freely moving constituent inside the proton:

$$\varkappa^\alpha = \langle P|J^\alpha|X\rangle = \sum_q e_q \bar{\chi}(p')\gamma^\alpha \chi(p)\,. \tag{8.53}$$

In writing Eq. (8.53) I am assuming that the parton, beside being free, is a fermion described by a spinor $\chi(p)$ obeying the Dirac equation. The hadron current becomes exactly what I have for the lepton current except for the momentum p, which is not well defined since I do not know how the proton momentum P is distributed among the parton constituents. Not knowing, I will give it a random fraction wP of the nucleon momentum P. (See Figure 8.13.) The random fraction follows a statistical distribution $q(w)$ over which I will eventually integrate, so that[8.10]

8.10 Neglecting the quark masses.

$$\begin{aligned}\frac{1}{2}\sum_{\text{pol}} \varkappa^\alpha \varkappa^{\beta *} &= 2\sum_q e_q^2 \int \mathrm{d}\omega\, q(w)\left[p'^\alpha p^\beta + p'^\beta p^\alpha - g_{\alpha\beta}(p'\cdot p)\right] \\ &= 2\sum_q e_q^2 \int \mathrm{d}\omega\, q(w)\Big\{w(wP+r)^\alpha P^\beta + w(wP+r)^\beta P^\alpha \\ &\qquad - g^{\alpha\beta}\Big[w(wP+r)\cdot P\Big]\Big\}\,.\end{aligned} \tag{8.54}$$

The phase space is given by

$$\frac{1}{2Mw}\int \mathrm{d}^3\vec{p}\,'\,\delta^{(3)}\underbrace{(\vec{k}\,'+\vec{p}\,'-\vec{k}-\vec{p})}_{\vec{k}\,'+\vec{p}\,'=\vec{p}+\vec{r}}\frac{\delta(\omega'-p^0-\omega+p'^0)}{2p'^0}\,, \tag{8.55}$$

reduced mass of parton

where the δ-function in the energy can be brought into a simple form by the following transformations:

$$\frac{\delta(\omega' - p^0 - \omega + p'^0)}{2p'^0} = \theta(p_0')\delta\Big[(p_0' - p_0 - r_0)(p_0' + p_0)\Big]$$
$$= \theta(p_0') \underbrace{\delta(r^2 + 2p\cdot r)}_{-Q^2+2wP\cdot r = 2wM\nu - Q^2} = \frac{1}{2M\nu}\delta(w - x). \tag{8.56}$$

By inserting this result into Eq. (8.44), I obtain

$$W^{\alpha\beta} = \sum_q \frac{e_q^2}{2M^2\nu}\int \mathrm{d}w\, q(w)\,\delta(w-x)$$
$$\times \left\{(wP+r)^\alpha P^\beta + (wP+r)^\beta P^\alpha - g^{\alpha\beta}\left[(wP+r)\cdot P\right]\right\}. \tag{8.57}$$

I can insert the expression into Eq. (8.57) into Eq. (8.51) and Eq. (8.52) to find[8.11]

8.11 Neglecting again terms like $w^2P^2 = m_q^2$.

$$\frac{4x^2}{\nu} P^\alpha P^\beta W_{\alpha\beta} = 0 \tag{8.58}$$

$$2Mg^{\alpha\beta}W_{\alpha\beta} = -4M^2\nu x \tag{8.59}$$

and then solve for F_1 and F_2 to find that

$$\frac{F_1}{2M} = \sum_q \frac{e_q^2}{2M}\int \mathrm{d}w\, q(w)\,\delta(w-x), \tag{8.60}$$

which gives

$$F_1 = \sum_q e_q^2 q(x) \tag{8.61}$$

and

$$\frac{F_2}{M^2\nu} = \sum_q \frac{e_q^2}{2M}\int \mathrm{d}w\, q(w)\,\delta(w-x)\frac{2w}{M\nu}, \tag{8.62}$$

which in turn gives

$$F_2 = \sum_q e_q^2 x\, q(x). \tag{8.63}$$

The result of the parton model computation is that the structure functions F_i satisfy (see Figures 8.14 and 8.15)

1. $\boxed{F_2 = xF_1}$ (Callan–Gross);

2. $\boxed{F_i(x, Q^2) \quad \text{independent of } Q^2.}$

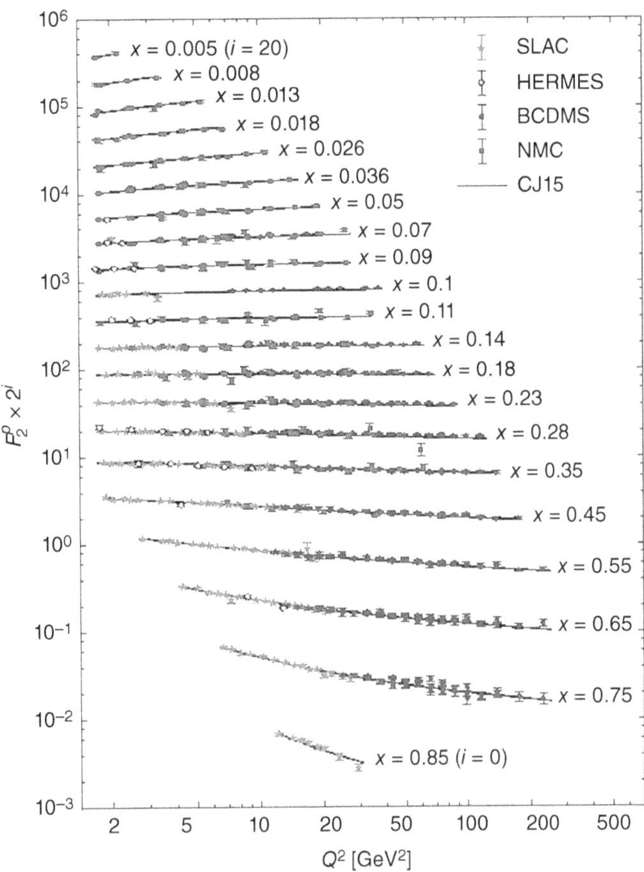

Figure 8.14 Relation between F_2 and Q^2. In the region $Q^2 < 5$ GeV^2, for values of x between 0.28 and 0.11, the form factor is approximately independent of Q^2. [R.L. Workman *et al.* (Particle Data Group), *Prog. Theor. Exp. Phys.* **2022** (2022) 083C01 http://pdg.lbl.gov.]

The **Callan–Gross relation** holds if the partons are taken, as I did, to be fermions. This is reasonably confirmed by the data (as shown in Figure 8.15) thus providing an important clue to the nature of the partons inside the nucleons.

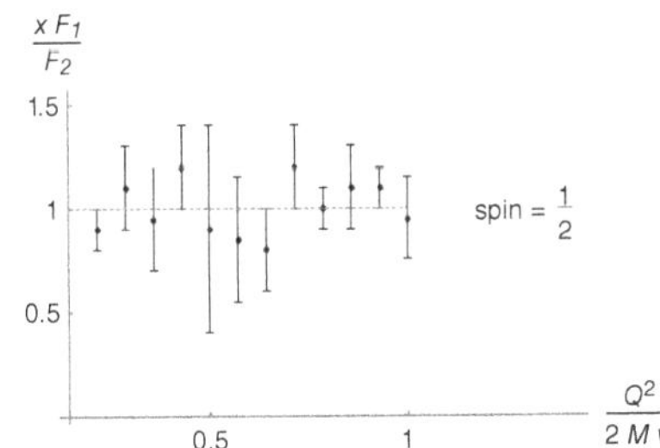

Figure 8.15 Early data at various values of Q^2 for the ratio xF_1/F_2.

The analogy I like to have in mind here is that of a little dog being walked on a long leash. As long as the dog runs within the length of the leash, he is free to do whatever he feels he wants to do. Yet, as he reaches the maximum length of the leash, he is brusquely stopped and recalled toward his owner.

The dependence of the structure function F_2 on Q^2 is shown in Figure 8.14. The energy range explored at the time ($Q^2 < 5$ GeV, $x > 0.2$) suggests that, indeed, the structure functions do not depend on Q^2. This is embodied in the parton model, where the quarks are bounded inside the nucleon but the energy scale of deep inelastic scattering is so high that the interaction takes place over time scales much shorter than those responsible for bounding the quarks.

In the parton model the nucleons are made out of quarks. The distribution functions $q_q(x)$ are given for each of the valence quarks u and d and also for those in the sea ($\bar{u}$, $\bar{d}$) as well as for the strange quark s. According to this model, the proton is described by the structure function

$$\begin{aligned} \frac{F_2^{ep}(x)}{x} &= \sum_i e_q^2 q_q(x) \\ &= \frac{4}{9}\left[u^p(x) + \bar{u}^p(x)\right] + \frac{1}{9}\left[d^p(x) + \bar{d}^p(x)\right] + \left[s^p(x) + \bar{s}^p(x)\right], \end{aligned} \tag{8.64}$$

while the neutron is described by

$$\begin{aligned} \frac{F_2^{en}(x)}{x} &= \sum_i e_q^2 q_q(x) \\ &= \frac{4}{9}\left[u^n(x) + \bar{u}^n(x)\right] + \frac{1}{9}\left[d^n(x) + \bar{d}^n(x)\right] + \frac{1}{9}\left[s^n(x) + \bar{s}^n(x)\right], \end{aligned} \tag{8.65}$$

where the isospin symmetry gives that

$$\begin{aligned} u^n &= d^p = d(x) \quad \text{and} \quad d^n = u^p = u(x) \\ s^n &= s^p = s(x) \quad \text{and} \quad \bar{q}^n = q^p = \bar{q}(x)\,. \end{aligned} \tag{8.66}$$

The distribution functions $q_q(x)$ have a part coming from the valence quarks $q_v(x)$ and one from the quarks in the "sea" $q_s(x)$ They must obey sum rules in such a way as to reproduce the nucleon charges:

$$\int_0^1 u_v(x)\mathrm{d}x = 2\,, \quad \int_0^1 d_v(x)\mathrm{d}x = 1\,, \quad \int_0^1 [q_s(x) - \bar{q}_s(x)]\mathrm{d}x = 0 \tag{8.67}$$

for the proton, and

$$\int_0^1 u_v(x)\mathrm{d}x = 1\,, \quad \int_0^1 d_v(x)\mathrm{d}x = 2\,, \quad \int_0^1 [q_s(x) - \bar{q}_s(x)]\mathrm{d}x = 0 \tag{8.68}$$

for the neutron.

The distribution functions look the same when the probing is by electromagnetic or weak interactions. This equivalence is shown by the ratio

$$\frac{F^{eN}(x)}{F^{\nu N}(x)} = \frac{18}{5}\frac{x[q(x) + \bar{q}(x)]}{x[q(x) + \bar{q}(x)]} = \frac{18}{5}\,, \tag{8.69}$$

the only difference coming from the overall factor 18/5 due to the quark electric charges.

The average parton momentum can be studied by means of the structure function F_2^{eN} in the scattering of electrons off nucleons,

$$F_2^{eN} = \frac{1}{2}\left[F_2^{ep} + F_2^{en}\right] = \frac{5}{18}\, x\left[u(x) + \bar{u}(x) + d(x) + \bar{d}(x)\right]; \qquad (8.70)$$

it is found to be

$$\frac{5}{18}\int_0^1 F_2^{eN}(x)\mathrm{d}x \simeq 0.5. \qquad (8.71)$$

This is the rather surprising result that quarks carry on average only 50% of a nucleon momentum. The other half must come from the gluons.

8.4 Another Take on the Parton Model*

The first time I studied the parton model I had the impression that it was somewhat grafted onto QCD. One had to assume the free partons in the nucleons in addition to the QCD Lagrangian. As it turns out, that assumption is not necessary: the parton model can be derived directly from the QCD Lagrangian by going into the deep inelastic kinematic regime and studying the behavior of the relevant products of fields.

I need to circle back and consider again the hadron tensor in Eq. (8.44),

$$W^{\alpha\beta} = \frac{1}{4\pi M}\int \frac{\mathrm{d}^3\vec{P}_X}{(2\pi)^3 2E_X} \times \frac{1}{2}\sum_{\text{pol}}\langle P|J^\alpha(0)|X\rangle\langle P|J^\beta(0)|X\rangle(2\pi)^4\delta^{(4)}(P_X - P - r), \qquad (8.72)$$

where as before $r = k' - k$, and rewrite this tensor as[8.12]

8.12 Recall: $(2\pi)^4\delta^{(4)}(P_X - P - r) = \int \mathrm{d}^4 z e^{-i(P_X - P - r)\cdot z}$.

$$W^{\alpha\beta} = \frac{1}{2\pi M}\int \mathrm{d}^4 z \int \frac{\mathrm{d}^3\vec{P}_X}{(2\pi)^3 2E_X}\frac{1}{2}\sum_{\text{pol}}\langle P|J^\alpha(z)|X\rangle\langle P|J^\beta(0)|X\rangle e^{ir\cdot z} \qquad (8.73)$$

by using the definition of the δ-function and after shifting the current as follows:

$$\langle P(P)|J^\alpha(z)|X(P_X)\rangle = \langle P(P)|J^\alpha(0)|X(P_X)\rangle e^{-i(P_X-P)\cdot z}. \qquad (8.74)$$

It can be written as a product of currents by using the completeness[8.13] of the states $|X\rangle$:

8.13 Recall: $\int \frac{\mathrm{d}^3\vec{P}_X}{(2\pi)^3 2E_X}|X\rangle\langle X| = 2M$.

$$W^{\alpha\beta} = \frac{1}{2\pi}\int \mathrm{d}^4 z \exp(ir\cdot x)\,\frac{1}{2}\sum_{\text{pol}}\langle P|J^\alpha(z)J^\beta(0)|X\rangle. \qquad (8.75)$$

The integration over z is dominated by those values that make the exponent $O(1)$, because otherwise the rapidly oscillating exponent makes the integration vanish. In the deep inelastic region,

$$r = \left(r^0 = \nu, 0, 0, r^3 = \sqrt{\nu^2 + Q^2}\right) \tag{8.76}$$

in the rest frame $p = (M, 0, 0, 0)$ of the nucleon. Therefore[8.14]

8.14 With ν, $Q^2 \gg M^2$ and $x = Q^2/2\nu M$ fixed.

$$\begin{aligned} r \cdot z &= \nu\left[z^0 - \frac{1}{\nu}\sqrt{\nu^2 + Q^2}\, z^3\right] \\ &= i\nu(z^0 - z^3) - i\frac{Q^2}{2\nu}z^3 = i\nu(z^0 - z^3) - ixMz^3 + O\left(\frac{1}{\nu}\right), \end{aligned} \tag{8.77}$$

which identifies the regions

$$|z^0 - z^3| \lesssim \frac{1}{\nu} \quad \text{and} \quad z^3 \lesssim \frac{1}{\nu}. \tag{8.78}$$

This gives

$$(z^0)^2 \lesssim \left(z^3 + \frac{1}{\nu}\right)^2 \simeq (z^3)^2 + 2\frac{z^3}{\nu} < \vec{z}^{\,2} + \frac{1}{x\nu M}, \tag{8.79}$$

that is, the deep inelastic region is the region where

$$z^2 \lesssim \frac{\text{const}}{Q^2} \to 0, \tag{8.80}$$

and the hadron tensor must be evaluated on the light-cone, where $z^2 \to 0$.

Before plunging into this estimate, I notice that Eq. (8.81) can be rewritten as

$$W^{\alpha\beta} = \frac{1}{2\pi}\int d^4z \exp(ir \cdot z)\frac{1}{2}\sum_{\text{pol}}\langle P|[J^\alpha(z), J^{\beta*}(0)]|X\rangle\,, \tag{8.81}$$

because the commuted currents give zero contribution.[8.15]

8.15 There is no state $|X\rangle$ with baryon number 1 and $E_X = P_X^0 - r^0 = M_N - \nu/M_N > M_N$.

Consider now the currents for free quarks as described by the spinor fields χ:

$$J_\alpha(z) = \bar{\chi}(z)\gamma_\alpha\chi(z)\,. \tag{8.82}$$

This is the essential point: the partons are described by free, fermionic fields. I can write the commutator as

$$\left[J_\alpha(z), J_\beta(0)\right] = \bar{\chi}_i(z)\gamma_\alpha\chi_i(z)\,\bar{\chi}_j(0)\gamma_\beta\chi_j(0) - \bar{\chi}_j(0)\gamma_\beta\chi_j(0)\,\bar{\chi}_i(z)\gamma_\beta\chi_i(z) \tag{8.83}$$

i,j Dirac indices

add zero to that, that is,

$$\left[\bar{\chi}(z)\gamma_\alpha\right]_i\bar{\chi}_j(0)\chi_i(z)\left[\gamma_\beta\chi(0)\right]_j - \bar{\chi}_j(0)\left[\bar{\chi}(z)\gamma_\alpha\right]_i\left[\gamma_\beta\chi(0)\right]_j\chi_i(z), \tag{8.84}$$

and find that

$$\left[J_\alpha(z), J_\beta(0)\right] = \left[\bar{\chi}(z)\gamma_\alpha\right]_i\Sigma_{ij}\left[\gamma_\beta\chi(0)\right]_j - \left[\bar{\chi}(0)\gamma_\beta\right]_j\Sigma_{ji}\left[\gamma_\alpha\chi(z)\right]_i, \tag{8.85}$$

where[8.16]

$$\Sigma_{ij} = \{\chi_i(z), \chi_j(0)\} = (i\not\partial + m)_{ij} i\Delta(z) \xrightarrow[z^2\to 0]{} \not\partial \frac{1}{2\pi}\epsilon(z_0)\,\delta(z^2). \quad (8.86)$$

8.16 Here I use one of the Green functions defined in quantum field theory. It is that describing the anticommutator of fermionic free fields – of which I retain only the singular part on the light-cone.

Collecting the various pieces together,

$$\begin{aligned}&\Big[J_\alpha(z), J_\beta(0)\Big]\\ &\to \Big\{\bar\chi(z)\gamma_\alpha\gamma_\lambda\gamma_\beta\chi(0) - \bar\chi(0)\gamma_\beta\gamma_\lambda\gamma_\alpha\chi(z)\Big\} \times \frac{1}{2\pi}\partial^\lambda\epsilon(z_0)\delta(z^2),\end{aligned} \quad (8.87)$$

which I can write as

$$\begin{aligned}&\Big[J_\alpha(z), J_\beta(0)\Big]\\ &= (g_{\alpha\lambda}g_{\beta\sigma} + g_{\alpha\sigma}g_{\beta\lambda} - g_{\alpha\beta}g_{\lambda\sigma})\Big[\underbrace{\bar\chi(z)\gamma^\sigma\chi(0) - \bar\chi(0)\gamma^\sigma\chi(z)}_{\text{finite matrix element}}\Big]\\ &\quad\times\frac{1}{2\pi}\underbrace{\partial^\lambda\epsilon(z_0)\delta(z^2)}_{\text{singular}}\end{aligned} \quad (8.88)$$

using the identity

$$\gamma_\alpha\gamma_\lambda\gamma_\beta = \Big(g_{\alpha\lambda}g_{\beta\sigma} + g_{\alpha\sigma}g_{\beta\lambda} - g_{\alpha\beta}g_{\lambda\sigma} + i\epsilon_{\alpha\lambda\beta\sigma}\gamma^5\Big)\gamma^\sigma \quad (8.89)$$

and dropping the parity-violating part proportional to γ^5.

The form of the current commutator in Eq. (8.88) is that of an **operator product expansion** in which the short-distance (divergent) part is factorized from the finite matrix element.

By performing a Taylor expansion around $z = 0$,

$$\bar\chi(z)\,\chi(0) = \sum_n \frac{1}{n!} z^{\mu_1}\cdots z^{\mu_n} \times \bar\chi(0)\,\overleftrightarrow{\partial}_{\mu_1}\cdots\overleftrightarrow{\partial}_{\mu_n}\chi(0) \quad (8.90)$$

(where the double-headed arrows indicate partial derivatives acting to the left and to the right). I can define

$$\hat\Psi^n_{\mu_1\cdots\mu_n\sigma} \equiv \bar\chi(0)\,\overleftrightarrow{\partial}_{\mu_1}\cdots\overleftrightarrow{\partial}_{\mu_n}\gamma_\sigma\chi(0) \quad (8.91)$$

and write

$$\begin{aligned}W^{\alpha\beta} &= \frac{1}{2\pi}\int d^4z\exp(ir\cdot z)\sum_n\frac{1}{n!}z^{\mu_1}\cdots z^{\mu_n}(g_{\alpha\lambda}g_{\beta\sigma} + g_{\alpha\sigma}g_{\beta\lambda} - g_{\alpha\beta}g_{\lambda\sigma})\\ &\quad\times\frac{1}{2\pi}\sum_{\text{pol}}\langle P|\hat\Psi^n_{\mu_1\cdots\mu_n\sigma}|P\rangle\partial^\lambda[\epsilon(x_0)\delta(x^2)].\end{aligned} \quad (8.92)$$

The last term in the equation above can be written, by exploiting its Lorentz structure, as

$$\frac{1}{2\pi}\sum_{\text{pol}}\langle P|\hat\Psi^n_{\mu_1\cdots\mu_n\sigma}|P\rangle = \mathbb{Q}^n P_{\mu_1}\cdots P_{\mu_n}P_\sigma + \text{terms with }\underbrace{M^2 g_{\mu_1\mu_2}\cdots}_{\text{higher twist}}, \quad (8.93)$$

where $\mathbb{Q}^n$ are coefficients to be determined. Neglecting higher-order terms, it follows that

$$W^{\alpha\beta} = \frac{1}{2\pi}\int d^4z \exp(ir\cdot z)\sum_n \frac{1}{n!}[z\cdot P]^n\,(g_{\alpha\lambda}g_{\beta\sigma} + g_{\alpha\sigma}g_{\beta\lambda} - g_{\alpha\beta}g_{\lambda\sigma})P^\sigma\,\mathbb{Q}^n \times \frac{1}{\pi}\partial^\lambda[\epsilon(z_0)\delta(z^2)]\,. \tag{8.94}$$

If I define

$$\sum_n [z\cdot P]^n\frac{\mathbb{Q}^n}{n!} \equiv \int dw\, e^{iwP\cdot z} q(w) \tag{8.95}$$

then this connects the coefficients $\mathbb{Q}^n$ to the probability distribution of the partons:[8.17]

8.17 The integration

$$\int d^4 z e^{iz\cdot(r+wP)}\epsilon(z_0)\delta(x^2) = 4\pi^2 i\delta\Big[(r+wP)^2\Big]\epsilon(r^0+wP^0)$$

is not obvious and is left as an exercise.

$$\begin{aligned} W^{\alpha\beta} &= \frac{1}{\pi}\int d^4z\int dw\, e^{iz\cdot(r+wP)}(g_{\alpha\lambda}g_{\beta\sigma} + g_{\alpha\sigma}g_{\beta\lambda} - g_{\alpha\beta}g_{\lambda\sigma})P^\sigma q(w) \\ &\quad\times \frac{1}{\pi}\partial^\lambda[\epsilon(z_0)\delta(z^2)] \\ &= \int dw(g_{\alpha\lambda}g_{\beta\sigma} + g_{\alpha\sigma}g_{\beta\lambda} - g_{\alpha\beta}g_{\lambda\sigma})P^\sigma q(w) \\ &\quad\times (r+wP)^\lambda\delta\Big[(r+wP)^2\Big]\epsilon(r^0+wP^0)\,, \end{aligned} \tag{8.96}$$

where

$$\delta\Big[(r+wP)^2\Big] = \frac{x}{Q^2}\delta(w-x) \tag{8.97}$$

kills the integration, leaving

$$\begin{aligned} W^{\alpha\beta} &= (g_{\alpha\lambda}g_{\beta\sigma} + g_{\alpha\sigma}g_{\beta\lambda} - g_{\alpha\beta}g_{\lambda\sigma})P^\sigma q(x)\frac{x}{Q^2}(r+xP)^\lambda \\ &\to -2\frac{x}{Q^2}q(x)\Big[-\underbrace{g_{\alpha\beta}(xM^2+P\cdot r)}_{=M\nu\,\propto\, F_1} + \underbrace{(r+xP)_\alpha P_\beta + P_\alpha(r+xP)_\beta}_{\propto\, F_2}\Big]. \end{aligned} \tag{8.98}$$

At this point, from the expression of the hadron tensor in Eq. (8.98) I can identify the coefficients (for a single parton of unit charge) F_1 and F_2 and find that[8.18]

8.18 $\dfrac{x}{Q^2} = \dfrac{1}{2\nu M}$.

$$F_1 = q(x) \quad\text{and}\quad F_2 = xF_1, \tag{8.99}$$

which is the parton model result: the independence of Q^2 and the Callan–Gross relation. As advertised, the QCD Lagrangian on its own leads to the parton model result for the hadron tensor in deep inelastic scattering.

8.5 Deep Inelastic Scattering *

"The secret of boring people lies in telling them everything." A. Chekhov

The structure functions F_1 and F_2 in the parton model do not depend on the exchanged momentum Q^2. What happens to the parton model when

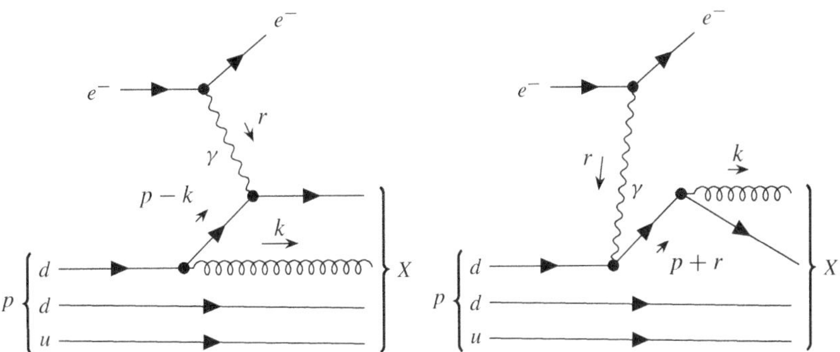

Figure 8.16 Feynman diagrams for deep inelastic scattering with the first QCD correction (gluon emission) included.

I turn on QCD? Some dependence on Q^2 is restored. This dependence is the hallmark of QCD, the fingerprint of its being the theory underlying the strong interactions.[8.19]

8.19 "A fox knows many things, but a hedgehog knows one big thing."

Archilochus

Practitioners of the electroweak part of the Standard Model are foxes, those of QCD are hedgehogs.

The first correction to the deep inelastic scattering process that is introduced by QCD is shown by the Feynman diagrams in Figure 8.16, in which the parton, before being struck by the photon, emits a gluon. A similar diagram has the parton emitting the gluon after being struck by the photon. The hadronic current – which in the parton model is given by Eq. (8.53) – now becomes

$$\varkappa^{\alpha} = \hat{e}_q g_s e \left[\frac{\bar{u}(p')\,\gamma_\alpha(\not{p} - \not{k})\not{\epsilon}^*(k)\,u(p)}{(p-k)^2} + \frac{\bar{u}(p')\,\not{\epsilon}^*(k)(\not{p} + \not{r})\gamma_\alpha\,u(p)}{(p+r)^2} \right] t^A_{ij}, \tag{8.100}$$

which must be inserted into the cross section for inelastic scattering. For simplicity, the overall function can be written as

$$W_{\alpha\beta} = \sum_q \hat{e}_q^2 \int_x^1 \frac{\mathrm{d}w}{w} q(w) \hat{W}_{\alpha\beta}, \tag{8.101}$$

in which

$$\hat{W}_{\alpha\beta} = \frac{1}{4\pi w M} \int \frac{\mathrm{d}^3\vec{p}\,'}{(2\pi)^3 2p_0'} \int \frac{\mathrm{d}^3\vec{k}'}{(2\pi)^3 2k_0'} \times \frac{1}{2} \sum_{\substack{\text{pol. hel.}\\ \text{color}}} \varkappa^{\alpha}\varkappa^{\beta*}\,(2\pi)^4\delta^{(4)}(p' - p + r - k) \tag{8.102}$$

depends only on the parton-level quantities.[3]

[3] The next couple of pages contain a rather involved computation. I am following the lead of *Quantum Chromodynamics* by W. Greiner, S. Schramm and E. Stein (Springer, 2002) where I learned to do it this way.

The sum over polarizations, helicities and colors of the product of the hadronic currents in Eq. (8.100) gives the tensor[8.20]

[8.20] Recall:

$$\frac{1}{3}\sum_{\text{color}} t^A t^A = \frac{4}{3}$$

$$\sum_{\text{hel.}} \varepsilon^{\mu}(k)\varepsilon^{\nu *}(k) = -g^{\mu\nu}$$

and

$$\sum_{\sigma} u(p)\bar{u}(p) = \not{p}\,.$$

$$\begin{aligned}\Xi_{\alpha\beta} &\equiv \frac{1}{2}\sum_{\substack{\text{pol. hel.}\\ \text{color}}} \varkappa^{\alpha}\varkappa^{\beta *}\\ &= \operatorname{Tr}\Big\{\not{p}\left[\gamma_\alpha(\not{p}-\not{k})\gamma_\beta+\gamma_\beta(\not{p}+\not{r})\gamma_\alpha\right]\\ &\quad\times(\not{p}-\not{r}-\not{k})\left[\gamma_\beta(\not{p}-\not{k})\gamma_\alpha+\gamma_\alpha(\not{p}+\not{r})\gamma_\beta\right]\Big\}.\end{aligned} \tag{8.103}$$

Since I am going to use Eq. (8.51) and Eq. (8.52) to extract the structure functions, I only need to compute the longitudinal part and trace of the tensor $\Xi_{\alpha\beta}$, which are, respectively,

$$\begin{aligned}p^{\alpha}p^{\beta}\,\Xi_{\alpha\beta} &= \frac{-2}{(p-k)^4}\operatorname{Tr}\left[\not{p}(\not{p}-\not{k})\not{p}(\not{p}-\not{k}+\not{r})\not{p}(\not{p}-\not{k})\right]\\ &= \frac{-32}{(p-k)^4}(p\cdot k)^2\,p\cdot(r-k) = 4\hat{u}\end{aligned} \tag{8.104}$$

and

$$\begin{aligned}\Xi^{\alpha}_{\alpha} =& \frac{4}{(p-k)^2}\operatorname{Tr}\left[\not{p}(\not{p}-\not{k})(\not{p}-\not{k}+\not{r})(\not{p}-\not{k})\right]\\ &- \frac{16}{(p-k)^2(p+q)^2}(p+r)\cdot(p-k)\operatorname{Tr}\left[\not{p}(\not{p}-\not{k}+\not{r})\right]\\ &+ \frac{4}{[(p-k)]^4}\operatorname{Tr}\left[\not{p}(\not{p}+\not{k})(\not{p}-\not{k}+\not{r})(\not{p}+\not{r})\right]\\ =& \,2(p\cdot r)(p\cdot k)\left(\frac{4}{(p\cdot k)^2}+\frac{4}{[k\cdot(p+r)]^2}\right)\\ &+ \frac{16}{(p\cdot k)[k\cdot(p+r)]}(p+r)\cdot(p-k)\,p\cdot(r-k)\\ =& -8\left(\frac{\hat{s}}{\hat{t}}+\frac{\hat{t}}{\hat{s}}-2\frac{Q^2\hat{u}}{\hat{s}\hat{t}}\right).\end{aligned} \tag{8.105}$$

The Mandelstam variables with a hat refer to the parton-level sub-process and are given by

$$\hat{t} = (p-k)^2 = 2p\cdot k \tag{8.106}$$

$$\hat{s} = (r+p)^2 = 2p\cdot r \tag{8.107}$$

$$\hat{u} = (r-k)^2 = -2p\cdot(r-k)\,, \tag{8.108}$$

and furthermore

$$Q^2 = -r^2 = 2(r+p)\cdot(p-k)\,. \tag{8.109}$$

The phase-space integration requires, as it often does, some work. It is best first to transform the three-dimensional integrations into Lorentz-covariant four-dimensional integrations by using the trick of writing

$$\frac{\mathrm{d}^3\vec{k}}{2k_0} = \frac{1}{2k_0}\underbrace{\delta(k_0 - |\vec{k}|)\mathrm{d}k_0}_{=1}\mathrm{d}^3\vec{k}$$
$$= \delta[2k_0(k_0 - |\vec{k}|)]\mathrm{d}^3\vec{k}\mathrm{d}k_0 = \theta(k_0)\delta^{(4)}(k^2)\mathrm{d}^4k, \qquad (8.110)$$

so that the phase-space integration can be written as

$$I = \int \frac{\mathrm{d}^4p'}{(2\pi)^3}\int \frac{\mathrm{d}^4k}{(2\pi)^3}(2\pi)^4\delta^{(4)}(p' - p + r - k)\theta(p'_0)\delta(p'^2)\theta(k_0)\delta(k^2). \qquad (8.111)$$

The integration over p' is eliminated by the presence of the Dirac δ-function and so

$$I = \frac{1}{(2\pi)^2}\int \mathrm{d}^4k\theta(k_0)\delta(k^2)\theta(p_0 + r_0 - k_0)\delta[(p + r - k)^2]$$
$$= \frac{1}{(2\pi)^2}\int_0^\infty \frac{|\vec{k}|^2\mathrm{d}|\vec{k}|}{2|\vec{k}|}\int_1^{+1}\mathrm{d}\cos\theta\ \theta(p_0 + r_0 - k_0)\delta[(p + r - k)^2]. \qquad (8.112)$$

The Mandelstam variable $\hat{t}$ can be used instead of the scattering angle θ:

$$\hat{t} = -2p\cdot k = 2|\vec{p}||\vec{k}|(1 + \cos\theta) \quad \text{and} \quad \mathrm{d}\cos\theta = -\frac{\mathrm{d}\hat{t}}{2|\vec{p}||\vec{k}|}. \qquad (8.113)$$

The integral I is computed most easily in the Breit frame of reference where[8.21]

8.21 Remember that r is purely space-like and pointing in the z-direction. $k = k_0 = |\vec{k}|$.

$$k = (|\vec{k}|, 0,\ |\vec{k}|\sin\theta,\ |\vec{k}|\cos\theta)$$
$$p = (|\vec{p}\,|, 0,\ 0,\ -|\vec{p}\,|)$$
$$r = (0, 0,\ 0,\ \sqrt{-Q^2}), \qquad (8.114)$$

and the momenta of the initial and final partons are back to back, as in Figure 8.17.

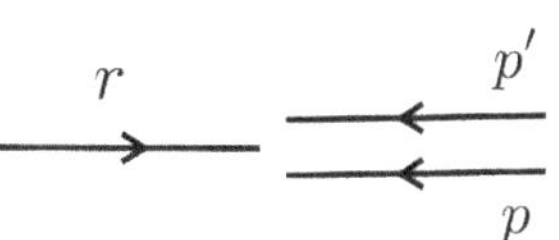

Figure 8.17 Momentum configuration in the Breit frame.

In the Breit frame I have that the argument of the δ-function in Eq. (8.112) is

$$(p + r - k)^2 = (|\vec{p}| - |\vec{k}|,\ 0,\ -|\vec{k}|\sin\theta,\ -|\vec{p}| + \sqrt{Q^2} - |\vec{k}|\cos\theta)^2$$
$$= \hat{t} + \hat{s} - 2|\vec{k}|\sqrt{Q^2} + 2|\vec{k}|\frac{Q^2}{\sqrt{Q^2}}(1 + \cos\theta)$$
$$= \hat{t}\left(1 - \frac{2Q^2}{\hat{s} + Q^2}\right) + \hat{s} - 2|\vec{k}|\sqrt{Q^2}, \qquad (8.115)$$

and, gathering together the various steps,[8.22]

8.22 $|\vec{p}|\sqrt{Q^2} = \hat{s} + Q^2$.

$$I = \frac{1}{2\pi}\int \mathrm{d}\hat{t}\int_0^\infty \frac{\mathrm{d}|\vec{k}|}{-4\sqrt{Q^2}|\vec{p}\,|}\theta(p_0 + q_0 - k)\,\delta\left[|\vec{k}| - \frac{1}{2\sqrt{Q^2}}\left(\hat{t} - \frac{2Q^2}{\hat{s} + Q^2}\right) + \hat{s}\right]$$
$$= \int \frac{\mathrm{d}\hat{t}}{8\pi(\hat{s} + Q^2)}. \qquad (8.116)$$

To obtain the limits in the integration over the Mandelstam variable $\hat{t}$, I must solve Eq. (8.115) for $\hat{t}$ and assign the two limiting values to $\cos\theta$:

$$\hat{t}\Big|_{t_{min}}^{t_{max}} = \frac{-\hat{s}}{1 - \frac{Q^2}{Q^2+\hat{s}} + \frac{2Q^2}{(1+\cos\theta)(\hat{s}+Q^2)}}\Bigg|_{\cos\theta=-1}^{\cos\theta=+1} \tag{8.117}$$

to obtain that $\hat{t}$ must go from 0 to $-\hat{s} - Q^2$.

I can use Eq. (8.51) and Eq. (8.52) to write

$$\begin{aligned} 2Mg^{\alpha\beta}\hat{W}_{\alpha\beta} &= \left[-3\hat{F}_1 + \frac{1}{x}\hat{F}_2\right] \\ &= \frac{16}{3}\alpha_s \int \frac{d\hat{t}}{8\pi(\hat{s}+Q^2)} \frac{\hat{t}^2+\hat{s}^2-2\hat{u}Q^2}{\hat{s}\hat{t}} \end{aligned} \tag{8.118}$$

and

$$\begin{aligned} \frac{4x^2}{\nu}p^\alpha p^\beta \hat{W}_{\alpha\beta} &= \left[\hat{F}_2 + x\hat{F}_1\right] \\ &= \frac{16}{3}\alpha_s \int \frac{d\hat{t}}{8\pi(\hat{s}+Q^2)} \frac{-\hat{u}}{2w^2}. \end{aligned} \tag{8.119}$$

These two relations can be solved, for instance to find the structure function F_2:[8.23]

8.23 $\hat{u} = -\hat{s} - \hat{t} - Q^2$.

$$\begin{aligned} \frac{F_2(x,Q^2)}{x} = & \sum_q 4\hat{e}_q^2\alpha_s \int_x^1 \frac{dw}{w} q(w) \\ & \times \int_{-Q^2-\hat{s}}^0 \frac{d\hat{t}}{8\pi(\hat{s}+Q^2)} \frac{4}{3} \underbrace{\left[\frac{\hat{s}}{-\hat{t}} + \frac{-\hat{t}}{\hat{s}} - 2\frac{Q^2\hat{u}}{\hat{s}\hat{t}} + O\left(\frac{1}{Q^2}\right)\right]}_{\simeq \frac{\hat{s}^2+2Q^2(\hat{s}+Q^2)}{(-\hat{t})\hat{s}} + O\left(\frac{1}{Q^2}\right)}. \end{aligned} \tag{8.120}$$

By introducing the variable

$$z = \frac{x}{w} = \frac{Q^2}{\hat{s}+Q^2} \tag{8.121}$$

so that $\hat{s} = Q^2(1-z)/z$, I can rewrite Eq. (8.120) in a compact form as

$$\begin{aligned} & \frac{F_2(x,Q^2)}{x} \\ & = \sum_q \frac{\hat{e}_q^2\alpha_s}{2\pi} \int_x^1 \frac{dz}{z} q\left(\frac{x}{z}\right) \int_0^{Q^2/z} d(-\hat{t})\frac{4}{3}\left[\frac{1+z^2}{1-z}\left(\frac{1}{-\hat{t}}\right) + O\left(\frac{1}{Q^2}\right)\right]. \end{aligned} \tag{8.122}$$

The expression in Eq. (8.122) contains two singularities:

- when $z \to 1$, which is an infrared divergence. I can regularize it by cutting the integration at a value $z_{\max}$;

- when $\hat{t} \to 0$, which is a mass singularity. I can regularize it by imposing that $-\mu^2 < \hat{t}$.

After this regularization, the integrals are well defined. I can put Eq. (8.122) together with the leading parton result of Eq. (8.63). The sum of the two terms gives

$$\frac{F_2(x, Q^2)}{x} = \sum_i e_i^2 \int_0^{z_{\max}} \frac{\mathrm{d}w}{w} q_i(x) \Big[\underbrace{\delta\left(1 - \frac{x}{w}\right)}_{\text{naive parton model}} + \underbrace{\frac{\alpha_s}{2\pi} P_{qq}\left(\frac{x}{w}\right) \ln \frac{Q^2}{\mu^2}}_{\substack{O(\alpha_s)\ QCD \\ \text{(long computation)}}} \Big], \tag{8.123}$$

where

$$P_{qq}(z) = \frac{4}{3}\left(\frac{1+z^2}{1-z}\right) \tag{8.124}$$

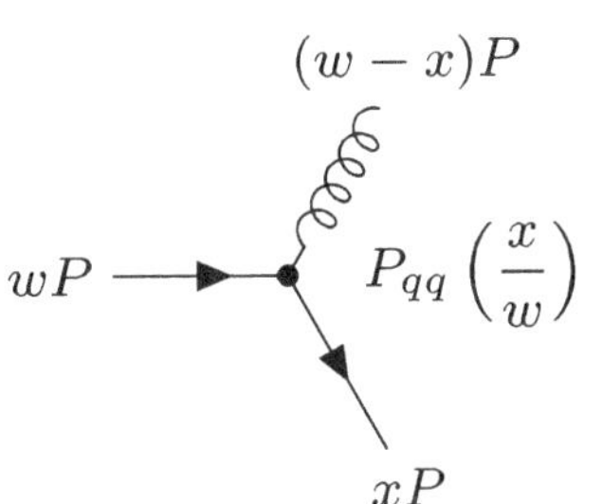

Figure 8.18 Feynman diagram for the splitting function in Eq. (8.124).

is the **splitting function**, which gives the probability for the initial quark to emit a gluon and then carry a fraction z of the original momentum. The corresponding Feynman diagram is depicted in Figure 8.18. The effects of some of the QCD corrections on the structure function $F_2(x)$ are shown in Figure 8.19.

The expression in Eq. (8.123) still contains the (regularized) collinear ($z \to 1$) and mass ($\hat{t} \to 0$) divergencies. They have to be renormalized, as the charge and mass parameters are in quantum field theory. For this reason, I need to introduce the renormalized distribution

$$q(x, Q^2) = q(x) + \frac{\alpha_s}{2\pi} \ln \frac{Q^2}{\mu^2} \int \frac{\mathrm{d}w}{w} q(w) P_{qq}\left(\frac{x}{w}\right), \tag{8.125}$$

which contains the cut off μ and an appropriate definition of the splitting function to remove the singularity for $z \to 1$.

The structure function is the same as in the parton model but the distribution function depends now on Q^2:

$$F_2(x, Q^2) = \sum_i e_i^2 x\, q_i(x, Q^2), \tag{8.126}$$

which is the effect of the QCD corrections.

The scale μ in Eq. (8.125) is the energy at which I factorize the short and the long-distance physics in this process. In QCD I can always factorize the short-distance physics (described by perturbative QCD) from the long-distance physics (which is non-perturbative and taken from the data). As shown in Figure 8.20, the cross section $\hat{\sigma}$ at the partonic level does not depend on the particular hadron from which the parton has been taken. On the other hand, the distribution functions $q_i(x)$ do depend on the hadron but they are universal, that is, they do not depend on the process in which the hadron is taking part.

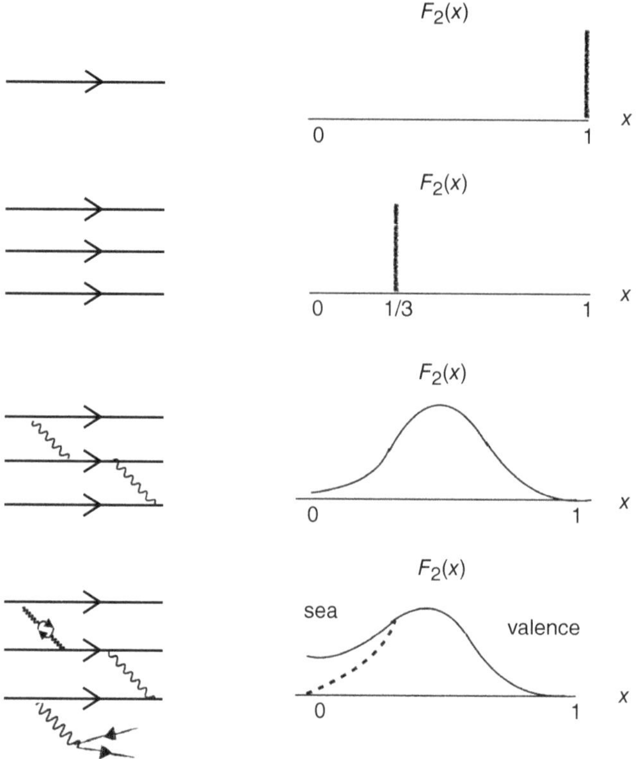

Figure 8.19 The nucleon structure function growing in shape from a δ-function (as in the parton model) to the curve giving the actual dependence on x after the inclusion of QCD corrections.

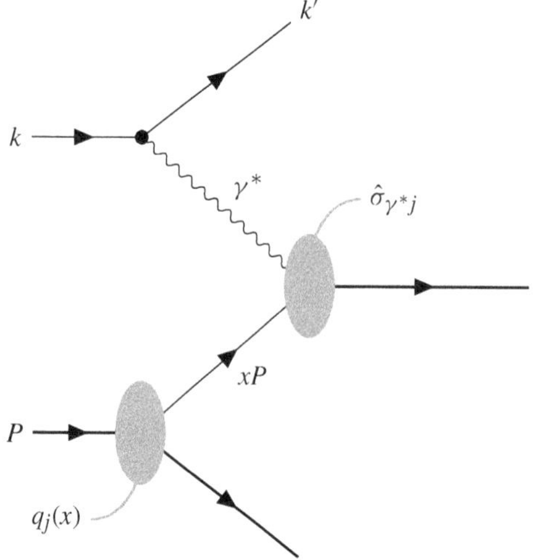

Figure 8.20 Factorization: the parton cross section $\hat{\sigma}_{\gamma^* j}$ is folded into the distribution functions $q_j(x)$ of the hadron taking part in the scattering.

This framework makes QCD a predictive theory because

- after taking the parton distribution functions from one (or more) processes, QCD tells me their contribution in another (different) process;
- given the parton distribution functions at a given scale Q_0^2, QCD tells me how they vary in going to a different scale Q^2.

By taking the logarithmic derivative of Eq. (8.125) I find the equation

$$\boxed{\frac{\mathrm{d}q(x, Q^2)}{\mathrm{d}\ln Q^2} = \frac{\alpha_s(Q^2)}{2\pi}\int_x^1 \frac{\mathrm{d}w}{w} q(w, Q^2)\, P_{qq}\left(\frac{x}{w}\right) + O(\alpha_s^2),} \tag{8.127}$$

which is the Dokshitzer–Gribov–Lipatov–Altarelli–Parisi equation. It describes how the parton distribution functions vary as Q^2 is varied. This variation is the analog of the running of the coupling constant $\alpha_s(Q^2)$. (See Figure 8.21.)

The actual equations describing the evolution of the parton distribution functions are in fact more complicated than Eq. (8.127) because I need to take into account the fact that, inside the nucleon also, the gluon can be the initial or final parton. These functions are controlled by the corresponding splitting functions, whose Feynman diagrams are shown in Figure 8.22.

Therefore, there is a whole set of coupled equations that must to be solved simultaneously to obtain the parton distribution functions. The final results of this fit to the experimental data together with the QCD corrections provide sets of parton distribution functions, like those shown in Figure 8.21, to be used in the physics analysis.

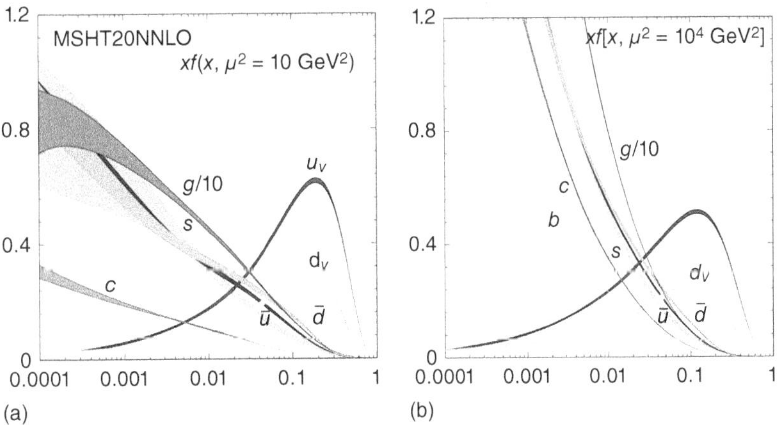

Figure 8.21 (a) Parton distribution functions as determined by the experiments at fixed $Q^2 = 10$ GeV^2 according to the MSHT20 collaboration global analysis. g, gluon distribution; u_v, valence quark distribution. (b) The Q^2 dependence the same distribution functions at $Q^2 = 10^4$ GeV^2 [R.L. Workman *et al.* (Particle Data Group), *Prog. Theor. Exp. Phys.* **2022** (2022) 083C01 http://pdg.lbl.gov].

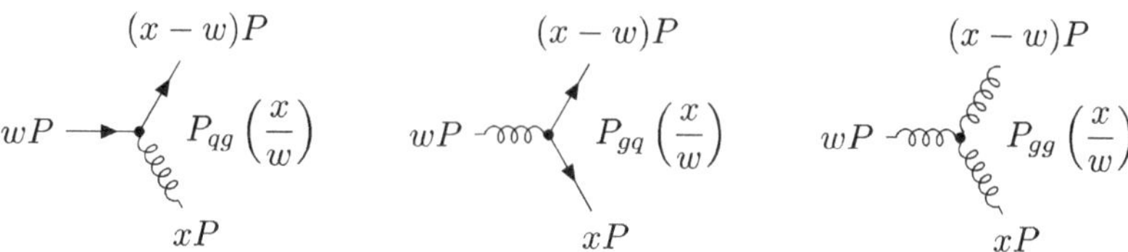

Figure 8.22 Feynman diagrams for the splitting functions P_{qg}, P_{gq} and P_{gg}.

8.6 Collider Physics

Collider physics has a language of its own in terms of kinematic variables. The momenta of the incoming protons (proton and antiproton) have a very large component in the direction of the beam, which can be identified with the z direction. The components of the momentum $p^\mu = (E, p'_x, p'_y, p'_y)$ are best written (as in Problem Session 7.2) as

$$p^\mu = (p^+, p_-, p_x, p_y), \tag{8.128}$$

where

$$p^\pm = \frac{E \pm p'_z}{\sqrt{2}} \quad \text{and} \quad \begin{cases} p_x = p_T \sin\phi \\ p_y = p_T \cos\phi \end{cases} \tag{8.129}$$

for $\vec{p}_T^{\,2} = p_x'^2 + p_y'^2$. The angle ϕ is called **azimuthal**. The on-shell condition is now written as

$$p^2 = 2p^+p^- - \vec{p}_T^{\,2}. \tag{8.130}$$

The **rapidity** is defined as

$$y = \frac{1}{2}\log\frac{p^+}{p_-} = \frac{1}{2}\log\frac{E+p'_z}{E-p'_z}. \tag{8.131}$$

The rapidity is related to the scattering angle. For a massless particle,

$$\begin{aligned} y &= \frac{1}{2}\log\frac{E+p'_z}{E-p'_z} = \frac{1}{2}\log\frac{|\vec{p}\,|+p'_z}{|\vec{p}\,|-p'_z} \\ &= \frac{1}{2}\log\frac{1+\cos\theta}{1+\cos\theta} = \frac{1}{2}\log\frac{1}{\tan^2\theta/2} \\ &= -\log\tan\frac{\theta}{2} = \eta\,, \end{aligned} \tag{8.132}$$

where η is called the **pseudorapidity**. For a massive particle,

$$\sinh\eta = \sqrt{1+\frac{m^2}{|\vec{p}|^2}}\,\sinh y\,. \tag{8.133}$$

After switching to the new variables, the phase-space volume element becomes

$$\frac{\mathrm{d}^3\vec{p}}{E} = p_T\,\mathrm{d}p_T\,\mathrm{d}\phi\,\mathrm{d}y = E_T\,\mathrm{d}E_T\,\mathrm{d}\phi\,\mathrm{d}y\,, \tag{8.134}$$

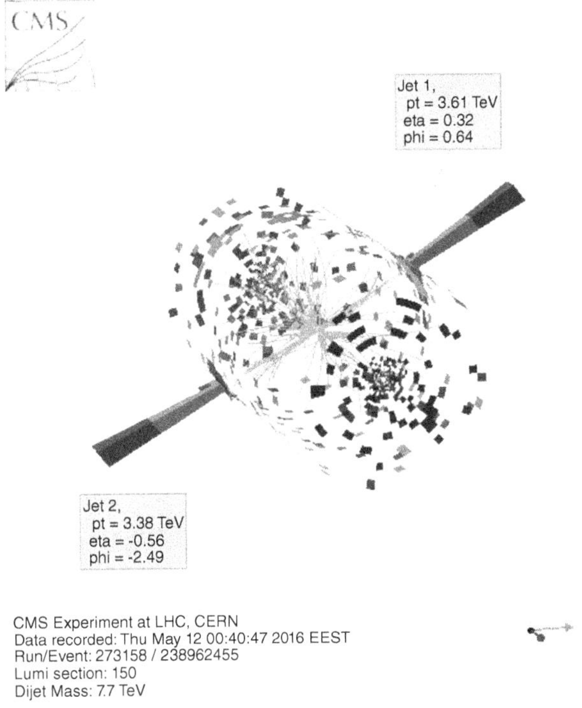

Figure 8.23 Lego plots for a two-jet event. The small shapes represent cells in the detector hit by the particles. The azimuthal variable ϕ is shown wrapped around to follow the cylindrical shape of the detector. The pseudorapidity η follows the axis of the cylinder. [The CMS Collaboration, *Journal of High Energy Physics* **02** (2019) 179.]

in which the scattering angle has been replaced by the rapidity. An event will be plotted in terms of the two variables η and ϕ in a so-called **lego plot** like that in Figure 8.23.

The distance between two events, for instance two jets, is measured by

$$\Delta R = \sqrt{\Delta\eta^2 + \Delta\phi^2}\,, \tag{8.135}$$

and the jet profile will also be described in terms of E_T, the rapidity and the azimuthal angles for the sub-components.

Interactions at colliders are the next level in the description of strong interactions. Whereas there are no hadrons in the initial states in electron-pair annihilation, and only one in deep inelastic scattering, there are two hadrons in the initial states of the interactions at colliders. At the leading order, there is an interaction between a parton in one nucleon and a parton in the other nucleon.

The Drell–Yan process (Figure 8.24) is the simplest process, in which the two partons annihilate into a photon which successively decays into a

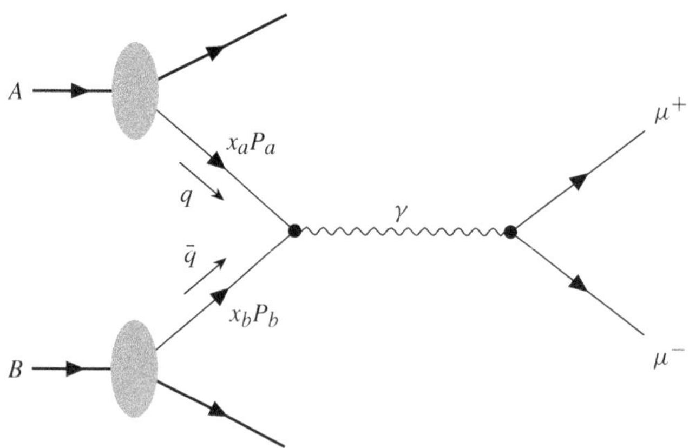

Figure 8.24 Drell–Yan process for the production of a muon pair.

pair of leptons (two muons, for instance). The partons have a momentum described by their distribution functions. They can be either a quark or an antiquark, to this order. Assuming factorization, the differential cross section is given by the convolution of the parton-level cross section, indicated by a hat, and the distribution functions of the two partons:

$$\frac{d\sigma_{DY}}{d\Omega\, d\hat{s}} = \iint \left[q_{a/A}(x_a)\bar{q}_{b/B}(x_b) + q_{b/A}(x_b)\bar{q}_{a/B}(x_a) \right] \\ \times \frac{d\hat{\sigma}}{d\Omega}(q + \bar{q} \to \gamma^* \to \mu^+\mu^-)\, \delta(m_{q\bar{q}}^2 - x_a x_b s)\, dx_a dx_b\,, \tag{8.136}$$

where $\hat{s} = m_{q\bar{q}}^2$ and $d\Omega = d\phi \sin\theta\, d\theta$ is the solid angle defined by the scattering angle between the initial quark and the final muon. The invariant mass is given by

$$m_{q\bar{q}}^2 = (p_a + p_b)^2 \simeq 2\, p_a \cdot p_b = 2\, x_a x_b P_a P_b \simeq x_a x_b (P_a + P_b)^2 = x_a x_b s\,, \tag{8.137}$$

as enforced by the δ-function. I can define the variable

$$\tau^2 = x_a x_b = \frac{m_{q\bar{q}}^2}{s}\,, \tag{8.138}$$

to rewrite the the differential cross section as[8.24]

8.24 Use $d\hat{s} = 2\, m_{q\bar{q}}\, dm_{q\bar{q}}$.

$$\frac{d\sigma_{DY}}{d\Omega\, dm_{q\bar{q}}} = \frac{4\pi\alpha^2}{9\hat{s}} \sum_q e_q^2\, L^{q\bar{q}}(\tau)\,, \tag{8.139}$$

using the parton-level cross section

$$\frac{d\hat{\sigma}}{d\Omega}(q + \bar{q} \to \gamma^* \to \mu^+\mu^-) = \left(\frac{1}{3}\right)\left(\frac{4}{3}\right)\frac{\pi\alpha^2}{\hat{s}}\,. \tag{8.140}$$

Here the symmetry and color factors are in evidence and also the **parton luminosity function** for the quark-like partons q and $\bar{q}$,

$$L^{q\bar{q}} = \frac{4\tau}{\sqrt{s}} \int_{\tau}^{1/\tau} \frac{\mathrm{d}z}{z} q_q(\tau z) q_{\bar{q}} \left(\frac{\tau}{z}\right), \tag{8.141}$$

which is obtained by doing one of the two integrals in Eq. (8.136) after using the δ-function. The luminosity function gives the amount with which each parton partakes in the cross section.

I expect corrections from QCD to this leading-order result. They are usually quantified as a κ-factor rescaling the cross section. At $O(\alpha_s)$,

$$\kappa = 1 + \frac{2\alpha_s}{3\pi}\left(\frac{4\pi^2}{3} - \frac{7}{2}\right) \simeq 2, \tag{8.142}$$

a rather large (100%!) correction.

The differential cross section for the production of two jets is computed in a similar manner. For example,

$$\frac{\mathrm{d}\sigma_{jj}}{\mathrm{d}\Omega\,\mathrm{d}\hat{s}} = \iint \left[q_{a/A}(x_a)\bar{q}_{b/B}(x_b) + q_{b/A}(x_b)\bar{q}_{a/B}(x_a)\right] \times \frac{\mathrm{d}\hat{\sigma}}{\mathrm{d}\Omega}(u + u \to u + u)\,\delta(m_{q\bar{q}}^2 - x_a x_b s)\,\mathrm{d}x_a \mathrm{d}x_b\,, \tag{8.143}$$

for two jets (with additional final states X carrying less p_T) produced by taking as initial partons inside the proton two u quarks. The parton-level cross section is computed by including the three diagrams in Figure 8.25.

These diagrams replace the s-channel diagram with the photon propagator in the Drell–Yan case. The parton-level cross section is given by

$$\frac{\mathrm{d}\hat{\sigma}}{\mathrm{d}\Omega}(u + u \to u + u) = \frac{4\pi\alpha^2}{9\hat{s}^2}\left[\frac{\hat{s}^2 + \hat{u}^2}{\hat{t}^2} + \frac{\hat{t}^2 + \hat{u}^2}{\hat{s}^2} - \frac{2}{3}\frac{\hat{u}}{\hat{s}\hat{t}}\right], \tag{8.144}$$

with $\hat{s}$, $\hat{t}$ and $\hat{u}$ the Mandelstam variables for the parton sub-process in Figure 8.25. The two jets can also be produced by the quarks going into gluons or the gluons being the initial partons. These cross sections must be added to obtain that for the proton-level process.

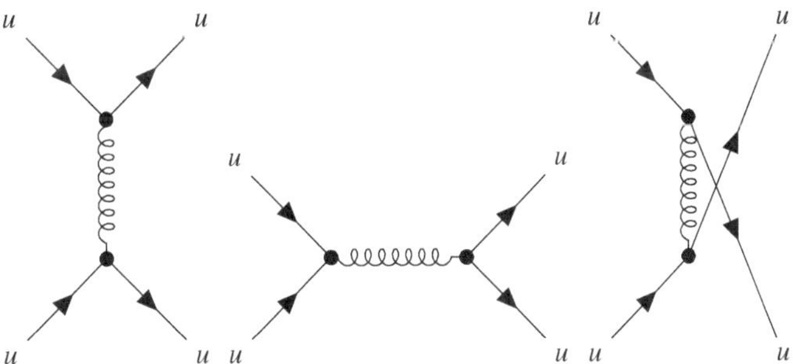

Figure 8.25 The three parton sub-processes contributing to $p + p \to j + j + X$.

8.7 *Problem Session*: Higgs Production

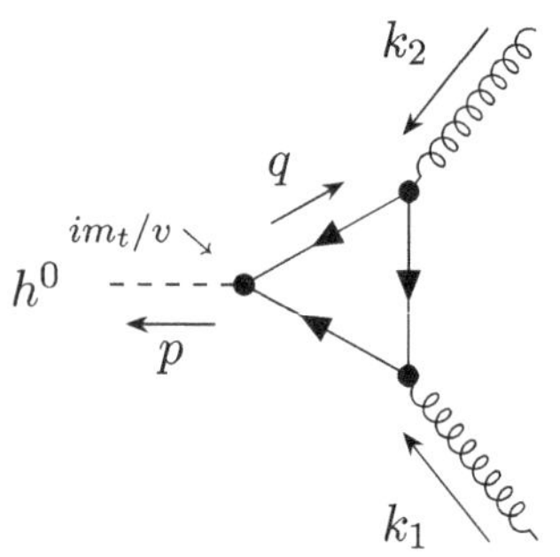

Figure 8.26 The one-loop Feynman diagram for the coupling of the Higgs boson to two gluons.

There is no direct interaction between the Higgs boson, which is colorless, and the gluons. An interaction arises from the one-loop diagram in Figure 8.26.

The magnitude of the interaction is

$$g_{ggh^0} h^0 G^{\mu\nu} G_{\mu\nu} , \tag{8.145}$$

whose effective coupling can be computed.

To find g_{ggh^0} I need to evaluate the amplitude

$$\int \frac{\mathrm{d}^4 q}{(2\pi)^4} \frac{im_t}{v} \frac{(-ig_s)^2(-i)^3 \mathrm{Tr}\,(T^A T^B)\, T^{\mu\nu}}{[q^2 - m_t^2][(q+k_1)^2 - m_t^2][(q+k_1+k_2)^2 - m_t^2]} , \tag{8.146}$$

where

$$T^{\mu\nu} = \mathrm{Tr}\left[(\not{q} + m_t)\gamma^\mu(\not{q} + \not{k}_1 + m_t)\gamma^\nu(\not{q} + \not{k}_1 + \not{k}_2 + m_t)\right] . \tag{8.147}$$

Let me denote by 1 and 2 the two gluons in the operator in Eq. (8.145). Then

$$\begin{aligned} G^{\mu\nu} G_{\mu\nu} &= i(k^1_\mu A^1_\nu - k^1_\nu A^1_\mu)(k^2_\mu A^2_\nu - k^2_\nu A^2_\mu) + O(A^2) \\ &= -2\Big[(k_1 \cdot k_2)(A_1 \cdot A_2) - (k_1 \cdot A_2)(k_1 \cdot A_1)\Big] \\ &= -2(k_1 \cdot k_2)\, A^1_\mu A^2_\nu \left(g^{\mu\nu} - \frac{k_1^\mu k_2^\nu}{(k_1 \cdot k_2)}\right) \\ &= -\sqrt{2} m_h^2\, A^1_\mu A^2_\nu\, P_T^{\mu\nu} , \end{aligned} \tag{8.148}$$

where

$$P_T^{\mu\nu} = \frac{1}{\sqrt{2}} \left(g^{\mu\nu} - \frac{k_1^\mu k_2^\nu}{(k_1 \cdot k_2)}\right) . \tag{8.149}$$

The tensor $P_T^{\mu\nu}$ satisfies

$$P_T^{\mu\nu} P_T^{\mu\nu} = 1 \quad \text{and} \quad P_T^{\mu\nu} k^1_\mu = P_T^{\mu\nu} k^2_\nu = 0 , \tag{8.150}$$

and can be used to simplify the computation. The tensor in Eq. (8.147) is proportional to $P_T^{\mu\nu}$ and

$$P_T^{\mu\nu} T_{\mu\nu} = \zeta P_T^{\mu\nu} P_T^{\mu\nu} = \zeta , \tag{8.151}$$

where ζ is a constant of proportionality. In this way, I can project the tensor in Eq. (8.147) and do a simpler computation to obtain that

$$P_T^{\mu\nu} T_{\mu\nu} = \frac{4m_t}{\sqrt{2}}\Big[-m_h^2 + 3m_t^2 - 2(k_1 \cdot q) - \frac{8}{m_h^2}(k_1 \cdot q)(k_2 \cdot q) + q^2\Big], \tag{8.152}$$

which gives

$$\zeta = \int \frac{\mathrm{d}^4 q}{(2\pi)^4} \frac{[-m_h^2 + 3m_t^2 - 2(k_1 \cdot q) - \frac{8}{m_h^2}(k_1 \cdot q)(k_2 \cdot q) + q^2]}{[q^2 - m_t^2][(q+k_1)^2 - m_t^2][(q+k_1+k_2)^2 - m_t^2]} . \tag{8.153}$$

The integral in Eq. (8.153) can be done either by hand or with software such as FEYNCALC. From its value, I obtain

$$\zeta = \frac{m_t}{4\sqrt{2}\pi^2}\left[1 + (1-\tau)f(\tau)\right] , \tag{8.154}$$

where $\tau = 4m_t^2/m_h^2$ and

$$f(\tau) = \left[\arcsin\frac{1}{\sqrt{\tau}}\right]^2 \xrightarrow[\tau\to\infty]{} \frac{1}{\sqrt{\tau}} + \cdots . \tag{8.155}$$

After gathering together the various pieces, the final result is an effective coupling between the gluons and the Higgs boson given by

$$g_{ggh^0} = -i\frac{\alpha_s}{8\pi}\frac{1}{v}\tau\left[1 + (1-\tau)f(\tau)\right] . \tag{8.156}$$

In the limit of an infinitely large top quark mass,

$$g_{ggh^0} \to -i\frac{\alpha_s}{12\pi v} , \tag{8.157}$$

from which I have the effective Lagrangian interaction

$$L_{ggh^0} = G^{\mu\nu}G_{\mu\nu}\left(\frac{\alpha_s}{\pi}\right)\frac{h^0}{12v} + \cdots . \tag{8.158}$$

The effective vertex gives the cross section $\hat{\sigma}(GG \to h^0)$, which can be inserted into the diagram shown in Figure 8.27 for gluon fusion.

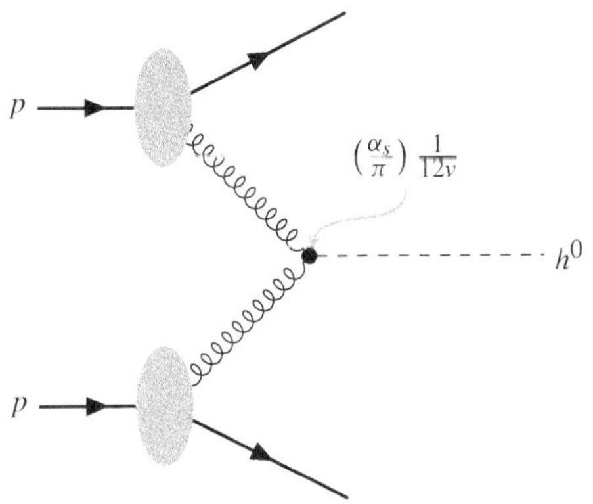

Figure 8.27 Feynman diagram representing gluon fusion into a Higgs boson.

The differential cross section is

$$d^2\sigma = q_{G/p}(x_1)\, q_{G/p}(x_2)\, \hat{\sigma}(GG \to h^0)\, dx_1\, dx_2 \,. \tag{8.159}$$

pdf of gluons inside proton

The cross section in Eq. (8.159) was used, together with the amplitude for the Higgs boson to decay into two photons in the search for and discovery of the Higgs boson.

8.8 QCD Meets Flavor Physics*

Fermi's theory of weak interactions is an example of an effective theory in which the propagator of the intermediate gauge boson is neglected and only the four-fermion operators are kept. This is the best strategy when one is computing processes at energies significantly lower than the mass of the gauge bosons – which is the case for most of flavor physics – for which it is indeed reasonable to use an effective theory of the Standard Model.

To be specific, consider the flavor-changing, non-leptonic $\Delta S = 1$ transition for a neutral kaon to decay into two pions. The effective theory (Figure 8.28) is given by the Lagrangian

$$\mathcal{L}_{\text{eff}} = \frac{G_F}{4\sqrt{2}} V_{us}^* V_{ud} \Big[\bar{s}(1-\gamma_5)\gamma_\mu u \Big] \Big[\bar{u}(1-\gamma_5)\gamma^\mu d \Big] \,. \tag{8.160}$$

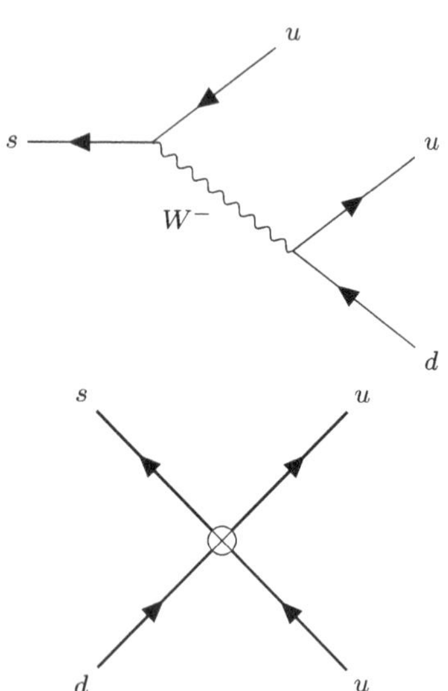

Figure 8.28 The effective vertex in Eq. (8.160) obtained in the effective theory (below) from the true diagram (above).

The Lagrangian in Eq. (8.160) is obtained from the Standard Model Lagrangian for charged currents by neglecting the momentum of the W-boson and making the identification $g^2/8m_W^2 = 1/\sqrt{2}G_F$. The structure in Eq. (8.160) can be generalized to

$$\mathcal{L}_{\text{eff}} = \frac{G_F}{4\sqrt{2}} \sum_i V_{\text{CKM}}^i C_i(\mu) Q_i \,, \tag{8.161}$$

assuming that there is more than just one operator and that other relevant operators Q_i enter the Hamiltonian with weights given by the coefficients C_i.

The effective Lagrangian in Eq. (8.161) is a series of operators Q_i multiplied by effective coupling constants C_i. This series is known under the name of the **operator product expansion**. Owing to the interplay of the electroweak and strong interactions there are many possible local operators. They can be classified with respect to their Dirac structure, color structure and the type of quarks and leptons relevant for a given decay.

The coefficients $C_i(\mu)$ contain the physics contributions from scales higher than μ and they can be calculated in perturbation theory as long as μ is not too small. They include the top quark contributions and contributions from other heavy particles such as the electroweak gauge bosons, and they depend on the masses of these heavy particles. This dependence can be found by evaluating the Feynman diagrams with W-, Z- and t-exchanges. The value of μ can be chosen arbitrarily. Its value serves to separate the physics contributions to a given decay amplitude into short-distance contributions at scales higher than μ and long-distance contributions corresponding to scales lower than μ. It is customary to choose μ to be of the order of the mass of the decaying hadron.

Coming back to Eq. (8.160), what happens to it when QCD is turned on? The first effect of having QCD is that more than one operator has to be included. Color is represented by the insertion of the Gell-Mann matrices t^a between the quark fields; this insertion can be written as

$$t^A_{\alpha\beta} t^A_{\gamma\delta} = \frac{1}{6}\delta_{\alpha\beta}\delta_{\gamma\delta} + \frac{1}{2}\delta_{\alpha\delta}\delta_{\beta\gamma}\,, \tag{8.162}$$

to make explicit that the saturation of the color indices can be done either across or between the quark pairs. The two cases yield two operators,

$$\begin{aligned} Q_1 &= \left[\bar{s}_\alpha(1-\gamma_5)\gamma_\mu u_\delta\right]\left[\bar{u}_\alpha(1-\gamma_5)\gamma^\mu d_\delta\right] \\ Q_2 &= \left[\bar{s}_\alpha(1-\gamma_5)\gamma_\mu u_\alpha\right]\left[\bar{u}_\delta(1-\gamma_5)\gamma^\mu d_\delta\right]. \end{aligned} \tag{8.163}$$

The effective Lagrangian is now given by the sum of the two operators:

$$\mathcal{L}_{\text{eff}} = \frac{G_F}{4\sqrt{2}} V^*_{us} V_{ud}\left[C_1(\mu)Q_1 + C_2(\mu)Q_2\right]. \tag{8.164}$$

In Eq. (8.164) I have written two coefficients C_1 and C_2 to give weights to the two operators. Turning QCD off, and including only the tree-level computation, I have $C_1 = 0$ and $C_2 = 1$ and am back to Fermi's theory. Yet, in general, these two coefficients are different and non-vanishing.

In an overall view, the physics of kaons comes about only after several layers of Standard Model physics have been crossed. Figure 8.29 shows three layers that can be identified by the gauge boson mass m_W, the c quark mass m_c and the kaon mass m_K. Following Fermi's approach to the effective interaction, around the scale m_W all the states of the Standard Model are propagating; below at the scale m_c some of these states are frozen because the energy is not sufficient to let them be on-shell and propagate. The effective Lagrangian contains only the propagating degrees of freedom. The same happens at the lower scale m_K where more of the Standard Model states stop propagating.

Below each of the scales, the effective theory contains only the lighter degrees of freedom. The effect of the heavier states is encoded by the matching of the two effective theories above and below the chosen scale. This is done by comparing the amplitudes above and below and imposing that they are the same at the matching scale. This matching gives us

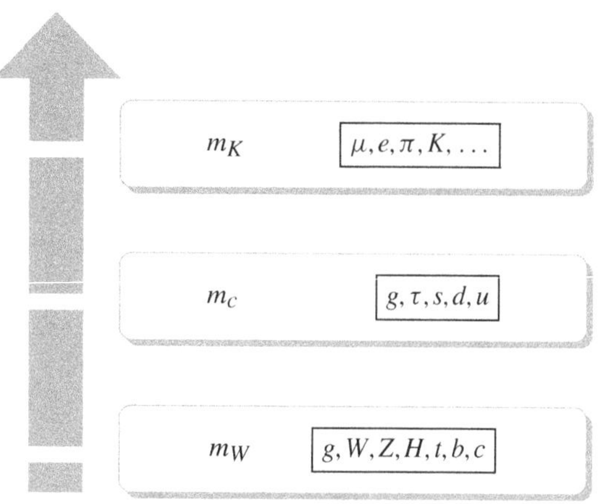

Figure 8.29 The ladder of effective theories, showing the degrees of freedom active at the various scales as the energy is decreased (going up with the big arrow).

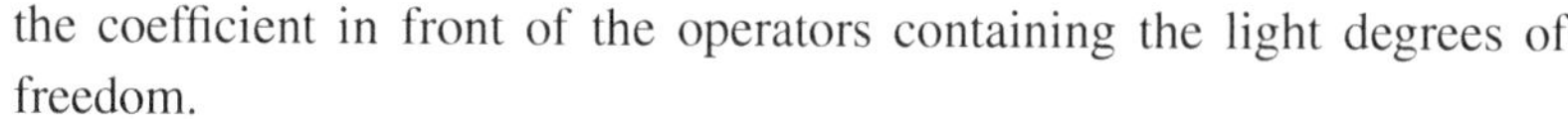

the coefficient in front of the operators containing the light degrees of freedom.

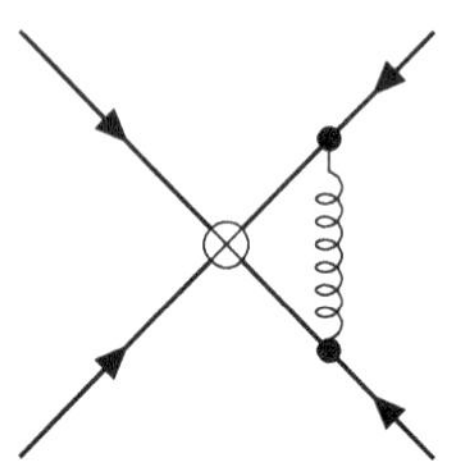

Figure 8.30 Simple QCD loop correction to the effective vertex in Figure 8.28.

In the case of Eq. (8.164) the two coefficients C_1 and C_2 are obtained. The coefficients are functions of the scale μ because the QCD corrections – which come in loop diagrams like the one in Figure 8.30 – are divergent and require a renormalization that introduces the scale μ.

The computation of the effect of these loop diagrams is best done by writing the two operators as

$$Q_{\pm} = \frac{1}{2}(Q_2 \pm Q_1), \tag{8.165}$$

because they factorize the renormalization group equations

$$\frac{\mathrm{d}C_{\pm}}{\mathrm{d}\log\mu} = \gamma_{\pm}^{(0)}(\alpha_s)C_{\pm}\,, \tag{8.166}$$

where $\gamma_{\pm}^{(0)}$ are the anomalous dimensions of the operators $Q_{\pm}$ at leading order. They are given by

$$\gamma_{\pm}^{(0)}(\alpha_s) = \frac{\alpha_s}{4\pi}\left[\pm 6\left(\frac{3\mp 1}{3}\right)\right] = \begin{cases} +4 \\ -8\,. \end{cases} \tag{8.167}$$

The solution to Eq. (8.166) is[8.25]

[8.25] $\beta_0 = 11 - \frac{2}{3}N_f$; here take $N_f = 4$.

$$C_{\pm}(\mu) = \left[\frac{\alpha_s(m_W)}{\alpha_s(\mu)}\right]^{\gamma_{\pm}^{(0)}/2\beta_0} C_{\pm}(m_w)\,, \tag{8.168}$$

where I have started the running from the energy scale given by the W-boson mass m_W, where $C_{\pm}(m_w) = 1$. The parameter β_0 is the same as in the β-function of the running of the coupling constant.

Writing the running coupling α_s at the leading order as

$$\alpha_s(\mu) = \alpha_s^0 \left(1 + \beta_0 \frac{\alpha_s^0}{4\pi} \ln \frac{m_W^2}{\mu^2}\right), \tag{8.169}$$

the coefficients are given by

$$C_\pm(\mu) = \left(1 + \beta_0 \frac{\alpha_s}{4\pi} \ln \frac{m_W^2}{\mu^2}\right)^{-\gamma_\pm^{(0)}/2\beta_0} . \tag{8.170}$$

Since $\gamma_+^{(0)} > 0$, C_+ decreases as we "run down" from the scale m_W. The opposite is true for C_-, which has $\gamma_-^{(0)} < 0$ and increases.

These coefficients quantify the impact of QCD on flavor physics in the case of the $\Delta S = 1$ transitions in the kaon system. The operator Q_- mediates transitions in which the isospin is changed by $\Delta I = 1/2$, while the operator Q_+ mediates transitions with $\Delta I = 3/2$. The two transitions start out as equal ($C_\pm(m_W) = 1$) but are modified by the QCD corrections. From Eq. (8.170) I do expect an enhancement of the isospin $\Delta I = 1/2$ transition over that for $\Delta I = 3/2$. Indeed, the experimental data show that the ratio between the amplitudes for these two transitions is[8.26]

8.26 The $\Delta I = 1/2$ rule.

$$\frac{A(\Delta I = 1/2)}{A(\Delta I = 3/2)} \simeq 22, \tag{8.171}$$

thus with a significant enhancement (of almost 500 in the decay widths!). The computation above gives a more modest factor of 3. There must be some additional enhancement coming from other operators in addition to Q_1 and Q_2 or from the estimates of the matrix elements or from both sources.

The same procedure can be extended to any other flavor-changing process that is best described in terms of an effective Lagrangian and running coefficients. In fact, this is the way in which all computations in flavor physics are actually done by people working on them.

Having the effective Lagrangian, I can compute the hadronic amplitude, for instance, for $K \to \pi\pi$:

$$\langle \pi\pi | \mathcal{L}_{\text{eff}} | K \rangle = \frac{G_F}{\sqrt{2}} V_{us}^* V_{ud} \Big[C_1(\mu) \langle \pi\pi | Q_1 | K \rangle + C_2(\mu) \langle \pi\pi | Q_2 | K \rangle \Big], \tag{8.172}$$

where $\langle \pi\pi | Q_i | K \rangle$ are the hadronic matrix elements of the operators Q_i between the kaon and the two-pion final state. The operator product expansion makes it possible to separate the problem of calculating a given amplitude into two distinct parts: the short-distance (perturbative) calculation of the couplings $C_i(\mu)$ and the long-distance (generally non-perturbative) calculation of the matrix elements $\langle Q_i \rangle$. The scale μ separates the physics contributions into short- and long-distance contributions.

The latter cannot be dealt with in perturbative QCD. They require non-perturbative techniques such as lattice QCD.

8.9 Exercises

1. **Structure functions.** Check Eq. (8.51) and Eq. (8.52).
2. **Scalar partons.** Derive the structure functions $F_1(x)$ and $F_2(x)$ in the case when the partons are scalar particles (instead of fermions, like the quarks).
   [Hint: You should find that $F_1(x)$ vanishes in this case.]
3. **Rapidity.** Show that the rapidity y transforms, under a boost ω, like the nonrelativistic velocity under a change of the frame of reference velocity.
   [Hint: Write the boost as $p^+ \to s^\omega p^+, p^- \to s^{-\omega} p^-$ and $\vec{p}_T \to \vec{p}_T$.]
4. **Jets.** Compute the parton-level cross sections for $\mathrm{d}\hat{\sigma}(qq \to gg)$ and $\mathrm{d}\hat{\sigma}(gg \to gg)$.
   [Hint: Draw the relevant Feynman diagrams. Pay attention to the color group factors.]
5. **Hadronic matrix elements.** How would you go about estimating the hadronic matrix elements in Eq. (8.172)?
   [Hint: Try to search the literature. Two possible approaches go under the key words “lattice QCD” and “$1/N$ expansion”.]

9 Finding Things Out

Contents

"Marvelous is the working of our world!" N. Gogol, *Nevsky Prospect*, 1835

Should this have been the first chapter? Where to start from in teaching the Standard Model? The data painfully collected by experiments are the basis of everything – but they would be mute were it not for our models and theories. On the other hand, it seems that our brain is not very good at working all by itself. It needs the gentle prodding of experimental data, of finding out how (real) things are. Without external inputs, our mind is easily led astray, going round in circles (often of narrower and narrower radius).

The need to look into the structure of atoms, nuclei and nucleons has led the high-energy community toward the construction of special machines designed to accelerate these particles and collide them against each other to make their structure available to our examination. The history of high-energy physics is mostly the history of these machines: the **accelerators**. After having been accelerated, particles are smashed against each other. This smashing is done either against a fixed target or between two beams of particles, each of which has been accelerated. In recent years, accelerators have become **colliders**, where particles are made to run around a circular ring several times before colliding.

Colliders have been compared to medieval cathedrals. Like those, they embody a synthesis of the best knowledge of the time. Whereas theology and masonry were brought together in the building of cathedrals, accelerators are built out of engineering and electronics. Accelerators have also been compared to microscopes – for the power they provide to see what is invisibly small. In a microscope we use light to look at the details of small samples. Ordinary light has too long a wavelength to be useful in the case of the minuscule elementary particles. Only other particles can probe elementary particles.

Accelerating particles to smash them is kind of fun but it is only the first step in the study of fundamental interactions. We also need to be able to observe these collisions and record what takes place. The particles produced by the collisions are so small as not to be directly visible – we can only see faint tracks resulting from their passage through matter. And even to be able to see these tracks requires an amplification process that has to be carried out by huge machines: the **detectors**.

The construction of accelerators and detectors, and their running, has spurred a multifold increase in the size of scientific collaborations – with a momentous shift in the sociology of research, which is now carried out by collaborations with hundreds of members.

9.1 Accelerating

A collider is defined by two parameters: energy and **luminosity**.

The energy parameter is taken to be the maximum energy available in the center of mass of the colliding particles, which is equal to the square root of the Mandelstam variable s:

$$E_{CM} = \sqrt{s}\,. \tag{9.1}$$

This energy is imparted by electromagnetic cavities in which strong electric fields accelerate the particles in their trajectory around the collider.

The particles are accelerated in bunches. The luminosity is given by the numbers, n_1 and n_2 respectively, of particles in the bunches of the two beams taking part in the collision, as they come to interact in an area of size a, the collisions taking place with frequency f:

$$\mathcal{L} = \frac{f n_1 n_2}{a}\,. \tag{9.2}$$

The luminosity is measured in squared centimeters per second.[9.1]

[9.1] The energy in the collision between two house flies is enormous (something like 10^{19} GeV) compared to that at colliders but the luminosity is very, very low.

In practice, what is interesting is the rate, the number of events per second, for a given scattering process. This is given by the cross section σ for that process to take place times the luminosity:

$$\underbrace{N}_{\text{rate (\# events/s)}} = \underbrace{\mathcal{L}}_{\text{luminosity}}\overbrace{\sigma}^{\text{cross section}}\,, \tag{9.3}$$

which leads to the total number of events observed over a given period of time:

$$N_{obs} = \sigma \times \underbrace{\int \mathcal{L}\, \mathrm{d}t}_{\substack{\text{total}\\ \text{luminosity}}} \times \epsilon \times A\,. \tag{9.4}$$

(a)

(b)

Figure 9.1 The Large Hadron Collider (LHC) at CERN, Geneva, Switzerland. It is probably the largest and most precisely designed machine ever built. (a) The path of the collider against a map of the region near Geneva where it is located. (b) The interior of the tunnel [© CERN (a) and Valentin Flauraud/Getty images (b)].

The last two factors in Eq. (9.4) take into account our limited ability to include all the produced particles in our sample. The **acceptance** A keeps track of the fraction of particles ending up in the detector out of those that are produced. Together with the **efficiency** ϵ in the reconstruction of the collision, these two quantities give the number of events actually observed.

Cross sections are measured in **barns**.[9.2] The total luminosity is measured in inverse barns. Because the unit of measure, one barn, is very large (it originated in the early days of nuclear physics), a fraction of it is often used, something like 10 fb^{-1}, that is, 10 inverse femtobarns.

[9.2] We are dealing with very small numbers, and the units are given by the prefixes "nano" $= 10^{-9}$, "pico" $= 10^{-12}$ and "femto" $= 10^{-15}$.

To have a feeling for the energy and the luminosity of an accelerator, consider that at the Tevatron (Fermilab), which is a proton–antiproton collider, the energy is $\sqrt{s} = 1.96$ TeV for a luminosity $\mathcal{L} = 2.1 \times 10^{33}$ $\text{cm}^{-2}\,\text{s}^{-1}$, and the proton–proton collider LHC (CERN) has $\sqrt{s} = 13$ TeV for $\mathcal{L} = 2 \times 10^{34}$ $\text{cm}^{-2}\,\text{s}^{-1}$. The total luminosity accumulated in Run 2 (2016–2018) at the LHC was 139 fb^{-1}. The two photographs in Figure 9.1 show an aerial view of the path of the LHC from above and an inside view.

Colliders can accelerate hadrons (protons and antiprotons) or leptons (electrons and positrons) or even a combination of the two. Like most things in life, each type of collider has advantages as well as disadvantages:

- **Lepton colliders** use all the energy in the collisions and have a limited amount of background events. Unfortunately, the synchrotron radiation of the electrons is a limiting factor as it increases as the fourth power of the energy – which means that it becomes harder and harder to accelerate them.[9.3]

[9.3] The power emitted by a particle of charge q, mass m and energy E, which is going around a circle of radius R, in the direction of the particle's velocity is

$$P = \frac{q^2 \beta^4}{6\pi R^2} \frac{E^4}{m^4} .$$

- **Hadron colliders** produce less synchrotron radiation (simply because this radiation also goes as the fourth power of the inverse of the mass of the accelerated particles) and can thus reach higher energies. The energy is shared among the constituents of the accelerated hadrons (the initial state is not known) and there is a very large background of underlying events.

Most of the effort of the high-energy community in recent years has gone into building bigger and bigger colliders.[9.4] This choice has been motivated by the desire to discover new particles with large masses – for which the largest energy available is required. This trend may continue or, depending on the theoretical framework that prevails, it may move to lepton colliders for more precise measurements. Alternatively, new avenues may be explored by colliding, for example, muons instead of electrons or by using novel ways of accelerating the colliding beams. Given the increasing cost of building these machines, the availability of funding will play a crucial factor as well.

[9.4] The actual working of a collider is complicated. It is an interesting feat of engineering and technology. It is a broad subject in itself, well worth reading about.

9.2 Detecting

Detectors, like colliders, are a vast subject but what I want to do here is only to discuss how the particles are seen and their properties recorded – and even that only in broad brushstrokes.

Figure 9.2 summarizes the progress made in the detection of particles, from the first bubble chambers to the latest detectors.

The various particles that can be seen by the detectors can be organized according to whether you can "see" them or almost "see" them or just not "see" them. I put "see" between inverted commas because we do not actually see them but only infer their presence thanks to the amplification of their interactions as they coast through the detector and the different kinds of matter of which it is made. According to this classification, there can be:

- **Stable states**. These are protons and electrons (and their antiparticles) and γ photons;
- **Long-lived states**. These are neutrons, light-mass hyperons like the Λ, kaons, pions and muons;
- **Short-lived states**. These are mesons such as π^0, $\rho^{0,\pm}$, the gauge bosons Z, $W^\pm$, the top quark t and the Higgs boson h plus mesons such as $B^{0,\pm}$ and $D^{0,\pm}$ and the heavy leptons $\tau^\pm$;
- **Invisible states**. These are the various neutrinos ν_f of the Standard Model and many of the conjectured new particles with names like like $\tilde{\chi}^0$, G_{KK}, $\tilde{\gamma}$ or whatever you might come up with.

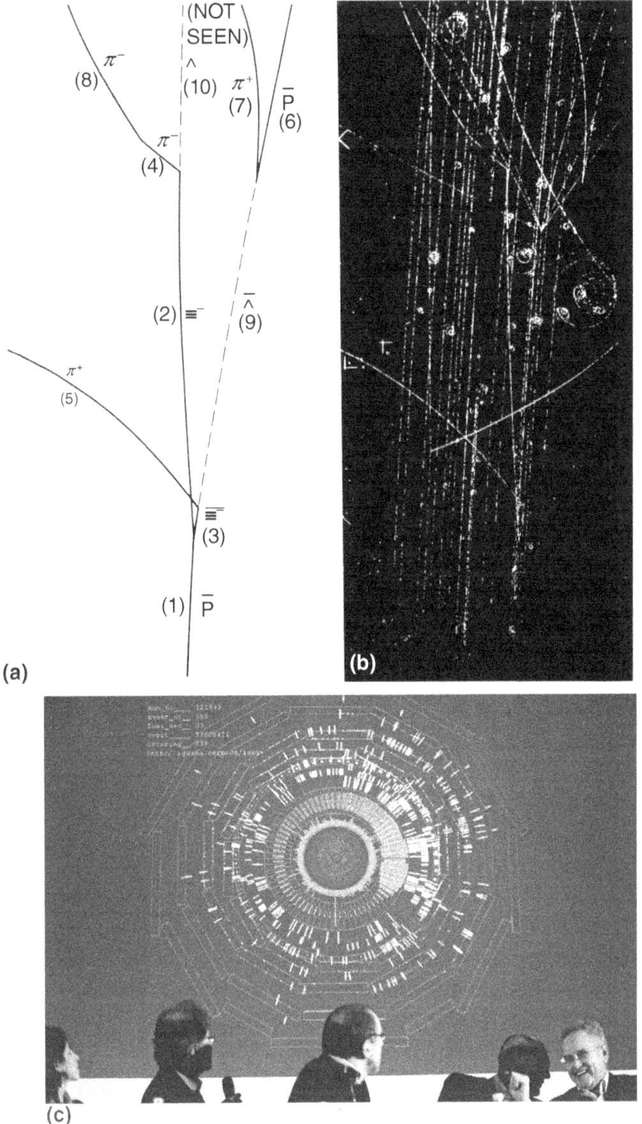

Figure 9.2 At the beginning most detectors were made of a gas in which the charged particles left a trail of bubbles. These were photographed and analyzed (a), (b). Modern detectors use electronics to record the particles. The tracks are visualized by sophisticated video imagining (c) [Smith Collection/Gado, Getty Images and Fabrice Coffini/Getty Images].

Short-lived particles are not visible (they decay inside the beam pipe) but their presence is deduced by their decays. Invisible particles are not seen at all and their presence is inferred by the missing energy.

Depending on their nature, different states will show up in different parts of the detector. Table 9.1 summarizes the various components of a typical detector and what states are designated to be captured.

Table 9.1 Components of a typical detector with a short note about the way in which they operate and the particles they can see.

Detector component	Operating method	Particles observed
Microvertex detector	High-resolution detector near the primary vertex	long-lived particles
Inner tracking chamber	Drift chamber to track charged particles	$\mu^{\pm}$, $e^{\pm}$, p, $\pi^{\pm}$ etc.
EM calorimeter	Energy deposited by electrons and photons	γ, $e^{\pm}$
Hadron calorimeter	Energy deposited by hadrons	n, p, $\pi^{\pm}$ etc.
Muon chamber	Drift chamber to register the trajectories of muons	$\mu^{\pm}$

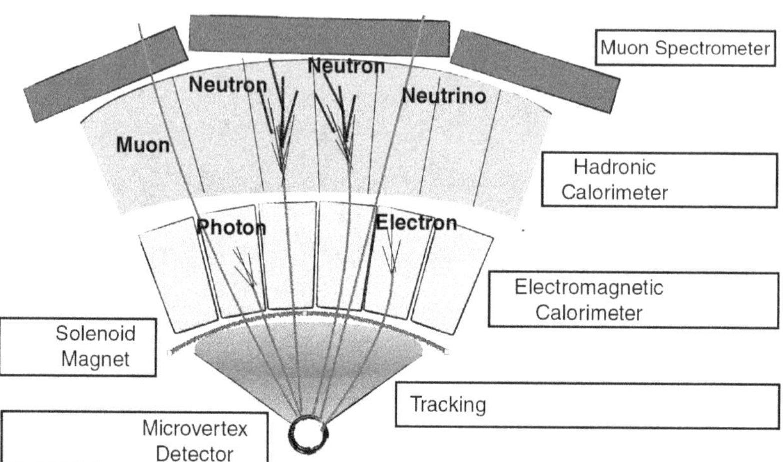

Figure 9.3 A sectional view of a typical detector showing the various layers of which it is composed.

The classification in Table 9.1 is useful whenever I ask myself what kind of signature the event in which I am interested will show. For example, I can try and tag a jet if it originates from a b quark (which is expected to show up in the microvertex detector) but not if it comes from a light quark.

Another way to visualize the same information is shown in Figure 9.3, which gives a more impressionistic view.

A huge structure is required to accommodate the various layers of which the detector is made and the electronics necessary to extract the information thus collected. The smaller the detail you wish to investigate, the bigger the apparatus you will need.

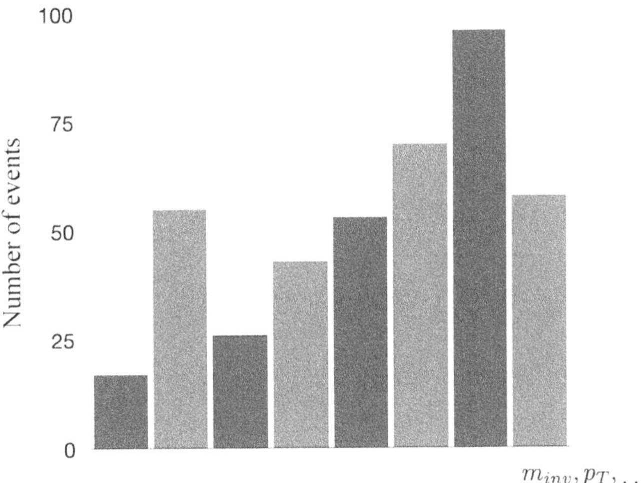

Figure 9.4 The data collected for a typical collision: a histogram of the number of events as a function of some kinematic variable.

The number of events produced by the collisions is staggering.[9.5] For instance, at the LHC, at each crossing of the beams there are approximately 25 collisions among the protons in the bunches. Given the number of crossings per second, it requires the acquisition of events at a frequency of about 25 MHz. Because each event contains about 1.5 Mb of information, there is an enormous production of information, which cannot be recorded in its entirety. For this reason, the recording of events in the detector is triggered only by certain kinds of events (mostly those containing particles ending up in the central section of the detector and carrying a large momentum). This series of triggers brings down the frequency of events to about 100 Hz – which are then duly recorded for future analysis.

[9.5] About 60 Gb of data are produced each second. This corresponds to about 60 movies in mpeg format.

An event in itself just consists of a number of particles, identified by their tracks in the detector, and the kinematic variables associated. These data can be visualized as histograms of the numbers of events versus the kinematic variable chosen to depict the data. Figure 9.4 shows one of these histograms.

Histograms are an effective way to visualize the available information. That is all there is. Out of the complicated process of accelerating and colliding protons, and then recording the particles produced by such a collision, we have a bunch of histograms.[9.6]

[9.6] These histograms of the data are often rendered as fancy pictures, like that in Figure 9.2, but histograms they do remain.

For the most elusive particles, such as neutrinos, and also Dark Matter candidates, another class of detectors has been used. These detectors are as big as those used at colliders (Figure 9.5). They are situated underground so as to screen them from unwanted particles, such as those originating from cosmic rays. The detector captures the particles within its enormous

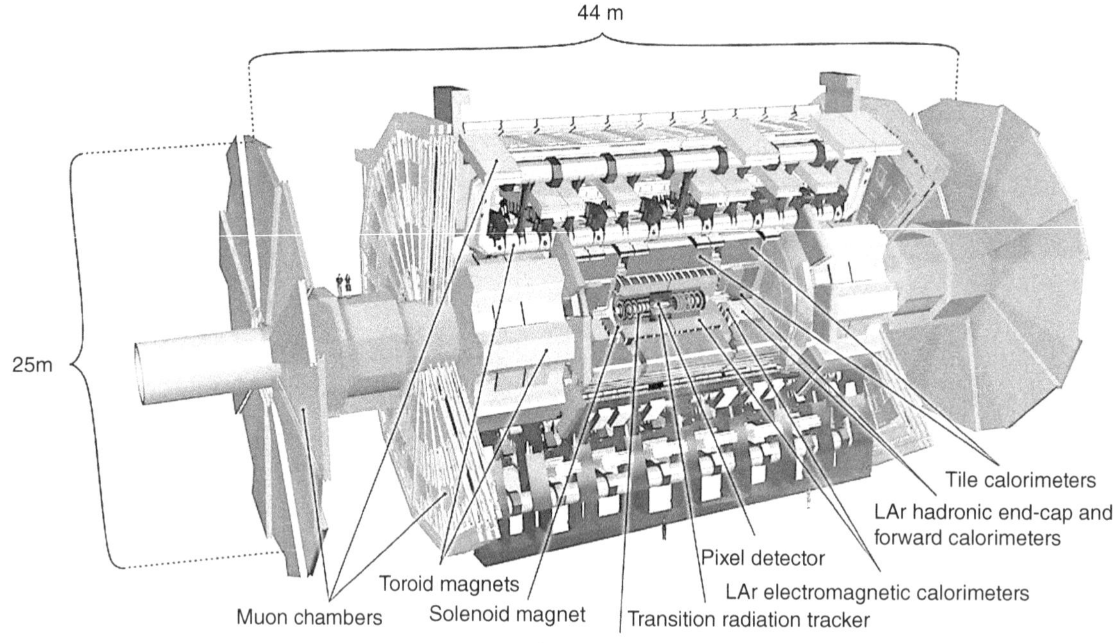

Figure 9.5 The illustrations gives some idea of how huge a detector can be. Look at the size of the two little figurines on the near side of the detector! The detector is as big as a five-storey building. [The ATLAS Collaboration, *Journal of Instrumentation* **3** (2008) S08003.]

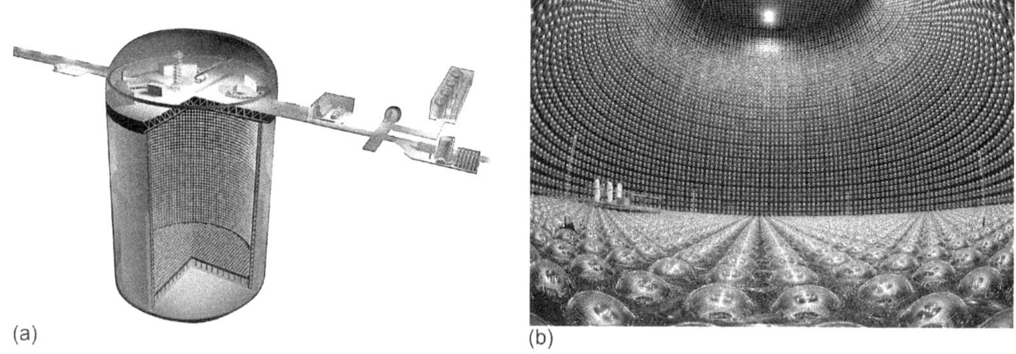

Figure 9.6 The SuperKamiokande detector. The photomultipliers cover the sides of the huge tank filled with water (a), which was empty at the time photograph (b) was taken. The huge size of the apparatus is shown by comparing that of the three white figures standing on its left side. [© Kamioka Observatory, ICRR (Institute for Cosmic Ray Research), The University of Tokyo.]

volume by various mechanisms which amplify their passage. For example, they can track photons emitted (by Cherenkov radiation, see the next section) as electrons, ionized by neutrinos – traveling in the water (or ice) of which the detector is made. A schematic of the SuperKamiokande detector and an inside view are shown in Figure 9.6.

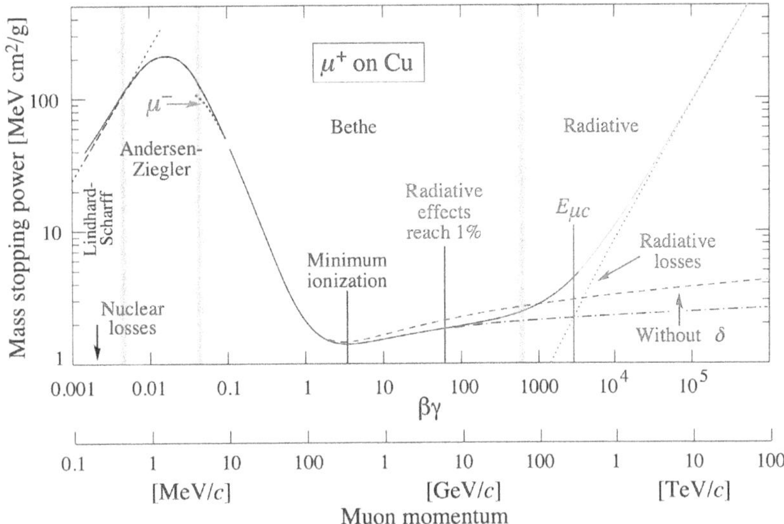

Figure 9.7 Energy loss of muons in copper [R.L. Workman *et al.* (Particle Data Group), *Prog. Theor. Exp. Phys.* **2022** (2022) 083C01 http://pdg.lbl.gov].

9.3 *Problem Session*: Particles Going through Matter

Our senses work on scales that are much bigger than those of these particles. In order to be able to see their behavior we must amplify their interactions. This is done by letting the particles fly through matter. When they do so, charged particles mostly do two things: ionize atoms along the way and emit radiation. Both of these processes are exploited to detect the particles and make their presence accessible to our senses.

- **Ionization** is the process through which a fast moving charged particle knocks electrons out of nearby atoms. In the process it loses energy. The ionization is detected by tracking the loss of energy E per unit distance x as[9.7]

[9.7] Bethe's formula for the energy loss by ionization (neglecting screening effects).

$$\frac{\mathrm{d}E}{\mathrm{d}x} = -4\pi\alpha^2 Q^2 \frac{N_A Z}{A}\frac{1}{\beta^2}\left[\frac{1}{2}\log\frac{2\gamma^2 m\beta^2}{\omega_{\text{atomic}}} - \beta^2\right]. \tag{9.5}$$

The energy loss depends on β, the velocity of the particle, and its mass m. It also depends on the properties of the medium: the atomic number A and charge Z, and the frequency of the atomic oscillations ω_{atomic}.

Figure 9.7 shows the loss of energy by ionization for muons. The energy loss depends only on the velocity of the particle and its charge. If the detector is located in a magnetic field so as to curve the particle

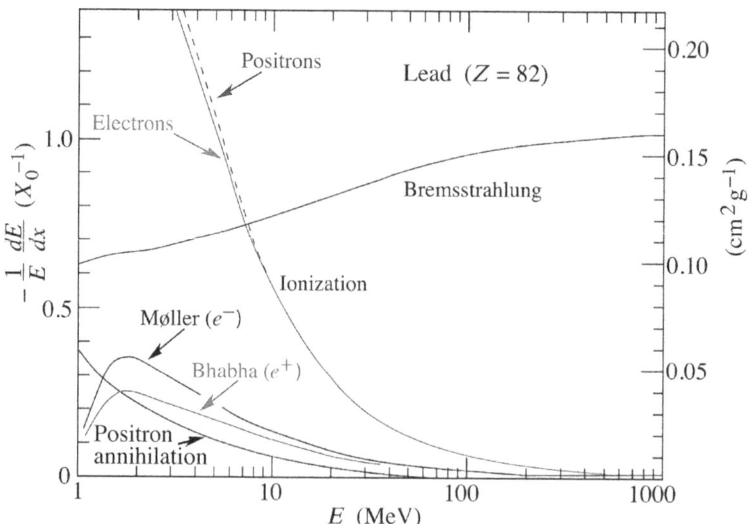

Figure 9.8 Energy losses for an electromagnetic shower [R.L. Workman *et al.* (Particle Data Group), *Prog. Theor. Exp. Phys.* **2022** (2022) 083C01 http://pdg.lbl.gov].

trajectory proportionally to its momentum, the mass of the particle can be established.

- Radiation emission is due to *Bremsstrahlung* – the radiation emitted as the incoming particle brakes. This process starts an **electromagnetic shower** of cascading processes that can be recorded in a calorimeter. The energy loss is proportional to the energy itself and is given by the equation

$$\frac{\mathrm{d}E}{\mathrm{d}x} = -\frac{1}{X_0}E, \tag{9.6}$$

where X_0 is the **radiation length**, which depends on the particle and the material of the detector.

The shower is started when an electron passes by an atom and emits a photon. The photon can successively decay into an electron–positron pair. And so on in a cascade of events.

Figure 9.8 shows the energy losses for electrons and positrons as a function of their energy. Notice how, for energies above about 10 MeV, *Bremsstrahlung* dominates over ionization losses. At relativistic velocities it is the dominant mechanism.

Another process utilized to make a charged particle visible is:

- **Cherenkov radiation**. This phenomenon occurs when a charged particle travels with a velocity that is larger than that of light in that medium. The particle moves ahead of its own radiation waves and, as with sound waves, the emitted radiation piles up producing a shock wave (Figure 9.9). For a medium with refraction index n, the characteristic angle θ_c of the shock wave is given by

$$\cos\theta_c = \frac{1}{\beta n}. \tag{9.7}$$

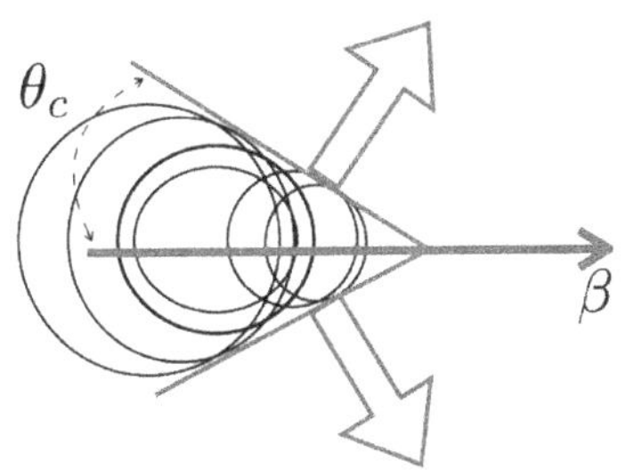

Figure 9.9 The geometry of Cherenkov radiation.

For hadrons, the interaction with the atomic nuclei produces a shower in which, in addition to photons, there are light hadrons. The shower is controlled by QCD and the hadronization process. The energy lost in this case is recorded by another layer in the detector: the **hadronic calorimeter**.

Different processes are at work in the case of neutral particles. These (mostly photons) are detected as the intensity of their flux is attenuated,

$$I(x) = I(0)\,e^{-\mu x}, \tag{9.8}$$

by the interaction with matter in the electromagnetic calorimeter. The attenuation factor μ depends on the cross section of three processes with which photons and electrons (or nuclei) interact:

- the knocking out of electrons from atoms by the **photoelectric effect**;
- **Compton scattering** and **pair production**, which were discussed in Chapter 1.

Figure 9.10 shows which of these three processes is the most important for photons of different energies.

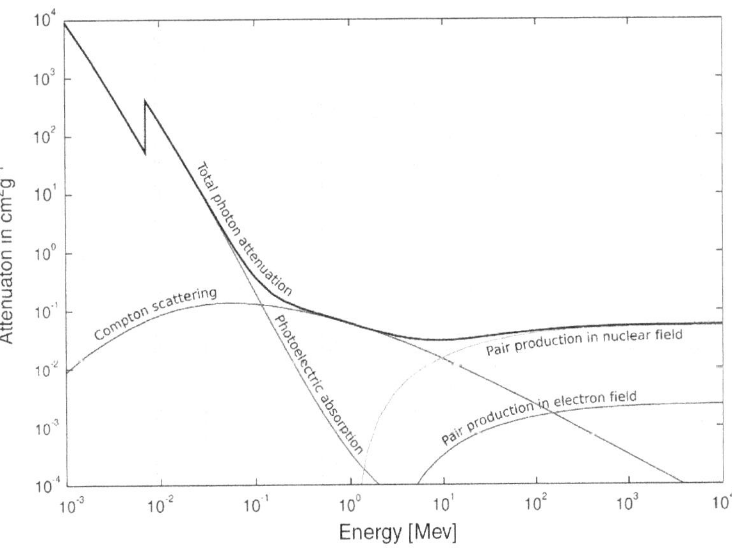

Figure 9.10 Energy loss (attenuation) for photons in matter. The photoelectric absorption line becomes the total photon attenuation line at low energies. Compton scattering dominates at intermediate energies while pair production dominates at higher energies. [Wikipedia Commons (Public domain).]

9.4 The Art of Physical Analysis

The problem facing any analysis of the events at modern colliders such as the LHC is simple to state. The typical cross section at the LHC for events with large transverse momentum is around 100 mb. It is mostly made by jets from the decays of gauge bosons and b quarks. The various processes are produced together and the total cross section is rather large.

Against this, the cross section for "interesting" processes is a small fraction of the total number of events. It can be as small as 10 pb, that is, 10^8 times smaller than the typical cross section. I have to qualify what I mean by "interesting." What is interesting is what I want to study. This can be a particular Standard Model process that occurs in the collision with some probability. It could also be a process that cannot take place within the Standard Model, the presence of which would signal physics beyond the Standard Model.

This means that for each interesting event, the **signal**, there many events in which I am not interested. They are called **background**. Statistical analysis is necessary to disentangle signal from background.[9.8]

9.8 "Today's events turn into tomorrow's background." Unattributed

This work of analysis requires a good deal of ingenuity.

Figure 9.11 shows the data for an event. The data are represented by the points with error bars. The background is fitted by the solid curve. If I believe that a signal (the dashed curve) is present then I must convince myself that the difference between the bold and dashed curves is statistically significant.

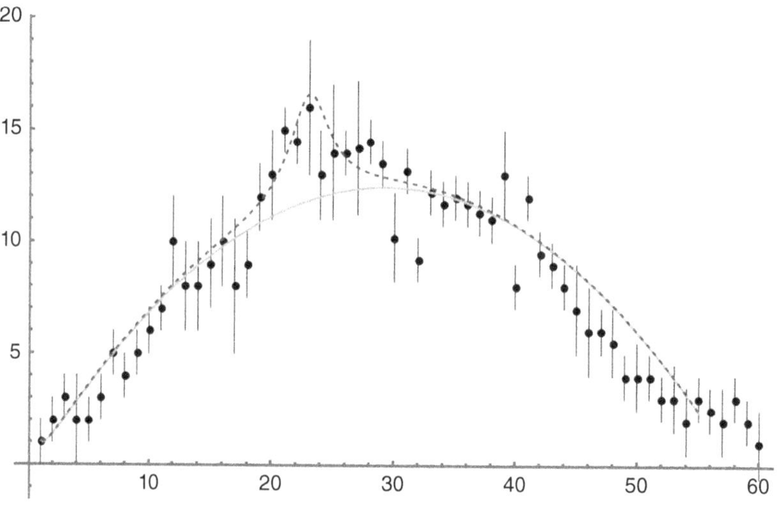

Figure 9.11 The number of data events (points with error bars) as a function of some variable. Also shown are the possible signal (dashed line) and background (solid line).

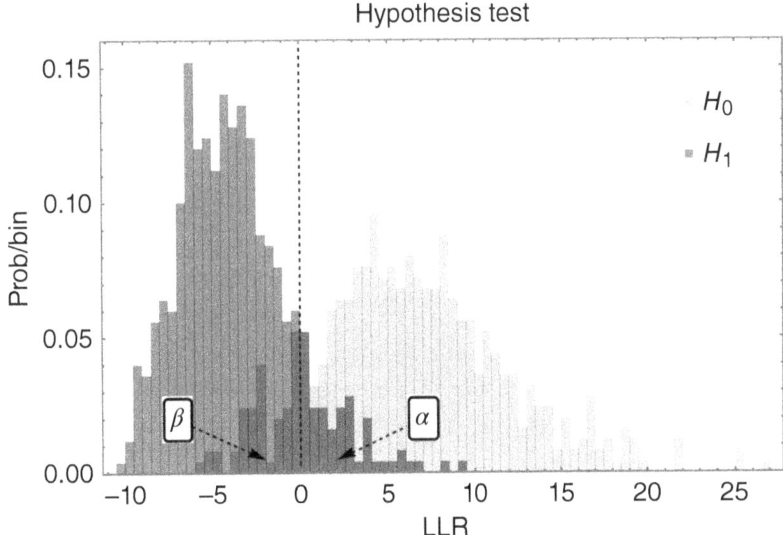

Figure 9.12 Hypothesis test (Neyman–Pearson) by means of the LLR.

This discrimination cannot be done by unaided eye. We need mathematical tools. The usual way to frame the question is by means of a statistical **hypothesis test**. In this kind of test a hypothesis is stated, the null hypothesis H_0, and compared with another hypothesis, H_1. In particle physics the null hypothesis is usually chosen to be that we only have the background. The opposite hypothesis would then be the signal.

The terminology illustrated in Figure 9.12 is that for H_0 as the null hypothesis (the data is all background), the area α is the probability of rejecting H_0 when it is true (Type I error) and the area β is the probability of accepting H_0 when it is false (Type 2 error). I have to steer between these errors to correctly estimate the significance of my signal.

In order to test the hypothesis that I have seen the signal over the background, I need the probability distribution $f(x|s+b)$ of a certain dataset with some characteristics x to be seen in the signal plus background and the probability distribution $f(x|b)$ for the same characteristics to be seen only in the background. The set of characteristics for the dataset can be anything of interest in the analysis: particle momenta, number of leptons or the energy of jets. To be definite, think of the invariant mass of the final states in the decay of a particle. I could compare these two distributions directly but statistics tells me that the most efficient quantity to use is the **logarithmic likelihood ratio**, LLR for short.

There is a total number n of events which is the sum of n_s events from the signal and n_b from the background. They are independent and occur at a fixed rate in time. Their statistics is modeled by a Poisson distribution.

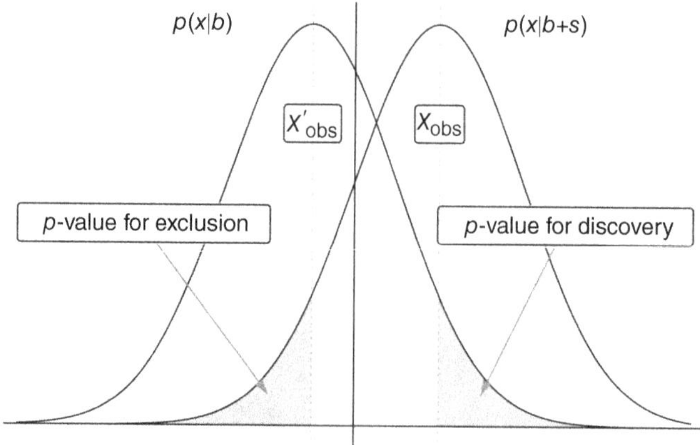

Figure 9.13 The *p*-values (introduced by R. Fisher) are the highlighted areas under the corresponding curves.

Accordingly, the probabilities of finding n such events in the signal and in the signal plus the background are given by

$$P(n|b) = \frac{b^n}{n!}e^{-b} \quad \text{and} \quad P(n|s+b) = \frac{(s+b)^n}{n!}e^{-(s+b)}, \tag{9.9}$$

respectively. The likelihood of the background is then defined to be

$$L_b = \frac{b^n}{n!}e^{-b}\prod_{i=1}^{n} f(x_i|b), \tag{9.10}$$

where x_i are the characteristics I introduced before – the index i stands here for the observation of those variables in that particular event out of the n events. The Poisson distribution is modulated by the joint distribution of the characteristics given by the product of their probability distributions. Similarly, the likelihood of the background plus signal is given by

$$L_{s+b} = \frac{(s+b)^n}{n!}e^{-(s+b)}\prod_{i=1}^{n} f(x_i|s+b). \tag{9.11}$$

The LLR is defined as[9.9]

$$\text{LLR} = -2\ln\frac{L_{s+b}}{L_b}. \tag{9.12}$$

It is the probability per unit bin of this quantity that is plotted in Figure 9.12. A specific event will give a certain value for the LLR that, according to where it falls, will define a value for α, β and the confidence level with which the null hypothesis has been rejected.

Often the result of a statistical analysis is expressed in terms of the ***p*-value**. The meaning of this parameter is shown in Figure 9.13. It represents

9.9 If the probability distributions are Gaussians, the likelihood ratio method becomes equivalent to the method of least squares.

the probability of the observed value (plus more extreme values) if the null hypothesis is true. The result expressed as p-values is easily translated into $\mathcal{Z}$, the **significance** (the number of standard deviations) by

$$\mathcal{Z} = \frac{1}{2} \operatorname{erf}^{-1}(1 - 2p) \tag{9.13}$$

which gives $\mathcal{Z} = 2$ for $p \simeq 0.023$ and $\mathcal{Z} = 5$ for $p = 2.87 \times 10^{-7}$. In most fields outside physics, the value $p = 0.05$ has some magical significance according to which experimental results are accepted or rejected. For the standards of high-energy physics a value of $p = 0.05$ is too large to make a definite decision.

If you know that your signal has some feature that the background lacks, you can try to select that feature in the event to make the signal stand out better. This selection is implemented by specific **cuts** in the range of the kinematic variables so as to bring the desired feature into evidence. For instance, in Figure 9.11 it is clear that retaining only the events between 20 and 30 (of the variable on the x-axis) will magnify the signal over the background. At the same time, the more cuts you decide to enforce, the fewer the events surviving the selection and the less robust the statistics of your analysis.

The art of physical analysis is a balancing act between these two requirements.

9.5 *Problem Session*: A Little Help (MADGRAPH, PYTHIA and DELPHES)

Figure 9.14 shows a typical event at a collider. The complexity is at first sight staggering. There are five processes simultaneously going on:

1. **The event proper**. The parton (quarks, leptons, gluons and electroweak gauge bosons) interaction is described by the Standard Model Lagrangian.
2. **The underlying events**. These are interactions taking place at the same time as the event proper.
3. **The shower**. The subsequent evolution of the partons generates decays and showers (both electromagnetic and strong). The showers originate from both the initial and the final states.
4. **The hadronization of the partons**. Eventually all the colored partons hadronize.
5. **The decays of the final hadrons**. The hadrons produced by the hadronization may also decay further.

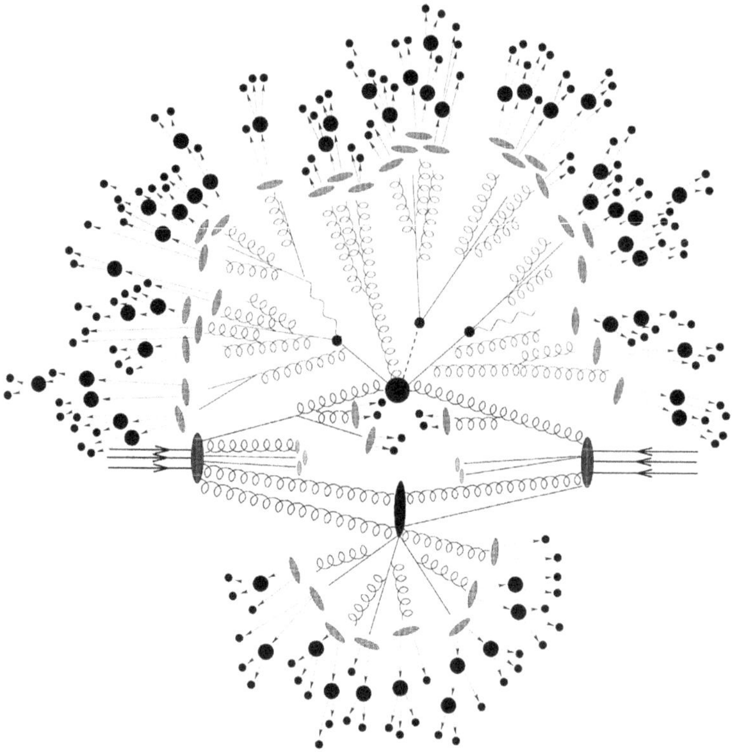

Figure 9.14 The awesome complexity of a typical event at a collider. This picture is not a Feynman diagram but an attempt at depicting the actual process. The initial partons in the incoming protons are the three arrows in the blobs on the left and the right). The event proper and the shower it produces is represented by the black blob in the center. The underlaying event is the disk-like blob below. The hadronization and the successive decays are the many little blobs on the rim of the figure [A. Scälicke *et al.*, *Progress in Particle and Nuclear Physics* **53** (2004) 329].

Because events at typical colliders are a messy superposition of various collisions and decays, dedicated software is employed to simulate the processes. The Lagrangian of the Standard Model gives us the matrix element for the event proper. This is easily computed in the simplest cases but is often a rather involved computation, mainly because of the integration over the final-state phase space, which becomes complicated if there are more than three final states. For this reason, this matrix element is computed by a dedicated software.[9.10] Here, I have used MadGraph.[9.11] After having been summoned up on your computer, as in Figure 9.15, MadGraph does three things. First, the program has the probability distribution functions for the partons inside protons. Second, it computes the cross sections at the partonic level as perturbative series in QCD and QED. Last, it does the phase-space integrals by a Monte Carlo integration.

[9.10] The programs I have chosen are those I normally use but many others are available and you should shop around to find those that best suit your needs and taste.

[9.11] Website: https://madgraph.phy.ucl.ac.be. The current distribution also contains Pythia and Delphes.

Figure 9.15 Screen shot of the initialization page of MadGraph.

The output of MadGraph is fed into another software package, Pythia, where the shower and the hadronization are simulated. Finally, the simulation of the detector is performed by Delphes, another piece of software, which simulates how the event is actually recorded in the detector by taking into account the acceptance of the experimental geometry and the detection efficiency.

To see how this chain of steps works, Let us start with the actual signal. In this example I am taking the production of the Higgs boson and its decay into two W-bosons (Figure 9.16). The W-bosons in turn decay into leptons. The final states actually reconstructed by the detector are the leptons and two jets.

The same final states of my signal can also be produced by many other processes. For instance, I can have a top-quark pair being produced and then decaying into W-bosons, or a Z-boson produced and decaying into two leptons (Figure 9.17). These two processes are the background.

All these processes (signal and background) are simulated through MadGraph and the result fed into Pythia. What is computed by Pythia is modulated by Delphes to approximate the actual tracking of the particles by the detector. The final result of this simulation is an output like the one depicted in Figure 9.18, where a table of particles and their kinematic variables is presented. The simulation is like having your very own collider (and detector!) to play with.

Signal and background are mixed together in the events as I come to plot their features in a series of histograms. In Figure 9.19 the number of events with a certain value of the transverse mass of the Higgs boson (related to the energies of the two leptons and the missing energy) and with properly defined invariant masses of the two leptons are shown for the signal (Higgs) and two possible backgrounds (Z and tt).

The signal is hidden below the background and I need to work the events through various cuts in order to make it stand out. For instance, by looking

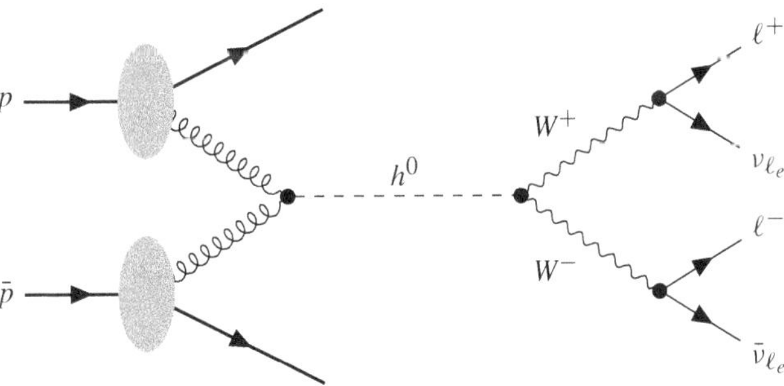

Figure 9.16 Feynman diagram of the signal: $h^0 \to W^+W^- \to \ell^+\nu_\ell\ell^-\bar{\nu}_\ell$.

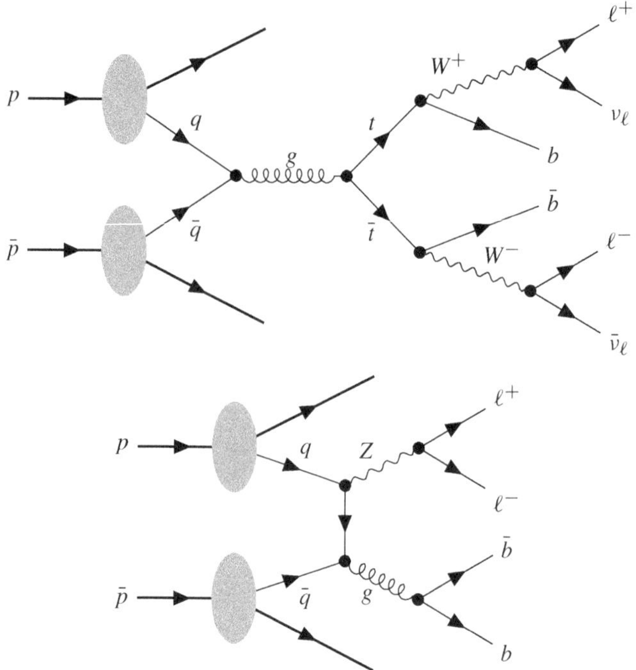

Figure 9.17 Feynman diagrams for two background processes: $t\bar{t} \to \ell^+ \nu_\ell\, b\, \ell^- \bar{\nu}_\ell\, \bar{b}$ and $Z \to \ell^+\, \ell^-$.

Event listing (HEP format) Event: 1

I	particle/jet	ISTHEP	IDHEP	JMOHEP		JDAHEP		PHEP(1,I)	PHEP(2,I)	PHEP(3,I)	PHEP(4,I)	PHEP(5,I)
1	!p+!	3	2212	0	0	0	0	0.00000	0.00000	6999.99994	7000.00000	0.93827
2	!p+!	3	2212	0	0	0	0	0.00000	0.00000	-6999.99994	7000.00000	0.93827
3	!d~!	3	-1	1	0	0	0	-0.88896	2.21221	1408.04177	1408.04379	0.00000
4	!u!	3	2	2	0	0	0	-0.39068	-1.52522	-2347.70817	2347.70870	0.00000
5	!d~!	3	-1	3	0	0	0	-0.72121	2.90451	1346.06386	1346.06718	0.00000
6	!d!	3	1	4	0	0	0	-5.17986	-14.13227	-24.99882	29.18035	0.00000
7	!t~!	3	-6	5	6	0	0	-38.59376	-4.05286	905.38580	922.96046	175.00476
8	!t!	3	6	5	6	0	0	32.69269	-7.17491	415.67924	452.28707	175.08306
9	!W-!	3	-24	7	0	0	0	-57.20538	46.33481	401.47429	416.17140	81.22594
10	!b~!	3	-5	7	0	0	0	18.61162	-50.38767	503.91151	506.78906	4.80000
11	!W+!	3	24	8	0	0	0	47.21011	26.36519	401.21912	412.69126	80.08357
12	!b!	3	5	8	0	0	0	-14.51741	-33.54010	14.46011	39.59581	4.80000
13	!tau-!	3	15	9	0	0	0	-12.88802	36.79518	325.22815	327.56143	1.77700
14	!nu_tau~!	3	-16	9	0	0	0	-43.74165	8.90700	75.19418	87.44616	0.00000
15	!mu+!	3	-13	11	0	0	0	39.97687	-13.99708	97.81917	106.59582	0.10566
16	!nu_mu!	3	14	11	0	0	0	6.59991	39.85240	298.86394	301.58154	0.00000
17	(W-)	2	-24	9	0	19	20	-56.62967	45.70218	400.42233	415.00760	81.22594
18	(W+)	2	24	11	0	21	22	46.57677	25.85532	396.68310	408.17736	80.08357
19	tau-	1	15	13	0	0	0	-12.88802	36.79518	325.22815	327.56143	1.77700
20	nu_tau~	1	-16	14	0	0	0	-43.74165	8.90700	75.19418	87.44616	0.00000
21	mu+	1	-13	15	0	0	0	39.97687	-13.99708	97.81917	106.59582	0.10566
22	nu_mu	1	14	16	0	0	0	6.59991	39.85240	298.86394	301.58154	0.00000
23	p+	1	2212	1	0	0	0	0.34195	-1.55888	1816.84702	1816.84796	0.93827
24	(s)	2	3	4	0	129	129	2.29667	-1.46586	-1.06702	2.96849	0.50000
25	(g)	2	21	4	0	129	129	-0.27958	0.22894	-1.74236	1.77944	0.00000
[...]												

Initial State (rows 1–6); Hard Event (rows 6–16); Stable Particles; Unstable Particles, Partons

Figure 9.18 The output of the machine-generated event [courtesy of M. Pinamonti].

at the histograms in Figure 9.19, it is tempting to cut out all events with $m_{\ell\ell} > 80$ GeV. In this way, half the top-quark pair events are removed while all the Higgs boson events are retained.

So, I do that and re-plot to see whether the signal has gained over the background. And then I cut again, and again, until I am able to discern the

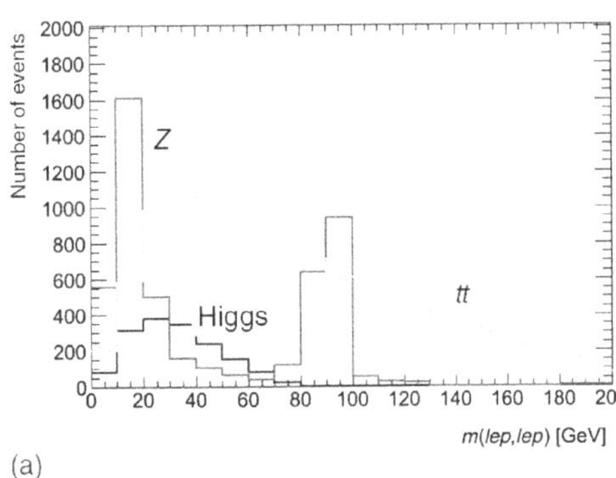

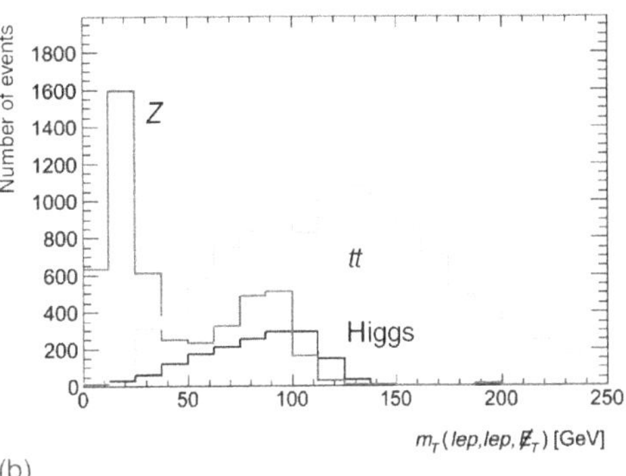

Figure 9.19 Histograms for the events produced by the Monte Carlo simulation. Depicted are the signal (Higgs) and two possible backgrounds (Z and tt) as functions of the lepton invariant mass $m_{\ell\ell}$ (a) and of the lepton transverse mass $m^T_{\ell\ell}$ (b) [courtesy of M. Pinamonti].

signal out of the background, hoping that in the process a sufficiently large number of the signal events survive to lend significance to the measure.

The simulation of the events prepares the way for a physical analysis of the data proper. It points the way toward what is possible and what is not possible to measure, and the sensitivity that can eventually be achieved.

9.6 Errors Lurking

The first thing that is taught to physics students: a measurement has no meaning if not given together with its error. Our measurements always have errors. Not because we are sloppy but because of the way measurements work. It is because of this ingrained idea that physicists – as opposed to, say, economists – look with great suspicion at any number that comes without an error.

There are two kinds of errors:

- **Statistical (random) errors**, which cause one measurement to differ slightly from the next. They come from unpredictable changes during an experiment, thus they too are unpredictable.
- **Systematic errors**, which always affect measurements by the same amount or by the same proportion. They come from our incomplete knowledge of the parameters or of the apparatus. They are in principle predictable.

Errors affect our ability to tell the signal apart from the background in the high-energy experiments at the colliders. The significance of a certain difference between signal and background goes like this:

$$\text{significance} \propto \frac{N_{sig}}{\sqrt{N_{sig} + N_{bkg}}} \simeq \sqrt{\mathcal{L}}\,. \tag{9.14}$$

The second relation holds because the number of events for both signal and background scale as the luminosity. For this reason we only have to wait for the luminosity to grow to improve the significance of our signal.

The square root law does not applies to systematic errors. In this case,

$$\text{significance} \propto \frac{N_{sig}}{N_{bkg}} \simeq \text{const}\,, \tag{9.15}$$

with no improvement for larger luminosities. On the other hand, systematic errors are predictable and are reduced as the apparatus becomes better known.

And then, there is a third kind of error:

- **Theoretical errors**[9.12] are not statistically well defined. They are mainly accounted for by uncertainties in the computations linking the data to the model, for example those coming from higher-order terms in the perturbative expansion of QCD.

9.12 Theoretical errors are different from the sign errors made by theorists in their computations.

All published measurements of observables come with their respective statistical and systematic errors and, often, with the theoretical error as well.

9.7 *Problem Session*: Statistics

There is more than a hypothesis test in getting right the statistics of high-energy experimental data. In this problem session I want to discuss two issues that are sometime neglected.

The comparison between the data and the theory (actually the model, in most cases) is based on the conditional probabilities of events. These probabilities depends on whether one set of data depends on the theory or *vice versa*. The example often presented in statistics textbooks is:

$$\underbrace{P(A|B)}_{\text{probability of being pregnant if a woman}} \neq \underbrace{P(B|A)}_{\text{probabilty of being a woman if pregnant}}\,. \tag{9.16}$$

In experimental physics, this example is translated into the probability

$$\underbrace{P(\text{DATA}|\text{THEORY})}_{\text{probability of data, assuming the theory}}\,, \tag{9.17}$$

which is what is called the **likelihood** and is different from

$$P(\text{THEORY}\,|\,\text{DATA})\,, \tag{9.18}$$

probability of theory for the given data

which is what we really have in mind when we say that a theory has been confirmed (verified).

These two probabilities are connected by **Bayes' theorem**:

$$\boxed{P(\text{THEORY}\,|\,\text{DATA}) = \frac{P(\text{DATA}\,|\,\text{THEORY})\,P(\text{THEORY})}{P(\text{DATA})}\,,} \tag{9.19}$$

which shows that when going from the likelihood to the posterior probability it is necessary to know what are $P(\text{THEORY})$, the prior probability, or **prior** for short, of the model and $P(\text{DATA})$, the prior probability of the data.

To make this point clearer consider the following question borrowed from the medical sciences:

> The Elisa test for HIV has a sensibility and specificity both of 99%. What is the probability of being infected if the test turns out positive?

One might be tempted to say 99% but it would be the wrong answer. The probability of actually being infected after testing positive depends on the prevalence of HIV in the population you belong to, and you need to apply Bayes' theorem.

This is done by first making explicit what sensibility and specificity mean. The sensibility of a test is defined as the probability $P(\text{T+}|\text{HIV})$ of testing positive if infected; the specificity is the probability $P(\text{T}-|\text{NOHIV})$ of testing negative if not infected. As a general rule, tests having a high sensibility are used for testing large populations which are expected to be mostly not ill; tests with high specificity are instead used after a first-level screening to rule out false positive results.

First of all, I need the incidence of the infection for which I am testing in the population. For HIV it is about 0.2% for a generic sample of people not at risk.

The probability of having a positive test result, $P(\text{T+})$, is given by

$$\begin{aligned} P(\text{T+}) &= P(\text{T+}|\text{HIV})P(\text{HIV}) + P(\text{T+}|\text{NOHIV})P(\text{NOHIV}) \\ &= 0.99 \times 0.002 + 0.01 \times 0.999 = 0.012\,, \end{aligned} \tag{9.20}$$

which includes both true positives and false positives. In Eq. (9.20) I have used $P(\text{T+}|\text{NOHIV}) = 1 - P(\text{T}-|\text{NOHIV}) = 0.01$.

Table 9.2 Test on a million patients. T+ indicates a positive result, T−, a negative result.

Test	Ill	Healthy	Total
T+	1980	9980	11 960
T−	20	988 020	998 040
Total	2000	998 000	1 000 000

The posterior probability $P(\mathrm{HIV}|\mathrm{T+})$ of being found to be ill after a positive test is obtained by Bayes' theorem to be

$$P(\mathrm{HIV}|\mathrm{T+}) = \frac{P(\mathrm{T+}|\mathrm{HIV})\, P(\mathrm{HIV})}{P(\mathrm{T+})} = \frac{0.99 \times 0.002}{0.01} = 0.165 \simeq 17\% \,. \tag{9.21}$$

This result is best understood by writing a table of the various probabilities where, for a sample of a million patients I expect 2% of them, that is, 2000, to actually be infected given the incidence of the infection; 1980 of these are true and 20 false positives. Among the remaining 998 000 patients, who are healthy, I will have 988 020 with a negative test but also 9980 patients with a positive test result (because of the specificity of 99%) that is a false positive (that is, a healthy patient who nevertheless has a positive test). These numbers can be set out as in Table 9.2.

I find that there are 11 960 people who have a positive test but of these only 1980 are actually ill. The posterior probability is the ratio of these two numbers, that is, about 17%.

The probabilities $P(\mathrm{T+})$ and $P(\mathrm{HIV})$ are the prior probabilities.[9.13] Unless they are close to 1, their effect can dramatically change our estimate in going from the likelihood to the posterior probability. $P(\mathrm{HIV})$, for instance, contains some estimate of how many individuals are infected in the population at large. In the case of HIV this number is given by the incidence of the viral infection. If I had been testing a higher-risk population with, say, an incidence of 5% instead of 0.2%, the effect of the priors would have been much less pronounced and the difference between likelihood and posterior probability less dramatic.

The problem is, the priors in physics are often hard to estimate. They are really about our prejudice regarding the result of a possible experiment. Framed in the language of Bayesian statistics, performing the experiment modifies our estimate of the likelihood of the outcome.

The prior probabilities are not the only problem of which we must be aware in a statistical analysis. Another possibly troublesome feature arises when I look at more than one observable at the same time.

[9.13] The saying "Extraordinary claims require extraordinary evidence" is a nice way to stress the role of prior probabilities in our search for physical evidence. If the prior probability of having a particle, for instance a neutrino, go faster then light is extremely low – after all we know that relativity has been verified many, many times – the burden of proof lies heavily on any experiment advancing such a claim.

Consider the following question:

> What is the probability for two people at a party with 23 people to be born the same day of the year? Make a guess!

And now let us check that guess (whatever it may be). Instead of computing the probability that they are born on the same day, it is easier first to compute the probability that they are not. For two people that probability is given by the ratio of all the dates that are not the same over the days of a year, that is,

$$\frac{364}{365}\,. \tag{9.22}$$

This is for just two people. For three people then I would have the same factor between the first two people, another between the second and the third and another between the third and the first. In general, for N people I have $N \times (N-1)/2$ factors. This number is the probability for not being born on the same day. The probability of being born on the same day is then just

$$1 - \left(\frac{364}{365}\right)^{(23\times 22)/2} \simeq 50\%\,. \tag{9.23}$$

As you might have guessed, I have chosen the number of 23 people to start with so as to have the nice (and surprisingly large) 50% result. Is this result useful beside being a nice party trick? It carries over to the problem of simultaneously estimating a number of parameters.

Consider the case of having, say, 20 parameters.[9.14] Each of their values ranges within its Gaussian distribution around the average value. What is the probability that at least one of them will assume a value at more than 2σ from the average value? Again, I do it backward. The probability for one parameter to be within the 2σ range is by definition 95%. If I have 20 parameters then the probability for all of them to be within the 2σ range is

$$P_1 P_2 \cdots P_{20} = \left(\frac{95}{100}\right)^{20} = 36\%\,, \tag{9.24}$$

and therefore the probability for at least one to be outside that range is

$$100\% - 36\% = 64\%\,. \tag{9.25}$$

This is the basis of what has been called the **look elsewhere effect**. The fluctuation of the background is estimated where the signal seems to be found but, generally speaking, the same signal could have been anywhere in the range under consideration. I must include these possibilities, whose probabilities sum up in the same way as those for the multiple parameters in the example.[9.15]

9.14 If you do a blood test and have 20 parameters measured, what is the probability that at least one is outside the 2σ range the medical community has decided to define to be normal?

9.15 Think how many papers on the measurement of different observables are produced by the experimental collaborations at the LHC every year.

Mostly because of these unavoidable uncertainties, the high-energy community has resolved to use the so-called **5σ rule** before calling out a possible discovery. The significance of the factor 5 comes from the experience in the past, when there have been various "discoveries" based on evidence at the level of 3 or 4σ which have eventually gone away.[9.16]

[9.16] The presence of possible systematic errors is also tamed by the very strict requirement of the 5σ rule.

By requiring such a large deviation from the mean, one hopes that both the priors and the look-elsewhere effect can be tamed.

9.8 Exercises

1. **Length of tracks.** Estimate the length of the track left by a particle with a lifetime of $\tau = 10^{-12}$ s.
   [Hint: The distance is given by $d = (\beta\tau)$, with β the particle's velocity.]
2. **Penetration depth.** What are the typical penetration depths for muons in various materials?
   [Hint: Search the web.]
3. **Muon identification.** The efficiency with which muons are identified in a detector is $\epsilon = 95\%$. A few pions are (unavoidably) erroneously identified as muons. Take that fraction to be $\delta = 5\%$. What is the probability that an event identified as a muon is actually a muon?
   [Hint: It depends on how many muons are in the sample considered. Take a case in which the muons are rare ($P(\mu) = 4\%$) and most of the particles are pions ($P(\pi) = 96\%$).]
4. **MadGraph.** Set up MadGraph and generate the events for the process $pp > h > WW$ together with the backgrounds as discussed in the text. Discuss the distributions of various kinematical variables and try to enforce the best cuts to raise the signal above the background.
   [Hint: You will need to load a model providing the effective vertex between the Higgs boson and the gluons.]
5. ***p*-value.** Comment on the following statement: if the p-value is equal to 5%, the null hypothesis has only a 5% probability of being true.
   [Hint: The null hypothesis is always assumed; it cannot be proved right or wrong.]

10 Things That Go Bump in the Night

Contents

Even though the physics of coupled two-state systems is something we learn about in a first class on quantum mechanics, it is still a fascinating subject and, as a matter of fact, important in the study of particle physics.

A quantum system with two (to begin with, non-interacting) energy levels is described by the Schrödinger equation[10.1]

$$H_0|\psi(t)\rangle = i\frac{\partial}{\partial t}|\psi(t)\rangle\,, \tag{10.1}$$

where the Hamiltonian is given by

$$H_0 = \begin{pmatrix} E_1 & 0 \\ 0 & E_2 \end{pmatrix}\,. \tag{10.2}$$

The eigenvectors (at time $t = 0$) are given by

$$|1\rangle = \begin{pmatrix} 1 \\ 0 \end{pmatrix} \quad \text{and} \quad |2\rangle = \begin{pmatrix} 0 \\ 1 \end{pmatrix}\,, \tag{10.3}$$

and are energy (or mass) eigenvectors. Their time evolution is dictated by Eq. (10.1) to be[10.2]

$$|i(t)\rangle = |i\rangle\, e^{-iE_it/2}\,. \tag{10.4}$$

If now I add a (perturbatively small) coupling between the levels, as well as an energy shift of the two levels themselves,

$$W = \begin{pmatrix} W_{11} & W_{12} \\ W_{12}^* & W_{22} \end{pmatrix}\,, \tag{10.5}$$

the Hamiltonian for the system becomes

$$H = H_0 + W = \begin{pmatrix} E_1 + W_{11} & W_{12} \\ W_{12}^* & E_2 + W_{22} \end{pmatrix}\,, \tag{10.6}$$

the eigenvalues of which can be found as usual by computing the determinant

$$\begin{vmatrix} E_1 + W_{11} - \lambda & W_{12} \\ W_{12}^* & E_2 + W_{22} - \lambda \end{vmatrix} = 0\,, \tag{10.7}$$

10.1 Units: $\boxed{\hbar = 1,\, c = 1}$

10.2 Since there is no coupling yet, the system is described just by two independent sub-systems, each evolving independently.

that is, by solving the equation

$$(E_1 + W_{11} - \lambda)(E_2 + W_{22} - \lambda) - |W_{12}|^2 = 0\,. \tag{10.8}$$

Equation (10.8) is a second-order algebraic equation:

$$\lambda^2 + \lambda(E_1 + W_{11} + E_2 + W_{22}) + |W_{12}|^2 - (E_1 + W_{11})(E_2 + W_{22}) = 0\,. \tag{10.9}$$

I know the solution. It is like meeting a friend from old times. There are two eigenvalues:

$$E_\pm = \frac{1}{2}\Bigg[E_1 + W_{11} + E_2 + W_{22} \pm \underbrace{\sqrt{(E_1 + W_{11} + E_2 + W_{22})^2 - 4(E_1 + W_{11})(E_2 + W_{22}) + 4|W_{12}|^2}}_{\sqrt{(E_1 + W_{11} - E_2 - W_{22})^2 + 4|W_{12}|^2}}\Bigg]\,. \tag{10.10}$$

The corresponding eigenvectors are defined by

$$H|\pm\rangle = E_\pm|\pm\rangle\,. \tag{10.11}$$

These two eigenvectors $|\pm\rangle$ provide a second set of eigenvectors for the system – according to which the system evolves after becoming coupled and perturbed. Their explicit form is found by solving the linear system (starting with E_+)

$$\begin{pmatrix} E_1 + W_{11} & W_{12} \\ W_{12}^* & E_2 + W_{22} \end{pmatrix}\begin{pmatrix} x \\ y \end{pmatrix} = E_+\begin{pmatrix} x \\ y \end{pmatrix}, \tag{10.12}$$

that is,

$$(E_1 + W_{11})x + W_{12}y = E_+ x\,, \tag{10.13}$$

which gives

$$\frac{y}{x} = \frac{E_1 - E_+ + W_{11}}{W_{12}}\,. \tag{10.14}$$

These expressions look horrible. So, define

$$\tan\theta = \frac{2\,|W_{12}|}{E_1 + W_{11} - E_2 - W_{22}} \tag{10.15}$$

with

$$W_{12} = |W_{12}|\,e^{-i\phi}\,, \tag{10.16}$$

and write the eigenvalue E_+ as

$$E_+ = \frac{1}{2}(E_1 + W_{11}) + \frac{1}{2}(E_2 + W_{22}) + |W_{12}|\underbrace{\sqrt{1 + \frac{1}{\tan^2\theta}}}_{1/\sin\theta}\,, \tag{10.17}$$

which makes it possible to write

$$\begin{aligned} E_1 - E_+ + W_{11} &= \frac{1}{2}(E_1 + W_{11} - E_2 - W_{22}) - \frac{|W_{12}|}{\sin\theta} \\ &= \frac{|W_{12}|}{\tan\theta} - \frac{|W_{12}|}{\sin\theta} = \underbrace{\left(\frac{1}{\tan\theta} - \frac{1}{\sin\theta}\right)}_{\tan\theta/2} |W_{12}| \end{aligned} \tag{10.18}$$

and (by using Eq. (10.14), Eq. (10.16) and Eq. (10.17))

$$\frac{y}{x} = \tan\frac{\theta}{2}\, e^{i\phi}\,. \tag{10.19}$$

The eigenvector $|+\rangle$ is now given by the components

$$\begin{cases} y = \sin\dfrac{\theta}{2}\, e^{i\phi/2} \\ x = \cos\dfrac{\theta}{2}\, e^{-i\phi/2}\,, \end{cases} \tag{10.20}$$

while the eigenvector $|-\rangle$ is obtained by the transformation

$$\begin{cases} x \to y^* \\ y \to x^*\,. \end{cases} \tag{10.21}$$

A backward glance shows that the eigenvectors are

$$\begin{cases} |+\rangle = \cos\dfrac{\theta}{2}\, e^{-i\phi/2}|1\rangle + \sin\dfrac{\theta}{2}\, e^{i\phi/2}|2\rangle \\ |-\rangle = -\sin\dfrac{\theta}{2}\, e^{-i\phi/2}|1\rangle + \cos\dfrac{\theta}{2}\, e^{i\phi/2}|2\rangle\,, \end{cases}$$

or, in terms of the mass basis,

$$\begin{cases} |1\rangle = e^{i\phi/2}\left(\cos\dfrac{\theta}{2}\,|+\rangle - \sin\dfrac{\theta}{2}\,|-\rangle\right) \\ |2\rangle = e^{-i\phi/2}\left(\sin\dfrac{\theta}{2}\,|+\rangle + \cos\dfrac{\theta}{2}\,|-\rangle\right), \end{cases}$$

which is the form that gives the time evolution. For instance, for an initial state $|\psi(t=0)\rangle = |1\rangle$,

$$|\psi(t)\rangle = e^{i\phi/2}\left(\cos\frac{\theta}{2}\,|+\rangle\, e^{-iE_+t/2} - \sin\frac{\theta}{2}\,|-\rangle\, e^{-iE_-t/2}\right). \tag{10.22}$$

I can now readily compute[10.3] the probability that a state starting out as $|2\rangle$ will go into $|1\rangle$:

10.3 Intermediate steps: $\langle 2|\psi(t)\rangle = \sin\frac{\theta}{2}\cos\frac{\theta}{2}\left(e^{-iE_+t/2} - e^{-iE_-t/2}\right) = \frac{1}{4}\sin^2\frac{\theta}{2}\, e^{-iE_+t/2}\left(1 - e^{-i(E_+-E_+)t/2}\right)$.

$$\begin{aligned} P_{1\to 2} &= |\langle 2|\psi(t)\rangle|^2 = \sin^2\theta\,\sin^2\frac{\overbrace{E_+ - E_-}^{=\omega}}{2}t \\ &= \frac{4|W_{12}|^2}{4|W_{12}|^2 + (E_1 - E_2)^2}\sin^2\frac{\sqrt{4|W_{12}|^2 + (E_1 - E_2)^2}}{2}t\,. \end{aligned} \tag{10.23}$$

The physical system starts out in the state $|2\rangle$ and it goes into the other state $|1\rangle$ and then back again, oscillating between the two states with frequency $\omega = (E_+ - E_-)/2$.

Equation (10.23) shows that, in oder to have oscillations between the two states of the system, two conditions must be satisfied:

1. **There is mixing**: $|W_{12}| \neq 0$.
2. **Eigenvalues are not degenerate**: $E_+ \neq E_-$.

The picture is slightly more complicated if the states I am considering can decay – which is the case for many elementary particles. The Hamiltonian acquires an extra term introducing the finite widths of the states:

$$H = M - \frac{i}{2}\Gamma = \begin{pmatrix} M_{11} & M_{12} \\ M_{12}^* & M_{22} \end{pmatrix} - \frac{i}{2}\begin{pmatrix} \Gamma_{11} & \Gamma_{12} \\ \Gamma_{12}^* & \Gamma_{22} \end{pmatrix}, \tag{10.24}$$

where the matrix elements satisfy

$$M_{21} = M_{12}^*, \quad \Gamma_{21} = \Gamma_{12}^* \tag{10.25}$$

because of the hermiticity of the operators, and

$$M_{22} = M_{11}, \quad \Gamma_{22} = \Gamma_{11}^* \tag{10.26}$$

because of CPT symmetry.

Now the eigenvalues can be written as

$$\begin{aligned} E_\pm &= \left(M_{11} - \frac{i}{2}\Gamma_{11}\right) \pm \sqrt{|M_{12}|^2 + \frac{1}{4}|M_{12}|^2 + \frac{i}{2}(M_{12}\Gamma_{12}^* - M_{12}^*\Gamma_{12})} \\ &= \left(M_{11} - \frac{i}{2}\Gamma_{11}\right) \pm \frac{1}{2}(\Delta m - \frac{i}{2}\Delta\Gamma), \end{aligned} \tag{10.27}$$

where

$$\frac{1}{4}\left[\Delta m^2 - \left(\frac{\Delta\Gamma}{2}\right)^2\right] = |M_{12}|^2 - \frac{1}{4}|\Gamma_{12}|^2$$

$$\Delta m \Delta\Gamma = 4\,\mathrm{Re}\, M_{12}\Gamma_{12}^*. \tag{10.28}$$

The time evolution is given, as before, by

$$|\psi(t)\rangle = e^{i\phi/2}\left(\cos\frac{\theta}{2}|+\rangle e^{-iE_+t/2} - \sin\frac{\theta}{2}|-\rangle e^{-iE_-t/2}\right) \tag{10.29}$$

but now with

$$E_\pm = m_\pm - \frac{i}{2}\gamma_\pm, \tag{10.30}$$

where

$$m_\pm = M_{11} \pm \frac{1}{2}\Delta m \quad \text{and} \quad \gamma_\pm = \Gamma_{11} \mp \Delta\Gamma. \tag{10.31}$$

The two-level system is a template for numerous physical problems. Every time the propagating eigenstates do not coincide with those that interact, there is the possibility of mixing and oscillation between (or among, if there are more than two) the sets of eigenstates.

Oscillations in strangeness in the kaon system and in flavor among massive neutrinos are two examples of this phenomenon in particle physics.

10.1 *Problem Session*: Two Coupled Pendulums

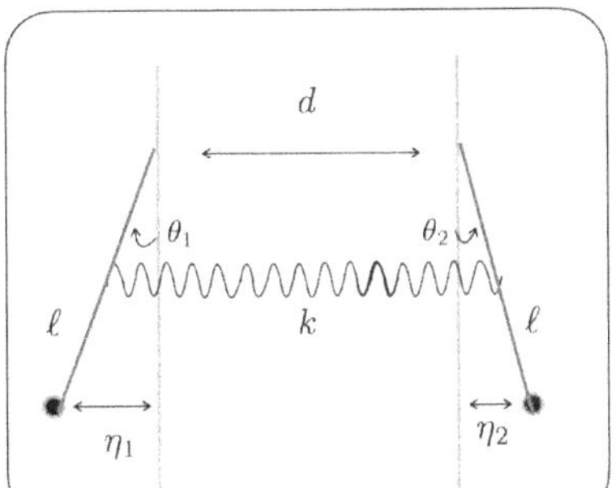

Figure 10.1 The two coupled pendulums.

A simple classical system that shares the essential features of the quantum two-level system consists of two pendulums coupled by a spring. The contraption is drawn in Figure 10.1. Taking their lengths to be equal to ℓ and masses to m, denote by η_i the horizontal displacements of the masses from the equilibrium position, θ_i the angles from the vertical position and k the spring constant. As usual, in the small-oscillation approximation I can assume that $\eta_i = \ell \sin\theta_i \simeq \ell\theta_i$.

The energy of the system is (in units of $m\ell^2/2$)

$$E = \dot{\theta}_1^2 + \dot{\theta}_2^2 + \frac{g}{\ell}\theta_1^2 + \frac{g}{\ell}\theta_2^2 + \frac{k}{m}(\theta_2 - \theta_1)^2 . \tag{10.32}$$

The free part of the Hamiltonian is just the potential energy of the two pendulums in the gravitation field – which I can write as a matrix acting on the two coordinates θ_1 and θ_2:

$$H_0 = \begin{pmatrix} g/\ell & 0 \\ 0 & g/\ell \end{pmatrix} , \tag{10.33}$$

while the spring that couples the two pendulums brings in the interaction

$$W = \begin{pmatrix} k/m & -k/m \\ -k/m & k/m \end{pmatrix} . \tag{10.34}$$

In the language of the two-level system, I have that

$$E_1 = E_2 = \frac{g}{\ell} \quad \text{and} \quad W_{11} = W_{22} = \frac{k}{m},\ W_{12} - W_{21}^* = -\frac{k}{m} \tag{10.35}$$

The full Hamiltonian can be diagonalized to give the eigenvalues

$$E_\pm = \frac{g}{\ell} + \frac{k}{m} \pm \frac{k}{m} , \tag{10.36}$$

with

$$\tan\theta = \frac{2k/m}{0} , \quad \text{that is,} \quad \theta = \frac{\pi}{2} \quad (\text{and } \phi = 0) , \tag{10.37}$$

in Eq. (10.15) and Eq. (10.16).

The eigenvectors are

$$\begin{cases} |+\rangle = \cos\frac{\pi}{4}\,|1\rangle + \sin\frac{\pi}{4}\,|2\rangle = \frac{1}{\sqrt{2}}\Big[|1\rangle + |2\rangle\Big] \\ |-\rangle = -\sin\frac{\pi}{4}\,|1\rangle + \cos\frac{\pi}{4}\,|2\rangle = \frac{1}{\sqrt{2}}\Big[|2\rangle - |1\rangle\Big], \end{cases}$$

which correspond to the two normal modes

$$|+\rangle = \begin{pmatrix} 1 \\ 1 \end{pmatrix} \qquad \Big[\bullet\rightarrow \quad \bullet\rightarrow \quad \text{moving in the same direction}\Big] \tag{10.38}$$

and

$$|-\rangle = \begin{pmatrix} -1 \\ 1 \end{pmatrix} \qquad \Big[\bullet\rightarrow \quad \leftarrow\bullet \quad \text{moving in opposite directions}\Big], \tag{10.39}$$

into which the motion of the two pendulums can be decomposed.

The time evolution of the two-pendulum system is described by the function

$$\begin{aligned} |\psi(t)\rangle &= \frac{1}{2}\begin{pmatrix} 1 \\ 1 \end{pmatrix} e^{-iE_+t/2} + \begin{pmatrix} -1 \\ 1 \end{pmatrix} e^{-iE_-t/2} \\ &= \frac{1}{2}|1\rangle\left(e^{-iE_+t/2} + e^{-iE_-t/2}\right) + \frac{1}{2}|2\rangle\left(e^{-iE_+t/2} - e^{-iE_-t/2}\right). \end{aligned} \tag{10.40}$$

Although I am discussing a classical problem, I can use the convenient trick of writing the time evolution of an oscillating system in complex space as a phase.

The result in Eq. (10.40) can be used to compute the time evolution of the system. Suppose that the pendulums start out with, for instance, only the first one out of the equilibrium position, in the eigenstate $|\psi(t=0)\rangle = |1\rangle$; then the evolution in time is

$$\begin{aligned} \mathcal{A}(t) = \text{Re}\,\langle 1|\psi(t)\rangle &= \frac{1}{2}\left(\cos\frac{E_+t}{2} + \cos\frac{E_-t}{2}\right) \\ &= \cos\underbrace{\frac{E_+ - E_-}{4}}_{\omega}t\,\cos\underbrace{\frac{E_+ + E_-}{4}}_{\Omega}t, \end{aligned} \tag{10.41}$$

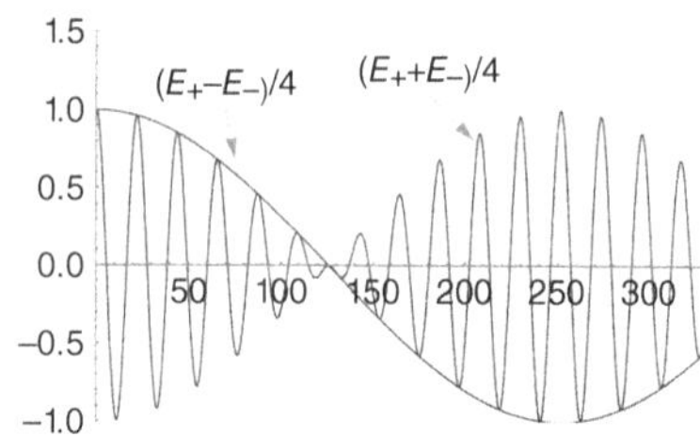

Figure 10.2 The basic and beat frequencies in the coupled system of two pendulums.

which shows a characteristic **beat frequency** ω emerging from the modulation of the basic frequency Ω. (See Figure 10.2.) The oscillation is transferred from one pendulum to the other and then back.

The two basic frequencies combine to produce the beat frequency, which is heard on top of the basic frequency. This is the effect using which an orchestra tunes up before a performance. The beat sound disappears when all instruments play the same note at the same frequency.

10.2 Kaons

Kaons are mesons made of two quarks. The neutral kaons are defined in terms of their quark content as follows:

$$K^0: \quad (\bar{s}d) \qquad \bar{K}^0: \quad (s\bar{d}). \tag{10.42}$$

They are degenerate states under the strong interaction. Why then keep them separated? Because they have opposite strangeness and under the weak interactions they decay differently, namely $K^0 \to \pi^- e^+ \nu_e$ and $\bar{K}^0 \to \pi^+ e^- \bar{\nu}_e$.

Under a parity transformation,

$$\begin{aligned} \hat{P}|K^0\rangle &= -\,|K^0\rangle \\ \hat{P}|\bar{K}^0\rangle &= -\,|\bar{K}^0\rangle, \end{aligned} \tag{10.43}$$

which shows that the neutral kaons are pseudoscalar $J^P = 0^-$ particles.

The two states are exchanged under a charge conjugation:

$$\begin{aligned} \hat{C}|K^0\rangle &= |\bar{K}^0\rangle \\ \hat{C}|\bar{K}^0\rangle &= |K^0\rangle. \end{aligned} \tag{10.44}$$

Therefore, under the combined transformation CP they are changed as follows:

$$\begin{aligned} \widehat{CP}|K^0\rangle &= -\,|\bar{K}^0\rangle \\ \widehat{CP}|\bar{K}^0\rangle &= -\,|K^0\rangle, \end{aligned} \tag{10.45}$$

and they are not eigenstates of CP. To bypass this problem I can define two new states

$$\begin{cases} K_1 = \dfrac{1}{\sqrt{2}}\Big[|K^0\rangle - |\bar{K}^0\rangle\Big] \\ K_2 = \dfrac{1}{\sqrt{2}}\Big[|K^0\rangle + |\bar{K}^0\rangle\Big] \end{cases}$$

to obtain

$$\begin{aligned} \widehat{CP}|K_1\rangle &= +\,|K_1\rangle \\ \widehat{CP}|K_2\rangle &= -\,|K_2\rangle, \end{aligned} \tag{10.46}$$

which are indeed eigenstate of CP. They are distinguishable by means of their decays. They like to decay into pions. What is the CP property of the two-pion state? The pion is a pseudoscalar particle for which $\hat{P}(\pi) = -1$ and $\hat{C}(\pi^0) = 1$. Two pions have a parity that is the product of their individual parities times that of the system, which is given by $(-1)^L$ with L the angular momentum. The two pions originate from a scalar state $J^P = 0^-$. The system must have total angular momentum $L = 0$. Therefore

$$\hat{P}(\pi^0\pi^0) = \hat{P}(\pi^0)\hat{P}(\pi^0)(-1)^L = (-1)(-1) = 1. \tag{10.47}$$

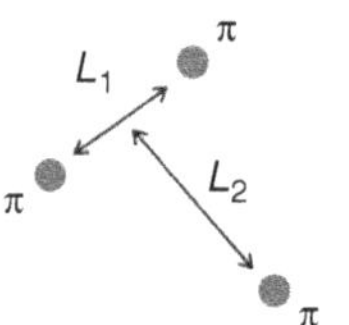

Figure 10.3 Angular momenta in the three-pion final state.

For three pions I have that

$$\hat{P}(\pi^0\pi^0\pi^0) = \hat{P}(\pi^0)\hat{P}(\pi^0)\hat{P}(\pi^0)(-1)^{L_1}(-1)^{L_2} = (-1)(-1) = -1\,, \tag{10.48}$$

where L_1 and L_2 are the angular momenta of the two subsystems consisting of the pions taken two by two, as shown in Figure 10.3. Again, by conservation of the total angular momentum it must follow that $L_1 = L_2$. The same is true for two neutral instead of charged pions.

The two-pion state therefore has

$$\widehat{CP}(\pi^0\pi^0) = \widehat{CP}(\pi^+\pi^-) = +\mathbf{1}\,, \tag{10.49}$$

while the three-pion state has

$$\widehat{CP}(\pi^0\pi^0\pi^0) = \widehat{CP}(\pi^0\pi^+\pi^-) = -\mathbf{1}\,. \tag{10.50}$$

If CP is conserved, I expect that

$$\begin{aligned} K_1 &\to 2\pi \quad (CP = +1) \\ K_2 &\to 3\pi \quad (CP = -1)\,. \end{aligned} \tag{10.51}$$

Because the decay into three pions has less final phase-space available ($m_K - 3m_\pi \simeq 80$ MeV), K_1 is called K_S for short-lived and K_2 is called K_L for long-lived.

Now – and this is what is interesting – kaons are produced by the strong interactions as K^0 or $\bar{K}^0$ but then they evolve in time according to their mass eigenvectors, which are K_L and K_S. It is like the general case of a two-state quantum system. Kaons are like coupled pendulums! The only catch is that I have to use the formulas for the case of states with a finite probability to decay. There are various rotations among the states to be done, we must keep our wits about us so as not to get confused.[10.4]

10.4 This discussion seems at first unrelated to the Standard Model. It was studied before the Standard Model was actually developed. It is a sort of **macro**scopic theory. Today we can (and must) understand it in terms of the **micro**scopic parameters of the the Standard Model.

The Hamiltonian of the system is

$$H|K_{L,S}\rangle = \left(m_{L,S} - \frac{i}{2}\Gamma_{L,S}\right)|K_{L,S}\rangle\,. \tag{10.52}$$

The mixing interaction can be understood in terms of the Standard Model fundamental (weak) interactions if I draw the Feynman diagram in Figure 10.4 for the transformation of a K^0 into a $\bar{K}^0$.

This diagram in Figure 10.4 (together with a diagram with the W-boson running vertically) shows how I can actually compute the parameters of the kaon system mixing. I just need to implement the Feynman rules and do the loop integral associated with the amplitude derived from the Feynman diagram in Figure 10.4 to find that

$$\Delta m_{LS} = \frac{G_F}{\sqrt{2}}\frac{\alpha}{6\pi}\frac{m_K f_K^2}{\sin^2\cos\theta_W}\mathrm{Re}\sum_{i,j} V_{is}V_{id}^* V_{js}V_{jd}^*\, F\left(\frac{m_i^2}{m_W^2}, \frac{m_j^2}{m_W^2}\right), \tag{10.53}$$

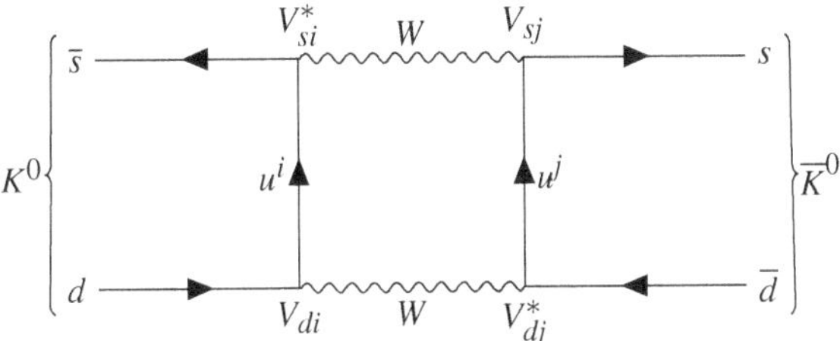

Figure 10.4 The Feynman box diagram between the neutral kaons K^0 and $\overline{K}^0$.

where $m_K f_K^2$ is a form factor coming from an estimate of the matrix element between the two kaons, and the function $F(x, y)$ is the result of the integration over the loop. Equation (10.53) shows that, in the Standard Model, the masses of the neutral kaons differ only because those of their constituent quarks differ. If I take $m_i = m_j$, then in Eq. (10.53) the factor involving the CKM matrices becomes $\sum_i |V_{is}V_{id}^*| = 0$ and the entire mass difference goes to zero.

The time evolution of an initial state $|\psi(t = 0)\rangle = |K_0\rangle$ is

$$|\psi(t)\rangle = \frac{1}{\sqrt{2}}\left[|K_S\rangle e^{-(im_S+\Gamma_S/2)t} + |K_L\rangle e^{-(im_L+\Gamma_L/2)t}\right]. \tag{10.54}$$

Let me write the exponential factors as

$$\xi_S(t) = e^{-(im_S + \Gamma_S/2)t} \quad \text{and} \quad \xi_L(t) = e^{-(im_L + \Gamma_L/2)t}, \tag{10.55}$$

so that I can rewrite Eq. (10.54) as

$$\begin{aligned}|\psi(t)\rangle &= \frac{1}{\sqrt{2}}\left[\xi_S(t)|K_S\rangle + \xi_L(t)|K_L\rangle\right] \\ &= \frac{1}{2}\left[\xi_S(t)(|K^0\rangle - |\bar{K}^0\rangle) + \xi_L(t)(|K^0\rangle + |\bar{K}^0\rangle)\right] \\ &= \frac{1}{2}\left[\xi_S(t) + \xi_L(t)\right]|K^0\rangle + \frac{1}{2}\left[\xi_L(t) - \xi_S(t)\right]|\bar{K}^0\rangle,\end{aligned} \tag{10.56}$$

which shows that K^0 evolves into $\bar{K}^0$. The probability for a K^0 to remain itself is

$$\Gamma(K^0 \to K^0) = \frac{1}{4}\left|\xi_S(t) + \xi_L(t)\right|^2, \tag{10.57}$$

and the probability to become a $\bar{K}_0$ is

$$\Gamma(K^0 \to \bar{K}^0) = \frac{1}{4}\left|\xi_L(t) - \xi_S(t)\right|^2, \tag{10.58}$$

which gives, after reintroducing the expressions in Eq. (10.55), that the probability for a K^0 to remain a K^0 is

$$\Gamma(K^0 \to K^0) = \frac{1}{4}\left[e^{-\Gamma_S t} + e^{-\Gamma_L t} + 2e^{-(\Gamma_L+\Gamma_S)t/2}\cos\left(\Delta m_{LS} t\right)\right] \tag{10.59}$$

and that to become a $\bar{K}^0$ is

$$\Gamma(K^0 \to \bar{K}^0) = \frac{1}{4}\left[e^{-\Gamma_S t} + e^{-\Gamma_L t} - 2e^{-(\Gamma_L+\Gamma_S)t/2}\cos(\Delta m_{\mathrm{LS}} t)\right]. \tag{10.60}$$

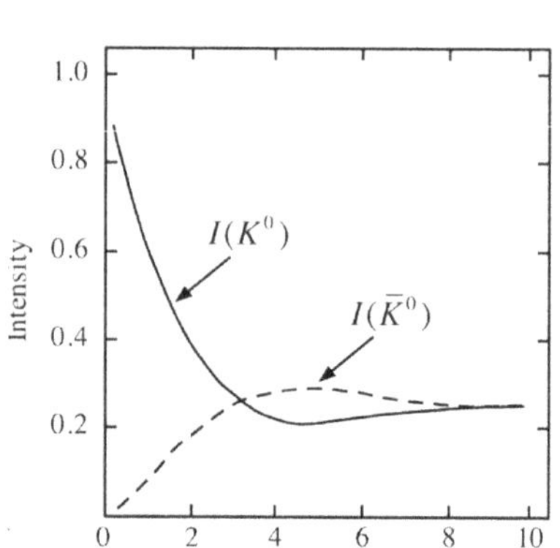

Figure 10.5 Oscillations in the number of neural kaons.

A schematic of the experimental result is shown in Figure 10.5.

In a typical experiment kaons K^0 are produced by collisions of protons and antiprotons ($p\bar{p} \to K^-\pi^+K^0$). The time dependence becomes the distance from the production point and it is possible to observe both the disappearance from the beam of the original K^0 (through its decay into $\pi^+e^-\bar{\nu}_e$) and the generation of $\bar{K}^0$ (through its subsequent decay into $\pi^-e^+\nu_e$).

To estimate the oscillation characteristic time, I need the value[1]

$$\Delta m_{LS} = (3.506 \pm 0.006) \times 10^{-15}\ \mathrm{GeV}. \tag{10.61}$$

The oscillation is visible because

$$\tau_{\mathrm{osc}} = \frac{2\pi}{\Delta m_{\mathrm{LS}}} \simeq 1.2 \times 10^{-9}\ \mathrm{s} \tag{10.62}$$

is sufficiently long compared with the lifetime $\tau_{K_S} \simeq 0.9 \times 10^{-10}$ s.

10.3 *CP* Violation

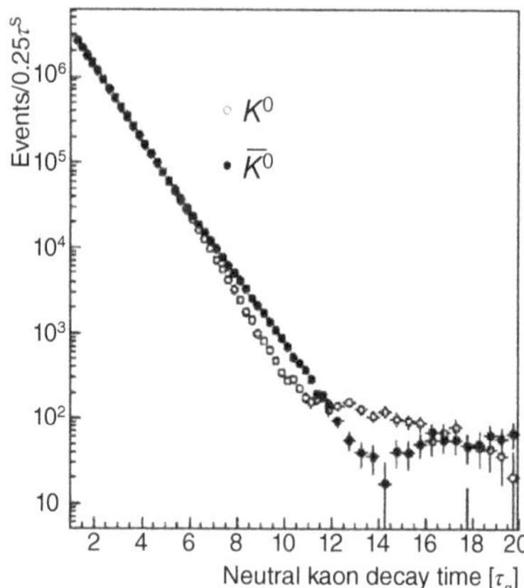

Figure 10.6 Data showing a difference in the decay times between K^0 and $\bar{K}^0$.

The physics of kaons had one more surprise up its sleeve. Observing the decays of the K_L, once in a while a decay into two pions is observed.

This odd decay is a very rare occurrence (45 out of 22 700 observed decays in the original experiment[2]). The only explanation is that K_S and K_L are not exactly the *CP* eigenstates K_1 and K_2: *CP* is not an exact symmetry of the model. Figure 10.6 shows the original data for this crucial experiment.

To take this into account I write (yes, one more time) the states as

$$\begin{cases} |K_S\rangle = \dfrac{1}{\sqrt{1+|\epsilon_K|^2}}\Big[|K_1\rangle + \epsilon_K|K_2\rangle\Big] \\ |K_L\rangle = \dfrac{1}{\sqrt{1+|\epsilon_K|^2}}\Big[|K_2\rangle + \epsilon_K|K_1\rangle\Big], \end{cases}$$

where ϵ_K is a (small) parameter taking into account the violation of *CP*.

Looking now at the K_L state,

$$|K_L\rangle = \frac{1}{\sqrt{1+|\epsilon_K|^2}}\Big[\underbrace{|K_2\rangle}_{\to\ \pi\pi\ \text{(direct)}} + \epsilon_K \underbrace{|K_1\rangle}_{\to\ \pi\pi\ \text{(indirect)}}\Big], \tag{10.63}$$

[1] PDG (2020) http://pdg.lbl.gov.

[2] J.H. Christienson, J.W. Cronin, V.L. Fitch and R. Turlay, *Physical Review Letters* **13** (1964) 138.

there are two possibilities for a $|K_L\rangle$ to go into two pions (violating *CP*):

- **indirectly** through the part of $|K_1\rangle$ (which naturally goes into two pions) that I now have, or
- **directly** through the $|K_2\rangle$ state – when the state itself goes into two pions.

Indirect *CP* violation. I have to go again through a maelstrom of rotations of the relevant eigenvectors. I need the strong interacting states in terms of the propagating states as

$$\begin{cases} |K^0\rangle = \sqrt{\dfrac{1+|\epsilon_K|^2}{2}}\dfrac{1}{1+\epsilon_K}\Big[|K_L\rangle + |K_S\rangle\Big] \\ |\bar{K}^0\rangle = \sqrt{\dfrac{1+|\epsilon_K|^2}{2}}\dfrac{1}{1+\epsilon_K}\Big[|K_L\rangle - |K_S\rangle\Big], \end{cases}$$

which is obtained by simply rewriting in Eq. (10.63) the $|K_{L,S}\rangle$ in terms of the $|K^0\rangle$ and $|\bar{K}^0\rangle$ and turning around the definitions. I can extract from the data[3] the *CP*-violating parameters:

$$|\epsilon_K| = (2.228 \pm 0.011) \times 10^{-3}$$
$$\phi_{\epsilon_K} = (43.19 \pm 0.75)^\circ .$$

The mass eigenstates can be made to evolve in time; for instance, for $|\psi(0)\rangle = |K^0\rangle$ I have

$$|\psi(t)\rangle = \sqrt{\frac{1+|\epsilon_K|^2}{2}}\frac{1}{1+\epsilon_K}\Big[\xi_L(t)|K_L\rangle + \xi_S(t)|K_S\rangle\Big], \qquad (10.64)$$

which can be written in terms of the states known to decay into two (K_1) or three (K_2) pions as

$$|\psi(t)\rangle = \frac{1}{\sqrt{2}}\frac{1}{1+\epsilon_K}\Big[(\xi_S(t)+\epsilon_K\xi_L(t))|K_1\rangle + (\xi_L(t)+\epsilon_K\xi_S(t))|K_2\rangle\Big]. \qquad (10.65)$$

In this form I can compute the probability for a kaon K^0 to decay into two pions by directly projecting out the component:[10.5]

10.5 $\left|\dfrac{1}{1+\epsilon_K^2}\right|^2 \simeq 1 - 2\,\mathrm{Re}\,\epsilon_K$, $|\xi_S + \epsilon_K\xi_L|^2 = \left|e^{-im_S-\Gamma_S t/2} + \epsilon_K^* e^{im_L-\Gamma_L t/2}\right| = e^{-\Gamma_S t} + |\epsilon_K|^2 e^{-\Gamma_L t} + 2\,\mathrm{Re}\left\{e^{-im_S-\Gamma_S t/2}\epsilon_K^* e^{im_L-\Gamma_L t/2}\right\}$ and $\epsilon_K = |\epsilon_K| e^{i\phi_{\epsilon_K}}$.

$$\begin{aligned} \Gamma(K^0 \to 2\pi) &= |\langle K_1|\psi(t)\rangle|^2 = \frac{1}{2}\left|\frac{1}{1+|\epsilon_K|^2}\right|^2 |\xi_S + \epsilon_K\xi_L|^2 \\ &= \frac{1}{2}(1-2\,\mathrm{Re}\,\epsilon_K)\Bigg\{\underbrace{e^{-\Gamma_S t}}_{K_S\text{ component}} + \underbrace{|\epsilon_K|^2 e^{-\Gamma_L t}}_{K_L\text{ component}} \\ &\qquad + \underbrace{2|\epsilon_K|e^{-(\Gamma_S+\Gamma_L)t/2}\cos(\Delta m_{LS} - \phi_{\epsilon_K})t}_{\text{interference}}\Bigg\}. \end{aligned} \qquad (10.66)$$

[3] PDG (2020) http://pdg.lbl.gov.

Direct *CP* violation. There is also the possibility of direct *CP* violation. This means that the violation is in the amplitudes themselves rather than in the mixing between states. The simplest way to parametrize this effect is by writing two amplitudes

$$\eta_{+-} = \frac{\langle \pi^+\pi^- | K_L \rangle}{\langle \pi^+\pi^- | K_S \rangle} \quad \text{and} \quad \eta_{00} = \frac{\langle \pi^0\pi^0 | K_L \rangle}{\langle \pi^0\pi^0 | K_S \rangle}, \tag{10.67}$$

which are both equal to ϵ_K if there is only indirect *CP* violation. In general, they can differ:

$$\eta_{+-} = \epsilon_K + \delta_{+-} \quad \text{and} \quad \eta_{00} = \epsilon_K + \delta_{00}, \tag{10.68}$$

where

$$\delta_{00} = -2\delta_{+-} \tag{10.69}$$

because of the *CPT* symmetry constraint

$$\Gamma(K^0 \to \pi^+\pi^-) + \Gamma(K^0 \to \pi^0\pi^0) = \Gamma(\bar{K}^0 \to \pi^+\pi^-) + \Gamma(\bar{K}^0 \to \pi^0\pi^0). \tag{10.70}$$

Finally, I obtain that (for $\epsilon' \equiv \delta_{+-}$)

$$\eta_{+-} = \epsilon_K + \epsilon' \quad \text{and} \quad \eta_{00} = \epsilon_K - 2\epsilon'. \tag{10.71}$$

In Eq. (10.71) I am neglecting small corrections due to differences in the isospin amplitudes. But never mind about that (a full explanation would require analysis in isospin components).

Experimentally, one measures the ratio[10.6]

10.6 Just expand

$$\left(\frac{\epsilon_K + \epsilon'}{\epsilon_K - 2\epsilon'}\right)\left(\frac{\epsilon_K + \epsilon'}{\epsilon_K - 2\epsilon'}\right)^*.$$

$$\left|\frac{\eta_{+-}}{\eta_{00}}\right| \simeq 1 - \mathrm{Re}\left(\frac{\epsilon'}{\epsilon_K}\right). \tag{10.72}$$

The most recent value for this parameter is[4]

$$\mathrm{Re}\,\frac{\epsilon'}{\epsilon_K} = (1.66 \pm 0.23) \times 10^{-3}, \tag{10.73}$$

which is one thousand times smaller than ϵ_K, which is itself rather small.

10.4 Neutrinos

The story of this evanescent particle begins, as so much of the Standard Model does, with β-decay. This decay, which was expected to have a sharp monochromatic line in the electron energy of a two-body decay, shows

4 PDG (2020) http://pdg.lbl.gov.

instead a broad spectrum, as can be seen in Figure 10.7, the fingerprint of a decay into three bodies: a proton, an electron and a new particle, the neutrino.[10.7]

Most new particles do not come alone. And indeed there are more flavors of neutrino. This has been determined by looking at a process like

$$\pi^+ \to \mu^+ + \nu_? \to \nu_? + n \to p + e^- . \tag{10.74}$$

If there were only one neutrino, the same for all lepton flavors, this process would be possible but it is found that is not: there is a neutrino for the weak interaction of the muon and another one for that of the electron. As a matter of fact (since the observation of ν_τ) there are as many neutrinos as there are leptons.

Neutrinos are the junior partners of the corresponding charged leptons in the electroweak doublets of the Standard Model. They were considered not to have a right-handed component because they were taken as massless. This assumption was supported by the decays of the charged pion:

$$\mu^+ \leftarrow \boxed{\pi^+} \to \underbrace{\nu_\mu}_{\text{LH}} \quad \text{and} \quad e^+ \leftarrow \boxed{\pi^+} \to \underbrace{\nu_e}_{\text{LH}} . \tag{10.75}$$

If the neutrinos are only left-handed, the charged μ^+ and e^+ must be left-handed as well by conservation of angular momentum. On the other hand, μ^+ and e^+ are antiparticles and right-handed in the relativistic limit. Therefore these decays are only possible because of the masses of these charged particles. The widths must be proportional to the square of their masses. The expectation is that

$$\frac{\Gamma(\pi^+ \to e^+\nu_e)}{\Gamma(\pi^+ \to \mu^+\nu_\mu)} \propto \left(\frac{m_e}{m_\mu}\right)^2 \simeq 10^{-4} , \tag{10.76}$$

which is indeed what is found.

This seemed nice at the time. There was even a good symmetry (chiral symmetry) to justify the vanishing of the neutrino masses. But it was not to be.

The actual measurement of the neutrino mass depends on the study of the endpoint of the energy spectrum of the β-decay (the Kurie[10.8] plot in Figure 10.8) where a non-zero mass would produce a faster fall-off of the spectrum. This determination is ongoing so that we still cannot say what are the values of the neutrino masses.

Never mind this – there are other reasons, as I am about to discuss, to believe in neutrinos being massive. And if they are indeed massive, I have to go back to the drawing board and think about the Lagrangian of the Standard Model. How do I add a mass term for the neutrinos to the Standard Model Lagrangian? There is no problem, I just do what I have already done for the up kind of quarks and write a term

[10.7] The origin of the neutrino hypothesis is an interesting instance of how physicists react to new experimental evidence, namely the electron spectrum in β-decay (Chadwick in 1914). The radical wing (Bohr) took the broad energy spectrum as a possible indication of energy non-conservation! The conservative wing (Pauli in 1930) suggested the existence of an undetectable new particle to make the two-body decay in to a three-body decay. The pragmatic wing (Fermi in 1933) named it "il neutrino". It will never be seen! said Bethe and Peierls in 1934. It was discovered by Cowan and Reines in 1956 at the nuclear plant of Savannah River by detection of the reaction $\bar{\nu} + {}^2_1\text{H} \to n + n + e^+$.

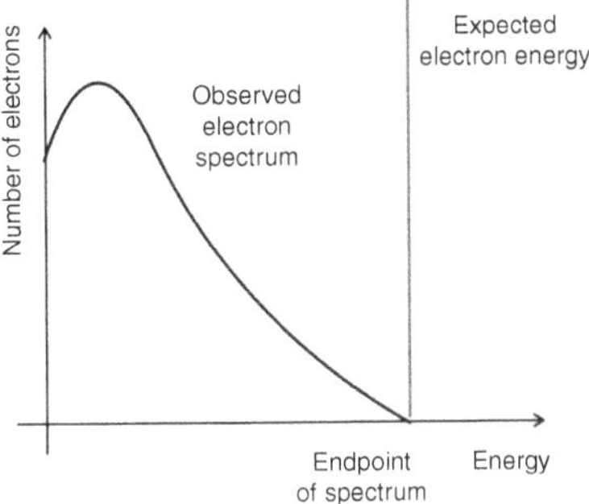

Figure 10.7 Electron energy spectrum in β decay.

[10.8] No misprint here: it is Kurie, not Curie as in "Madame Curie".

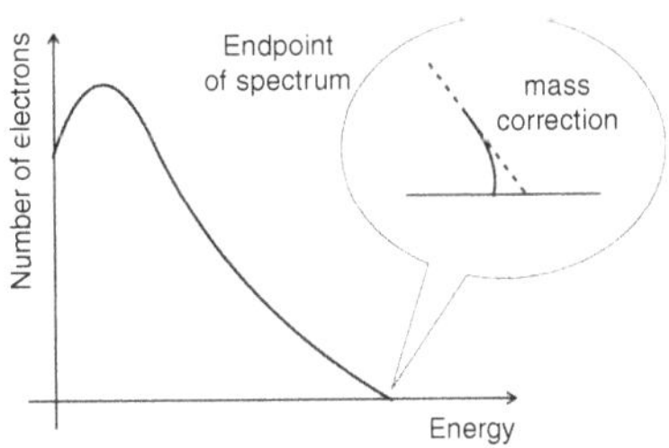

Figure 10.8 Kurie plot of the number of electrons as a function of their energy.

$$Y_{ij}^{\nu_i} \overline{L}_i \nu_{Rj} \tilde{H} + \text{h.c.}\,, \tag{10.77}$$

which, as the Higgs boson develops a vacuum expectation value, will generate the required mass terms:

$$m_D\, \bar{\nu}_R \psi\, \nu_L + m_D \bar{}\, \nu_L \nu_R\,. \tag{10.78}$$

This would be the end of it but for the fact that neutrinos are neutral fermions and as such could be Majorana particles.

Majorana particles can make a right-handed spinor simply by conjugation,[10.9]

$$(\psi_L)^c = (\psi^c)_R \quad \text{and} \quad (\psi_R)^c = (\psi^c)_L\,, \tag{10.79}$$

and I can write a mass term as

$$\frac{1}{2} m_L\, \overline{(\nu_L)^c} \nu_L + \frac{1}{2} m_R\, \overline{\nu_R} (\nu_R)^c + \text{h.c.}\,, \tag{10.80}$$

which is fine, except that neutrinos form a component of a weak isospin doublet, so that this term is a triplet. I need an isotriplet out of the Higgs boson fields, which is provided by

$$H^T i\sigma_2 H\,, \tag{10.81}$$

and the mass term for the Standard Model Lagrangian reads now

$$\frac{f}{M} (L^T \hat{C} i\sigma_2 L)(H^T i\sigma_2 H)\,. \tag{10.82}$$

In general, I can have both mass terms, the Dirac type and the Majorana type:

$$\begin{aligned} &\tfrac{1}{2} m_L\, \overline{(\nu_L)^c} \nu_L + \tfrac{1}{2} m_R\, \overline{\nu_R} (\nu_R)^c + \text{h.c.} && \text{(Majorana)} \\ &m_D\, \overline{\nu}_R \nu_L + \text{h.c.} && \text{(Dirac)}\,. \end{aligned} \tag{10.83}$$

This general structure gives rise to an interesting possibility. First I rewrite Eq. (10.83) in terms of the physical states

$$\chi_L = \frac{\nu_L + (\nu_L)^c}{\sqrt{2}} \quad \text{and} \quad \chi_R = \frac{\nu_R + (\nu_R)^c}{\sqrt{2}}\,, \tag{10.84}$$

so that now I have

$$(\bar{\chi}_L, \bar{\chi}_R) \begin{pmatrix} m_L & m_D \\ m_D & m_R \end{pmatrix} \begin{pmatrix} \chi_L \\ \chi_R \end{pmatrix}, \tag{10.85}$$

which I must diagonalize to find the mass eigenstates. The eigenvalues are

$$\lambda^{\pm} = \frac{m_R + m_L}{2} \pm \sqrt{\frac{(m_R - m_L)^2}{4} + 4m_D^2}\,. \tag{10.86}$$

10.9 This is a longish one:

$$\begin{aligned} (\psi_L)^c &= \left[\tfrac{1-\gamma^5}{2}\psi\right]^c = \hat{C}\,\overline{\left[\tfrac{1-\gamma^5}{2}\psi\right]}^T \\ &= \hat{C}\left[\left(\tfrac{1-\gamma^5}{2}\psi\right)^\dagger \gamma^0\right]^T \\ &= \hat{C}\left[\psi^\dagger \tfrac{1-\gamma^5}{2}\gamma^0\right]^T = \hat{C}\left[\gamma^0 \tfrac{1-\gamma^5}{2}\psi^{\dagger T}\right] \\ &= \hat{C}\left[\tfrac{1+\gamma^5}{2}\gamma^0 \psi^{\dagger T}\right] \\ &= \hat{C}\left[\tfrac{1+\gamma^5}{2}(\psi^\dagger \gamma^0)^T\right] \\ &= \tfrac{1+\gamma^5}{2}\hat{C}\overline{\psi} = \tfrac{1+\gamma^5}{2}\psi^c = (\psi^c)_R\,. \end{aligned}$$

10.10 How about a mass with negative sign? I can always rotate the field by $\gamma^5\nu$ and change it to a plus.

Assuming $m_L \simeq 0$ and $m_R \gg m_D$ I obtain[10.10]

$$\lambda^{\pm} = \begin{cases} -4\dfrac{m_D^2}{m_R} \\ m_R\,, \end{cases} \tag{10.87}$$

with one heavy Majorana neutrino $\nu_H = \chi_R$ and one naturally very light Majorana neutrino $\nu \simeq \chi_L$. This is called the **see-saw mechanism**.

So, is it Majorana or Dirac? The jury is still out – let alone that we do not know its mass by a direct measurement. To determine the nature of such mass I need a process where being a Majorana particle makes a difference.

For heavy (nonrelativistic) neutrinos the difference between Dirac and Majorana can be detected in decays. For light (relativistic) neutrinos the only process where there is a difference is the neutrinoless double β-decay, the Feynman diagram of which is shown in Figure 10.9. A Majorana neutrino is its own antiparticle and it is possible to have two neutrons decaying into two positrons without neutrinos in the final states. For this reason a decay like

$${}^{76}_{32}\mathrm{Ge} \rightarrow {}^{76}_{34}\,\mathrm{Se} + e^- + e^-, \tag{10.88}$$

which is depicted in Figure 10.10, is currently under close scrutiny in nuclear physics.

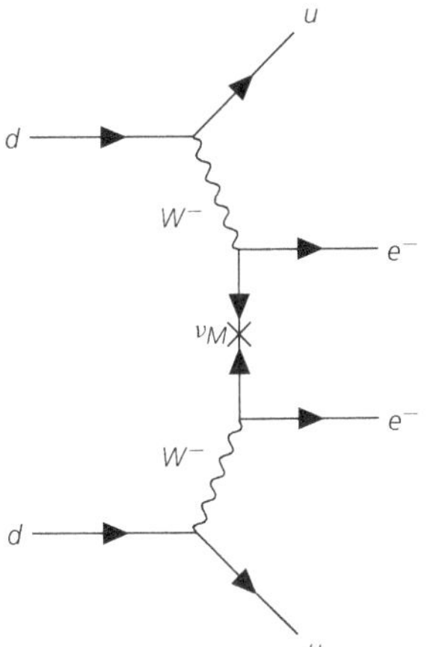

Figure 10.9 Feynman diagram for the neutrinoless double β-decay.

10.5 Flavor Oscillations

If neutrinos are indeed massive (and there is no lingering doubt about that) then they are another example of a two-state quantum system (actually three states, but never mind about that for the moment).

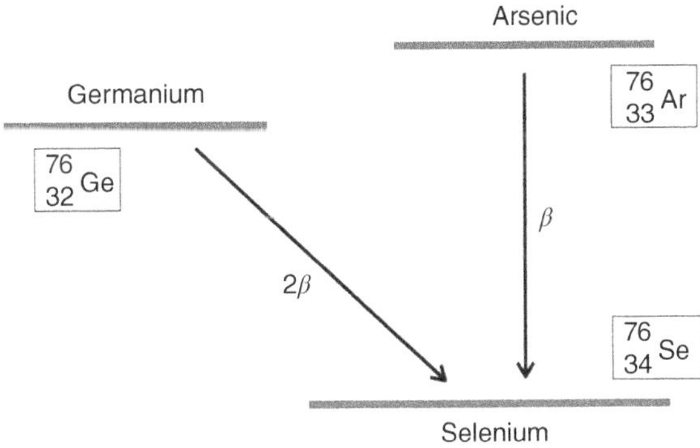

Figure 10.10 The neutrinoless 2β-decay from germanium to selenium. Also shown is the ordinary β-decay from arsenic to selenium.

I can write a flavor state, say $|\nu_e\rangle$, as a coherent superposition

$$|\nu_e\rangle = U_{e1}|\nu_1\rangle + U_{e2}|\nu_2\rangle\,, \tag{10.89}$$

which I can parametrize as

$$\begin{pmatrix} \nu_e \\ \nu_\mu \end{pmatrix} = \begin{pmatrix} \cos\theta & \sin\theta \\ -\sin\theta & \cos\theta \end{pmatrix} \begin{pmatrix} \nu_1 \\ \nu_2 \end{pmatrix}\,. \tag{10.90}$$

Neutrinos are produced as flavor eigenstates that I can write in this angular parametrization as

$$|\nu_e\rangle = \cos\theta|\nu_1\rangle + \sin\theta|\nu_2\rangle, \tag{10.91}$$

but they evolve in space-time as mass eigenstates

$$|\nu_i\rangle e^{-iE_it+i\vec{p}_i\cdot\vec{x}}\,. \tag{10.92}$$

An actual process consists in the production of an electron-neutrino flavor state at a certain location $\vec{x}_P$, its propagation as mass eigenstates ν_j and detection in $\vec{x}_D$ at a distance $\vec{L} = \vec{x}_P - \vec{x}_D$. To be definite, consider the production of electron neutrinos in the Sun as boron becomes beryllium. These neutrinos can be detected on Earth by inverse β-decay in a target detector made of chlorine, by measuring how much argon is produced. The process is depicted in Figure 10.11.

The amplitudes for production and detection are given by the usual Fermi effective Lagrangians, which can be used to write the vertex contributions at the production and detection points:[10.11]

10.11 Neglecting neutral current contributions.

$$\begin{aligned} V_{xp} &= -\frac{G_F}{\sqrt{2}}\sum_j U^*_{ej}\bar{u}_{\nu_j}(x)\gamma^\lambda(1-\gamma_5)v_e(x)\varkappa^\dagger_\lambda(x) \\ V_{xD} &= -\frac{G_F}{\sqrt{2}}\sum_j U_{ej}\bar{u}_e(x)\gamma^\lambda(1-\gamma_5)u_{\nu_j}(x)\varkappa_\lambda(x), \end{aligned} \tag{10.93}$$

with the nuclear contribution to the β-decay given by[10.12]

10.12 θ_c is the Cabibbo angle in the CKM matrix.

$$\varkappa_\lambda(x) = \cos\theta_c\bar{\psi}_p(x)\gamma_\lambda(1-\tilde{g}\gamma_5)\psi_n(x)\,. \tag{10.94}$$

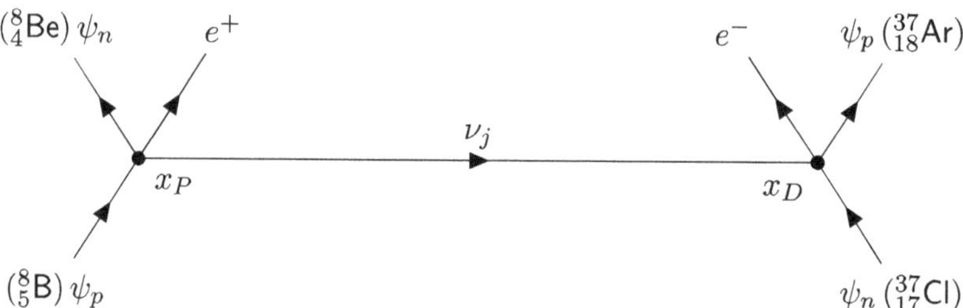

Figure 10.11 Diagram for the production, propagation and detection of neutrinos; ψ_n and ψ_p are neutron and proton wave functions.

If I were to take plane-wave solutions for all the particles then I would have the production and detection points spread over the entire space (the probability of finding a free particle is equal to 1 over all space-time). No oscillation would be possible because the momentum of the propagating neutrino would be fixed and accordingly also its mass eigenstate. Instead, what I need is to keep the coherent superposition of the mass eigenstate. This coherent propagation can only be accomplished by localizing the production and detection wave functions by writing

$$\begin{aligned}\psi_p(x) &= \Psi_p(\vec{x}-\vec{x}_P)\exp(-iE_p^P t)\\ \psi_n(x) &= \Psi_n(\vec{x}-\vec{x}_P)\exp(-iE_n^P t),\end{aligned} \tag{10.95}$$

for the nucleons in the nuclei of boron that are becoming beryllium and, similarly,

$$\begin{aligned}\psi_p(x) &= \Psi_p(\vec{x}-\vec{x}_D)\exp(-iE_p^D t)\\ \psi_n(x) &= \Psi_n(\vec{x}-\vec{x}_D)\exp(-iE_n^D t)\end{aligned} \tag{10.96}$$

for those in chlorine that are becoming argon. The wave functions $\Psi(\vec{x}-\vec{x}_P)$ and $\Psi(\vec{x}-\vec{x}_D)$ are wave packets centered around the production point $\vec{x}_P$ and the detection point $\vec{x}_D$. They can be taken to be Gaussians centered at the production and detection points with a spread given by σ:

$$\Psi(\vec{x}) = \frac{1}{(\sqrt{2\pi}\sigma)^{2/3}} e^{-\vec{x}^2/4\sigma^2}\,. \tag{10.97}$$

After inserting the propagator in space-time from x_1 to x_2 of the neutrino ν_j of mass m_j,

$$i\int \frac{\mathrm{d}^4 q}{(2\pi)^4} e^{-iq\cdot(x_1-x_2)}\frac{\not{q}+m_j}{q^2-m_j^2}\,, \tag{10.98}$$

and plane waves for the positron emission at the production site and for the electron at the detection site, I obtain the amplitude for the electron neutrino produced in x_P to be detected in x_D:

$$\begin{aligned}\mathcal{M}(\nu_e\to\nu_e) = \langle \nu_e(x_D)|\nu_e(x_P)\rangle &= i\frac{G_F^2}{2}\sum_j \int \mathrm{d}^4x_1\mathrm{d}^4x_2 \int \frac{\mathrm{d}^4 q}{(2\pi)^4} e^{-iq\cdot(x_1-x_2)}\\ &\times e^{-i(E_n^P-E_p^P)t_1} e^{-i(E_n^P-E_p^P)t_2} e^{ip_1\cdot x_1+ip_2\cdot x_2}\\ &\times \varkappa_\lambda^*(\vec{x}_1-\vec{x}_P)v_e(\vec{p}_1)\gamma^\lambda(1-\gamma_5)\\ &\times U_{ej}^*\frac{\not{q}+m_j}{q^2-m_j^2}U_{ej} \longleftarrow \nu_j \text{ propagator from } x_P \text{ to } x_D\\ &\times \gamma^\mu(1-\gamma_5)\bar{u}_e(\vec{p}_2)\varkappa_\mu(\vec{x}_2-\vec{x}_D),\end{aligned} \tag{10.99}$$

where the hadron current must now be written in terms of wave packets:

$$\varkappa_\lambda(x) = \cos\theta_c\bar{\Psi}_p(x)\gamma_\lambda(1-\tilde{g}\gamma_5)\Psi_n(x)\,. \tag{10.100}$$

10.13 $\int d^4x\, e^{-ik\cdot x} f(x+b) = e^{ik\cdot b}\tilde{f}(k)$.

The space-time integrals can be performed to find[10.13]

$$\mathcal{M}(\nu_e \to \nu_e) = i\frac{G_F^2}{2(2\pi)^2} e^{-i\vec{p}_1\cdot x_P - i\vec{p}_2\cdot x_D} \sum_j \int d^4q\, \delta(q_0+E_1)\delta(q_0+E_2)e^{-i\vec{q}\cdot\vec{L}}$$
$$\times\ \tilde{\varkappa}_\lambda^*(\vec{p}_1-\vec{q})\nu_e(\vec{p}_1)\gamma^\lambda(1-\gamma_5)$$
$$\times\ U_{ej}^*\frac{\not{q}+m_j}{q^2-m_j^2}U_{ej}$$
$$\times\ \bar{u}_e(\vec{p}_2)\gamma^\mu(1-\gamma_5)\tilde{\varkappa}_\mu(\vec{p}_2+\vec{q}), \tag{10.101}$$

where I have defined

$$E_1 = E_p^P - E_n^P - p_1^0, \qquad E_2 = E_p^D - E_n^D + p_2^0, \tag{10.102}$$

and the tilde indicates Fourier-transformed functions.

To perform the integration over q, I need a result that I am not going to prove, namely that[5]

$$\int d^3q\, F(\vec{q})e^{-i\vec{q}\cdot\vec{L}}\frac{1}{A-\vec{q}^{\,2}} \underset{L\to\infty}{\longrightarrow} -\frac{2\pi^2}{L}F\left(\frac{\sqrt{A}\vec{L}}{L}\right), \tag{10.103}$$

where F is an arbitrary function and $L = |\vec{L}|$. Using Eq. (10.103) (with $A = q_0^2 - m_j^2$) to integrate the amplitude in Eq. (10.101), I find (in the limit $L \to \infty$)[10.14]

10.14 The factor $1/L$ accounts for the diminishing flux of the produced neutrinos as they travel in space.

$$\mathcal{M}(\nu_e \to \nu_e) = -i\frac{G_F^2}{4}\frac{1}{L}\delta(E_1-E_2)e^{-i\vec{p}_1\cdot x_P - i\vec{p}_2\cdot x_D}\sum_j U_{ej}^* U_{ej} e^{iq_jL}$$
$$\times\ \tilde{\varkappa}_\lambda^*(\vec{p}_1+q_j\vec{l})\nu_e(\vec{p}_1)\gamma^\lambda(1-\gamma_5)(q_0\gamma^0+q_j\vec{l}\cdot\vec{\gamma}+m_j)$$
$$\times\ \bar{u}_e(\vec{p}_2)\gamma^\mu(1-\gamma_5)\tilde{\varkappa}_\mu(\vec{p}_2+q_j\vec{l}), \tag{10.104}$$

with $\vec{l} = \vec{L}/L$, $q_0 = -E_1 = -E_2$ and $q_j = \sqrt{q_0^2 - m_j^2}$.

I can rewrite the term

$$q_0\gamma^0 + q_j\vec{l}\cdot\vec{\gamma} + m_j = -\not{p}_j + m_j = -\sum_s u_{\nu_j}^s(\vec{p}_j)\bar{u}_{\nu_j}^s(\vec{p}_j), \tag{10.105}$$

so that the amplitude in Eq. (10.104) contains explicitly the product of the amplitudes A_j^P and A_j^D for the production and detection of the neutrino ν_j.

Making an approximation for small masses $E_j \gg m_j$ (the neutrino masses are expected in any case to be very small) I can write

$$q_j \simeq E_i\left(1-\frac{m_i^2}{E_i^2}\right)^{1/2} \simeq E_i - \frac{m_i^2}{2E_i}, \tag{10.106}$$

[5] I have mostly followed W. Grimus and P. Stockinger, arXiv: hep-ph/9603430, where they prove this formula as well.

and the amplitude as

$$M(\nu_e \to \nu_e) = \sum_j A_j^P A_j^D e^{iE_j} \underbrace{U_{ej}^* U_{ej} e^{im_j^2 L/2E_j}}_{\text{neutrino oscillation}}, \tag{10.107}$$

where the quantities

$$\begin{aligned} A_j^P &= \tilde{\varkappa}_\lambda^*(\vec{p}_1 + q_j\vec{l})v_e(\vec{p}_1)\gamma^\lambda(1-\gamma_5)\bar{u}_{\nu_j}(\vec{p}_j) \\ A_j^D &= u_{\nu_j}(\vec{p}_j)\gamma^\mu(1-\gamma_5)\bar{u}_e(\vec{p}_2)\tilde{\varkappa}_\mu(\vec{p}_2 + q_j\vec{l}) \end{aligned} \tag{10.108}$$

refer only to the neutrino production and detection mechanisms.

If the nuclear currents $\tilde{H}$ were completely localized,[10.15] the values of q_j would be all equal and there would be no oscillation since the neutrino would be forced to be in one, and only one, mass eigenstate. For this reason I must take the nuclear currents to be wave packets with some spread (as in Eq. (10.100)) so that the neutrino momenta q_1 and q_2 could be different as long as

10.15 Take $q_1\vec{l} = -\vec{p}_1$, which would be the case if the function $\tilde{H}$ had no spread and thus was collapsed to a δ-function. Then q_2 would have to satisfy $q_2\vec{l} + \vec{p}_1 = q_2\vec{l} - q_1\vec{l} = 0$ and so $q_1 = q_2$.

$$|q_1 - q_2| \leq \sigma_P \quad \text{and} \quad |q_1 - q_2| \leq \sigma_D, \tag{10.109}$$

where σ_P and σ_D are the spread of the functions $\tilde{H}$ at the production and detection points, respectively. The spread of the wave packets in space-time must be smaller than the oscillation length L for the oscillation to take place. These conditions are the pay-off from the computation above: as long as the conditions in Eq. (10.109) are satisfied, the nuclear currents are not exponentially suppressed, but if the conditions in Eq. (10.109) are not satisfied then the amplitude in Eq. (10.107) vanishes.

The square of this expression gives the probability of finding an electron neutrino at the detector. It is going to be less than 1 because some of the electron neutrinos have oscillated into muon neutrinos.

Since I will be interested only in the mixing and the phase, I can normalize the amplitude terms in such a way that the wave-packet part is reabsorbed (but remember that it vanishes if the conditions in Eq. (10.109) are not satisfied) and write the amplitude as[10.16]

10.16 Why not start from here? Localization of the production and detection is necessary in order to understand how to preserve the coherence of the neutrino mass eigeinstates. If you insist on starting from the amplitude in Eq. (10.110) then you will have to explain why a neutrino with fixed momentum can have different masses.

$$\tilde{\mathcal{M}}(\nu_e \to \nu_e) = \sum_j U_{ej}^* U_{ej} e^{i\phi_j}, \tag{10.110}$$

where $\phi_j = \dfrac{m_j^2}{2E_j} L$.

The neutrino, after traveling the distance L, has the wave function (writing the mixing matrices U_{ej} explicitly in terms of the angle θ)

$$\begin{aligned} |\nu_e(L)\rangle &= \cos\theta\Big[\cos\theta|\nu_e\rangle - \sin\theta|\nu_\mu\rangle\Big]e^{-i\phi_1} + \sin\theta\Big[\sin\theta|\nu_e\rangle + \cos\theta|\nu_\mu\rangle\Big]e^{-i\phi_2} \\ &= |\nu_e\rangle\Big[\cos^2\theta e^{-i\phi_1} + \sin^2\theta e^{-i\phi_2}\Big] \\ &\quad + |\nu_\mu\rangle\Big[\sin\theta\cos\theta(e^{-i\phi_1} - e^{-i\phi_2})\Big], \end{aligned} \tag{10.111}$$

and the probability of finding a neutrino ν_e from a beam of ν_e is

$$\tilde{\mathcal{M}}(\nu_e \to \nu_e) = \langle \nu_e | \nu_e(L) \rangle = \sum_{j=1,2} U^*_{ej} U_{ej} e^{i\phi_j} , \tag{10.112}$$

so that

$$\begin{aligned} P(\nu_e \to \nu_e) &= |\tilde{\mathcal{M}}|^2 = |\langle \nu_e | \nu_e(L) \rangle|^2 \\ &= \cos^4\theta + \cos^4\theta \\ &\quad + \sin^2\theta \cos^2\theta (e^{-i\phi_2} - e^{-i\phi_1})(e^{+i\phi_2} - e^{-i\phi_1}) \\ &= 1 - \frac{1}{2}\sin^2 2\theta \Big[1 - \cos(\phi_1 - \phi_2)\Big] \\ &= 1 - \underbrace{\sin^2 2\theta}_{\text{mixing}} \underbrace{\sin^2\left(\frac{\phi_1 - \phi_2}{2}\right)}_{\text{oscillation}} , \end{aligned} \tag{10.113}$$

which is often written in convenient units as

$$P(\nu_e \to \nu_e) = 1 - \sin^2 2\theta_{e\mu} \sin^2 \left[1.27 \, \frac{\Delta m^2_{21}(\text{eV}^2) L(\text{km})}{E(\text{GeV})} \right] . \tag{10.114}$$

In order to see the oscillation, the argument of the sine in Eq. (10.114) must be a number of order 1, that is,

$$\frac{\Delta m^2 L}{E} = O(1) , \tag{10.115}$$

which means that, for very small differences in the masses, rather large values of the ratio between the energy and the baseline of neutrino experiment are needed, and *vice versa*:

$$\begin{cases} \Delta m^2 \text{ small} \Longrightarrow L/E \text{ large} \\ \Delta m^2 \text{ large} \Longrightarrow L/E \text{ small} . \end{cases} \tag{10.116}$$

This simple argument is behind the search for neutrino oscillations from the Sun (the long-baseline case) and in the Earth's atmosphere (the short-baseline case).

Since, in general, I expect to have three flavors of neutrinos, the mixing is given by[10.17]

$$\begin{pmatrix} \nu_e \\ \nu_\mu \\ \nu_\tau \end{pmatrix} = \underbrace{\begin{pmatrix} U_{e1} & U_{e2} & U_{e3} \\ U_{\mu 1} & U_{\mu 2} & U_{\mu 3} \\ U_{\tau 1} & U_{\tau 2} & U_{\tau 3} \end{pmatrix}}_{\text{PMNS matrix}} \begin{pmatrix} \nu_1 \\ \nu_2 \\ \nu_3 \end{pmatrix} , \tag{10.117}$$

[10.17] The PMNS matrix here plays the role of the CKM matrix in the quark sector. The acronym comes from the names of the physicists B. Pontecorvo, Z. Maki, M. Nakagawa and S. Sakata.

with, for example, the probability for the transition $\nu_e \to \nu_\mu$ now given by a straightforward generalization of Eq. (10.113):

$$\begin{aligned}P(\nu_e \to \nu_\mu) = 2\,\mathrm{Re}\left\{\left(U_{e1}U^*_{\mu 1}U_{e2}U^*_{\mu 2}\right)\left[e^{-i(\phi_1-\phi_2)}-1\right]\right\} \\ +\,2\,\mathrm{Re}\left\{\left(U_{e1}U^*_{\mu 1}U_{e3}U^*_{\mu 3}\right)\left[e^{-i(\phi_1-\phi_3)}-1\right]\right\} \\ +\,2\,\mathrm{Re}\left\{\left(U_{e2}U^*_{\mu 2}U_{e3}U^*_{\mu 3}\right)\left[e^{-i(\phi_2-\phi_3)}-1\right]\right\},\end{aligned} \tag{10.118}$$

where

$$\mathrm{Re}\left[e^{-i(\phi_i-\phi_j)}-1\right] = \cos(\phi_j-\phi_i)-1 = 2\sin^2\frac{\Delta m^2_{ji}L}{4E}. \tag{10.119}$$

The evidence for neutrino flavor oscillations comes (originally) from two sets of experiments. There are now many more experiments but let me just limit myself to these historically crucial ones:

- **Solar neutrinos.** The Sun produces a lot of neutrinos as it goes about burning its nuclear fuel. The three most important reactions, which produce neutrinos with an energy spectrum shown in Figure 10.12, are

$$p+p \to d+e^+\nu_e \tag{10.120}$$

$$e^- +{}^7\,\mathrm{Be} \to {}^7\mathrm{Li}+\nu_e \tag{10.121}$$

$${}^8_5\mathrm{B} \to {}^8_4\mathrm{Be}^* + e^+\nu_e\,. \tag{10.122}$$

When I say a lot of neutrinos I mean about 10^{38} neutrinos per second.

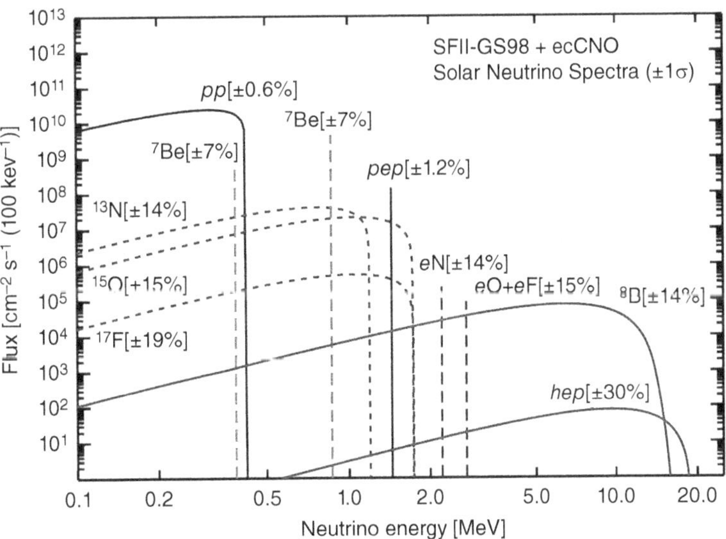

Figure 10.12 The energy spectra of solar neutrinos [R.L. Workman *et al.* (Particle Data Group), *Prog. Theor. Exp. Phys.* **2022** (2022) 083C01 http://pdg.lbl.gov].

This number gives a flux of neutrinos on Earth of about $6 \times 10^{10}/\mathrm{cm}^2$, which is indeed a huge number and it was only natural for someone (Ray Davis since 1964 at the Homestake mine) to try to measure it, the very small cross section notwithstanding. The detector consisted of chlorine and the reaction is an inverse β-decay:

$$\nu_e + {}^{37}_{17}\mathrm{Cl} \rightarrow {}^{37}_{18}\mathrm{Ar} + e^- . \tag{10.123}$$

This experiment, and those that followed (SNO at Sudbury since 2002 and Super-Kamiokande since 1996) have consistently found a flux roughly one-third smaller than what was expected.

- **Atmospheric neutrinos.** Cosmic rays beget pions, which beget muons, which beget neutrinos, through the decays

$$\begin{aligned} \pi^+ &\rightarrow \mu^+ + \nu_\mu \\ &\qquad \mu^+ \rightarrow e^+ + \nu_e + \bar{\nu}_\mu \end{aligned} \tag{10.124}$$

and

$$\begin{aligned} \pi^- &\rightarrow \mu^- + \bar{\nu}_\mu \\ &\qquad \mu^- \rightarrow e^- + \bar{\nu}_e + \nu_\mu . \end{aligned} \tag{10.125}$$

The results shown in Figure 10.13 are from the original work that reported these fluxes. A schematic of the experimental setup is shown in Figure 10.14. By simple arithmetic I expect the number of ν_μ to be twice as large as that of ν_e, that is,

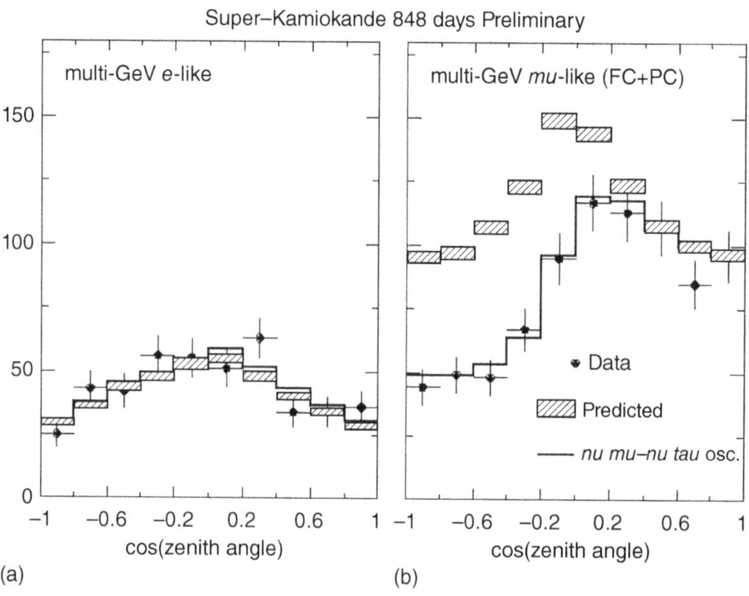

Figure 10.13 Neutrino flux as a function of the zenith angle above the detector. (a) The zenith angle distribution for ν_e. (b) The zenith angle distribution for ν_μ. [R.L. Workman *et al.* (Particle Data Group), *Prog. Theor. Exp. Phys.* **2022** (2022) 083C01 http://pdg.lbl.gov.]

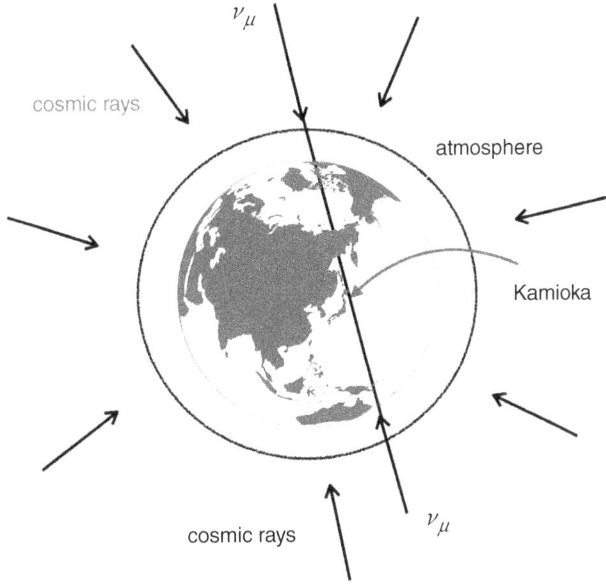

Figure 10.14 The experimental setup for the detection of atmospheric neutrinos at SuperKamiokonde. The muon neutrinos coming from above go only through the atmosphere while those coming from below also go through the Earth.

$$\frac{N_{\nu_\mu+\bar{\nu}_\mu}}{N_{\nu_e+\bar{\nu}_e}} \simeq 2\,. \tag{10.126}$$

Again, this expectation has been contradicted by experiments, which have found

$$\frac{N_{\nu_\mu+\bar{\nu}_\mu}}{N_{\nu_e+\bar{\nu}_e}} = 0.7 \pm 0.1\,. \tag{10.127}$$

Data on atmospheric neutrinos also provides information on the asymmetry

$$A = \frac{N^{\text{Up}}_{\nu_\mu} - N^{\text{Down}}_{\nu_\mu}}{N^{\text{Up}}_{\nu_\mu} + N^{\text{Down}}_{\nu_\mu}} = 0.3 \pm 0.05 \tag{10.128}$$

of the distribution of the neutrino flux from below and above the detector, that is, its variation with the zenith angle, which again can be used to study neutrino oscillations.

When all these data (plus the many more coming from terrestrial experiments with nuclear reactors) are combined, the neutrino parameters are constrained. The constraints on possible values of the first and third columns of the PMNS mixing matrix (10.117) are shown in the plot in Figure 10.15.

The neutrino sector has **six** parameters: two mass differences Δm_{21}, Δm_{32}, three angles θ_{12}, θ_{21}, θ_{32} and one phase δ_{CP} (**two** more if they are

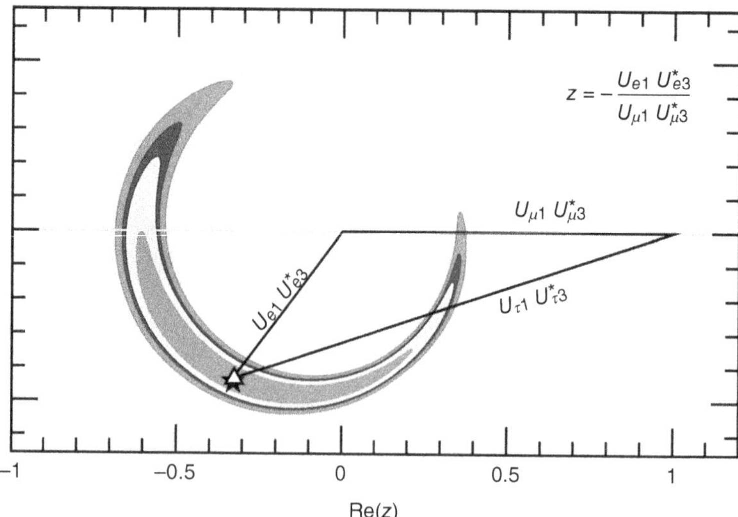

Figure 10.15 Leptonic unitarity triangle for the first and third columns of the PMNS mixing matrix *U*. [R.L. Workman *et al.* (Particle Data Group), *Prog. Theor. Exp. Phys.* **2022** (2022) 083C01 http://pdg.lbl.gov.]

Majorana particles) parametrizing the PNMS matrix in the standard form. All these parameters are known and are given by[6]

$$|\Delta m_{21}|^2 = (7.53 \pm 0.18) \times 10^{-5}\ \text{eV}^2 \quad \text{(Sun)} \tag{10.129}$$

$$|\Delta m_{32}|^2 = (2.453 \pm 0.033) \times 10^{-3}\ \text{eV}^2 \quad \text{(atmosphere)} \tag{10.130}$$

$$\sin^2 2\theta_{13} = 0.0218 \pm 0.0007 \quad \text{(SNO+KamLAND)} \tag{10.131}$$

$$\sin^2 2\theta_{12} = 0.307\,^{+0.013}_{-0.012} \quad \text{(KamLAND+Sun)} \tag{10.132}$$

$$\sin^2 2\theta_{23} = 0.539 \pm 0.022 \tag{10.133}$$

$$\delta_{CP} = 1.36 \pm 0.17. \tag{10.134}$$

This is a major achievement. All the parameters of the leptonic sector of the Standard Model are almost as well known as those of the quark sector. The PNMS matrix elements seem to follow a pattern without the hierarchical structure that, in the case of the quarks, made it possible to write the mixing matrix in the Wolfenstein form.

The data from oscillations only give values for the squares of the mass differences in Eq. (10.129) and Eq. (10.130). Therefore two possible scenarios are possible for these masses, as illustrated in Figure 10.16. Either there are two very light (and very close) neutrino masses to explain the solar data and a slightly heavier one that enters in the atmospheric data (the normal hierarchy), or only one very light neutrino mass, and the other two above it (the inverted hierarchy). The absolute value of the masses is still unknown though bounded by cosmological constraints.

6 PDG (2020) http://pdg.lbl.gov.

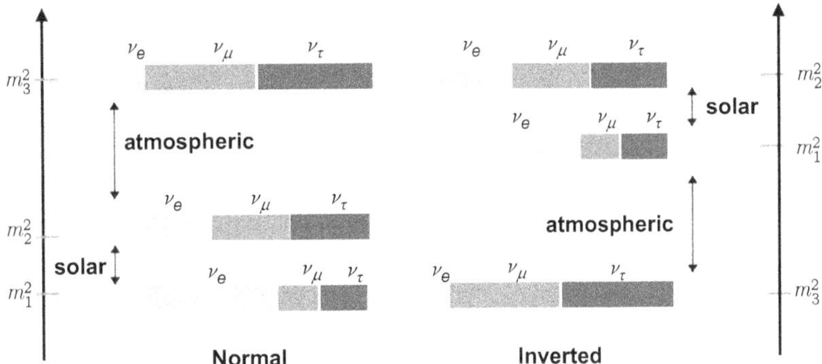

Figure 10.16 Normal and inverted hierarchies for the neutrino masses. The lengths of the rectangles represent the weights of the different flavor components in the mass eigenstates.

10.6 *Problem Session*: Oscillations in Matter

Let me start again. Massive neutrinos propagate following the equation

$$i\frac{\mathrm{d}}{\mathrm{d}t}\begin{pmatrix}\nu_1\\ \nu_2\end{pmatrix} = H\begin{pmatrix}\nu_1\\ \nu_2\end{pmatrix}, \tag{10.135}$$

where I have the free Hamiltonian

$$H = \frac{1}{2E}\begin{pmatrix}m_1^2 & 0\\ 0 & m_2^2\end{pmatrix}. \tag{10.136}$$

The Hamiltonian can be rewritten in terms of the flavor eigenstates by rotating the mass states with the matrix

$$U = \begin{pmatrix}\cos\theta & \sin\theta\\ -\sin\theta & \cos\theta\end{pmatrix}, \tag{10.137}$$

so that Eq. (10.138) becomes

$$i\frac{\mathrm{d}}{\mathrm{d}t}\begin{pmatrix}\nu_e\\ \nu_\mu\end{pmatrix} = UHU^\dagger\begin{pmatrix}\nu_e\\ \nu_\mu\end{pmatrix}, \tag{10.138}$$

where

$$UHU^\dagger = \frac{m_1^2+m_2^2}{4E}\begin{pmatrix}1 & 0\\ 0 & 1\end{pmatrix} + \frac{\Delta m^2}{4E}\begin{pmatrix}\cos 2\theta & \sin 2\theta\\ \sin 2\theta & \cos 2\theta\end{pmatrix}. \tag{10.139}$$

Neutrinos have charged and neutral current interactions. When they propagate in matter, these interactions add a term to Eq. (10.138), which becomes

$$i\frac{\mathrm{d}}{\mathrm{d}t}\begin{pmatrix}\nu_e\\ \nu_\mu\end{pmatrix} = \underbrace{\frac{\Delta m^2}{4E}\begin{pmatrix}-\cos 2\theta & \sin 2\theta\\ \sin 2\theta & \cos 2\theta\end{pmatrix}}_{\text{in vacuum}}\begin{pmatrix}\nu_e\\ \nu_\mu\end{pmatrix} + \underbrace{\begin{pmatrix}V_e & 0\\ 0 & V_\mu\end{pmatrix}}_{\text{in matter}}\begin{pmatrix}\nu_e\\ \nu_\mu\end{pmatrix}, \tag{10.140}$$

where the terms

$$V_e = \sqrt{2}G_F\left(N_e - \frac{N_n}{2}\right) \quad \text{and} \quad V_\mu = \sqrt{2}G_F\left(-\frac{N_n}{2}\right) \tag{10.141}$$

take into account the propagation in matter. The two quantities in Eq. (10.141) require a bit of explanation. The effective Lagrangian for the interaction of neutrinos in matter is given by the charged current

$$-\mathcal{L}_{CC} = -\frac{G_F}{\sqrt{2}}\bar{\nu}_e(1-\gamma^5)\gamma^\mu\nu_e\bar{e}(1-\gamma^5)\gamma_\mu e \tag{10.142}$$

and the neutral current

$$-\mathcal{L}_{NC} = -\frac{G_F}{\sqrt{2}}\bar{\nu}_e(1-\gamma^5)\gamma^\mu\nu_e\bar{e}[\gamma_\mu(1-\gamma^5)g_L^e+(1+\gamma^5)g_R^e]e, \tag{10.143}$$

with $g_L^e = -1/2 + \sin^2\theta_W$ and $g_R^e = \sin^2\theta_W$.

A number of simplifications to these interactions can be made because I am in the low-energy regime. First I can neglect terms containing a γ^5 because matter can be assumed to be unpolarized. I can take the product of the remaining currents at rest, where only the γ^0 component survives (the total momentum of matter is zero), and set $\bar{e}\gamma^0 e = e^\dagger e = N_e$, the electron density.[10.18] I can also do the same for the other fermions in matter (namely, protons and neutrons). I will also neglect $\sin^2\theta_W$. Under these simplifications I have that

10.18 The electron number density varies widely. To provide some context: N_e is approximately 10^{-16} eV^3 averaged over the galaxy and 10^{11} eV^3 averaged inside the Sun.

$$-\mathcal{L}_{CC} \simeq \bar{\nu}V_e^{CC}\nu_e \quad \text{and} \quad -\mathcal{L}_{NC} \simeq \bar{\nu}V_e^{NC}\nu_e, \tag{10.144}$$

in which $V_e^{CC} = \sqrt{2}G_F N_e$ and $V_e^{CC} = -G_F N_e/\sqrt{2}$. To obtain Eq. (10.141) I have to remember that the only charged current affecting the propagation of neutrinos is the one for electron neutrinos while the muon neutrinos see only the neutral current (there are no muons in matter). Final point: the overall neutrality of matter makes the contribution of electrons and protons cancel out in the neutral interactions.

Equation (10.140) is best rewritten by introducing a new mixing angle and mass difference as follows:

$$\frac{\Delta m_m^2}{4E}\begin{pmatrix} -\cos 2\theta_m & \sin 2\theta_m \\ \sin 2\theta_m & \cos 2\theta_m \end{pmatrix}, \tag{10.145}$$

where

$$\Delta m_m^2 = \sqrt{(\Delta m^2\cos 2\theta - 2\sqrt{2}G_F N_e E)^2 + (\Delta m^2 \sin 2\theta)^2} \tag{10.146}$$

and

$$\tan 2\theta_m = \frac{\tan 2\theta}{\underbrace{\frac{\Delta m^2}{2E}\cos 2\theta - \sqrt{2}G_F N_e}_{\text{resonance when } = 0}} \tag{10.147}$$

are the mass difference and mixing angle for neutrinos propagating in matter.

Equation (10.147) gives the resonance condition for which the mixing is maximal, $\theta_m = \pi/4$, in a manner similar to the two coupled pendulums for which $\tan\theta = \infty$ corresponds to the mixing being maximal. The maximal mixing amplifies the neutrino oscillations and can be used to improve the fit to the experimental results.

As it often happens, things are not so simple when we try to compare formulas and experimental results. The medium through which neutrinos propagate – be it the Sun or the Earth's atmosphere and core – is not uniform and the equation describing neutrino propagation is thus more complicated and must be solved numerically. Yet the main idea is there and is beautiful.

10.7 Exercises

1. ***B*-mesons.** Consider the mass difference between the neutral *B*-mesons. What is the typical scale for $\Delta B = 2$ oscillations compared with that for $\Delta S = 2$ of the kaon system?
2. ***T* non-conservation.** Why does *CP* violation imply *T* non-conservation?
   [Hint: Think of the *CPT* theorem of quantum field theory.]
   What do you reckon: is the violation of *CP* invariance in the kaon system sufficient to explain the thermodynamical arrow of time?
3. **μ Collider.** In a μ collider muons are accelerated instead of electrons. This has the advantage of reduced Bremsstrahlung radiation and the possibility of reaching very large center of mass energies. As the muons go around the collider, some of them decay. Given the muon lifetime and velocity (for a 10 TeV machine with a radius of 3 km), how many times would a typical muon go around the collider? How many neutrinos do you expect to have because of μ-decays? Should we be concerned?
4. **Majorana neutrinos.** How would you use flavor oscillations to distinguish between Majorana and Dirac neutrinos?
   [Hint: You cannot; explain why.]
5. **Quark oscillations.** Forgetting that quarks are confined inside hadrons, why cannot flavor oscillations between the two first generations of quarks be observed?
   [Hint: Use the values of the masses of the quarks *s* and *d* to estimate the oscillation length for an average energy of about 100 MeV.]

APPENDIX A

Numbers, Formulas and Feynman Rules

For ease of perusal, I collect in this appendix the values of some parameters, the Feynman rules and other useful formulas given in the main text and necessary in computations.

A.1 Parameters and Observables

Numerical values from CODATA, website: https://physics.nist.gov/cuu/Constants/index.html; PDG (2020), website: https://pdg.lbl.gov/.

$$\begin{aligned}
\alpha(0) &= e^2/4\pi = 1/137.035999084... \\
G_F &= 1.1663787(6) \times 10^{-5} \text{ GeV}^{-2} \\
\sin^2\theta_W(m_Z) &= 0.23122(4) \quad (\overline{MS}) \\
\alpha_s(m_Z) &= 0.1181(11)
\end{aligned}$$

$$\begin{aligned}
m_e &= 0.51099899461 \pm 0.000000031 \\
m_\mu &= 105.6583715 \pm 0.0000024 \\
m_\tau &= 1776.86 \pm 0.12 \qquad \text{(in MeV)}
\end{aligned}$$

$$|\Delta m_{21}|^2 = (7.53 \pm 0.18) \times 10^{-5} \quad |\Delta m_{32}|^2 = (2.453 \pm 0.033) \times 10^{-3} \quad (\text{in eV}^2)$$

$$\begin{aligned}
&m_u = 2.16^{+0.49}_{-0.26} \quad m_d = 4.67^{+0.48}_{-0.17} \quad m_s = 93^{+11}_{-5} \quad (\overline{MS}, \text{in MeV}) \\
&m_c = 1.27 \pm 0.02 \quad m_b = 4.18^{+0.03}_{-0.02} \quad m_t = 172.76 \pm 0.30 \quad (\overline{MS}, \text{in GeV})
\end{aligned}$$

$$m_W = 80.379(12) \quad m_Z = 91.1876(21) \quad m_h = 125.25 \pm 0.17 \quad \text{(in GeV)}$$

$$\begin{aligned}
m_p &= 938.27200813 \pm 0.0000058 & m_n &= 939.5654133 \pm 0.0000058 \\
m_{\pi^0} &= 134.9768 \pm 0.0005 & m_{\pi^\pm} &= 139.57039 \pm 0.00017 \\
m_\rho &= 775.26 \pm 0.23 & & \\
m_{K^0} &= 497.611 \pm 0.013 & m_{K^\pm} &= 497.611 \pm 0.013 \quad \text{(in MeV)}
\end{aligned}$$

$$\boxed{V_{\text{CKM}}:}$$

$$\sin\theta_{12} = 0.22650 \pm 0.00048$$

$$\sin\theta_{13} = 0.00361^{+0.00011}_{-0.00009}$$

$$\sin\theta_{23} = 0.04053^{+0.00083}_{-0.00061}$$

$$\delta_{CP} = 1.196^{+0.045}_{-0.043}$$

$$\boxed{V_{\text{PMNS}}:}$$

$$\sin^2 2\theta_{13} = 0.0218 \pm 0.0007$$

$$\sin^2 2\theta_{12} = 0.307\,^{+0.013}_{-0.012}$$

$$\sin^2 2\theta_{23} = 0.539 \pm 0.022$$

$$\delta_{CP} = 1.36 \pm 0.17$$

A.2 Diracology

A.2.1 Dirac Equation

$$(\not{p} - m)_{ab} u^b(p,\sigma) = 0 \quad \text{and} \quad (\not{p} + m)_{ab} v^b(p,\sigma) = 0$$

with wave functions given by

$$u^b(p,\sigma) = \sqrt{E+m}\begin{pmatrix} \chi_\sigma \\ \dfrac{\vec{\sigma}\cdot\vec{p}}{E+m}\chi_\sigma \end{pmatrix}$$

and

$$v^b(p,\sigma) = \sqrt{E+m}\begin{pmatrix} \dfrac{\vec{\sigma}\cdot\vec{p}}{E+m}\chi_\sigma \\ \chi_\sigma \end{pmatrix},$$

and with the spinors

$$\chi_{+1/2} = \begin{pmatrix} 1 \\ 0 \end{pmatrix} \quad \text{and} \quad \chi_{-1/2} = \begin{pmatrix} 0 \\ 1 \end{pmatrix}.$$

A.2.2 Dirac Matrices

$$\{\gamma^\mu, \gamma^\nu\} = 2g^{\mu\nu}\,\mathbb{1}$$

with metric tensor $g = \text{diag}(1, -1, -1, -1)$.

Also

$$\gamma_5 = -\frac{i}{4!}\epsilon_{\mu\nu\rho\sigma}\gamma^\mu\gamma^\nu\gamma^\rho\gamma^\sigma$$

with

$$\{\gamma_5, \gamma^\mu\} = 0$$

and (the Levi–Civita tensor)

$$\epsilon^{\mu\nu\rho\sigma} = \begin{cases} 1 & \text{even permutation of } (\mu\nu\rho\sigma) \\ -1 & \text{odd permutation of } (\mu\nu\rho\sigma) \\ 0 & \text{otherwise} \end{cases}$$

and

$$\begin{aligned} \epsilon_{\mu\nu\rho\sigma} &= -\epsilon^{\mu\nu\rho\sigma} \\ \epsilon^{\alpha\beta\gamma\delta}\epsilon_{\alpha\beta\gamma\delta} &= -24 \\ \epsilon^{\alpha\beta\gamma\mu}\epsilon_{\alpha\beta\gamma\nu} &= -6\,\delta^{\mu}_{\nu} \\ \epsilon^{\alpha\beta\mu\nu}\epsilon_{\alpha\beta\rho\sigma} &= -2\,(\delta^{\mu}_{\rho}\delta^{\nu}_{\sigma} - \delta^{\mu}_{\sigma}\delta^{\nu}_{\rho})\,. \end{aligned}$$

A.2.3 Gordon Identity

$$\bar{u}(p')\gamma^{\mu}u(p) = \frac{1}{2m}\bar{u}(p')(p+p')^{\mu}u(p) + \frac{i}{2m}\bar{u}(p')\sigma^{\mu\nu}(p-p')_{\nu}u(p)$$

A.2.4 Useful Relations

$$\begin{aligned} \gamma^{\mu}\gamma_{\mu} &= 4 \\ \gamma^{\mu}\gamma_{\alpha}\gamma_{\mu} &= -2\gamma_{\alpha} \\ \gamma^{\mu}\gamma^{\nu}\gamma^{\rho}\gamma_{\mu} &= 4\,g^{\nu\rho} \\ \gamma^{\mu}\gamma_{\alpha}\gamma_{\beta}\gamma_{\mu} &= -2\,\gamma_{\alpha}\gamma_{\beta} + 4\,g_{\alpha\beta} \\ \gamma^{\mu}\gamma_{\alpha}\gamma_{\beta}\gamma_{\delta}\gamma_{\mu} &= -2\,\gamma_{\delta}\gamma_{\beta}\gamma_{\alpha}\,. \end{aligned}$$

Also used:

$$\sigma_{\mu\nu} = \frac{i}{2}[\gamma^{\mu}, \gamma^{\nu}]$$

with space components

$$\sigma_{ij} = \Sigma_k = \begin{pmatrix} \sigma_k & 0 \\ 0 & \sigma_k \end{pmatrix}.$$

A.2.5 Hermitian Conjugation

$$\begin{aligned} \gamma^{0}\gamma^{\mu}\gamma^{0} &= \gamma^{\mu\dagger} \\ \gamma^{0}\gamma_{5}\gamma^{0} &= -\gamma_{5}^{\dagger} = \gamma_{5} \\ \gamma^{0}\sigma^{\mu\nu}\gamma^{0} &= \sigma^{\mu\nu\dagger} \\ (\bar{u}_a\Gamma u_b)^{*} &= \bar{u}_b(\gamma^{0}\Gamma^{\dagger}\gamma^{0})u_a \quad \text{for any product of Dirac matrices } \Gamma\,. \end{aligned}$$

A.2.6 Fierz Identities

$$\left[(\mathbb{1}\pm\gamma^{5})\gamma^{\mu}\right]_{ij}\left[(\mathbb{1}\pm\gamma^{5})\gamma^{\mu}\right]_{kl} = -\left[(\mathbb{1}\pm\gamma^{5})\gamma^{\mu}\right]_{il}\left[(\mathbb{1}\pm\gamma^{5})\gamma^{\mu}\right]_{kj}$$

$$\left[(\mathbb{1}\pm\gamma^{5})\gamma^{\mu}\right]_{ij}\left[(\mathbb{1}\mp\gamma^{5})\gamma^{\mu}\right]_{kl} = 2(\mathbb{1}\pm\gamma^{5})_{il}(\mathbb{1}\mp\gamma^{5})_{kj}$$

plus more complicated identities.

A.2.7 Traces

$$
\begin{aligned}
\mathrm{Tr}\,\mathbb{1} &= 4 \\
\mathrm{Tr}\ (\text{odd number of } \gamma s) &= 0 \\
\mathrm{Tr}\,\gamma^\mu\gamma^\nu &= 4\,g^{\mu\nu} \\
\mathrm{Tr}\,\gamma^\mu\gamma^\nu\gamma^\rho\gamma^\sigma &= 4\,(g^{\mu\nu}g^{\rho\sigma} - g^{\mu\rho}g^{\mu\sigma} + g^{\mu\sigma}g^{\mu\rho}) \\
\mathrm{Tr}\,\gamma^\mu\gamma^\nu\gamma^\rho\gamma^\sigma\gamma_5 &= -4i\,\epsilon^{\mu\nu\rho\sigma} \\
\mathrm{Tr}\,\gamma_5 = \mathrm{Tr}\,\gamma_5\gamma^\mu\gamma^\nu &= 0
\end{aligned}
$$

A.3 *SU*(*N*) Group Factors

$$
\begin{aligned}
[t^A, t^B] &= if^{ABC}t^C \\
\mathrm{Tr}\,t^A_{ij}t^A_{jk} &= \frac{N^2-1}{2N}\delta_{ik} \\
\mathrm{Tr}\,t^At^B &= \frac{1}{2}\delta^{AB} && SU(2):\ \mathrm{Tr}\,\sigma_j\sigma_k = 2\delta_{jk}\mathbb{1} \\
\mathrm{Tr}\,t^At^Bt^C &= \frac{i}{4}f^{ABC} + \frac{1}{4}d^{ABC} && SU(2):\ \mathrm{Tr}\,\sigma_i\sigma_j\sigma_k = 2i\epsilon_{ijk} \\
\{t^A, t^B\} &= \frac{1}{N}\delta^{AB} + d^{ABC}t^C && SU(2):\ \{\sigma_j, \sigma_k\} = 2\delta_{jk}\mathbb{1} \\
t^At^B &= \frac{1}{2}\left[\frac{1}{N}\delta^{AB} + (d^{ABC} + if^{ABC})t^C\right] && SU(2):\ \sigma_j\,\sigma_k = \delta_{jk}\mathbb{1} + i\epsilon_{jki}\sigma_j \\
f^{ACD}f^{BCD} &= N\delta^{AB} \\
f^{ABC}t^Bt^C &= i\frac{N}{2}t^A \\
f^{ACD}f^{BCD} &= N\delta^{AB} \\
f^{ACD}f^{ACD} &= N(N^2-1)
\end{aligned}
$$

A.3.1 Fierz Identity

$$
t^A_{ab}t^A_{cd} = \frac{1}{2}\left(\delta_{ad}\delta_{bc} - \frac{1}{N}t^A_{ad}t^A_{bc}\right)
$$

A.4 Cross Sections and Decays

A.4.1 Units

$$
\begin{aligned}
\hbar c &= 197.3269804\ldots\ \mathrm{MeV\ fm} \\
1\ \mathrm{barn} &= 10^{-28}\ \mathrm{m}^2\,,
\end{aligned}
$$

where 1 fm = 10^{-15} m $\simeq 5$ GeV^{-1}.

Prefixes: atto (a) 10^{-18}	femto (f) 10^{-15}	pico (p) 10^{-12}
mega (M) 10^{6}	giga (G) 10^{9}	tera (T) 10^{12}

A.4.2 Two-Body Cross Section

$$\left(\frac{\mathrm{d}\sigma}{\mathrm{d}\Omega}\right)_{\mathrm{CM}} = \frac{1}{64\pi^2 E_{\mathrm{CM}}^2}\frac{p_f}{p_i}|\overline{\mathcal{M}_{fi}}|^2\theta(E_{\mathrm{CM}} - m_3 - m_4).$$

Elastic case (same particles in the initial and final states!), with $p_f = p_i$. For non-elastic scattering, with different masses in the initial ($m_1 = m_2 = m_i$) and final states ($m_3 = m_4 = m_f$), the factors $p_f = \sqrt{1 - 4m_f^2/E_{\mathrm{CM}}^2}$ and $p_f = \sqrt{1 - 4m_i^2/E_{\mathrm{CM}}^2}$ must be retained.

A.4.3 Width

$$\frac{\mathrm{d}\Gamma}{\mathrm{d}\Omega} = \frac{p^*}{32\pi^2 M^2}|\overline{\mathcal{M}_{fi}}|^2\,\theta(M - m_1 - m_2),$$

with $p^* = \sqrt{[M^2 - (m_1 + m_2)^2][M^2 - (m_1 - m_2)^2]}/2M$.

A.4.4 Matrix Element

$$|\overline{\mathcal{M}_{fi}}|^2 = \prod_i \frac{1}{2s_i + 1}\sum_{\mathrm{pol}}|\langle f|\hat{\mathcal{M}}|i\rangle|^2$$

with s_i the spins of the initial particles.

A.4.5 Breit–Wigner Formula

$$\sigma(A + B \to R \to f) = \left[\frac{(2s_R + 1)c_R}{(2s_A + 1)(2s_B + 1)C_A c_B}\right]\frac{16\pi\ \Gamma_{AB}\Gamma_R}{(s - m_R^2)^2 + m_R^2\Gamma^2},$$

where s_X and c_X are the spin and color of particle X, Γ_{AB} is the width of the decay $R \to AB$ and Γ_R is the total width of the decay $R \to f$.

A.4.6 Sum Over Polarizations

$$\sum_{\sigma=1,2} u_a^\sigma(p)\bar{u}_b^\sigma(p) = (\not{p} + m)_{ab}$$

$$\sum_{\sigma=1,2} v_a^\sigma(p)\bar{v}_b^\sigma(p) = (\not{p} - m)_{ab}$$

$$\sum_{\lambda=1,2} \varepsilon_\mu^\lambda(k)\varepsilon_\nu^{\lambda *}(k) = -g_{\mu\nu} + \frac{k_\mu n_\nu + n_\mu k_\nu}{k \cdot n} - n^2 \frac{k_\mu k_\nu}{(p \cdot n)^2} \quad \text{(massless)}$$

$$\sum_{\lambda=1,2,3} \varepsilon_\mu^{\lambda *}(k)\varepsilon_\nu^\lambda(k) = -g_{\mu\nu} + \frac{k_\mu k_\nu}{M^2} \quad \text{(massive)}$$

In the massless case, n^μ is a (non-covariant) fixed four-vector that one can take to be light-like, $n^2 = 0$.

A.4.7 Projector Over Definite Polarization

$$u^\sigma(p) \otimes \bar{u}^\sigma(p) = (\not{p} + m)\frac{(\mathbb{1} - \gamma_5 \not{\sigma})}{2}$$
$$v^\sigma(p) \otimes \bar{v}^\sigma(p) = (\not{p} - m)\frac{(\mathbb{1} - \gamma_5 \not{\sigma})}{2}$$

with σ the fermion's polarization.

$$\begin{aligned}
\varepsilon_\mu^\lambda(k)\varepsilon_\nu^{\lambda' *}(k) &= \frac{1}{2} e_\mu^{(\lambda)} \left(\mathbb{1} + \vec{\xi} \cdot \vec{\sigma}\right) e_\nu^{(\lambda')} \quad \text{(massless:} \quad \lambda, \lambda' = 1, 2) \\
&= \frac{1}{2}\left(e_\mu^{(1)} e_\nu^{(1)} + e_\mu^{(2)} e_\nu^{(2)}\right) + \frac{\xi_1}{2}\left(e_\mu^{(1)} e_\nu^{(2)} + e_\mu^{(2)} e_\nu^{(1)}\right) \\
&\quad - \frac{\xi_2}{2}\left(e_\mu^{(1)} e_\nu^{(2)} - e_\mu^{(2)} e_\nu^{(1)}\right) + \frac{\xi_3}{2}\left(e_\mu^{(1)} e_\nu^{(1)} - e_\mu^{(2)} e_\nu^{(2)}\right)
\end{aligned}$$

with ξ_i the Stokes parameters and σ_i the Pauli matrices. The four-vectors $e_\mu^{(\lambda)}$ provide a basis; they are two orthonormal four-vectors orthogonal to the momentum $e^{(\lambda)} \cdot e^{(\lambda')} = -\delta^{\lambda\lambda'}$, $e_\mu^{(\lambda)} \cdot k = 0$.

$$\begin{aligned}
\varepsilon_\mu^\lambda(k)\varepsilon_\nu^{\lambda' *}(k) &= \frac{1}{3}\Big(- g_{\mu\nu} + \frac{k_\mu k_\nu}{M^2}\Big)\delta_{\lambda'\lambda} + \frac{i}{2M}\epsilon^{\mu\nu\sigma\tau} k_\sigma \zeta_\tau^i (S_i)_{\lambda'\lambda} \\
&\quad - \frac{1}{2}\zeta_i^\mu \zeta_j^\nu (S^{ij})_{\lambda'\lambda} \quad \text{(massive:} \quad \lambda', \lambda = 1, 2, 3),
\end{aligned}$$

where $(\zeta_1^\mu, \zeta_2^\mu, \zeta_3^\mu)$ are the boosted (to the momentum $\vec{k}$) four-vectors of the triad $(\hat{\theta}, \hat{\phi}, \hat{s})$, defined in the particle rest frame, where $\vec{s}$ is the spin vector; S_i are 3×3 spin-1 matrices, $S_{ij} = \{S_i, S_j\} - (4/3)\mathbb{1}\delta_{ij}$ is a symmetric and traceless 3×3 tensor. Notice the transposition of λ with λ' on the right-hand side.

A.5 Integrals

A.5.1 Parametrization

$$\frac{1}{A_1A_2\cdots A_n}=\int_0^1 \mathrm{d}x_1\mathrm{d}x_2\cdots\mathrm{d}x_n\,\delta\left(\sum_i x_i-1\right)\frac{(n-1)!}{x_1A_1+x_2A_2+\cdots+x_nA_n]^n}$$

Special case:

$$\frac{1}{A^\alpha B^\beta}=\frac{\Gamma(\alpha+\beta)}{\Gamma(\alpha)\Gamma(\beta)}\int_0^1\mathrm{d}x\frac{x^{\alpha-1}(1-x)^{\beta-1}}{[Ax+(1-x)B]^{\alpha+\beta}}\,.$$

A.5.2 Integration in D Dimensions ($D=4-\epsilon$)

Loop integrals are first defined in $4-\epsilon$ space-time dimensions so as to make them well defined even in the presence of divergent contributions in $D=4$. The result of the integration is written by means of the following formulas:

$$\int \mathrm{d}^Dk\frac{1}{(m^2-2p\cdot k-k^2)^\alpha}=\frac{i\pi^{D/2}\Gamma(\alpha-D/2)}{(m^2+p^2)^{\alpha-D/2}\Gamma(\alpha)}$$

$$\int \mathrm{d}^Dk\frac{k_\mu}{(m^2-2p\cdot k-k^2)^\alpha}=\frac{i\pi^{D/2}\Gamma(\alpha-D/2)}{(m^2+p^2)^{\alpha-D/2}\Gamma(\alpha)}(-p_\mu)$$

$$\int \mathrm{d}^Dk\frac{k^2}{(m^2-2p\cdot k-k^2)^\alpha}=\frac{i\pi^{D/2}}{(m^2+p^2)^{\alpha-D/2}\Gamma(\alpha)}\Big[\Gamma(\alpha-D/2)p^2 - \Gamma(\alpha-1-D/2)(D/2)(m^2+p^2)\Big]$$

$$\int \mathrm{d}^Dk\frac{k_\mu k_\nu}{(m^2-2p\cdot k-k^2)^\alpha}=\frac{i\pi^{D/2}}{(m^2+p^2)^{\alpha-D/2}\Gamma(\alpha)}\Big[\Gamma(\alpha-D/2)p_\mu p_\nu - \Gamma(\alpha-1-D/2)(1/2)g_{\mu\nu}(m^2+p^2)\Big]$$

$$\int\frac{\mathrm{d}^Dk}{(2\pi)^D}\frac{(k^2)^r}{[k^2-R^2]^m}=i\frac{(-1)^{r-m}}{(16\pi^2)^{D/4}}\frac{\Gamma(r+D/2)}{\Gamma(D/2)}\frac{\Gamma(m-r-D/2)}{\Gamma(m)(R^2)^{m-r-D/2}}$$

$$\int \mathrm{d}^Dk\,k^\mu k^\nu f(k^2)=\frac{g^{\mu\nu}}{D}\int \mathrm{d}^Dk\,k^2 f(k^2)$$

$$\Gamma(1+\epsilon)=1-\gamma\epsilon+O(\epsilon)$$

$$(R^2)^{\epsilon/2}=1+\frac{\epsilon}{2}\log R^2+O(\epsilon^2)$$

$$\frac{\Gamma(2-D/2)}{(4\pi)^{D/2}}\left(\frac{1}{R}\right)^{2-D/2}=\frac{1}{(4\pi)^2}\left(\frac{2}{\epsilon}-\log R^2-\gamma+\log(4\pi)+O(\epsilon)\right)$$

with (Euler–Mascheroni constant) $\gamma\simeq 0.5772$.

The singular parts, proportional to the poles in ϵ, are absorbed in the redefinition of the bare parameters. The finite parts give the loop corrections.

A.6 Feynman Rules

- Multiply by i the terms in the Lagrangian to obtain the corresponding Feynman rules. Derivatives ∂_μ go into $-ip_\mu$.
- External states (color and weak isospin indices are not shown):

	IN	OUT
fermion	$u^s(p)$	$\bar{u}^s(p)$
antifermion	$\bar{v}^s(p)$	$v^s(p)$
boson	$\varepsilon^\mu_\lambda(k)$	$\varepsilon^{\mu *}_\lambda(k)$

- Multiply each diagram by:

1. a symmetry factor $1/S$, where $S = p\prod_{n=2,3,\ldots} 2^\beta (n!)^{\alpha_n}$ and where p is the number of permutations of vertices leaving the diagram unchanged with fixed external lines, β is the number of lines connecting a vertex to itself and α_n is the number of pairs of vertices connected by n identical lines;
2. a factor (-1) for each fermion loop;
3. a factor (-1) between diagrams which differ by the interchange of two identical external fermions; and
4. an overall sign for the external fermion lines, coming from their permutation with respect to Green functions.

A.6.1 QED

QED Feynman rules

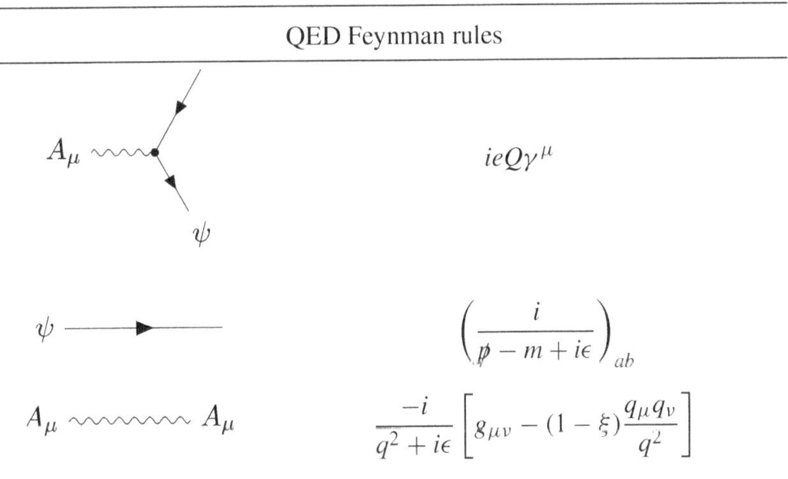

A.6.2 Electroweak Theory (Unitary Gauge)

In this gauge there are no ghosts and the gauge boson propagators are those of massive spin-1 particles.

Electroweak Feynman rules

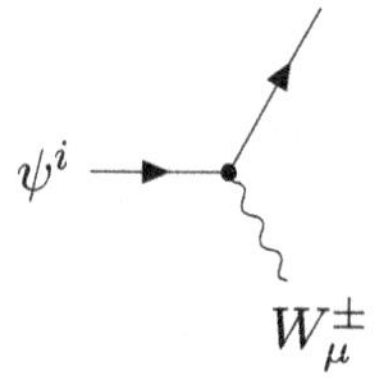

ψ^i, $W^\pm_\mu$

$$i\frac{g}{2\sqrt{2}}\gamma^\mu(1-\gamma_5)T^\pm$$

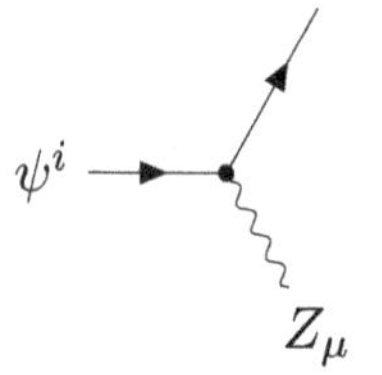

ψ^i, Z_μ

$$i\,\frac{g}{2\cos\theta_W}\gamma^\mu\left[g^i_L(1-\gamma_5)+g^i_R(1+\gamma_5)\right]$$

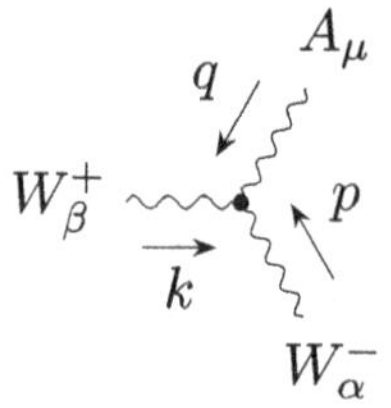

A_μ, q, W^+_β, p, k, W^-_α

$$-ie\left[g_{\alpha\beta}(p-k)_\mu+g_{\beta\mu}(k-q)_\alpha+g_{\mu\alpha}(q-p)_\beta\right]$$

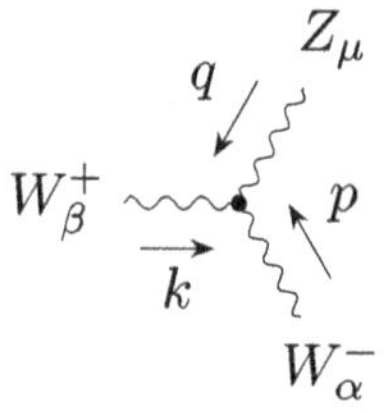

Z_μ, q, W^+_β, p, k, W^-_α

$$-ig\cos\theta_W\left[g_{\alpha\beta}(p-k)_\mu+g_{\beta\mu}(k-1)_\alpha+g_{\mu\alpha}(q-p)_\beta\right]$$

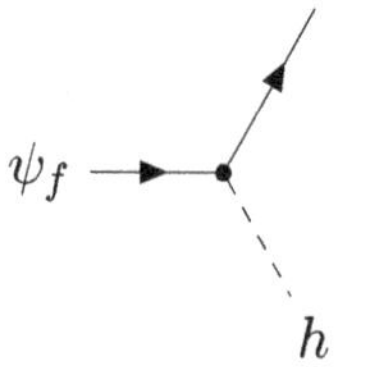

ψ_f, h

$$-i\frac{g}{2}\frac{m_f}{m_W}$$

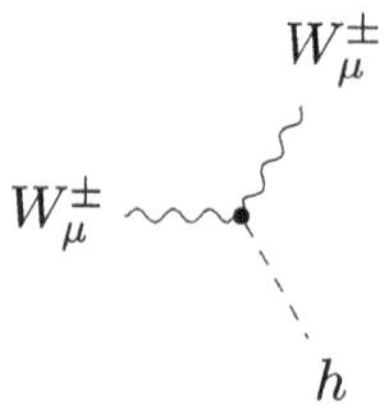

$W^\pm_\mu$, $W^\pm_\mu$, h

$$-i\,g\,m_W\,g_{\mu\nu}$$

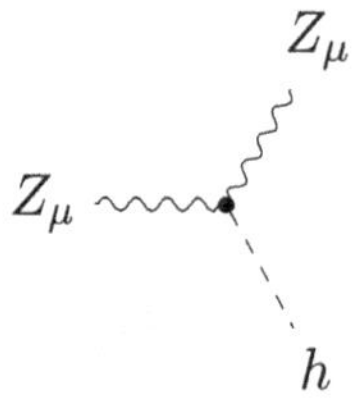

Z_μ, Z_μ, h

$$-i\,\frac{g}{\cos\theta_W}m_Z\,g_{\mu\nu}$$

continued on next page

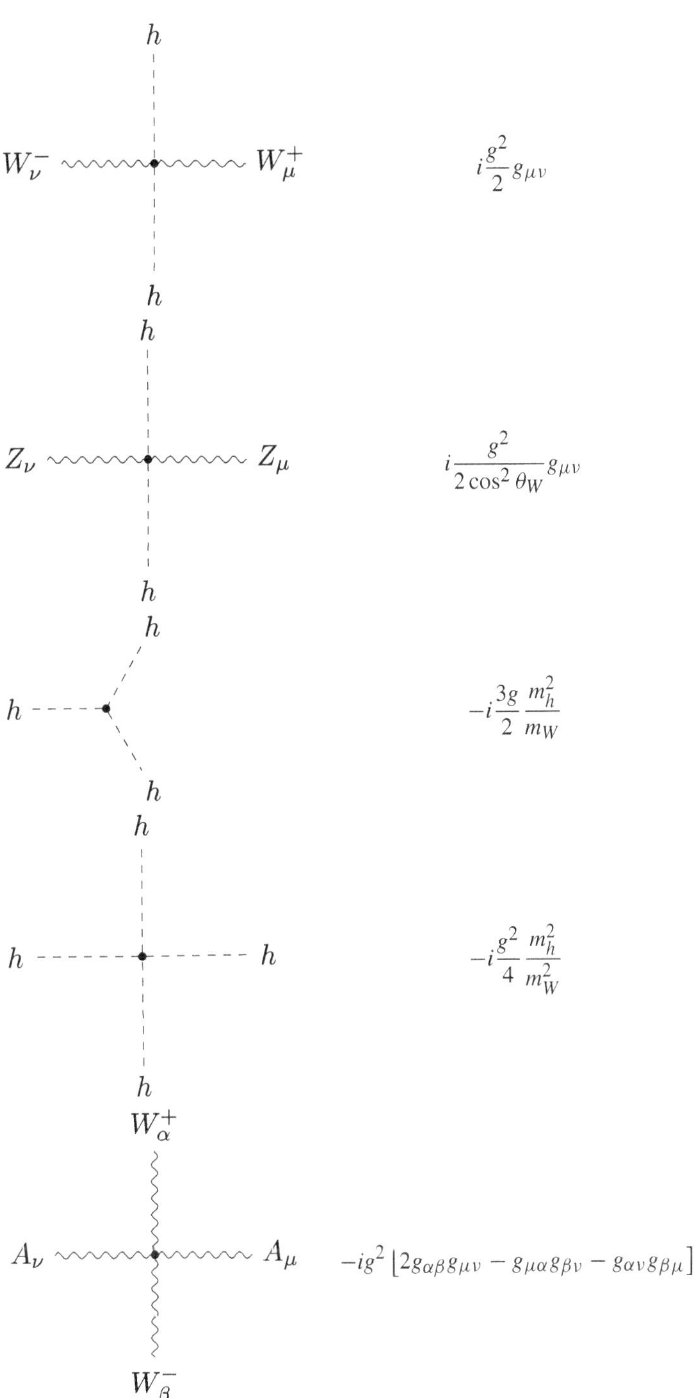

continued on next page

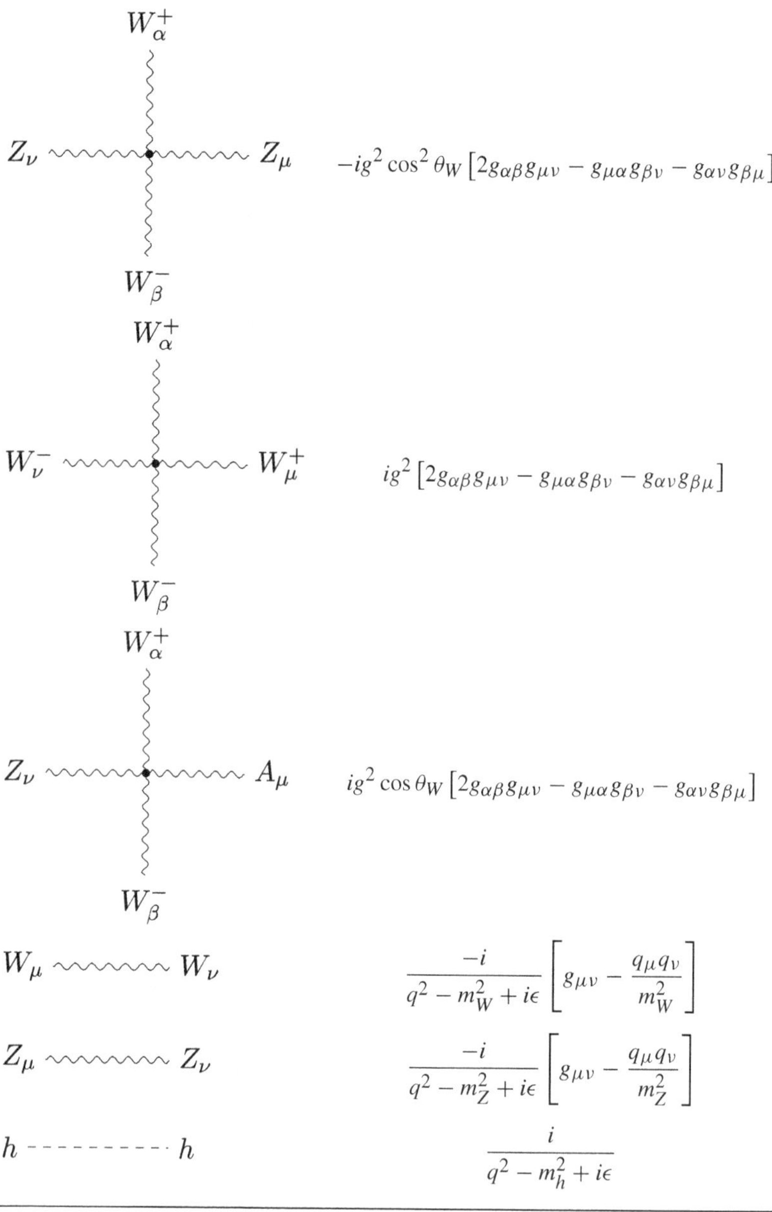
W^+_α
Z_ν
Z_μ
W^-_β
$-ig^2 \cos^2\theta_W \left[2g_{\alpha\beta}g_{\mu\nu} - g_{\mu\alpha}g_{\beta\nu} - g_{\alpha\nu}g_{\beta\mu}\right]$
W^+_α
W^-_ν
W^+_μ
W^-_β
$ig^2 \left[2g_{\alpha\beta}g_{\mu\nu} - g_{\mu\alpha}g_{\beta\nu} - g_{\alpha\nu}g_{\beta\mu}\right]$
W^+_α
Z_ν
A_μ
W^-_β
$ig^2 \cos\theta_W \left[2g_{\alpha\beta}g_{\mu\nu} - g_{\mu\alpha}g_{\beta\nu} - g_{\alpha\nu}g_{\beta\mu}\right]$
W_μ
W_ν
$\frac{-i}{q^2 - m_W^2 + i\epsilon}\left[g_{\mu\nu} - \frac{q_\mu q_\nu}{m_W^2}\right]$
Z_μ
Z_ν
$\frac{-i}{q^2 - m_Z^2 + i\epsilon}\left[g_{\mu\nu} - \frac{q_\mu q_\nu}{m_Z^2}\right]$
h
h
$\frac{i}{q^2 - m_h^2 + i\epsilon}$

A.6.3 QCD

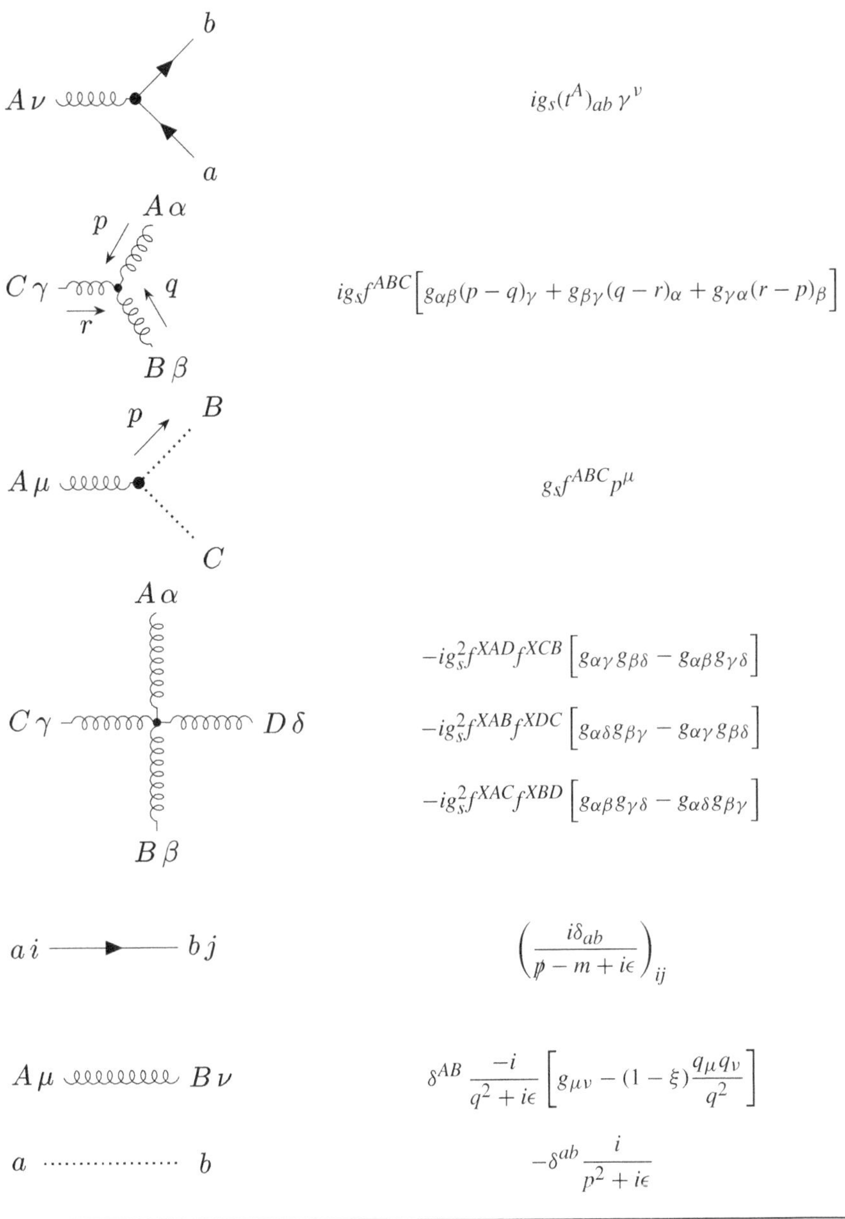

QCD Feynman rules	
$A\,\nu$; a; b	$ig_s(t^A)_{ab}\,\gamma^\nu$
$A\,\alpha$; $B\,\beta$; $C\,\gamma$; p, q, r	$ig_s f^{ABC}\Big[g_{\alpha\beta}(p-q)_\gamma + g_{\beta\gamma}(q-r)_\alpha + g_{\gamma\alpha}(r-p)_\beta\Big]$
$A\,\mu$; B; C; p	$g_s f^{ABC} p^\mu$
$A\,\alpha$; $B\,\beta$; $C\,\gamma$; $D\,\delta$	$-ig_s^2 f^{XAD} f^{XCB}\left[g_{\alpha\gamma}g_{\beta\delta} - g_{\alpha\beta}g_{\gamma\delta}\right]$ $-ig_s^2 f^{XAB} f^{XDC}\left[g_{\alpha\delta}g_{\beta\gamma} - g_{\alpha\gamma}g_{\beta\delta}\right]$ $-ig_s^2 f^{XAC} f^{XBD}\left[g_{\alpha\beta}g_{\gamma\delta} - g_{\alpha\delta}g_{\beta\gamma}\right]$
$a\,i$ — $b\,j$	$\left(\dfrac{i\delta_{ab}}{\not{p} - m + i\epsilon}\right)_{ij}$
$A\,\mu$ — $B\,\nu$	$\delta^{AB}\,\dfrac{-i}{q^2+i\epsilon}\left[g_{\mu\nu} - (1-\xi)\dfrac{q_\mu q_\nu}{q^2}\right]$
a ······ b	$-\delta^{ab}\,\dfrac{i}{p^2+i\epsilon}$

APPENDIX B

Primary Sources

It is a pity that reading the original papers is not currently part of the curriculum in physics. This choice stands in stark contrast to what happens in the humanities, where it is almost inconceivable to have a class without reading the original texts under discussion. A class on Victorian literature without reading Trollope's "Can you forgive her?" I do not think so.

Collected in this appendix are some of the primary sources of the Standard Model. They are often surprisingly accessible. I think that it is useful to read them directly (but you have to to go to the editor's web site for the full text because of the copyright restriction). They give a flavor of the original work. Only by confronting these works directly can we truly appreciate both the ingenuity and the breadth of the physicists who first revealed the structure of the Standard Model. It is not a question of priority disputes. It is a question about how history develops through many variants and also missteps. Enjoy!

B.1 Fermi's Paper

TENTATIVO DI UNA TEORIA DEI RAGGI β

Nota [1] di ENRICO FERMI

Sunto. - *Si propone una teoria quantitativa dell'emissione dei raggi β in cui si ammette l'esistenza del «neutrino» e si tratta l'emissione degli elettroni e dei neutrini da un nucleo all'atto della disintegrazione β con un procedimento simile a quello seguito nella teoria dell'irradiazione per descrivere l'emissione di un quanto di luce da un atomo eccitato. Vengono dedotte delle formule per la vita media e per la forma dello spettro continuo dei raggi β, e le si confrontano coi dati sperimentali.*

Figure B.1 From E. Fermi, Tentativo di una theoris dei raggi β, *Il Nuovo Cimento* **11** (1934). 1. English translation: F.L. Wilson, *American Journal of Physics* **36** (1968) 1150. Abstract reproduced with permission from Springer Nature.

B.2 Yang's and Mills' Paper

PHYSICAL REVIEW VOLUME 96, NUMBER 1 OCTOBER 1, 1954

Conservation of Isotopic Spin and Isotopic Gauge Invariance*

C. N. YANG † AND R. L. MILLS
Brookhaven National Laboratory, Upton, New York
(Received June 28, 1954)

It is pointed out that the usual principle of invariance under isotopic spin rotation is not consistent with the concept of localized fields. The possibility is explored of having invariance under local isotopic spin rotations. This leads to formulating a principle of isotopic gauge invariance and the existence of a **b** field which has the same relation to the isotopic spin that the electromagnetic field has to the electric charge. The **b** field satisfies nonlinear differential equations. The quanta of the **b** field are particles with spin unity, isotopic spin unity, and electric charge $\pm e$ or zero.

Figure B.2 From C.N. Yang and R.L. Mills, Conservation of isotopic spin and isotopic gauge invariance, *Physical Review* **96** (1954) 191. Abstract reproduced with permission from the American Physical Society.

B.3 Parity Non-Conservation

Experimental Test of Parity Conservation in Beta Decay*

C. S. WU, *Columbia University, New York, New York*

AND

E. AMBLER, R. W. HAYWARD, D. D. HOPPES, AND R. P. HUDSON,
National Bureau of Standards, Washington, D. C.
(Received January 15, 1957)

Figure B.3 From C.S. Wu, F. Ambler, R.W. Hayward, D.D. Hoppes and R.P. Hudson, Experimental test of parity conservation in beta decay, *Physical Review* **105** (1957) 1413. Article title reproduced with permission from American Physical Society.

B.4 The *V–A* Model

Chirality Invariance and the Universal Fermi Interaction*

E. C. G. SUDARSHAN, *Harvard University, Cambridge, Massachusetts*

AND

R. E. MARSHAK, *University of Rochester, Rochester, New York*
(Received January 10, 1958)

Figure B.4 From E.C.G. Sudarshan and R.E. Marshak, Chirality invariance and the universal Fermi interaction, *Physical Review* **109** (1958) 1860. Article title reproduced with permission from the American Physical Society.

PHYSICAL REVIEW VOLUME 109, NUMBER 1 JANUARY 1, 1958

Theory of the Fermi Interaction

R. P. FEYNMAN AND M. GELL-MANN
California Institute of Technology, Pasadena, California

(Received September 16, 1957)

The representation of Fermi particles by two-component Pauli spinors satisfying a second order differential equation and the suggestion that in β decay these spinors act without gradient couplings leads to an essentially unique weak four-fermion coupling. It is equivalent to equal amounts of vector and axial vector coupling with two-component neutrinos and conservation of leptons. (The relative sign is not determined theoretically.) It is taken to be "universal"; the lifetime of the μ agrees to within the experimental errors of 2%. The vector part of the coupling is, by analogy with electric charge, assumed to be not renormalized by virtual mesons. This requires, for example, that pions are also "charged" in the sense that there is a direct interaction in which, say, a π^0 goes to π^- and an electron goes to a neutrino. The weak decays of strange particles will result qualitatively if the universality is extended to include a coupling involving a Λ or Σ fermion. Parity is then not conserved even for those decays like $K\rightarrow 2\pi$ or 3π which involve no neutrinos. The theory is at variance with the measured angular correlation of electron and neutrino in He^6, and with the fact that fewer than 10^{-4} pion decay into electron and neutrino.

Figure B.5 From R.P. Feynman and M. Gell-Mann, Theory of the Fermi interaction, *Physical Review* **109** (1958) 193. Abstract reproduced with permission from the American Physical Society.

B.5 Bell's Paper

Physics Vol. 1, No. 3, pp. 195–200, 1964 Physics Publishing Co. Printed in the United States

ON THE EINSTEIN PODOLSKY ROSEN PARADOX*

J. S. BELL†
Department of Physics, University of Wisconsin, Madison, Wisconsin

(*Received* 4 *November* 1964)

Figure B.6 From J.S. Bell, On the Einstein Podolsky Rosen paradox, *Physics* **1** (1964) 195. Article title reproduced with permission from the American Physical Society.

B.6 Quark Model

Volume 8, number 3 PHYSICS LETTERS 1 February 1964

A SCHEMATIC MODEL OF BARYONS AND MESONS *

M. GELL-MANN
California Institute of Technology, Pasadena, California

Received 4 January 1964

Figure B.7 From M. Gell-Mann, A schematic model of baryons and mesons, *Physics Letters* **8** (1964) 214. Article title reproduced with permission from Elsevier B.V.

B.7 *CP* Violation

VOLUME 13, NUMBER 4 PHYSICAL REVIEW LETTERS 27 JULY 1964

EVIDENCE FOR THE 2π DECAY OF THE K_2^0 MESON*†

J. H. Christenson, J. W. Cronin,‡ V. L. Fitch,‡ and R. Turlay§
Princeton University, Princeton, New Jersey
(Received 10 July 1964)

Figure B.8 From J.H. Christenson, J.W. Cronin, V.L. Fitch and R. Turlay, Evidence for the 2π decay of the K_2^0 meson, *Physical Review Letters* **13** (1964) 138. Article title reproduced with permission from the American Physical Society.

PHYSICAL REVIEW VOLUME 97, NUMBER 5 MARCH 1, 1955

Behavior of Neutral Particles under Charge Conjugation

M. GELL-MANN,* *Department of Physics, Columbia University, New York, New York*
AND
A. PAIS, *Institute for Advanced Study, Princeton, New Jersey*
(Received November 1, 1954)

Some properties are discussed of the θ^0, a heavy boson that is known to decay by the process $\theta^0 \to \pi^+ + \pi^-$. According to certain schemes proposed for the interpretation of hyperons and K particles, the θ^0 possesses an antiparticle $\bar{\theta}^0$ distinct from itself. Some theoretical implications of this situation are discussed with special reference to charge conjugation invariance. The application of such invariance in familiar instances is surveyed in Sec. I. It is then shown in Sec. II that, within the framework of the tentative schemes under consideration, the θ^0 must be considered as a "particle mixture" exhibiting two distinct lifetimes, that each lifetime is associated with a different set of decay modes, and that no more than half of all θ^0's undergo the familiar decay into two pions. Some experimental consequences of this picture are mentioned.

Figure B.9 From M. Gell-Mann and A. Pais, Behavior of neutral particles under charge conjugation, *Physical Review* **97** (1955) 1387. Abstract reproduced with permission from the American Physical Society.

B.8 Glashow's and Weinberg's Papers

PARTIAL-SYMMETRIES OF WEAK INTERACTIONS

SHELDON L. GLASHOW †
Institute for Theoretical Physics, University of Copenhagen, Copenhagen, Denmark

Received 9 September 1960

Abstract: Weak and electromagnetic interactions of the leptons are examined under the hypothesis that the weak interactions are mediated by vector bosons. With only an isotopic triplet of leptons coupled to a triplet of vector bosons (two charged decay-intermediaries and the photon) the theory possesses no partial-symmetries. Such symmetries may be established if additional vector bosons or additional leptons are introduced. Since the latter possibility yields a theory disagreeing with experiment, the simplest partially-symmetric model reproducing the observed electromagnetic and weak interactions of leptons requires the existence of at least four vector-boson fields (including the photon). Corresponding partially-conserved quantities suggest leptonic analogues to the conserved quantities associated with strong interactions: strangeness and isobaric spin.

Figure B.10 From S.L. Glashow, Partial-symmetries of weak interactions, *Nuclear Physics* **22** (1961) 579. Abstract reproduced with permission from Elsevier B.V.

A MODEL OF LEPTONS*

Steven Weinberg†
Laboratory for Nuclear Science and Physics Department,
Massachusetts Institute of Technology, Cambridge, Massachusetts
(Received 17 October 1967)

Figure B.11 From S. Weinberg, A model of leptons, *Physical Review Letters* **19** (1967) 1264. Article title reproduced with permission from the American Physical Society.

B.9 Anderson's and Higgs' Papers

PHYSICAL REVIEW VOLUME 130, NUMBER 1 1 APRIL 1963

Plasmons, Gauge Invariance, and Mass

P. W. Anderson
Bell Telephone Laboratories, Murray Hill, New Jersey
(Received 8 November 1962)

Schwinger has pointed out that the Yang-Mills vector boson implied by associating a generalized gauge transformation with a conservation law (of baryonic charge, for instance) does not necessarily have zero mass, if a certain criterion on the vacuum fluctuations of the generalized current is satisfied. We show that the theory of plasma oscillations is a simple nonrelativistic example exhibiting all of the features of Schwinger's idea. It is also shown that Schwinger's criterion that the vector field $m\neq 0$ implies that the matter spectrum before including the Yang-Mills interaction contains $m=0$, but that the example of superconductivity illustrates that the physical spectrum need not. Some comments on the relationship between these ideas and the zero-mass difficulty in theories with broken symmetries are given.

Figure B.12 From P.W. Anderson, Plasmons, gauge invariance, and mass, *Physical Review* **130** (1963) 439. Abstract reproduced with permission from the American Physical Society.

Volume 13, Number 16 PHYSICAL REVIEW LETTERS 19 October 1964

BROKEN SYMMETRIES AND THE MASSES OF GAUGE BOSONS

Peter W. Higgs
Tait Institute of Mathematical Physics, University of Edinburgh, Edinburgh, Scotland
(Received 31 August 1964)

Figure B.13 From P. Higgs, Broken symmetries and the masses of gauge bosons, *Physical Review Letters* **13** (1964) 508. Article title reproduced with permission from the American Physical Society.

B.10 Triangle Anomaly

PHYSICAL REVIEW VOLUME 177, NUMBER 5 25 JANUARY 1969

Axial-Vector Vertex in Spinor Electrodynamics

Stephen L. Adler
Institute for Advanced Study, Princeton, New Jersey 08540
(Received 24 September 1968)

Working within the framework of perturbation theory, we show that the axial-vector vertex in spinor electrodynamics has anomalous properties which disagree with those found by the formal manipulation of field equations. Specifically, because of the presence of closed-loop "triangle diagrams," the divergence of axial-vector current is not the usual expression calculated from the field equations, and the axial-vector current does not satisfy the usual Ward identity. One consequence is that, even after the external-line wave-function renormalizations are made, the axial-vector vertex is still divergent in fourth- (and higher-) order perturbation theory. A corollary is that the radiative corrections to $\nu_l l$ elastic scattering in the local current-current theory diverge in fourth (and higher) order. A second consequence is that, in massless electrodynamics, despite the fact that the theory is invariant under γ_5 transformations, the axial-vector current is not conserved. In an Appendix we demonstrate the uniqueness of the triangle diagrams, and discuss a possible connection between our results and the $\pi^0 \to 2\gamma$ and $\eta \to 2\gamma$ decays. In particular, we argue that as a result of triangle diagrams, the equations expressing partial conservation of axial-vector current (PCAC) for the neutral members of the axial-vector-current octet must be modified in a well-defined manner, which completely alters the PCAC predictions for the π^0 and the η two-photon decays.

Figure B.14 From S.L. Adler, Axial-vector vertex in spinor electrodynamics, *Physical Review* **177** (1969) 2426. Abstract reproduced with permission from the American Physical Society.

IL NUOVO CIMENTO VOL. LX A, N. 1 1° Marzo 1969

A PCAC Puzzle: $\pi^0 \to \gamma\gamma$ in the σ-Model.

J. S. BELL
CERN - Geneva

R. JACKIW (*)
CERN - Geneva

Jefferson Laboratory of Physics, Harvard University - Cambridge, Mass.

(ricevuto l'11 Settembre 1968)

Summary. — The effective coupling constant for $\pi^0 \to \gamma\gamma$ should vanish for zero pion mass in theories with PCAC and gauge invariance. It does not so vanish in an explicit perturbation calculation in the σ-model. The resolution of the puzzle is effected by a modification of Pauli-Villars-Gupta regularization which respects both PCAC and gauge invariance.

Figure B.15 From J.S. Bell and R. Jackiw, A PCAC puzzle: π^0 –> $\gamma\gamma$ in the sigma-model, *Il Nuovo Cimento* **LX** (1969) 47. Abstract reproduced with permission from Springer Nature.

B.11 Deep Inelastic Scattering

VOLUME 23, NUMBER 16 PHYSICAL REVIEW LETTERS 20 OCTOBER 1969

OBSERVED BEHAVIOR OF HIGHLY INELASTIC ELECTRON-PROTON SCATTERING

M. Breidenbach, J. I. Friedman, and H. W. Kendall
Department of Physics and Laboratory for Nuclear Science,*
Massachusetts Institute of Technology, Cambridge, Massachusetts 02139

and

E. D. Bloom, D. H. Coward, H. DeStaebler, J. Drees, L. W. Mo, and R. E. Taylor
Stanford Linear Accelerator Center,† Stanford, California 94305
(Received 22 August 1969)

Results of electron-proton inelastic scattering at 6° and 10° are discussed, and values of the structure function W_2 are estimated. If the interaction is dominated by transverse virtual photons, νW_2 can be expressed as a function of $\omega = 2M\nu/q^2$ within experimental errors for $q^2 > 1$ (GeV/c)^2 and $\omega > 4$, where ν is the invariant energy transfer and q^2 is the invariant momentum transfer of the electron. Various theoretical models and sum rules are briefly discussed.

Figure B.16 From M. Breidenbach *et al.*, Observed behavior of highly inelastic electron-proton scattering, *Physical Review Letters* **23** (1969) 935. Abstract reproduced with permission from the American Physical Society.

VERY HIGH-ENERGY COLLISIONS OF HADRONS

Richard P. Feynman
California Institute of Technology, Pasadena, California
(Received 20 October 1969)

Proposals are made predicting the character of longitudinal-momentum distributions in hadron collisions of extreme energies.

Figure B.17 From R.P. Feynman, Very high-energy collisions of hadrons, *Physical Review Letters* **23** (1969) 1415. Abstract reproduced with permission from the American Physical Society.

PHYSICAL REVIEW VOLUME 185, NUMBER 5 25 SEPTEMBER 1969

Inelastic Electron-Proton and γ-Proton Scattering and the Structure of the Nucleon*

J. D. Bjorken and E. A. Paschos
Stanford Linear Accelerator Center, Stanford University, Stanford, California 94305
(Received 10 April 1969)

A model for highly inelastic electron-nucleon scattering at high energies is studied and compared with existing data. This model envisages the proton to be composed of pointlike constituents ("partons") from which the electron scatters incoherently. We propose that the model be tested by observing γ rays scattered inelastically in a similar way from the nucleon. The magnitude of this inelastic Compton-scattering cross section can be predicted from existing electron-scattering data, indicating that the experiment is feasible, but difficult, at presently available energies.

Figure B.18 From J.D. Bjorken and E.A. Paschos, Inelastic electron-proton and γ-proton scattering and the structure of the nucleon, *Physical Review* **185** (1969) 1975. Abstract reproduced with permission from the American Physical Society.

B.12 Neutral Currents

Volume 46B, number 1 PHYSICS LETTERS 3 September 1973

OBSERVATION OF NEUTRINO-LIKE INTERACTIONS WITHOUT MUON OR ELECTRON IN THE GARGAMELLE NEUTRINO EXPERIMENT

F.J. HASERT, S. KABE, W. KRENZ, J. Von KROGH, D. LANSKE, J. MORFIN, K. SCHULTZE and H. WEERTS
III. Physikalisches Institut der Technischen Hochschule, Aachen, Germany

G.H. BERTRAND-COREMANS, J. SACTON, W. Van DONINCK and P. VILAIN*1
Interuniversity Institute for High Energies, U.L.B., V.U.B. Brussels, Belgium

U. CAMERINI*2, D.C. CUNDY, R. BALDI, I. DANILCHENKO*3, W.F. FRY*2, D. HAIDT, S. NATALI*4, P. MUSSET, B. OSCULATI, R. PALMER*4, J.B.M. PATTISON, D.H. PERKINS*6, A. PULLIA, A. ROUSSET, W. VENUS*7 and H. WACHSMUTH
CERN, Geneva, Switzerland

V. BRISSON, B. DEGRANGE, M. HAGUENAUER, L. KLUBERG, U. NGUYEN-KHAC and P. PETIAU
Laboratoire de Physique Nucléaire des Hautes Energies, Ecole Polytechnique, Paris, France

E. BELOTTI, S. BONETTI, D. CAVALLI, C. CONTA*8, E. FIORINI and M. ROLLIER
Istituto di Fisica dell'Università, Milano and I.N.F.N. Milano, Italy

B. AUBERT, D. BLUM, L.M. CHOUNET, P. HEUSSE, A. LAGARRIGUE, A.M. LUTZ, A. ORKIN-LECOURTOIS and J.P. VIALLE
Laboratoire de l'Accélérateur Linéaire, Orsay, France

F.W. BULLOCK, M.J. ESTEN, T.W. JONES, J. McKENZIE, A.G. MICHETTE*9 G. MYATT* and W.G. SCOTT*6,*9
University College, London, England

Received 25 July 1973

Events induced by neutral particles and producing hadrons, but no muon or electron, have been observed in the CERN neutrino experiment. These events behave as expected if they arise from neutral current induced processes. The rates relative to the corresponding charged current processes are evaluated.

Figure B.19 From F.J. Hasert *et al.*, Observation of neutrino-like interactions without muon or electron in the Gargamelle neutrino experiment, *Physics Letters* **46B** (1973) 138. Abstract reproduced with permission from Elsevier B.V.

B.13 QCD

Volume 47B, number 4 PHYSICS LETTERS 26 November 1973

ADVANTAGES OF THE COLOR OCTET GLUON PICTURE☆

H. FRITZSCH*, M. GELL-MANN and H. LEUTWYLER**
California Institute of Technology, Pasadena, Calif. 91109, USA

Received 1 October 1973

It is pointed out that there are several advantages in abstracting properties of hadrons and their currents from a Yang–Mills gauge model based on colored quarks and color octet gluons.

Figure B.20 From H. Fritzsch, M. Gell-Mann and H. Leutwyler, Advantages of the color octet gluon picture, *Physics Letters* **47B** (1973) 365. Abstract reproduced with permission from Elsevier B.V.

Non-Abelian Gauge Theories of the Strong Interactions*

Steven Weinberg
Lyman Laboratory of Physics, Harvard University, Cambridge, Massachusetts 02138
(Received 30 May 1973)

A class of non-Abelian gauge theories of strong interactions is described, for which parity and strangeness are automatically conserved, and for which the nonconservations of parity and strangeness produced by weak interactions are automatically of order α/m_W^2 rather than of order α. When such theories are "asymptotically free," the order-α weak corrections to natural zeroth-order symmetries may be calculated ignoring all effects of strong interactions. Speculations are offered on a possible theory of quarks.

Figure B.21 From S. Weinberg, Non-Abelian gauge theories of the strong interactions, *Physics Review Letters* **31** (1973) 494. Copyright 1973 by the American Physical Society.

B.14 Asymptotic Freedom

VOLUME 30, NUMBER 26 PHYSICAL REVIEW LETTERS 25 JUNE 1973

Ultraviolet Behavior of Non-Abelian Gauge Theories*

David J. Gross† and Frank Wilczek
Joseph Henry Laboratories, Princeton University, Princeton, New Jersey 08540
(Received 27 April 1973)

It is shown that a wide class of non-Abelian gauge theories have, up to calculable logarithmic corrections, free-field–theory asymptotic behavior. It is suggested that Bjorken scaling may be obtained from strong-interaction dynamics based on non-Abelian gauge symmetry.

Figure B.22 From D. Gross and F. Wilczek, Ultraviolet behavior of non-Abelian gauge theories, *Physical Review Letters* **30** (1973) 1343. Abstract reproduced with permission from the American Physical Society.

Reliable Perturbative Results for Strong Interactions?*

H. David Politzer
Jefferson Physical Laboratories, Harvard University, Cambridge, Massachusetts 02138
(Received 3 May 1973)

An explicit calculation shows perturbation theory to be arbitrarily good for the deep Euclidean Green's functions of any Yang-Mills theory and of many Yang-Mills theories with fermions. Under the hypothesis that spontaneous symmetry breakdown is of dynamical origin, these symmetric Green's functions are the asymptotic forms of the physically significant spontaneously broken solution, whose coupling could be strong.

Figure B.23 From H.D. Politzer, Reliable perturbative results for strong interactions?, *Physical Review Letters* **30** (1973) 1346. Abstract reproduced with permission from the American Physical Society.

B.15 Renormalization Group Equations

PHYSICAL REVIEW VOLUME 95, NUMBER 5 SEPTEMBER 1, 1954

Quantum Electrodynamics at Small Distances*

M. Gell-Mann† and F. E. Low
Physics Department, University of Illinois, Urbana, Illinois
(Received April 1, 1954)

The renormalized propagation functions D_{FC} and S_{FC} for photons and electrons, respectively, are investigated for momenta much greater than the mass of the electron. It is found that in this region the individual terms of the perturbation series to all orders in the coupling constant take on very simple asymptotic forms. An attempt to sum the entire series is only partially successful. It is found that the series satisfy certain functional equations by virtue of the renormalizability of the theory. If photon self-energy parts are omitted from the series, so that $D_{FC}=D_F$, then S_{FC} has the asymptotic form $A[p^2/m^2]^n[i\gamma\cdot p]^{-1}$, where $A=A(e_1^2)$ and $n=n(e_1^2)$. When all diagrams are included, less specific results are found. One conclusion is that the *shape* of the charge distribution surrounding a test charge in the vacuum does not, at small distances, depend on the coupling constant except through a scale factor. The behavior of the propagation functions for large momenta is related to the magnitude of the renormalization constants in the theory. Thus it is shown that the unrenormalized coupling constant $e_0^2/4\pi\hbar c$, which appears in perturbation theory as a power series in the renormalized coupling constant $e_1^2/4\pi\hbar c$ with divergent coefficients, may behave in either of two ways:
(a) It may really be infinite as perturbation theory indicates;
(b) It may be a finite number independent of $e_1^2/4\pi\hbar c$.

Figure B.24 From M. Gell-Mann and F.E. Low, Quantum electrodynamics at small distances, *Physical Review* **95** (1954) 1300. Abstract reproduced with permission from the American Physical Society.

PHYSICAL REVIEW B VOLUME 4, NUMBER 9 1 NOVEMBER 1971

Renormalization Group and Critical Phenomena.
I. Renormalization Group and the Kadanoff Scaling Picture*

Kenneth G. Wilson
Laboratory of Nuclear Studies, Cornell University, Ithaca, New York 14850
(Received 2 June 1971)

The Kadanoff theory of scaling near the critical point for an Ising ferromagnet is cast in differential form. The resulting differential equations are an example of the differential equations of the renormalization group. It is shown that the Widom-Kadanoff scaling laws arise naturally from these differential equations if the coefficients in the equations are analytic at the critical point. A generalization of the Kadanoff scaling picture involving an "irrelevant" variable is considered; in this case the scaling laws result from the renormalization-group equations only if the solution of the equations goes asymptotically to a fixed point.

Figure B.25 From K.G. Wilson, Renormalization group and critical phenomena. I. Renormalization group and the Kadanott scaling picture, *Physical Review B* **4** (1971) 3174. Abstract reproduced with permission from the American Physical Society.

B.16 Neutrino Oscillations

Volume 28B, number 7 PHYSICS LETTERS 20 January 1969

NEUTRINO ASTRONOMY AND LEPTON CHARGE

V. GRIBOV * and B. PONTECORVO
Joint Institute for Nuclear Research, Dubna, USSR

Received 20 December 1968

It is shown that lepton nonconservation might lead to a decrease in the number of detectable solar neutrinos at the earth surface, because of $\nu_e \rightleftarrows \nu_\mu$ oscillations, similar to $K^0 \rightleftarrows \bar{K}^0$ oscillations. Equations are presented describing such oscillations for the case when there exist only four neutrino states.

Figure B.26 From V. Gribov and B. Pontecorvo, Neutrino astronomy and lepton charge, *Physics Letters* **B28** (1969) 493. Abstract reproduced with permission from Elsevier B.V.

VOLUME 81, NUMBER 8 PHYSICAL REVIEW LETTERS 24 AUGUST 1998

Evidence for Oscillation of Atmospheric Neutrinos

Y. Fukuda,[1] T. Hayakawa,[1] E. Ichihara,[1] K. Inoue,[1] K. Ishihara,[1] H. Ishino,[1] Y. Itow,[1] T. Kajita,[1] J. Kameda,[1] S. Kasuga,[1] K. Kobayashi,[1] Y. Kobayashi,[1] Y. Koshio,[1] M. Miura,[1] M. Nakahata,[1] S. Nakayama,[1] A. Okada,[1] K. Okumura,[1] N. Sakurai,[1] M. Shiozawa,[1] Y. Suzuki,[1] Y. Takeuchi,[1] Y. Totsuka,[1] S. Yamada,[1] M. Earl,[2] A. Habig,[2] E. Kearns,[2] M.D. Messier,[2] K. Scholberg,[2] J.L. Stone,[2] L.R. Sulak,[2] C.W. Walter,[2] M. Goldhaber,[3] T. Barszczxak,[4] D. Casper,[4] W. Gajewski,[4] P.G. Halverson,[4,*] J. Hsu,[4] W.R. Kropp,[4] L.R. Price,[4] F. Reines,[4] M. Smy,[4] H.W. Sobel,[4] M.R. Vagins,[4] K.S. Ganezer,[5] W.E. Keig,[5] R.W. Ellsworth,[6] S. Tasaka,[7] J.W. Flanagan,[8,†] A. Kibayashi,[8] J.G. Learned,[8] S. Matsuno,[8] V.J. Stenger,[8] D. Takemori,[8] T. Ishii,[9] J. Kanzaki,[9] T. Kobayashi,[9] S. Mine,[9] K. Nakamura,[9] K. Nishikawa,[9] Y. Oyama,[9] A. Sakai,[9] M. Sakuda,[9] O. Sasaki,[9] S. Echigo,[10] M. Kohama,[10] A.T. Suzuki,[10] T.J. Haines,[11,4] E. Blaufuss,[12] B.K. Kim,[12] R. Sanford,[12] R. Svoboda,[12] M.L. Chen,[13] Z. Conner,[13,‡] J.A. Goodman,[13] G.W. Sullivan,[13] J. Hill,[14] C.K. Jung,[14] K. Martens,[14] C. Mauger,[14] C. McGrew,[14] E. Sharkey,[14] B. Viren,[14] C. Yanagisawa,[14] W. Doki,[15] K. Miyano,[15] H. Okazawa,[15] C. Saji,[15] M. Takahata,[15] Y. Nagashima,[16] M. Takita,[16] T. Yamaguchi,[16] M. Yoshida,[16] S.B. Kim,[17] M. Etoh,[18] K. Fujita,[18] A. Hasegawa,[18] T. Hasegawa,[18] S. Hatakeyama,[18] T. Iwamoto,[18] M. Koga,[18] T. Maruyama,[18] H. Ogawa,[18] J. Shirai,[18] A. Suzuki,[18] F. Tsushima,[18] M. Koshiba,[19] M. Nemoto,[20] K. Nishijima,[20] T. Futagami,[21] Y. Hayato,[21,§] Y. Kanaya,[21] K. Kaneyuki,[21] Y. Watanabe,[21] D. Kielczewska,[22,4] R.A. Doyle,[23] J.S. George,[23] A.L. Stachyra,[23] L.L. Wai,[23,||] R.J. Wilkes,[23] and K.K. Young[23]
(Super-Kamiokande Collaboration)

[1]*Institute for Cosmic Ray Research, University of Tokyo, Tanashi, Tokyo, 188-8502, Japan*
[2]*Department of Physics, Boston University, Boston, Massachusetts 02215*
[3]*Physics Department, Brookhaven National Laboratory, Upton, New York 11973*
[4]*Department of Physics and Astronomy, University of California at Irvine, Irvine, California 92697-4575*
[5]*Department of Physics, California State University, Dominguez Hills, Carson, California 90747*
[6]*Department of Physics, George Mason University, Fairfax, Virginia 22030*
[7]*Department of Physics, Gifu University, Gifu, Gifu 501-1193, Japan*
[8]*Department of Physics and Astronomy, University of Hawaii, Honolulu, Hawaii 96822*
[9]*Institute of Particle and Nuclear Studies, High Energy Accelerator Research Organization (KEK), Tsukuba, Ibaraki 305-0801, Japan*
[10]*Department of Physics, Kobe University, Kobe, Hyogo 657-8501, Japan*
[11]*Physics Division, P-23, Los Alamos National Laboratory, Los Alamos, New Mexico 87544*
[12]*Department of Physics and Astronomy, Louisiana State University, Baton Rouge, Louisiana 70803*
[13]*Department of Physics, University of Maryland, College Park, Maryland 20742*
[14]*Department of Physics and Astronomy, State University of New York, Stony Brook, New York 11794-3800*
[15]*Department of Physics, Niigata University, Niigata, Niigata 950-2181, Japan*
[16]*Department of Physics, Osaka University, Toyonaka, Osaka 560-0043, Japan*
[17]*Department of Physics, Seoul National University, Seoul 151-742, Korea*
[18]*Department of Physics, Tohoku University, Sendai, Miyagi 980-8578, Japan*
[19]*The University of Tokyo, Tokyo 113-0033, Japan*
[20]*Department of Physics, Tokai University, Hiratsuka, Kanagawa 259-1292, Japan*
[21]*Department of Physics, Tokyo Institute of Technology, Meguro, Tokyo 152-8551, Japan*
[22]*Institute of Experimental Physics, Warsaw University, 00-681 Warsaw, Poland*
[23]*Department of Physics, University of Washington, Seattle, Washington 98195-1560*
(Received 6 July 1998)

We present an analysis of atmospheric neutrino data from a 33.0 kton yr (535-day) exposure of the Super-Kamiokande detector. The data exhibit a zenith angle dependent deficit of muon neutrinos which is inconsistent with expectations based on calculations of the atmospheric neutrino flux. Experimental biases and uncertainties in the prediction of neutrino fluxes and cross sections are unable to explain our observation. The data are consistent, however, with two-flavor $\nu_\mu \leftrightarrow \nu_\tau$ oscillations with $\sin^2 2\theta > 0.82$ and $5 \times 10^{-4} < \Delta m^2 < 6 \times 10^{-3}$ eV^2 at 90% confidence level. [S0031-9007(98)06975-0]

Figure B.27 From Y. Fukuda *et al.*, Evidence for oscillation of atmospheric neutrinos, *Physical Review Letters* **81** (1998) 1562. Abstract reproduced with permission from the American Physical Society.

APPENDIX C
Time Lines

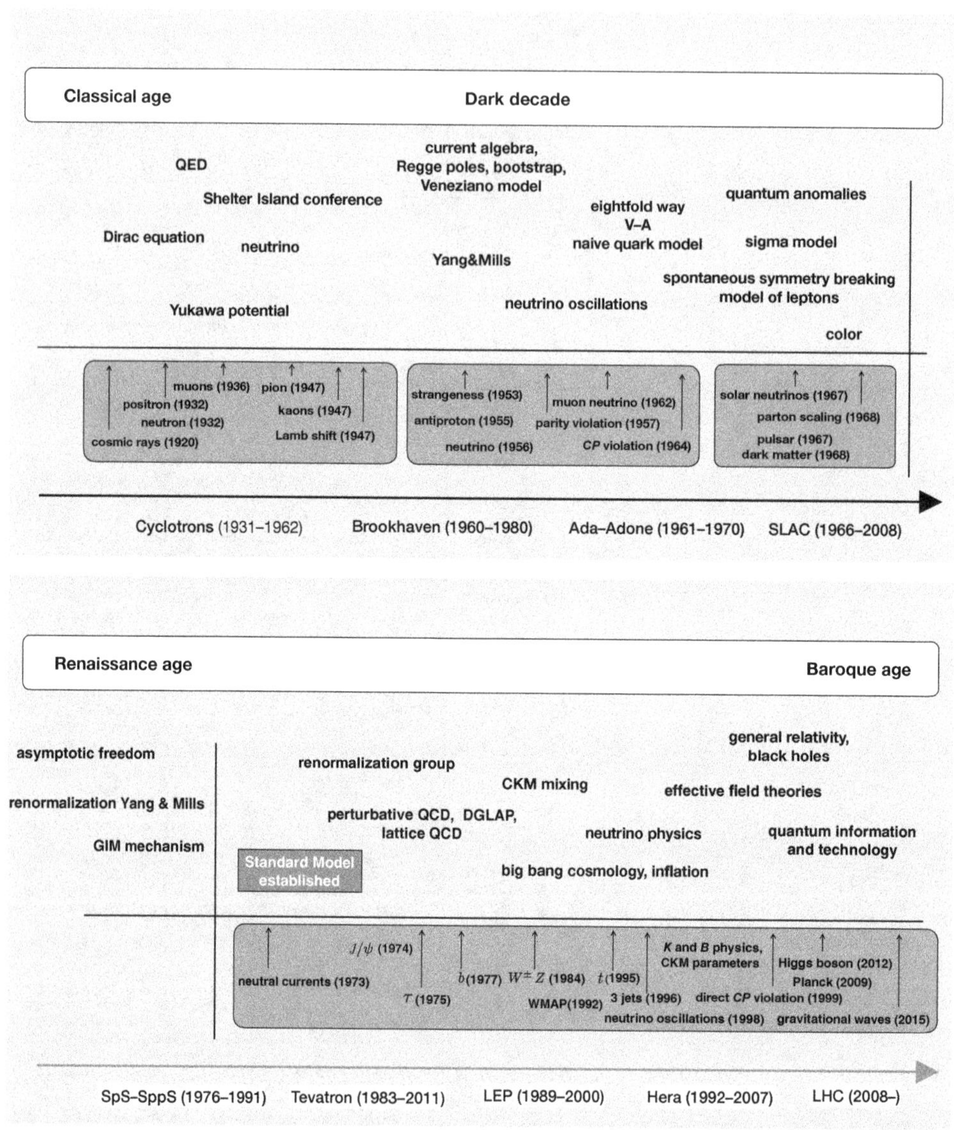

Figure C.1 Development of high-energy physics.

APPENDIX D
Dirac Equation

There exist various ways to derive the Dirac equation. Here I use group theory to obtain first the Weyl equations and from them the Dirac equation. It is perhaps the most direct and fastest way.

The idea is to look at the basis of the spin representations $(1/2, 0)$ and $(0, 1/2)$ of the Lorentz group. They are introduced in Section 2.6, Table 2.3. The two-dimensional generators of the group are

$$K_1 = \frac{1}{2}\begin{pmatrix} 0 & i \\ i & 0 \end{pmatrix} \quad K_2 = \frac{1}{2}\begin{pmatrix} 0 & 1 \\ -1 & 0 \end{pmatrix} \quad K_3 = \frac{1}{2}\begin{pmatrix} i & 0 \\ 0 & -i \end{pmatrix} \tag{D.1}$$

$$J_1 = \frac{1}{2}\begin{pmatrix} 0 & 1 \\ 1 & 0 \end{pmatrix} \quad J_2 = \frac{1}{2}\begin{pmatrix} 0 & -i \\ i & 0 \end{pmatrix} \quad J_3 = \frac{1}{2}\begin{pmatrix} 1 & 0 \\ 0 & -1 \end{pmatrix}. \tag{D.2}$$

The basis of this representation is two two-dimensional spinors u_α and v_α whose transformations are[D.1]

D.1 The dagger of a spinor means conjugation and transposition.

opposite sign!

$$u_\alpha \to e^{i\vec{\theta}\cdot\frac{\vec{\sigma}}{2}} u_\alpha \qquad u_\alpha^\dagger \to u_\alpha^\dagger \; e^{-i\vec{\theta}\cdot\frac{\vec{\sigma}}{2}} \text{ (rotations)} \tag{D.3}$$

$$v_\alpha \to e^{\vec{\chi}\cdot\frac{\vec{\sigma}}{2}} v_\alpha \qquad v_\alpha^\dagger \to \; v_\alpha^\dagger e^{\vec{\chi}\cdot\frac{\vec{\sigma}}{2}} \text{ (boosts)} \tag{D.4}$$

same sign!

Under rotations, the product $u^\dagger u$ is invariant and $u^\dagger\vec{\sigma}u$ transforms as a three-vector. Instead, an arbitrary boost – because of the same sign in the exponent of the transformation factor of the spinors u and $u^\dagger$ – can bring a rotation into a boost, as in

$$u^\dagger u \to u^\dagger e^{\vec{\chi}\cdot\frac{\vec{\sigma}}{2}} e^{\vec{\chi}\cdot\frac{\vec{\sigma}}{2}} u = u^\dagger e^{\vec{\chi}\cdot\vec{\sigma}} u \simeq u^\dagger(\mathbb{1}+\vec{\chi}\cdot\vec{\sigma})u = u^\dagger u + \vec{\chi}\cdot u^\dagger\vec{\sigma}u\,. \tag{D.5}$$

The transformation in Eq. (D.5) mixes the $u^\dagger u$ part with the three-vector part $u^\dagger\vec{\sigma}u$. It means that a covariant description must deal with a four-vector, in which the two parts are different components of the same object:

$$\omega^\mu = (u^\dagger u,\; u^\dagger\vec{\sigma}u), \tag{D.6}$$

in which

$$\sigma^\mu = (\mathbb{1}, \vec{\sigma}), \tag{D.7}$$

with $\mathbb{1}$ the 2×2 unit matrix. Now that I have identified the relevant four-vector, I can write a scalar Lagrangian by taking the product of the partial derivative and ω^μ. There are no other four-vectors. Since $\partial_\mu \omega^\mu$ is a total derivative, I am left with the other possibility, namely

$$\mathcal{L} = i u_\alpha^\dagger \sigma^\mu \partial_\mu u_\alpha \,, \tag{D.8}$$

and the equation of motion is

$$\boxed{\sigma^\mu \partial_\mu u_\alpha = \left(\frac{\partial}{\partial t} + \vec{\sigma} \cdot \vec{\nabla} \right) u_\alpha = 0 \,.} \tag{D.9}$$

Equation (D.9) is a **Weyl equation**. It describes the evolution in time and space of a two-component spinor, that is, a massless fermion.

The Lagrangian for the other spinor in the basis is similar,[D.2]

D.2 $\bar{\sigma}^\mu$ is defined as $(\mathbb{1}, -\vec{\sigma})$.

$$\mathcal{L} = i v_\alpha^\dagger \bar{\sigma}^\mu \partial_\mu v_\alpha, \tag{D.10}$$

and the equation of motion is also similar:

$$\boxed{\bar{\sigma}^\mu \partial_\mu v_\alpha = \left(\frac{\partial}{\partial t} - \vec{\sigma} \cdot \vec{\nabla} \right) v_\alpha = 0 \,.} \tag{D.11}$$

Equation (D.11) too is a **Weyl equation**, this time for the other two-component spinor v. It describes the evolution of a massless fermion.

A parity transformation sends $u \to v$ and *vice versa* – which means that for parity to be conserved I have to join these two bases somehow. This is done by taking the tensor product of the two spinor representations, $(0, 1/2) \otimes (1/2, 0)$. The two Weyl equations in Eq. (D.9) and Eq. (D.11) are combined in a single equation:

$$\begin{pmatrix} 0 & \bar{\sigma}^\mu \partial_\mu \\ \sigma^\mu \partial_\mu & 0 \end{pmatrix} \begin{pmatrix} u_\alpha \\ v_\alpha \end{pmatrix} = 0 \,. \tag{D.12}$$

Equation (D.12) is a 4×4 matrix equation.

The Weyl equation describes, as discussed above, massless fermions. A mass term in the equation of motion arises by adding

$$\mathcal{L}' = -m(u^\dagger v + v^\dagger u) \tag{D.13}$$

to the Lagrangian. The new terms enter and modify the equation of motion in Eq. (D.12) to

$$\begin{pmatrix} -m & i\bar{\sigma}^\mu \partial_\mu \\ i\sigma^\mu \partial_\mu & -m \end{pmatrix} \begin{pmatrix} u_\alpha \\ v_\alpha \end{pmatrix} = 0 \,. \tag{D.14}$$

If I introduce four 4×4 matrices (in the Weyl representation)

$$\gamma^\mu = \begin{pmatrix} \mathbb{0} & \bar{\sigma}^\mu \\ \sigma^\mu & \mathbb{0} \end{pmatrix} \tag{D.15}$$

then Eq. (D.14) can be written as[D.3]

D.3 Here $\mathbb{1}$ is the identity matrix in four dimensions.

$$\boxed{(i\gamma^\mu \partial_\mu - \mathbb{1}\, m)\Psi(x) = 0\,,} \tag{D.16}$$

which is the **Dirac equation** for a free massive fermion described by the four-dimensional spinor

$$\Psi(x) = \begin{pmatrix} u_\alpha \\ v_\alpha \end{pmatrix}. \quad \longleftarrow \text{4 components} \tag{D.17}$$

If I add to Eq. (D.16) a minimal coupling to an electromagnetic four-potential $A^\mu(x) = (\phi(x), A^i(x))$ and write the Dirac equation keeping the time component separate from the space components, I obtain

$$\boxed{\left\{\gamma^0\left[i\frac{\partial}{\partial t} - e\phi(\vec{x},t)\right] - \gamma^i\left[i\frac{\partial}{\partial x^i} - eA_i(\vec{x},t)\right] - \mathbb{1}\, m\right\}\Psi(\vec{x},t) = 0\,,} \tag{D.18}$$

which is the equation used in Section 2.8.

The plane-wave solutions of Eq. (D.16) in momentum space are given in appendix section A.2. They can be derived by first going to Fourier space, where the differential equation in Eq. (D.16) becomes algebraic and therefore simpler to solve. Thus, I can write a generic plane wave as[D.4]

D.4 The spinor $u(\vec{p}, \sigma)$ is just the Fourier transform of $\Psi^+_\sigma(x)$.

$$\Psi^+_\sigma(x) = \int \frac{\mathrm{d}^3\vec{p}}{(2\pi)^3}\, u(\vec{p}, \sigma)\, e^{-ip\cdot x} \tag{D.19}$$

in which $p \cdot x = Et - \vec{p}\cdot\vec{x}$. The four-dimensional spinor $u(\vec{p}\,)$ must satisfy the equation[D.5]

D.5 $\not{p} = \gamma^\mu p_\mu$.

$$(\not{p} - m)_{ab}\, u^b(\vec{p}, \sigma) = 0\,. \tag{D.20}$$

In Eq. (D.20), I have written out explicitly the four components and retained the traditional notation even though there is a clash with Eq. (D.9), in which u denotes one of the two Weyl spinors.

At $\vec{p} = 0$,

$$u^a(\vec{0}, \sigma) = \begin{pmatrix} \chi_\sigma \\ \mathbb{0} \end{pmatrix} \tag{D.21}$$

in which χ_σ ($\sigma = 1/2, \; -1/2$) are two-component spinors that can be written as

$$\chi_{1/2} = \begin{pmatrix} 1 \\ 0 \end{pmatrix} \quad \text{and} \quad \chi_{-1/2} = \begin{pmatrix} 0 \\ 1 \end{pmatrix}, \tag{D.22}$$

by identifying the z-axis with the direction of $\vec{p}$. They describe the spin of the fermion being up or down, respectively. Their expression in an arbitrary direction is obtained by a rotation given by the $M^{(1/2)}$ Wigner D-matrix:

$$\chi_{1/2} = \begin{pmatrix} e^{-i\phi/2}\cos\theta/2 \\ e^{i\phi/2}\sin\theta/2 \end{pmatrix} \quad \text{and} \quad \chi_{-1/2} = \begin{pmatrix} e^{-i\phi/2}\sin\theta/2 \\ e^{i\phi/2}\cos\theta/2 \end{pmatrix}. \tag{D.23}$$

Equation (D.20) is trivially satisfied for[D.6]

D.6 N_p is a momentum dependent normalization.

$$u^a(\vec{p}, \sigma) = N_p(\not{p} + m)\, u^a(\vec{0}, \sigma), \tag{D.24}$$

because

$$(\not{p} - m)\, u^a(\vec{p}, \sigma) = (\not{p} - m)(\not{p} + m)\, u^a(\vec{0}, \sigma) = 0 \tag{D.25}$$

since

$$(\not{p} - m)(\not{p} + m) = p^2 - m^2 = 0\,. \tag{D.26}$$

If I write out explicitly the matrix form of $\not{p} + m$, I have that

$$\begin{pmatrix} E + m & -\vec{\sigma} \cdot \vec{p} \\ \vec{\sigma} \cdot \vec{p} & m - E \end{pmatrix} \tag{D.27}$$

and the spinor $u^a(\vec{p}\,)$ can be written as

$$u^a(\vec{p}, \sigma) = N_p\, (\not{p} + m) \begin{pmatrix} \chi_\sigma \\ \mathbb{0} \end{pmatrix} = N_p\, (E + m) \begin{pmatrix} \chi_\sigma \\ \frac{\vec{\sigma} \cdot \vec{p}}{E+m}\, \chi_\sigma \end{pmatrix}. \tag{D.28}$$

The normalization factor N_p is readily obtained by requiring the covariant normalization

$$\sum_\sigma u^\dagger(\vec{p}, \sigma) u(\vec{p}, \sigma) = 2E\,, \tag{D.29}$$

which gives

$$N_p = \frac{1}{\sqrt{E + m}}\,. \tag{D.30}$$

The solutions of the Dirac equation are therefore given by Eq. (D.19) with

$$\boxed{u^a(\vec{p}, \sigma) = \sqrt{E + m} \begin{pmatrix} \chi_\sigma \\ \frac{\vec{\sigma} \cdot \vec{p}}{E+m}\, \chi_\sigma \end{pmatrix}.} \tag{D.31}$$

The other set of solutions,

$$\Psi^-_\sigma(x) = \int \frac{\mathrm{d}^3\vec{p}}{(2\pi)^3}\, v(\vec{p}, \sigma)\, e^{ip \cdot x}\,, \tag{D.32}$$

is found by following the same steps, starting with

$$v^a(\vec{p}, \sigma) = N_p(\not{p} - m) v^a(\vec{0}, \sigma)\,, \tag{D.33}$$

in which

$$v^a(\vec{0}, \sigma) = \begin{pmatrix} \mathbb{0} \\ \chi^\sigma \end{pmatrix}. \tag{D.34}$$

The final result is

$$\boxed{v^a(\vec{p}, \sigma) = \sqrt{E + m} \begin{pmatrix} \frac{\vec{\sigma} \cdot \vec{p}}{E+m}\, \chi^\sigma \\ \chi^\sigma \end{pmatrix}.} \tag{D.35}$$

The solutions in Eq. (D.19) describe a wave in time

$$\Psi^{+} \propto e^{-iEt} \tag{D.36}$$

with $E = \sqrt{\vec{p}^{\,2} + m^2}$. If I compute the energy of this wave I will find that it is positive. The opposite happens for the solutions in Eq. (D.32), which go as

$$\Psi^{-} \propto e^{+iEt} \tag{D.37}$$

and have negative energy. These negative-energy solutions are disconcerting at first, but only as far as I insist in working within the relativistic theory of a single fermion – which does not exist, as discussed in Section 2.10. Once the solutions $\Psi^{\pm}_{\sigma}(x)$ are promoted to operators in quantum field theory, the Dirac field operator is given by

$$\hat{\Psi}(x) = \sum_{\sigma=1,2} \int \frac{d^3\vec{p}}{(2\pi)^3} \frac{1}{\sqrt{2E_{\vec{p}}}} \left[a(\vec{p},\sigma)\, u(\vec{p},\sigma)\, e^{-i\vec{p}\cdot\vec{x}} + b^{\dagger}(\vec{p},\sigma)\, v(\vec{p},\sigma)\, e^{+i\vec{p}\cdot\vec{x}} \right] \tag{D.38}$$

in which $a^{\dagger}(\vec{p},\sigma)$ creates a fermion and $b^{\dagger}(\vec{p},\sigma)$ an antifermion. The solutions with positive energy in Eq. (D.19) are associated with fermions while those in Eq. (D.32) to antifermions.[D.7]

D.7 The antifermion can be thought as moving backward in time if I want its energy to be positive. Again, see the discussion in Section 2.10.

The results in Appendix A for the normalization of the Dirac spinors and the sum over their polarizations can be derived from these solutions. For instance, the sum over the spin of the fermion can be computed by starting from[D.8]

D.8 $\bar{u}(\vec{p},\sigma) = \gamma^0 \bar{u}^{\dagger}(\vec{p},\sigma)$.

$$\sum_{\sigma=1,2} u(\vec{p},\sigma)\bar{u}(\vec{p},\sigma) = (E+m) \sum_{\sigma=1,2} \begin{pmatrix} \chi^{\sigma} \\ \frac{\vec{\sigma}\cdot\vec{p}}{E+m}\chi^{\sigma} \end{pmatrix} \left(\chi^{\sigma\dagger},\ -\frac{\vec{\sigma}\cdot\vec{p}}{E+m}\chi^{\sigma\dagger}\right). \tag{D.39}$$

In Eq. (D.39) we have

$$\sum_{\sigma=1,2} \chi^{\sigma}\chi^{\sigma\dagger} = 1\,, \tag{D.40}$$

and it follows that[D.9]

D.9 Recall that

$$(\vec{\sigma}\cdot\vec{p})^2 = \vec{p}^{\,2}$$

and

$$\vec{p}^{\,2} = (E-m)(E+m).$$

$$\begin{aligned} \sum_{\sigma=1,2} u(\vec{p},\sigma)\bar{u}^{\sigma}(\vec{p},\sigma) &= (E+m) \begin{pmatrix} \mathbb{1} & -\frac{\vec{\sigma}\cdot\vec{p}}{E+m} \\ \frac{\vec{\sigma}\cdot\vec{p}}{E+m} & -\frac{(\vec{\sigma}\cdot\vec{p})^2}{(E+m)^2} \end{pmatrix} \\ &= \begin{pmatrix} E+m & -\vec{\sigma}\cdot\vec{p} \\ \vec{\sigma}\cdot\vec{p} & m-E \end{pmatrix}. \end{aligned} \tag{D.41}$$

The last matrix is just $\not{p} + m$ and the final result is

$$\sum_{\sigma=1,2} u^{\sigma}(\vec{p})\bar{u}^{\sigma}(\vec{p}) = \not{p} + m\,. \tag{D.42}$$

A similar result for the antifermion fields $v(\vec{p},\sigma)$ is readily obtained by following the same steps.

Bibliography of Primary Sources

S.L. Adler, Axial-vector vertex in spinor electrodynamics, *Physical Review* **177** (1969) 2426.

P.W. Anderson, Plasmons, gauge invariance, and mass, *Physical Review* **130** (1963) 439.

J.S. Bell, On the Einstein Podolsky Rosen paradox, *Physics* **1** (1964) 195.

J.S. Bell and R. Jackiw, A PCAC puzzle: $\pi^0 \to \gamma\gamma$ in the sigma-model, *Il Nuovo Cimento* **LX** (1969) 47.

J.D. Bjorken and E.A. Paschos, Inelastic electron-proton and γ-proton scattering and the structure of the nucleon, *Physical Review* **185** (1969) 1975.

M. Breidenbach, *et al.* Observed behavior of highly inelastic electron-proton scattering, *Physical Review Letters* **23** (1969) 935.

J.H. Christenson, J.W. Cronin, V.L. Fitch and R. Turlay, Evidence for the 2π decay of the K_2^0 meson, *Physical Review Letters* **13** (1964) 138.

E. Fermi, Tentativo di una theoris dei raggi β, *Il Nuovo Cimento* **11** (1934) 1. English translation: F.L. Wilson, *American Journal of Physics* **36** (1968) 1150.

R.P. Feynman, Very high-energy collisions of hadrons, *Physical Review Letters* **23** (1969) 1415.

R.P. Feynman and M. Gell-Mann, Theory of the Fermi interaction, *Physical Review* **109** (1958) 193.

H. Fritzsch, M. Gell-Mann and H. Leutwyler, Advantages of the color octet gluon picture, *Physics Letters* **47B** (1973) 365.

Y. Fukuda *et al.*, Evidence for oscillation of atmospheric neutrinos, *Physical Review Letters* **81** (1998) 1562.

M. Gell-Mann, A schematic model of baryons and mesons, *Physics Letters* **8** (1964) 214.

M. Gell-Mann and F.E. Low, Quantum electrodynamics at small distances, *Physical Review* **95** (1954) 1300.

M. Gell-Mann and A. Pais, Behavior of neutral particles under charge conjugation, *Physical Review* **97** (1955) 1387.

S.L. Glashow, Partial-symmetries of weak interactions, *Nuclear Physics* **22** (1961) 579.

V. Gribov and B. Pontecorvo, Neutrino astronomy and lepton charge, *Physics Letters* **B28** (1969) 493.

D. Gross and F. Wilczek, Ultraviolet behavior of non-Abelian gauge theories, *Physical Review Letters* **30** (1973) 1343.

F.J. Hasert *et al.*, Observation of neutrino-like interactions without muon or electron in the Gargamelle neutrino experiment, *Physics Letters* **46B** (1973) 138.

P. Higgs, Broken symmetries and the masses of gauge bosons, *Physical Review Letters* **13** (1964) 508.

H.D. Politzer, Reliable perturbative results for strong interactions?, *Physical Review Letters* **30** (1973) 1346.

E.C.G. Sudarshan and R.E. Marshak, Chirality invariance and the universal Fermi interaction, *Physical Review* **109** (1958) 1860.

S. Weinberg, A model of leptons, *Physical Review Letters* **19** (1967) 1264.

S. Weinberg, Non-Abelian gauge theories of the strong interactions, *Physical Review Letters* **31** (1973) 494.

K. G. Wilson, Renormalization group and critical phenomena. I. Renormalization group and the Kadanott scaling picture, *Physical Review B* **4** (1971) 3174.

C.S. Wu *et al.*, Experimental test of parity conservation in beta decay, *Physical Review* **105** (1957) 1413.

C.N. Yang and R.L. Mills, Conservation of isotopic spin and isotopic gauge invariance, *Physical Review* **96** (1954) 191.

Index

For EU product safety concerns, contact us at Calle de José Abascal, 56–1°,
28003 Madrid, Spain or eugpsr@cambridge.org.

www.ingramcontent.com/pod-product-compliance
Lightning Source LLC
LaVergne TN
LVHW082020150826
845671LV00005B/215